NEW FRONTIERS IN BARNACLE EVOLUTION

CRUSTACEAN ISSUES

10

General editor:
FREDERICK R. SCHRAM
Institute for Systematics and Population Biology, University of Amsterdam

A.A. BALKEMA / ROTTERDAM / BROOKFIELD / 1995

NEW FRONTIERS IN BARNACLE EVOLUTION

Edited by

FREDERICK R. SCHRAM

Institute for Systematics and Population Biology, University of Amsterdam

JENS T. HØEG

Departmentof Cell Biology and Anatomy, Institute of Zoology, University of Copenhagen

A.A. BALKEMA / ROTTERDAM / BROOKFIELD/ 1995

Published by
A.A. Balkema, P.O. Box 1675, 3000 BR Rotterdam, Netherlands (Fax: +31.10.4135947)
A.A. Balkema Publishers, Old Post Road, Brookfield, VT 05036, USA (Fax: 802.276.3837)

ISSN 0168-6356
ISBN 90 5410 626 3

Printed in the Netherlands

Table of contents

Preface

One day while sitting around the table in the cirripede laboratory in Copenehagen, we were brainstorming over how to establish a cooperative research program in barnacle biology. We had the magnificent monograph of Goeffrey Boxshall and Rony Huys before us, *Evolution of Copepoda* (London: Ray Soc. 1992), and we suggested to ourselves that cirripedology needed a volume like that. We immediately realized that a one or two author tome was not feasible, and we then further realized that even a multi-author book was not possible given the great gaps in our knowledge about many aspects of most of the known biological diversity of the Cirripedia. We concluded that before we could ever contemplate a book that could be entitled *Evolution of Cirripedia* a lot of research had to occur. Not only that, it would take a fairly concerted international effort to arrive at that point and would entail people moving off in new directions away from what had been done in the past.

'How might we facilitate this?' we asked ourselves. Aside from the obvious wish that we had unlimited sources of funds available on demand, eventually we came to the conclusion that what we needed was some initial 'push'. We decided that that push might take the form of a preliminary volume, or an international symposium, or both. And so was born this project.

Very quickly we came to believe that an effective focus of this would be to honor William Newman with a Festschrift. Not only is Bill a dear friend and close colleague of ours, he has been, in our opinion, one of the most productive and pioneering cirripedologists since Charles Darwin put down the group and went on the 'other matters'. After four decades of research on barnacles, Bill Newman is still hard at work and has no intention of retiring. However, in that four decades he has been at the heart of new discoveries and ideas concerning matters related to the systematics and evolution of the group.

In keeping with our original intent, we did not want to do the 'usual' kind of Festschrift, one with lots of people writing short papers and naming new species for the honoree. Nor did we intend to look backward in any way and summarize from where barnacle biology had come from. That was effectively done in 1987 in *Barnacle Biology* (A.J. Southward, ed., *Crustacean* Issues 5). We wanted to look forward, to honor one of the most forward looking cirripede workers we knew of, and set out some pathways for other researchers to follow (maybe some of the new and upcoming generation) to that 'new frontier' we wanted to reach.

Whether we will succeed only time will tell. We could not include everyone we might have liked. Time constraints and other commitments often caused people to drop out of the project. Nevertheless, we hope that what we have assembled is a book whose contents will help refocus research in the field.

This volume (drawing as it did a contribution via Prof. Joel W. Hedgpeth, a poetic effort, from that bard of biology, Jerome Tichenor) was linked with an international symposium presented at the annual meeting of The Crustacean Society in Washington, DC, December 1995. This symposium was co-sponsored by the society, the Division of Invertebrate Zoology of the American Society of Zoologists, and the Groupe d'Etudes et de Réflexion sur l'Evolution des Crustacés.

Frederick R. Schram, Amsterdam

Jens T. Høeg, Copenhagen

Ad majoriam cirripediae gloriam

Hail Newman, cardinal of the fringy footed creatures
Encrusting round the world on rocky oceansides
(It took Darwin years to master all their features),
And student of the torrid ocean vents besides.

Of barnacles there is no end of things to know:
Some live on pycnogonids in their latter stages,
But the editors of *JCB* have said 'No!
Those hosts must not be mentioned in our pages'.

There are some crusty balanoids who ride on whales,
And among the hosts in more congenial climes
Is dainty *Janthina* who harbors as she sails
A little *Lepas* sweeping water for its feeding times.

Cirripedes are monoecious, but alternate their sexes,
Sedentary ones extend penes to their nearest neighbor —
But parasitic *Sacculina*'s blob a crab sore vexes
By depriving it of life's joys, and that's no favor.

Their variety of life and form is great in ocean's range:
Borers into snail shells, and parasitic ways so nauseous,
And shadow dodging *Pollicipes* is passing strange,
Yet all begin their lives as a dainty nauplius!

Ah Newman, veteran of many phylogenetic frays:
How fortunate to remain within your chosen Class,
While I, confused by an ultrastructural maze*
Must remain perforce by ambiguity unclassed.

Jerome Tichenor

*see W.H. Fahrenbach: *J. Morphology* 222:33-48 (1994).

William Anderson Newman: An appreciation

Alan J. Southward
Marine Biological Association, Plymouth, UK

William Anderson (Bill) Newman was born in San Francisco, California, on 13 November 1927, 5 months before me. He is a rare-bird in that fast-growing state today, where most of the people living there were born elsewhere. His grandfather ran vineyards near Modesto and his father was also involved in the business for a while, so that Bill is a wine connoisseur by birth as well as by palate.

Both Bill and I experienced the war-time emergencies during our teens, and the immediate aftermath of World War II. In Bill's case this meant conscription into the army as soon as he finished high school. Army service was a form of 'adult education' denied me because of my disability, though I later experienced the milder forms of it at agricultural camps for students, based in ex-army hutments. Bill's and my first publication dates differ as a result of his army service, about which he talks very little. He does admit to 'goofing about' building sailing boats for a couple of years after leaving the army, before enrolling at the Berkeley Campus of the University of California. This was a departure from the rest of the family, all Stanford graduates. In those days before student riots, Berkeley was an enviable paradise to Europeans, especially to us British still struggling to throw off the war-time austerity that was extended unnecessarily into the days of peace by the government. Also at that time (1950's), biology at Berkeley provided splendid training in comparative and functional anatomy and behaviour of invertebrates, numbering among the faculty the late much-respected Ralph Smith. After his master's degree in 1954 Bill left Berkeley for two years of teaching at the Pacific Islands Central School, Truk (1954-1956), a stay that introduced him to coral reef ecology, a topic that still concerns him. On return to Berkeley, Bill resumed researches on estuarine ecology and barnacles, eventually presented as a doctoral dissertation in 1962, with a committee composed of Ralph Smith, Cadet Hand and Wyatt Durham. The ecological part of his work at Berkeley was published separately (Newman & Muscatine 1961; Newman 1963a) from the experimental part which appeared in a symposium volume from India (Newman 1967c). Such investigations were of interest to many Berkeley researchers in those days, studying the complex origins and evolution of the San Francisco Bay fauna. Conomos (1979) has edited a more recent review on the topic.

Bill soon found a more congenial niche in the morphology and systematics of barnacles. As a result he has so far published, excluding book reviews, over 120 papers and book chapters. This total corresponds to a rate of 3.3 publications a year, but in some years the number reached much higher, peaking at 10 in 1974, a proud record. Some of the papers have a very high citation value, and almost all represent work carried out at the Scripps Institution of Oceanography, La Jolla, where Bill was based in 1962-1963 and since 1965; this is now part of the extensive San Diego campus of the University of California. His stays at Harvard and Plymouth were not quite as productive as the years at La Jolla, but interestingly both resulted in papers on burrowing deep-sea acrothoracicans, among other things. The move to New England in 1963 could have been a major achievement for Bill, but his education in the democratic forms of higher education in California had not prepared him for the autocratic regime at the Museum of Comparative Zoology at Harvard, where an echinoderm specialist was in charge of invertebrates. I had first met Bill face-to-face, after much correspondence, in May 1963 in La Jolla, when he was living in the 'Bird Rock' part of the town, well south of the Scripps campus, and where he was a most gracious host. We had an interesting time investigating Californian barnacles, with a brief foray into northern Mexico, where Bill has been a frequent visitor for many years, especially for the fall shooting season. I and my wife had been staying at Westwood Village, near the Los Angeles campus of the University of California, with a friend from Plymouth who was working in Ted Bullock's Department, the late Trevor Shaw. Trevor was not a good freeway driver, often missing entrances and exits (sometimes twice when engaged in scientific thought), and it was a relief to be driven by someone as calm as Bill, but being asked to keep quiet and not flourish UK passports at the Mexican border was a bit nerve-wracking. In later years, we would come to know the same border quite well after forays into southern Baja California for tropical chthamalids.

In 1963, my wife and I travelled from California along the coast up to British Columbia and Alaska, and by late September, on the way back to England, we were visiting Howard Sanders at Woods Hole. We were quite surprised when Bill turned up in Howard's laboratory, in the process of taking up the post at Harvard. It just happened that Howard had found a peculiar animal in a sample from the famous Gay Head-Bermuda transect of epibenthic sled samples, and he showed it to me in the hope I might be interested. It was evidently an acrothoracican barnacle, a group with which I was not well acquainted, and on mentioning this fact I was an immediate centre of interest from Bill Newman and Bob Hessler who tussled over the specimen a bit before it was agreed Bill should describe it (Newman 1971). Bill returned to Scripps in 1965, and except for a sabbatical half year at Plymouth noted below, he has remained faithful to California.

A glance through the titles of Bill Newman's papers from Scripps shows that at first both biogeography (Newman 1960a) and stalked barnacles jostled for attention (Newman 1960b,c, 1961b,c) but acrothoracicans had already appeared (Tomlinson & Newman 1960). His interest in stalked barnacles appears to have been stimulated when a Californian spiny lobster was served up for supper in the Newman household, complete with *Octolasmis* in the gill cavity. Bill's studies became wider still and he began trying to find out how cirripeds have evolved (Newman 1962). His early introduction to the fauna of the Pacific Islands while at Truk in 1954-1956 was expanded into studies of morphology and distribution of Pacific barnacles, including a chthamalid (Newman 1961a) and more general studies on the Fanning Island fauna (Newman 1963b), followed by Hawaian (Matsui, Shane & Newman 1964) and Pacific guyot species (Zullo & Newman 1964). By this time an interest in geological evolution had involved him in collaborative studies on fossil cirripedes with Arnold Ross and the late Vic Zullo (Ross & Newman 1967; Shepard et al. 1967; Newman, Zullo & Withers 1969; Newman & Zullo 1969). This type of work also involved collaboration with geologists about changes in sea level and related matters (Curray et al. 1967; Newman, Veeh & Curray 1967; Shepard, Curray & Newman 1967; Shepard et al. 1967).

In the late 1960's and the early 1970's Bill Newman was collaborating on barnacle taxonomy with Arnold Ross, studying Antarctic and cold water sea areas (Ross & Newman 1969; Newman & Ross 1971) as well as tropical regions (Ross & Newman 1969; Newman et al. 1971; Newman 1972a, 1972c; Rao & Newman 1972). There were further forays into the geology and ecology of coral reefs, including some notable contributions on the problem of the coral-eating starfish, *Acanthaster* that was said to be destroying some reefs (Newman & Dana 1972; Dana & Newman 1972; Dana, Newman & Fager 1972; Dana & Newman 1973).

Bill had first shown interest in deep sea barnacles during his early years at Scripps (Newman 1961b). This continued into the 1970's with the contributions on the acrothoracicans that burrow into soft sedimentary rocks exposed along the continental slope (Newman 1971, 1974a). Much of the work involved in writing these last two papers was carried out while Bill and family were living in Plymouth, from July to December, 1972. Some of the typing was done on his wife Lynn's old college manual typewriter, which had been a gift from her parents on graduation from grammar school. She was then persuaded to leave it behind when they left for home, on the promise of a new electric machine; it still remains at Plymouth, being last used in the 1970's by another visiting Californian, Grover Stephens. Bill and family had many adventures during the half-sabbatical year at Plymouth, including taking an ageing English Ford Cortina (a rare version with automatic gear shift) around Europe. Among the culture-shocks received on his only sabbbatical, it appears to have been the long European lunch 'hour' that Bill found most uncongenial. An inveterate 'brown-bagger' at home, accustomed to eat at his desk to save time, he was eventually able to tolerate the lunch break and conversation in the Plymouth common room. Nevertheless, Bill never returned to Europe, whereas he followed up his research cruises with several working visits and symposia attendances around the Pacific.

With his extensive experience of the distribution of deep-sea barnacles (Newman 1980a; Dayton, Newman & Oliver 1982) and his situation at one of the founding institutions for study of the ocean ridges (Lonsdale 1977), it was natural for Bill to join in the investigations into the newly-discovered abundant communities of life around hydrothermal vents. His knowledge of the fossil barnacles was invaluable in determining the systematic position of the new species that turned up as successive hydrothermal vent regions were visited, first a Mesozoic relict lepadid (Newman 1979b), then a primitive verrucid (Newman & Hessler 1989; Newman 1989b), and more recently a living relative of the first sessile barnacles, the Brachylepadomorphs, that disappeared from the fossil record in the Miocene (Newman & Yamaguchi 1995). These and other contributions (Newman 1985a; Yamaguchi & Newman 1990) have strengthened Bill's views as to the antiquity of the deep-sea barnacle fauna, especially at the hydrothermal vents. The results support his ideas on the influence of competition on the distribution and evolution of the Cirripedia in general (Stanley & Newman 1980; Newman & Stanley 1981), in contrast to the opinions of ecologists who think other factors important (Paine 1981). The work on distribution of the hydrothermal vent barnacles was accompanied by other studies on the biogeography of the Pacific, mostly undertaken in collaboration with the late Brian Foster of New Zealand (Newman & Foster 1982, 1983, 1987; Newman 1986). Bill and Brian had earlier disagreed about Bill's interpretation of experiments on estuarine barnacles (Newman 1967c), but they came together again for biogeographical researches. The results of the biogeographical studies have to be viewed together with Bill's investigations on cirripede groups that have many fossil representatives, such as the species of *Concavus* (Newman 1982a) and the acrothoracicans (Grygier & Newman 1985). As a result he has been led to theorize about the origin and evolution of the barnacles (Newman 1987, 1990a,e, 1992b). Like many other cirripedologists, Bill has also pondered about the influence of Darwin on the way we study barnacles (Newman 1993).

As already remarked, Bill's interest in coral reefs began during his years on Truk and was extended to the barnacles associated with corals. This interest was further encouraged by several collaborators (Curray et al. 1967; Newman et al. 1971). The collaboration with Arnold Ross included coral barnacles (Ross & Newman 1973) and there were additional papers (Newman, Jumars & Ross 1974; Newman & Ladd 1974; Ladd, Newman & Sohl 1974) on coral reefs as well as barnacles. The results of these studies led Bill to formulate hypotheses and speculate about the neogene radiation of coral barnacles (Newman 1988).

After his move to the new biology building at Scripps in the 1970s, Bill developed a splendidly equipped laboratory and offices for barnacle studies, including ingeniously constructed storage cabinets for preserved material, sea-water tanks for live material, and facilities for shell studies. Bill is clever with his hands and skilled at sectioning shells and mounting mouthparts on slides; at home also, he shows a talent for wood and metal-working.

At Scripps Bill attracted a number of graduate students, nearly all of whom enjoyed talking barnacles into the small hours, and some of whom are contributing to this volume or are co-authors of papers quoted. Mark Grygier worked with Bill on acrothoracicans and lepadids; the late Larry Ritchie worked with Bill and Jens Hoeg on Rhizocephala; Bob Van Syoc has used molecular techniques to follow up earlier suggestions of latitudinal diversity in *Pollicipes*. Other graduate students have included Jorge Laguna who worked on the barnacles of Panama and Eduardo Gomez on *Conopea*. Bill has played host to many visiting scientists who joined him in fruitful collaboration. Some have already been listed, but we should also mention Y. Achituv (cirral activity), J. Buckeridge (fossils), D.J. Crisp (*Hemioniscus*), J. Høeg (Rhizocephala), J. Moyse (ascothoracicans) and T. Yamaguchi (deep-sea). It was the joint work with Arnold Ross that led to the highly-quoted contributions on the nomenclature and biogeography of barnacles, first the article on cirripedia in the Antarctic Research Series (Newman & Ross 1971), then a revision of the coral-inhabiting barnacles (Ross & Newman 1973), followed by a revision of the balanomorphs (Newman & Ross 1976). These monographs have extremely valuable bibliographies, and the barnacle world owes a lot to Bill and Arnold for their labours. Productivity, however one measures it, is a poor yardstick for judging taxonomists like

Bill. For example, the search for literature and material on the coral-inhabiting barnacles took Bill Newman and Arnold Ross almost 13 years before publication, whereas Darwin seems to have spent only 41 days on the group (de Beer 1959). However, the work on the coral-eating barnacle, *Hoekia*, took a mere 30 days plus another 22 being rejected by 'Science', compared with 6 days for acceptance by 'Pacific Science', the time required for air mail between Hawaii and continental USA

Some barnacle workers stubbornly refused to accept the much-needed revisions of Darwin's species by Bill Newman and Arnold Ross, changes that had been foreshadowed by Pilsbry (1916). This stubborness led eventually to somewhat acrimonious correspondence instigated by the late Dennis Crisp, about the necessity or not of nomenclatural changes (Crisp & Fogg 1989; Newman 1989). It was the changing of the name of the common boreo-arctic barnacle from *Balanus balanoides* to *Semibalanus balanoides* that rankled most in the minds of critics who were not themselves experienced taxonomists. Bill and others, including myself, thought that such disagreement might do taxonomy a disservice at a period when it was under attack by administrators and experimentalists, and we strove hard to overcome the problem.

Other important collaborative work by Bill was with southern hemisphere workers, notably Brian Foster. Bill developed new ideas about endemism and extinction of previously amphitropical taxa (Newman & Foster 1983, 1986, 1987; Foster & Newman 1986). The collaboration with John Killingley on shell carbonate isotope ratios (Killingley et al. 1980; Killingley & Newman 1982) led to an interesting account of how unusual outlying distributions can be verified or falsified by determination of the likely temperatures experienced during their life. Examination of microscopic fragments of a stalked barnacle *Pollicipes*, supposedly from India, showed that the shells had grown at cool temperatures consistent with those off California, and could not have lived in the tropical waters of the Bay of Bengal from where they were supposed to have been collected (Newman & Killingley 1985). An allied investigation of the apatite content of barnacle shell plates (Lowenstam, Weiner & Newman 1992) has palaeontological implications. Additional palaentological studies were continued by Bill in association with Fred Schram on fossil verrucids (Schram & Newman 1980) and with John Buckeridge (Buckeridge & Newman 1992). Bill's important review of fossil cirripedes in collaboration with the late Vic Zullo and the late Henry Withers was published in the Treatise on Invertebrate Palaeontology (Newman, Zullo & Withers 1968) and constitutes another much quoted reference.

Additional Calfornian students, with whom Bill was associated in some capacity, took part in an international workshop on barnacles held at Scripps in the fall of 1978. Those present in addition to Bill and myself were Dora Henry, Pat McLaughlin, Cindy Lewis, Denis Hedgecock, Arnold Ross, Henry Spivey and John Standing. This was a memorable meeting if only for the amazing mix of ages and topics. It was enlivened by Dora Henry's description of enzyme electrophoresis as 'hocus pocus' and by the unofficial competition between the participants from Florida and California as to who could produce the biggest home-grown avocado. Patsy McLaughlin won with examples from Florida, one of which was big enough to supply half a dozen people with canapes to go with the gin & tonic mixes superintended by Dora. However, Bill's skills as gardener are unsurpassed among cirripedologists, and at home on the hills close to Scripps he and his wife raise flowers and vegetables as well as superb fruit.

My collaboration with Bill originated in a common interest in the deep sea Bathylasmatidae. His own involvement in this group grew from examination of the extensive material from the 'Eltanin' cruises in the southern ocean, and occupied many years during which his research interest changed considerably. Our collaboration proceeded from the deep sea to chthamalids of the Indian Ocean (Newman 1967; Southward 1967) and the discovery of relict four-plated barnacles in the Atlantic, work published by Bill with Arnold Ross (Newman & Ross 1977). We have long been studying the respective distributions of chthamalids on both sides of the Isthmus of Panama, though early lumping of species (Southward & Newman 1977) has been outdated by work on proteins with other collaborators, including Paul Dando and Dennis Hedgecock, mostly still unpublished (Hedgecock 1979; Dando & Southward 1980). Bill and I are both pursuing possible Tethyan relict barnacles (Newman 1992; and unpublished).

These collaborative investigations led to a highly productive and entertaining 'collecting trip' cruise with R.V. 'Alpha Helix' in the spring of 1978 along the coast of Central America from Panama back to California. We anchored for the day off rocky and coral reefs, many of them not easily accessible by road, where the barnacles were investigated by landing from a small boat. We dredged at night on the way between the intertidal sites. Bill had masterly success in dealing with a ship's crew who wanted to get home quicker, and in overcoming unforeseen but not entirely unrelated breakdowns of ship's winches and generators. Investigations, fairly hurried, were even made in the Gulf of Fonseca, where El Salvador, Honduras and Nicaragua meet in uneasy truce, but other investigations in the Gulf of California were shortened by radio instructions from maritime HQ at Scripps; single sideband transmitters have destroyed the independence of ship's captains and chief scientists.

In his 34 years of publishing, Bill has made significant contributions to science, and his major works have been helpful to many. I mention particularly the chapter on Cirripedia in the revised edition of Light's Manual of Intertidal Invertebrates published from Berkeley (Newman & Abbott 1975), the later chapter in a book from Stanford University (Newman & Abbott 1980), an article in the Encyclopaedia Brittanica (Newman 1974b) and a contribution to the volume on crustacea edited by Abele (Newman 1982b). As mentioned above, not all cirripedologists have been willing to accept his iconoclastic attitude to Darwin's 'bible' (Darwin 1851, 1854a,b), even though many of the changes made to genera were well justified upgradings of natural divisions previously recognised (Pilsbry 1907, 1916). Bill's own achievements in barnacle systematics can be ranked at least equal with those of a previous American expert on barnacles, Henry Pilsbry, although his approach has been different. Among Bill's most endearing features is his own particular brand of urbanity and his willingness to discuss problems. These things have helped remove the misunderstandings that inevitably occur between colleagues from time to time whether interpreting the results of experiments, fossil findings or phylogenetic trees. In this respect Bill has a strong advantage compared with most present day cirripedologists. He has worked on the taxonomy and morphology of all the orders and most of the families of the class, including fossils. It is this experience that allows inductive reasoning to be followed by proper deductive assessment of the phylogeny and evolution of the barnacles.

Bill's powers in taxonomy have been acknowledged by many of his peers who

Plate 1. William Anderson Newman (photo compliments of Arnold Ross).

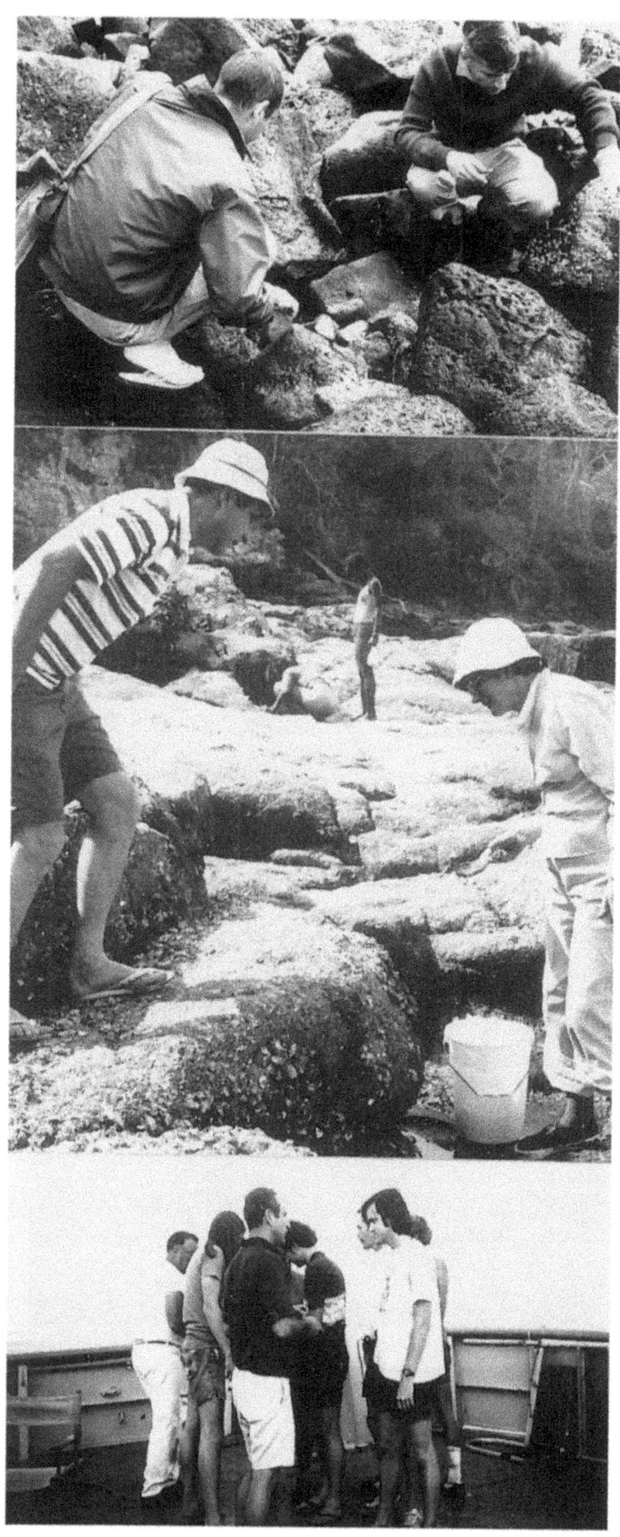

Plate 2. Upper photo. Bill Newman with Alan Southward collecting at Puerto Desconso, northern Baja California, Mexico, May 1963. Middle photo. Bill Newman with Eve Southward, studying Panamic cirripedes at Farfan Point near Panama City, March, 1978. Lower photo. Sunset conference of Bill Newman with his team, preparing for the nightly dredge haul aboard R.V. 'Alpha Helix', March 1978.

have named new species and higher taxa in his honour. These include five species: *Utinomia newmani* (Tomlinson 1963), *Cryptophialus newmani* (Tomlinson 1969), *Calantica (Paracalantica) newmani* (Rosell 1981), *Mesolasma newmani* (Buckeridge 1983), *Arossia newmani* (Zullo 1992), and three genera: *Newmanella* (Ross 1969), *Newmanilepas* (Zevina & Yakhontov 1987), *Wanella* (Anderson 1992). This volume is also dedicated to him.

ACKNOWLEDGEMENT

I would like to acknowledge the help given by Bill's family and friends, especially Lynn Newman and Arnold Ross, in completing this account. To Bill, himself, thanks are due for the labour-saving gift of a bibliography on disc.

REFERENCES

Anderson, D.T. 1992. Structure, function and phylogeny of coral inhabiting barnacles. *Zool. J. Linn. Soc. Lond.* 106: 277-339.

Buckeridge, J.S. 1983. Fossil barnacles of New Zealand and Australia. *N.Z. Geol Surv. Paleo. Bull.* 50: 1-151, 13 plates.

Conomos, J.T. (ed.) 1979. *San Francisco Bay. The urbanized estuary.* San Francisco: American Association for the Advancement of Science, Pacific Division.

Dando, P.R. & A.J. Southward, 1980. A new species of *Chthamalus* (Crustacea: Cirripedia) characterised by enzyme electrophoresis and shell morphology: with a revision of other species of *Chthamalus* from the western shores of the Atlantic Ocean. *J. Mar. Biol. Ass. UK* 60: 787-831.

Darwin, C., 1851. A Monograph on the Sub-class Cirripedia, with figures of all the species. The Lepadidae; or, pedunculated cirripedes. London: Ray Society.

Darwin,C., 1854. A monograph on the fossil Balanidae and Verrucidae of Great Britain. London: Palaeontographical Society.

Darwin, C., 1855. A Monograph on the Sub-class Cirripedia. The Balanadae, the Verrucidae etc. London: Ray Society

De Beer, G.R., 1959. Darwin's Journal. *Bull. Brit. Mus. (Nat. Hist.) Hist. Ser.* 2: 3-21.

Hedgecock, D., 1979. Biochemical genetic variation and evidence of speciation in *Chthamalus* barnacles of the tropical eastern Pacific. *Mar. Biol.* 54: 207-214.

Paine, R.T. (1981). Barnacle ecology: is competition important? The forgotten roles of disturbance and predation. *Palaeobiol.* 7: 553-560.

Pilsbry, H.A., 1907. The barnacles (Cirripedia) contained in thecollections of the U.S. National Museum. *Bull. U.S. Nat. Mus.* 60: 1-114.

Pilsbry, H.A., 1916. The sessile barnacles (Cirripedia) contained in the collections of the U.S. National Museum; including a monograph of the American species. *Bull U.S. Nat. Mus.* 93, 1-366.

Rosell, N.C., 1982. Resultats de Campagnes MUSORSTOM. 1. Phillipines (18-28 Mars, 1976). Crustacea: Cirripedia. *Mem. ORSTOM* 91: 277-307.

Ross, A., 1969. Studies on the Tetraclitidae (Cirripedia: Thoracica): revision of *Tetraclita. Trans. S. Diego Soc. Nat Hist.* 15: 237-251.

Southward, A.J., 1967 On the ecology and cirral behaviour of a new barnacle from the Red Sea and Indian Ocean. *J. Zool. Lond.* 153: 437-444.

Tomlinson, J.T., 1963. Two new Acrothoracican cirripeds from Japan. *Publ. Seto Mar. Biol. Lab.* 11: 263-280.

Tomlinson, J.T., 1969. The burrowing barnacles (Cirripedia: orderAscothoracica). *Bull. U.S. Nat. Mus.* 296: 1-162.

Zevina, G.B. & I.V. Yakhontova, 1987. A new barnacle genus of the family Scalpellidae (Crustacea, Cirripedia) from the North Atlantic. *Zool. Zh.* 66: 1262-1264 (in Russian).

PUBLICATIONS BY W.A. NEWMAN, EXCLUDING BOOK REVIEWS

1958

Newman, W.A. 1958. Sedimentary and biological characteristics of San Diego Bay floor in 1958. Marine Advisors, La Jolla, for the State of California Water Pollution Control Board, 38pp.

1959

Newman, W.A. 1959. Some observations on the geology of certain islands in Micronesia. *Veliger* 1: 28-31.

1960

Newman, W.A. 1960a. On the paucity of intertidal barnacles in the tropical Western Pacific. *Veliger* 2: 89-94.

Newman, W.A. 1960b. Five pedunculate cirripeds from the Western Pacific, including two new forms. *Crustaceana* 1: 100-116.

Newman, W.A. 1960c. *Octolasmis californiana*, a new pedunculate barnacle from the gills of the California Spiny Lobster. *Veliger* 3: 9-11.

Tomlinson, J.T. & W.A. Newman 1960. *Lithoglyptes spinatus*, a new burrowing barnacle from Jamaica. *Proc. U. S. Nat. Mus.* 112: 517-526.

1961

Newman, W.A. 1961a. On the nature of the basis in certain species of the *Hembeli* section of *Chthamalus* (Cirripedia, Thoracica). *Crustaceana* 2: 142-150.

Newman, W.A. 1961b. Notes on certain deep-sea species of *Octolasmis* (Cirripedia, Thoracica) from deep-sea Crustacea. *Crustaceana* 2: 326-329.

Newman, W.A. 1961c. On certain littoral species of *Octolasmis* (Cirripedia, Thoracica) symbiotic with decapod Crustacea from Australia, Hawaii and Japan. *Veliger* 4: 99-107.

Newman, W.A. & L. Muscatine 1961. Benthic invertebrates. A pilot study of physical, chemical and biological characteristics of waters and sediments of South San Francisco Bay. *Sanitary Engineering Laboratory, UC-Berkeley*: 215-254.

1962

Newman, W.A. 1962. Origin and evolution of Cirripedia, In I. Gordon (ed.). *Conference on the evolution of Crustacea. Crustaceana* 4: 166.

1963

Newman, W.A. 1963a. On the introduction of an edible oriental shrimp (Caridea, Palaemonidae) to San Francisco Bay. *Crustaceana* 5: 119-132.

Newman, W.A. 1963b. On Cirripedia. In: The Fanning Island Expedition-1963. *Nature*. 200: 325-326.

1964

Matsui, T., G. Shane & W.A. Newman 1964. On *Balanus eburneus* Gould (Cirripedia, Thoracica) in Hawaii. *Crustaceana* 7: 141-145.

Zullo, V.A. & W.A. Newman 1964. Thoracic Cirripedia from a Southeast Pacific Guyot. *Pac. Sci.* 18: 355-372.

1965

Newman, W.A. 1965a. Prospectus on larval cirriped setation formulae. *Crustaceana* 9: 51-56.

Newman, W.A. 1965b. Physiology and behavior of estuarine barnacles. *Abstracts: Symposium on Crustacea.* (3): 52. Mar. Biol. Assoc. India.

1967

Newman, W.A. 1967a. Shallow water versus deep sea *Octolasmis*(Cirripedia, Thoracica). *Crustaceana* 12: 13-32.

Newman, W.A. 1967b. A new genus of Chthamalidae (Cirripedia, Balanomorpha) from the Red Sea and Indian Ocean. *J. Zool. Lond.* 153: 423-435.

Newman, W.A. 1967c. On physiology and behaviour of estuarine barnacles. In: *Symposium on Crustacea.* (3): 1038-1066. Mar. Biol. Assoc. India.

Newman, W.A., H.H. Veeh & J.R. Curray 1967. Shallow submarine terraces of Micronesian oceanic reefs. *Abstracts: Annual Meeting Geol. Soc. Amer.* 1967: 163.

Newman, W.A., V.A. Zullo & S.A. Wainwright 1967. A critique on recent concepts of growth of Balanomorpha (Cirripedia, Thoracica). *Crustaceana* 12: 167-178.

Ross, A. & W.A. Newman 1967. Eocene Balanidae of Florida, including a new genus and species with a unique plan of `Turtle Barnacle' organization. *Amer. Mus. Novitates* 2288:1-21.

Curray, J.R., A.L. Bloom, W.A. Newman, N.D. Newell, F.P. Shepard, J.I. Tracey & H.H. Veeh 1967. Late Holocene sea-level fluctuations in Micronesia. *Abstracts: Annual Meeting Geol. Soc. Amer.* 1967: 40-41.

Shepard, F.P., J.R. Curray & W.A. Newman 1967. Lagoonal Topography of Caroline and Marshall Islands. p. 202. *Abstracts: Annual Meeting Geol. Soc. Amer.* 1967: 202.

Shepard, F.P., J.R. Curray, W.A. Newman, A.L. Bloom, N.D., Newell, J.I. Tracey & H.H. Veeh 1967. Holocene changes in sea level: Evidence in Micronesia. *Science* 157:542-544.1968. Same abstracts and pagination as 1967f-i. *Abstracts: Annual Meeting Geol. Soc. Amer.* 1967.

1969

Newman, W. A., V. A. Zullo & T. H. Withers 1969. Cirripedia. In: R.C. Moore (ed.), *Treatise on Invertebrate Paleontology*, Part R, Arthropoda 4, Vol. 1, pp. R206-295. Lawrence: Univ. Kansas & Geol. Soc. Amer.

Newman, W.A. & V.A. Zullo 1969. Addendum to Cirripedia. In: R.C. Moore (ed.), *Treatise on Invertebrate Paleontology*. Vol. 2, p. R628. Lawrence: Univ. Kansas & Geol. Soc. Amer.

Ross, A. & W.A. Newman 1969a. Cirripedia. In: J.W. Hedgpeth (ed.). *Distribution of selected groups of marine invertebrates in waters south of 35°S latitude. Antarctic Folio Series* 11:30-32 + pl. 17. Amer. Geograph. Soc.

Ross, A. & W.A. Newman 1969b. A coral-eating barnacle. *Pac. Sci.* 23: 252-256.

1970

Newman, W.A. 1970. *Acanthaster*: A disaster? *Science* 167: 1274-1275.

Lonsdale, P.F., W.R. Normark, W.A. Newman & E.C. Allison 1970.Submarine erosion of the crest of Horizon guyot. *Abstracts: Geol. Soc. Amer. 1970 Annual Meeting* 2: 610.

1971

Newman, W.A. 1971. A deep-sea burrowing barnacle (Cirripedia: Acrothoracica) from Bermuda. *J. Zool. Lond.* 165: 423-429.

Newman, W.A., J.R. Curray, P.J.S. Crampton, T.F. Dana, J.E. McCosker and M.O. Stallard 1971. Comparative coral reef investigations on the north and south coasts of Panama. In: *Alpha Helix Research Program 1969-1970,* 36-39. La Jolla: Scripps Inst. Oceanogr.

Newman, W.A. & A. Ross 1971. Antarctic Cirripedia. American Geophysical Union. *Antarctic Research Series* 14: 1-257.

1972

Newman, W.A. 1972a. Lepadids from the Caroline Islands (Cirripedia, Thoracica). *Crustaceana* 22: 31-38.

Newman, W.A. 1972b. The National Academy of Science committee on the ecology of the Interoceanic Canal. *Bull. Biol. Soc. Wash.* 2: 247-260.

Newman, W.A. 1972c. An oxynaspid (Cirripedia, Thoracica) from the eastern Pacific. *Crustaceana* 23: 202-208.

Newman, W.A. & T.F. Dana 1972. Potential consequences of a Panamic sea-level canal. *Abstracts: Seventeenth International Congress of Zoology. Symposium* 3. Monaco.

Dana, T.F. & W.A. Newman 1972. Massive aggregation and migration of *Acanthaster*: Behavioral responses to severe food limitation. *Abstracts: Seventeenth International Congress of Zoology. Symposium* 3. Monaco.

Dana, T.F., W.A. Newman & E.W. Fager 1972. *Acanthaster* aggregations: Interpreted as primarily responses to natural disturbances. Pac. Sci. 16: 355-372.

Lonsdale, P, W.R. Normark & W.A. Newman 1972. Sedimentation and erosion of Horizon Guyot. *Bull Geol. Soc. Amer.* 83: 289-316.

Rao, L.M.V. & W.A. Newman 1972. Thoacic Cirripedia from guyots of the Mid-Pacific Mountains. *Trans. San Diego Soc. Nat. Hist.* 17: 69-94.

Zullo, V.A., W.A. Newman & A. Ross 1972. Kolosvary Gabor (Gabriel von Kolosvary), 1901-1968. *Crustaceana* 22: 95-102.

1973

Dana, T.F. & W.A. Newman 1973. Massive aggregations and migration of *Acanthaster:* Behavioral responses to severe food limitation. *Micronesica* 9: 205.

Gomez, E.D., D.J. Faulkner, W.A. Newman & C. Ireland 1973. Juvenile hormone mimics: Effect on cirriped crustacean metamorphosis. *Science* 179: 813-814.

Ladd, H.S. & W.A. Newman 1973. Geologic history of Horizon Guyot, Mid-Pacific Mountains: Discussion. *Bull. Geol. Soc. Amer.* 84: 1501-1504.

Ross, A. & W.A. Newman 1973. Revision of the coral-inhabiting barnacles (Cirripedia: Balanidae). *Trans. San Diego Soc. Nat. Hist.* 17: 137-173.

1974

Newman, W.A. 1974a. Two new deep-sea Cirripedia (Ascothoracica and Acrothoracica) from the Atlantic. *J. Mar.Biol. Ass. UK* 54: 437-456.

Newman, W.A. 1974b. Cirripedia. *Encyclopedia Britannica* 4: 641-643.

Newman, W.A. 1974c. Storm frequency and rainfall patterns: Effects on contemporary reef and reef land development. *Abstracts: Int. Symp. on Indo-Pacific Trop. Reef Biol., Guam*, June 23, 1974.

Newman, W.A. & T.F. Dana 1974. *Acanthaster*: Test of the time course of coral destruction. *Science* 183: 103.

Newman, W.A. & H.S. Ladd 1974. Occurrences of modern coral-inhabiting barnacles in the Miocene of Fiji and the Marshall Islands. *Abstracts: Int. Symp. Indo-Pac. Trop. Reef Biol., Guam*, p. 29. June 23, 1974.

Newman, W.A. & H.S. Ladd 1974. Origin of coral-inhabiting balanids (Cirripedia, Thoracica). *Verhandl. Naturf. Ges. Basel* 84: 381-396.

Newman, W.A. & J.T. Tomlinson 1974. Ontogenetic dimorphism in *Lithoglyptes* (Cirripedia, Acrothoracica). *Crustaceana* 27: 204-208.

Newman, W.A., P.A. Jumars & A. Ross 1974. Diversity trends in coral-inhabiting barnacles (Cirripedia, Pyrgomatinae):14. *Abstracts: Int. Symp. Indo-Pacific Trop. Reef Biol., Guam*, p. 14 June 23, 1974.

Keeley, L.S. & W.A. Newman 1974. The Indo-West Pacific genus *Pagurolepas* (Cirripedia, Poecilasmatidae) in Floridan waters. *Bull. Mar. Sci.* 24: 628-637.

Ladd, H.S., W.A. Newman & N.F. Sohl 1974. Darwin Guyot, the Pacific's oldest atoll. *Proc. Second Int. Coral Reef Symp. Brisbane, Australia.* 2: 513-522.

1975

Newman, W.A. 1975. Cirripedia. In: R.I. Smith & J.T. Carlton (eds.) *Intertidal Invertebrates of the Central California Coast*, 259-269. Berkeley: Univ. Cal. Press.

Hessler, R.R. and Newman, W.A 1975. A trilobitomorph origin for the Crustacea. *Fossils and Strata* 4: 437-459.

1976

Newman, W.A. & A. Ross 1976. Revision of the balanomorph barnacles; including a catalogue of the species. *Mem. San Diego Soc. Nat. Hist.* 9: 1-108.

Newman, W.A., P.A. Jumars & A. Ross 1976. Diversity trends in coral-inhabiting barnacles (Cirripedia, Pyrgomatinae). *Micronesica* 12: 69-82.

1977

Newman, W.A. & A. Ross 1977a. Emendation: Superfamilies of the Balanomorpha (Cirripedia, Thoracica). *Crustaceana* 32: 102.

Newman, W.A. & A. Ross 1977b. A living *Tesseropora* (Cirripedia: Balanomorpha) from Bermuda and the Azores: First records from the Atlantic since the Oligocene. *Trans. San Diego Soc. Nat. Hist.* 18: 207-216.

Southward, A.J. & W.A. Newman 1977. Aspects of the ecology andbiogeography of the intertidal and shallow-water balanomorph Cirripedia of the Caribbean and adjacent sea-areas. *FAO Fisheries Rep.* 200: 407-425.

1979

Newman, W.A. 1979a. On the biogeography of balanomorph barnacles of the southern ocean including new balanid taxa; a subfamily, two genera and three species. *Proc. Int. Symp. Mar. Biogeogr. Evol. Sthn. Hemisphere. New Zealand Department of Scientific Industrial Research Information, Ser. 137*, 1: 279-306.

Newman, W.A. 1979b. A new scalpellid (Cirripedia); a Mesozoic relic living near an abyssal hydrothermal spring. *Trans. San Diego Soc. Nat. Hist.* 19: 153-167.

Newman, W. A. 1979c. Californian transition zone: Significance ofshort-range endemics. In: J. Gray and A. J. Boucot (eds.), *Historical Biogeography, Plate Tectonics and the changingEnvironments*, 339-416. Corvallis: Oregon State Univ. Press.

1980

Newman, W.A. 1980a. A review of extant *Scillaelepas* (Cirripedia:Scalpellidae) including recognition of new species from the North Atlantic, Western Indian Ocean and New Zealand. *Tethys* 9: 379-398.

Newman, W.A. 1980b. Crustacea. *World Book Encyclopedia* 4:928-928b.

Newman, W.A. & D.P. Abbott. 1980. Cirripedia In: . R.H. Morris, D.P. Abbott & E.C. Haderlie (eds.), *Intertidal Invertebrates of California*, 504-535. Stanford Univ. Press.

Killingley, J.S., W.H. Berger, K.C. MacDonald & W.A. Newman 1980. $^{18}O/^{16}O$ variations in deep-sea carbonate shells from the Rise Hydrothermal field. *Nature* 288: 218-221.

Schram, F.R. & W.A. Newman 1980. *Verruca withersi* n. sp. (Crustacea: Cirripedia) from the middle Cretaceous of Colombia. *J. Paleo.* 54: 229-233.

Stanley, S.M. & W.A. Newman 1980. Competitive exclusion inevolutionary time: the case of the acorn barnacles. *Paleobiol.* 6: 173-183.

1981

Newman, W.A. 1981a. Huzio Utinomi, 24 February 1910 - 19 June 1979. *Crustaceana* 40: 100-108.

Newman, W.A. & S.M. Stanley. 1981. Competition wins out overall: Reply to Paine. *Paleobiol.* 7: 561-569.

Enright, J.T., W.A. Newman, R.R. Hessler & J.A. McGowan 1981. Deep-ocean hydrothermal vent communities. *Nature* 289: 219-221.

1982

Newman, W.A. 1982a. A review of extant taxa of the `Group of *Balanus concavus*' (Cirripedia; Thoracica) and a proposal for genus-group ranks. *Crustaceana* 43: 25-36.

Newman, W.A. 1982b. Cirripedia. In: L. Abele (ed.), *The Biology of Crustacea* 1:197-221. New York: Academic Press.

Newman, W.A. & B.A. Foster 1982. The Rapanuian faunal district (Easter & Sala y Gomez): Peripheral isolation and speciation across the East Pacific Rise. *Abstracts: Second Int. Symp. Marine Biogeography and Evolution in the Pacific. Sydney, Australia*. July 5-9, 1982.

Dayton, P.K., W.A. Newman and J. Oliver 1982. The vertical zonation of the deep-sea Antarctic acorn barnacle, *Bathylasma corolliforme* (Hoek): Experimental transplants from the shelf into shallow water. *J. Biogeogr.* 9: 95-109.

Killingley, J.S. & W. A. Newman 1982. ^{18}O fractionation in barnacle calcite: A barnacle paleotemperature equation. *J. Mar. Research* 40: 893-902.

1983

Newman, W.A. 1983. Origin of the Maxillopoda; Urmalacostracan ontogeny and progenesis. In: F. Schram (ed.) *Crustacean phylogeny, Crustacean Issues* 1: 105-120. Rotterdam: Balkema.

Newman, W.A. & B.A. Foster 1983. The Rapanuian faunal district (Easter and Sala y Gomez): In search of ancient archipelagos. *Bull. Mar. Sci.* 33: 633-644.

1984

Newman, W.A. & M.D. Knight 1984. The carapace and crustacean evolution - A rebuttal. *J. Crust. Biol.* 4: 682-687.

1985

Newman, W.A. 1985. The abyssal hydrothermal vent invertebrate fauna: A glimpse of antiquity? *Bull. Biol. Soc. Wash.* 6: 231-242.

Newman, W.A. & J.S. Killingley 1985. The north-east Pacificintertidal barnacle *Pollicipes polymerus* in India? Abiogeographical enigma elucidated by ^{18}O fractionation in barnacle calcite. *J. Nat. Hist.* 19: 1191-1196.

Grygier, M.J. and W.A. Newman 1985. Motility and calcareous parts in extant and fossil Acrothoracica (Crustacea; Cirripedia), based primarily upon new species burrowing in the deep-sea scleractinian coral *Enallopsammia. Trans. San Diego Soc. Nat. Hist.* 21: 1-22.

1986

Newman, W.A. 1986. Origin of the Hawaiian marine fauna: Dispersal and vicariance as indicated by barnacles and other organisms. In: R. Gore, R. & K. Heck (eds.) *Biogeography of Crustacea.* Crustacean Issues 4: 21-49. Rotterdam: Balkema.

Foster, B.A. & W.A. Newman 1986. Chthamalid barnacles of Easter Island, Southeast Pacific: Peripheral isolation of the Notochthamalinae subfam. nov. and the hembeli-group of Euraphiinae (Cirripedia: Chthamaloidea). *Program/Abstracts, p. 15. Second Int. Symp. on Indo-Pacific Marine Biol. University of Guam, Agana.*

Newman, W.A. & B.A. Foster 1986. The highly endemic barnacle fauna of the southern hemisphere (Crustacea: Cirripedia): Explained in part by extinction of northern hemisphere members of previously amphitropical taxa? *Program/Abstracts Second Int. Symp. on Indo-Pacific Marine Biol. University of Guam, Agana*, p. 24.

1987

Newman, W.A. 1987. Evolution of Cirripedes and their major groups. In: A.J. Southward (ed.) *Barnacle Biology. Crustacean Issues* 5: 3-42. Rotterdam: Balkema.

Foster, B.A. & W.A. Newman 1987. Chthamalid barnacles of Easter Island, Southeast Pacific: Peripheral Pacific Isolation of Notochthamalinae subfam. nov. and the hembeli-group of Euraphiinae (Cirripedia: Chthamaloidea). *Bull. Mar. Sci.* 41: 322-336.

Newman, W.A. & B.A. Foster 1987. The highly endemic barnacle fauna of the southern hemisphere (Crustacea; Cirripedia): Explained in part by extinction of northern hemisphere members of previously amphitropical taxa? *Bull. Mar. Sci.* 41: 361-377.

1988

Newman, W.A. 1988. The spectacular Neogene radiation of coralbarnacles: A triumph over a background of relictualization and extinction in numerous other groups. *Third International Symposium on Marine Biogeography & Evolution in the Pacific. Abstracts*, p. 19. University of Hong Kong, 26 June-3 July 1988.

Newman, W.A. & R.R. McConnaughey 1988. The tropical eastern Pacific barnacle, *Megabalanus coccopoma* (Darwin), in Southern California, following El Nino 1982-83. *Pac. Sci.* 41: 31-36.

1989

Newman, W.A. 1989a. Barnacle taxonomy. *Nature* 337: 23-24.

Newman, W.A. 1989b. Juvenile ontogeny and metamorphosis in the most primitive living sessile barnacle, *Neoverruca*, from an abyssal hydrothermal spring. *Bull. Mar. Sci.* 45: 467-477.

Newman, W.A. & R.R. Hessler 1989. A new abyssal hydrothermal Verrucomorphan (Cirripedia; Sessilia): The most primitiveliving sessile barnacle. *Trans. San Diego Soc. Nat. Hist.* 21: 259-273.

1990

Newman, W.A. 1990a. Biotic cognates of eastern boundary conditions in the Pacific and Atlantic; Relicts of Tethys and climatic change? *Abstracts and Program, Symposium on Crustacea of the Eastern Pacific. Estacion Mazatlan, Instituto de Ciencias del Mar y Limnologia, Universidad Nacional Autonoma de Mexico*. p. 32. March 1990.

Newman, W.A. 1990b. Origins of southern hemisphere endemism,especially among marine Crustacea. *Conference Handbook, Third International Crustacean Conference, Brisbane, Australia*, 2-6 July 1990. University of Queensland, p. 77.

Newman, W.A. 1990c. Carl August Nilsson-Cantell, 28 December 1893 - 14 January 1987. *Crustaceana* 59: 89-294.

Newman, W.A. 1990d. Cirripedes. *Encyclopedia Britannica, 15th Edition* Vol. 16: 49-854 & 859.

Newman, W.A. 1990e. The Neogene radiation of coral barnacles; a triumph over a background of reliction and extinction. *Abstracts: Bull. Mar. Sci.* 47: 258.

Yamaguchi, T. & W.A. Newman 1990. A new and primitive barnacle(Cirripedia; Balanomorpha) from the North Fiji Basin abyssal hydrothermal field, and its evolutionary implications. *Pac. Sci.*44: 135-155.

1991

Newman, W.A. 1991a. Capricorn Guyot and the Tonga Trench: The drowning of a Neogene coral reef by subduction. *Abstracts: Geol. Soc. Amer., 87th Annual Cordilleran Section, Abstracts with programs*, Abstract No. 18676, p. 83.

Newman, W.A. 1991b. Origins of Southern Hemisphere Endemism, especially among marine Crustacea. *Mem. Queensland Mus.* 31: 51-76.

Grygier, M.J. & W.A. Newman 1991. A new genus and two new species of Microlepadidae (Cirripedia: Pedunculata) found on western Pacific diadematid echinoids. *Galaxia* 9: 1-22.

Spivey, H.R., Newman & V.A. Zullo 1991. Norman Edward Weisbord (1901-1990). *Crustaceana* 61: 88-92.

1992

Newman, W.A. 1992a. Biotic cognates of eastern boundary conditions in the Pacific and Atlantic: Relicts of Tethys and climatic change. *Proc. San Diego Soc. Nat. Hist.* 16: 1-7.

Newman, W.A. 1992b. Origin of the Maxillopoda. *Acta Zoologica(Stockholm)* 73: 319-322.

Buckeridge, J.S. & W.A. Newman 1992. A reexamination of *Waikalasma* (Cirripedia: Thoracica) and its significance in balanomorph phylogeny. *J. Paleo.* 66: 341-345.

Lowenstam, H.A., S. Weiner & W. A. Newman 1992. Carbonate apatite-containing plates of a barnacle (Cirripedia). In: H.Slavkin & P. Price (eds.). *Chemistry and biology of mineralized tissues*, 73-84. New York: Elsevier.

Van Syoc, R.J. & W.A. Newman 1992. Genetic divergence between reproductive types in northern and southern populations of the edible goose barnacle, *Pollicipes. California Sea Grant Biennial Report of Completed Projects 1988-90. University of California, La Jolla.* Publ. No. R-CSGCP-033: 210-211.

1993

Newman, W.A. 1993. Darwin and Cirripedology. In: F.M. Truesdale (ed.), *History of Cirripedology*. Crustacean Issues 8: 349-434. Rotterdam: Balkema.

1994

Buckeridge, J.S. & W.A. Newman 1994. Brian Arthur Foster (10 July 1942-26 June 1992). *Crustaceana* 66: 247-252.

Newman, W.A., J.S. Buckeridge, R.W. Portell & H.R. Spivey 1994. Victor August Zullo (24 July 1936-16 July 1993). *J. Crust. Biol.* 14: 399-405.

1995

Newman, W.A. & T. Yamaguchi 1995. A new sessile barnacle (Cirripedia: Balanomorpha) from the Lau Back-Arc Basin, Tonga; first record of a living representative since the Miocene. *Bull. Mus. nat. d'Hist. natur., Paris* (in press).

Ross, A. & W.A. Newman 1995. A coral-eating barnacle revisited. *Contrib. to Zool.* (in press).

A revision of the scalpellomorph subfamily Calanticinae *sensu* Zevina (1978)

D.S. Jones
Western Australian Museum, Perth, Australia

N. S. Lander
W.A. Herbarium, Como, Western Australia

ABSTRACT

The confused taxonomy of the scalpellomorph subfamily Calanticinae is briefly reviewed. A preliminary revision of the scalpellomorph subfamily Calanticinae in the sense of Zevina (1978) is being undertaken, using the DELTA System to expand and manage the relevant character data. Some of the difficulties involved in preparation of data for manipulation by the DELTA System are discussed. The DELTA System facilitates the application of computer technology to the organisation and manipulation of taxonomic data. DELTA is a flexible, data-coding format for taxonomic descriptions with an associated set of programs for producing natural language descriptions and keys, for interactive identification and information retrieval, and for conversion of the data to formats required for phylogenetic and phenetic analysis.

1 INTRODUCTION

A satisfactory diagnosis of the scalpellid subfamily Calanticinae has yet to be produced. Several genera have been assigned to the subfamily, but the number of genera varies between different authors. Zevina (1978) defined the Calanticinae as follows:

> With 9-18 fully delineated plates; umbo of the carina apical; umbo of the scutum usually apical, although sometimes subcentral; hermaphrodites with complemental males clearly divided into a peduncle and a capitulum with six or more small capitular plates.

Zevina included five genera in the Calanticinae, namely *Calantica* Gray, 1825, *Euscalpellum* Hoek, 1907, *Paraclantica* Utinomi, 1949, *Scillaelepas* Seguenza, 1867 and *Smilium* Gray, 1825. These genera are distributed in tropical shallow waters, with some also occuring in deep and some in boreal waters.

More recently Buckeridge (1983) also includeded fossil genera in his consideration of the Calanticinae. He placed the genera *Calantica* and *Smilium,* together with *Pollicipes* Leach, 1817, *Pisiscalpellum* Utinomi, 1958, *Capitulum* Gray, Oken, 1815 and the fossil genera *Zeugmatolepas,* Withers, 1913 and *Titanolepas* Withers in the Calanticinae. Buckeridge recognised *Scillaelepas* Seguenza as a subgenus of *Calantica* and placed *Euscalpellum* in the Scalpellinae. He defined calanticinids as having:

> 12-34 plates, comprising a large rostrum, carina, and paired scuta, terga, carinolatera, upper latera, rostrolatera and (except with *Pisiscalpellum*) a subcarina; carinas with apical umbos; valves arranged in 2 more or less distinct whorls, with those in the lower whorl either overlapping or being over lapped by adjacent valves; with a cosmopolitan distribution, from the Upper Jurassic to Recent.

Buckeridge also suggested that *Smilium* was the most derived genus of the subfamily and furthermore that *Smilium, Calantica* and *Euscalpellum* may form a distinct group within the Calanticinae.

Recent acquisitions of calanticinid material from Australian waters uncovered new taxa which have proved difficult to determine as they fall outside the limitations presently attributed to recognised calanticinid species and genera. It has become clear that the Calanticinae is ambiguously defined and that several of the genera therein are not homogeneous. We are seeking to begin a resolution of this problem by an analysis of the subfamily Calanticinae using the DELTA System. Initially we are considering the extant species in the genera *Calantica, Euscalpellum, Paraclantica, Scillaelepas* and *Smilium* (subfamily Calanticinae *sensu* Zevina, 1978), but the analysis will be extended to include all the genera, extant and fossil, which have been considered within the subfamily.

2 THE DELTA SYSTEM

DELTA stands for DEscription Language for TAxonomy, a standardised format for coding taxonomic descriptions. DELTA-format data can be converted into natural language descriptions, and into formats required by programs for key generation, interactive identification and information retrieval, phenetic analysis, and phylogenetic analysis. Not only does the DELTA System facilitate the application of computer technology to the organisation and manipulation of taxonomic data, but it also encourages the conscientious acquisition of properly comparative data and ensures that due attention is paid to questions of characters and character-state definitions and their interdependence. Such rigor is essential for any subsequent rational classificatory argument or phylogenetic analysis.

The DELTA System was originally developed by Dallwitz (1980). The programs for processing the DELTA-format data have been prepared by the C.S.I.R.O. Division of Entomology, Canberra. The latest current edition of the DELTA Primer is Partridge et al. (1993), and of the DELTA User's Guide is Dallwitz et al. (1993).

All applications of the DELTA System are based on information contained in three files: the characters (CHARS), the items (ITEMS) and the specifications (SPECS). The CHARS file contains a numbered list of the characters and (where relevant) the states which are to be used to describe the taxa (see Appendix 1). Each character consists of a feature and a set of states. Five main types of characters are recognised: 1) unordered multistate (UM) – two or more states, with no relation of ordering between them; 2) ordered multistate (OM) – two or more states with a relation of ordering between them; 3) integer numeric (IN) – a measurement which is always a whole number; 4) real numeric (RN) – a measurement which may take fractional values; and 5) text (T) –-any text.

The ITEMS file contains descriptions of the taxa, coded in terms of the character and state numbers (see Appendix 2). An item is a taxon name and description in terms of the character set. The description consists of character numbers and state values (attributes). Alternative state values, ranges of state values, and states existing in combination can be accomodated, and combinations, ranges and alternatives may be combined. For integer or real numeric charcters, the outermost values of a range may be enclosed in parentheses, to denote rarely occurring extreme values. If the state of a character is unknown, then the character is omitted, or the state value coded as U. If the character is inapplicable, the state may be coded. Alternatively, the logical dependency between characters may be specified.

The SPECS file contains information about the characters and items, for example the total number of characters, the maximum number of states, the maximum number of items, or the dependent characters (see Appendix 3).

CONFOR is a format-conversion program that converts the coded descriptions into natural language, and into formats required by programs for key-generation, interactive identification, information retrieval, phylogenetic analysis, and phenetic analysis. For example, the DELTA-format data, prepared as indicated above, is converted to the format required by the key-generation program CONFOR. Then the program KEY (Dallwitz 1974) is run to produce a conventional key suitable for printing. CONFOR also converts the data to the form required by INTKEY (Watson et al. 1989). The program INTKEY can identify a specimen by comparing its attributes with stored descriptions of taxa. The program can also be used to interrogate the stored data. Natural language descriptions may also be produced by CONFOR.

3 CHARACTER DISCUSSION

Emphasis in our research has been placed on developing a comprehensive, detailed character list for each taxon of calanticine. This encompasses not only details of the number, shape and relative positions of the capitular plates (the standard means of calanticine species description by many previous authors), but also details where known of things like soft part morphology, ontogenetic development, infra-specific variation, complemental or dwarf males, and ova. Information has been amassed from the examination of actual specimens whenever possible. The various techniques that may be used to prepare comprehensive species descriptions have been described by Jones (1993). Details have also been included from all known published descriptions and illustrations for each taxon.

Comprehensive descriptions of extant species were prepared for *Calantica* (26 described species but by synonymy probably only 19 species, plus possibly 2 or 3 new species), *Euscalpellum* (6 described species), *Paraclantica* (3 described species), *Smilium* (18 described species but by synonymy probably 19 species, plus possibly 2 new species) and *Scillaelepas* (11 described species but by synonymy probably 9 actual species). From these descriptions 227 features were used to generate a list of characters and character-state definitions for organisation and manipulation using the DELTA System (Appendix 1).

Initially we scored a single species (*Scillaelepas arnaudi* Newman, 1980) from its written description, using the character list. However, when considering the species descriptions for all genera, we found that a great deal of additional characters were available, which we felt would be dangerous to neglect, especially as we had used a template of sorts for scoring. For the time being we have provisionally scored this additional data as comments, although this is hardly satisfactory.

Two descriptions of *S. arnaudi* were then produced by CONFOR. One omitted all comments (see Appendix 4); the other included them (see Appendix 5). By comparing them the magnitude of the uncoded data becomes obvious. We must also point out that, although such uncoded data can read quite well in natural language translations, in this form it will not be available for interrogation by INTKEY, or for use directly in keys, or for use in conversion to formats required by analytical programs.

This causes a dilemma on how to proceed with the data preparation. On the one hand, data for many of the characters in our current list seems unavailable; on the other hand, a great deal of additional data is available. Ideally we should fill in the missing data for our current character set and formulate the necessary characters for the additional data. On the other hand, we could simply score available data against the current character set, either with or without scoring the additional data as comments.

Therefore, assuming the missing data is unavailable at present, the three choices are as follows:

A) Score available data against the current character set *without* the additional data as comments;

B) Score available data against the current character set *with* the additional data as comments;

C) Formulate the necessary characters for the additional data and score all available data against the extended character set.

In terms of speed A would be the solution; B would take longer but give superior descriptions; C would take much longer but would result in a superior interactive system.

4 CONCLUSION

The thorough going adoption of the DELTA System in this ongoing revision of the scalpellomorph subfamily Calanticinae *sensu* Zevina (1978) will yield many advantages to all aspects of this study. Apart from the ability to output uniform, strictly comparative descriptions and to construct conventional identification keys, DELTA can be used to transform data to the various forms necessary for input into both phenetic and phyletic programs thus aiding in analysis and in the construction of a system of classification within the subfamily. Further, the powerful interactive program INTKEY will enable the researchers to explore the data at will interactively. In combination with phylogenetic programs, this could include assessment of the practicality of classifications suggested by various analyses performed by ourselves or by others (e.g., Zevina, 1978; Buckeridge, 1983). In particular, rigorous descriptions of putative taxa combining data from their components can be produced automatically. For example, a description of a putative genus derived from the data scored for its component species can be output by INTKEY and used in subsequent analysis, key construction, or exploration.

Additionally, INTKEY can serve as a sophisticated vehicle for the presentation of taxonomic data concerning this subfamily, combining descriptive material with computer generated maps, scanned illustrations and photographs, and even sounds, if relevant. A further refinement that we plan to implement is the use of DELTA to output descriptive text and images in HTML (Hyper Text Markup Language) format, thus making the results of our taxonomy available in a form suitable for presentation on Internet's World Wide Web where it can be read elegantly and intuitively by means of a hypertext browser such as MOSAIC.

We are convinced of the power, usefulness and flexibility of the DELTA approach to taxonomic studies of the Thoracica, and indeed for any taxonomic group. We are decidedly of the view that before further scoring is undertaken our character list should be extended to encompass all available morphological data concerning members of the Scalpellid subfamily Calanticinae, thus bringing the interactive system to bear on the full body of descriptive data available.

REFERENCES

Buckeridge, J.S. 1983. Fossil barnacles (Cirripedia: Thoracica) of New Zealand and Australia. New Zealand *Geol. Survey Paleontol. Bull.* 50: 1-151.

Dallwitz, M.J. 1974. A flexible computor program for generating identification keys. *Syst. Zool.* 23: 50-7.

Dallwitz, M.J. 1980. A general system for coding taxonomic descriptions. *Taxon* 29: 41-6.

Dallwitz, M.J., T.A. Paine, & E.J. Zurcher 1993. *'User's Guide to the DELTA System: a General System for Processing Taxonomic Descriptions.'* 4th edition. 136pp. Canberra: C.S.I.R.O. Division of Entomology.

Jones, D.S. 1993. Techniques to investigate the structure of scalpellomorph barnacles (Thoracica: Pedunculata). *J. Crust. Biol.* 13: 343-348.

Newman, W.A. 1980. A review of extant *Scillaelepas* (Cirripedia: Scalpellidae) including recognition of new species from the North Atlantic, western Indian Ocean, and New Zealand. *Tethys* 9:379-398.

Partridge, T.R., M.J. Dallwitz, & L. Watson 1993. *'A Primer for the DELTA System.'* 3rd edition. 15pp. Canberra: C.S.I.R.O. Division of Entomology.

Watson, L., G.E. Gibbs Russell, & M.J. Dallwitz 1989. Grass genera of southern Africa: interactive identification and information retrieval from an automated data bank. *S. Afr. J. Bot.* 55: 452-63.

Zevina, G.B. 1978. A new classification of the Scalpellidae (Cirripedia, Thoracica). 1. Subfamilies Lithotryinae, Calanticinae, Pollicipinae, Scalpellinae, Brochiinae, and Scalpellopsinae. *Zool. Zh.* 57: 998-1007. [In Russian].

APPENDIX 1

*SHOW Barnacle species character list. 21 July 1994. Revised Fri Jul 22 16:35:23 1994
*CHARACTER LIST
#1. <synonyms: 'species' included in the current description>/

#2. distribution <geographical>/
1. cosmopolitan/
2. tropical <23°27'N to 23°27'S>/
3. subtropical <23°27' to 34°00' in either hemisphere>/
4. temperate <34°00' to 60°00'in either hemisphere>/
5. subpolar <60°00' to 70°00' in either hemisphere>/

6. polar <70° or greater in either hemisphere>/

#3. distribution <depth>/
1. littoral <0 m>/
2. sublittoral <0-200 m>/
3. deep <200-1000 m>/
4. bathyal <1000-4000 m>/
5. abyssal <4000-6000 m>/

#4. <species, secondary sexual dimorphic type>/
1. simultaneous hermaphrodite <all individuals hermaphrodite>/
2. dioecious <large female, small attached (dwarf) males>/
3. androdioecious <large hermaphrodite, small attached (complemental) males>/

#5. capitulum <mature hermaphrodite or female, maximum height>/ mm high/

#6. capitulum <mature hermaphrodite or female, maximum width>/ mm wide/

#7. capitulum <hermaphrodite or female, form>/
1. triangular or almost so/
2. square/
3. ovate/
4. elongate/

#8. capitulum <hermaphrodite or female, shape>/
1. breadth much less than height/
2. breadth equal to height/
3. breadth much more than height/

#9. capitulum <hermaphrodite or female, form>/
1. laterally compressed/
2. not laterally compressed/

#10. capitulum <hermaphrodite or female, form>/
1. inflated/
2. not inflated/

#11. capitulum <hermaphrodite or female, inflated form>/
1. in lower half only/
2. in upper half only/
3. generally/

#12. <capitulum> carinal margin <hermaphrodite or female, shape>/
1. more strongly curved or angled than occludent/
2. as strongly curved or angled as occludent/
3. less strongly curved or angled than occludent/

#13. <capitulum> occludent margin <hermaphrodite or female, shape>/
1. straight or almost so/
2. curved/
3. sinuate/
4. angularly indented/
5. angularly indented/

#14. capitular plates <hermaphrodite or female, number>/

#15. capitular plates <hermaphrodite or female, color>/

#16. capitular plates <hermaphrodite or female, arrangement>/
1. in 2 whorls/
2. in 3 whorls/
3. in >3 whorls/

#17. capitular plates <hermaphrodite or female, placement>/
1. approximate/
2. slightly separated/
3. well separated/

#18. capitular plates <of lower (second) whorl, hermaphrodite or female, size>/
1. moderately large/
2. small/

#19. capitular plates <of lower (second) whorl, hermaphrodite or female, imbrication>/
1. overlapping bases of upper whorl/
2. not overlapping bases of upper whorl/

#20. capitular plates <hermaphrodite or female, thickness>/
1. thin <delicate>/
2. moderately thick/
3. very thick/

#21. capitular plates <hermaphrodite or female, surface>/
1. smooth/
2. with fine furrows/
3. with deep furrows/

#22. umbos <of capitular plates, hermaphrodite or female, position>/
1. all apical/
2. all apical except scutum <sub-apical or other>/
3. all apical except scutum and tergum <sub-apical or other>/
4. all apical except scutum, carina and upper latus <sub-apical or other>/
5. all apical except scutum, carina and 4 pairs of latera <sub-apical or other>/

#23. <capitular plates> chitinous membrane <hermaphrodite or female, presence>/
1. present/
2. absent/

#24. chitinous membrane <capitular plates, hermaphrodite or female, form>/
1. thin/
2. thick/

#25. chitinous membrane <capitular plates, hermaphrodite or female, form>/
1. naked <without setae or hairs>/
2. velvety or setose/

#26. carina <hermaphrodite or female, presence>/
1. present/
2. absent/

#27. carina <separation>/
1. widely separated from carinolateral/
2. narrowly separated from carinolateral/
3. closely approximate to carinolatera/

#28. carina <shape>/
1. straight or nearly so/
2. bowed <evenly bent>/
3. angular/

#29. <carina> angular bend <position>/
1. in middle of plate/
2. towards top of plate/
3. at top of plate/

#30. <carina> umbo <position>/
1. medial/
2. subapical/
3. apical/

#31. <carina> apex <form>/
1. pointed/
2. rounded/

#32. <carina> apex <length>/
1. reaching beyond apex of tergum/
2. reaching middle of tergum or beyond/
3. just reaching or slightly extending along lower tergum/

#33. <carina> apex <extent>/
1. extending between terga/
2. not extending between terga/

#34. <carina> roof <form>/
1. dorsally rounded or curved/
2. bordered dorsally by ridges/

#35. <carina> roof <width>/
1. narrow/
2. wide/

#36. <carina> base <form>/
1. rounded/
2. flat <not rounded>/

#37. tergum <hermaphrodite or female, presence>/
1. present/
2. absent/

#38. tergum <hermaphrodite or female, position>/
1. occupying whole space between scutum and carina/
2. not occupying whole space between scutum and carina/

#39. tergum <form>/
1. rhombiform or quadrangular/
2. triangular or almost so/
3. triangular with depression in basal margin/
4. bifurcate/

#40. tergum <size>/
1. large <most of length of capitulum>/
2. moderate <½ length of capitulum>/
3. small <less than ½ length of capitulum>/

#41. tergum <length>/
1. projecting above carinal apex/
2. not projecting above carinal apex/

#42. tergum <height>/
1. greater than width/
2. equal to width/

#43. <tergum> umbo <position>/
1. apical/
2. subapical/
3. marginal/
4. medial/
5. basal/

#44. <tergum> apex <form>/
1. recurved towards carina/
2. erect/

#45. <tergum> scutal margin <form>/
1. straight/
2. curved/
3. excavated/

#46. <tergum> carinal margin <form>/
1. straight/
2. curved/
3. excavated/

#47. scutum <hermaphrodite or female, presence>/
1. present/
2. absent/

#48. scutum <size relative to capitulum>/
1. large <most of length of capitulum>/
2. moderate <½ length of capitulum>/
3. small <less than ½ length of capitulum>/

#49. scutum <size relative to tergum>/
1. larger than tergum/
2. same size as tergum/
3. smaller than tergum/

#50. scutum <length>/
1. shorter than carina/
2. equal to carina/
3. longer than carina/

#51. scutum <shape>/
1. triangular/
2. square/
3. rectangular/
4. pentangular/

#52. scutum <height>/
1. more than 2 × width/
2. equal to width/
3. less than 2 × width/

#53. <scutum> umbo <position>/
1. apical/
2. subapical/
3. marginal/
4. medial/
5. basal/

#54. <scutum> apex <extent>/
1. produced into a flat, spear-shape projection/
2. not produced/

#55. <scutum> apex <form>/
1. slightly overlapping tergum/
2. not overlapping tergum/

#56. <scutum> basal margin <form>/
1. straight/
2. convex/

#57. <scutum> carino-basal margin <form>/
1. straight/
2. convex/
3. deeply excavated/

#58. <scutum> tergal margin <form>/
1. straight/
2. convex/
3. sinuate/
4. concave/
5. angularly indented/
6. irregular/

#59. occludent margins of scutum and tergum <form>/
1. forming an even curve/
2. straight or almost so/

#60. upper latus <hermaphrodite or female, presence>/
1. present/
2. absent/

#61. upper latus <form>/
1. large/
2. small/

#62. upper latus <form>/
1. triangular/
2. elongate/
3. banana-shaped/
4. pentagonal/

#63. <upper latus> umbo <position>/
1. apical/
2. subapical/
3. marginal/
4. medial/
5. basal/

#64. <upper latus> distal portion <form>/
1. covering lower parts of tergum and scutum/
2. not covering tergum and scutum/

#65. <upper latus> basal margin <form>/
1. overlapped by rostrolatus and carinolatus/
2. not overlapped by rostrolatus and carinolatus/

#66. <upper latus> scutal margin <form>/
1. straight/
2. concave/

#67. <upper latus> carinal margin <form>/
1. straight/
2. concave/

#68. subcarina <hermaphrodite or female, presence>/
1. present/
2. absent/

#69. subcarina <hermaphrodite or female, size>/
1. large/
2. small/

#70. subcarina <hermaphrodite or female, form>/
1. ½ cone-shaped <very curved>/
2. horn-shaped <moderately curved>/
3. triangular <barely or not curved>/
4. rectangular <barely or not curved>/

#71. subcarina <hermaphrodite or female, form>/
1. lying close beneath carina/
2. lying distantly beneath carina/

#72. subcarina <hermaphrodite or female, form>/
1. overlapping basal portion of carinolatus/
2. not overlapping basal portion of carinolatus/

#73. subcarina <hermaphrodite or female, form>/
1. projecting backwards/
2. erect/

#74. subcarina <hermaphrodite or female, form>/
1. strongly projecting beyond carinal margin/
2. projecting very little beyond carinal margin/

#75. <subcarina> umbo <hermaphrodite or female>/
1. apical/
2. subapical/
3. marginal/
4. medial/
5. basal/

#76. carinolatus <hermaphrodite or female, presence>/
1. present/
2. absent/

#77. carinolatus <hermaphrodite or female, size>/
1. large/
2. small/

#78. carinolatus <shape>/
1. triangular/
2. square/
3. rectangular/
4. pentagonal/

#79. carinolatus <shape>/
1. broader than high/
2. as broad as high/
3. higher than broad/

#80. carinolatus <position>/
1. obliquely set/
2. straight set/

#81. carinolatus <form>/
1. meeting behind carina/
2. not meeting behind carina/

#82. <carinolatus> umbo <position>/
1. apical/
2. subapical/
3. marginal/
4. medial/
5. basal/

#83. <carinolatus> apex <form>/
1. recurved/
2. erect/

#84. <carinolatus> apex <form>/
1. directed towards carina/
2. directed towards scutum/

#85. <carinolatus> apex <position>/
1. extending between carina and upper latus/
2. curving towards scutum and over-lying scutal baso-tergal margin/
3. erect and lying between scutal baso-tergal angle and tergal baso-scutal angle/

#86. carinolatus <form>/
1. extending freely out from capitulum/
2. just reaching edge of capitulum/
3. not extending out from capitulum/

#87. <carinolatus> upper angle <projection>/
1. produced outwards from capitulum/
2. not produced/

#88. inframedian latus <hermaphrodite or female>/
1. present/
2. absent/

#89. inframedian latus <size>/
1. large/
2. small/

#90. inframedian latus <size>/
1. larger than upper latus/
2. same size as upper latus/
3. smaller than upper latus/

#91. inframedian latus <shape>/
1. triangular/
2. 4-sided/
3. 5-sided/
4. 6-sided/
5. rounded/

#92. inframedian latus <size>/
1. reaching upper latus/
2. not reaching upper latus/

#93. inframedian latus <form>/
1. with small apical depression/
2. without small apical depression/

#94. <inframedian latus> upper parts <form>/
1. curving outwards/
2. erect/

#95. <inframedian latus> umbo <position>/
1. apical/
2. subapical/
3. marginal/

4. medial/
5. basal/

#96. rostrolatus <hermaphrodite or female, presence>/
1. present/
2. absent/

#97. rostrolatus <size>/
1. large/
2. small/

#98. rostrolatus <form>/
1. lanceolate/
2. square/
3. triangular/
4. irregular/

#99. rostrolatus <form>/
1. wider than high/
2. as wide as high/
3. higher than wide/

#100. rostrolatus <size>/
1. larger than carinolatus/
2. same size as carinolatus/
3. smaller than carinolatus/

#101. <rostrolatus> umbo <position>/
1. apical/
2. subapical/
3. marginal/
4. medial/
5. basal/

#102. rostrolatus <form>/
1. lying transversely/
2. not lying transversely/

#103. <rostrolatus> apex <form>/
1. directed towards scutum/
2. directed towards rostrum/

#104. rostrolatus <form>/
1. definitely produced backwards <extending below base of carinolatus for some distance>/
2. slightly produced backwards <just extending below base of carinolatus>/
3. not produced backwards below base of carino-latus/

#105. other latera <hermaphrodite or female, presence>/
1. present/
2. absent/

#106. other latera <hermaphrodite or female, number of pairs>/

#107. other latera <size>/
1. large/
2. small/

#108. other latera <position, in which whorl>/

#109. rostrum <hermaphrodite or female, presence>/
1. present/
2. absent/

#110. rostrum <form>/
1. wholly exposed/
2. slightly embedded/
3. deeply embedded/

#111. rostrum <form>/
1. well developed/
2. small but definite/
3. rudimentary/

#112. rostrum <form>/
1. lying transversely below scutal basal margin/
2. lying straight below scutal basal margin/

#113. rostrum <form>/
1. half cone-shaped <curved>/
2. triangular <slightly or not curved>/
3. pentagonal <slightly or not curved>/

#114. rostrum <form>/
1. overlapped by rostrolatera <not separating RL>/
2. not overlapped by rostrolatera <separating RL>/

#115. rostrum <form>/
1. higher than wide/
2. height equal to width/
3. wider than high/

#116. <rostrum> apex <form>/
1. recurved beneath base of capitular orifice/
2. erect/
3. projecting outwards from capitulum/

#117. <rostrum> umbo <position>/
1. apical/
2. subapical/
3. marginal/
4. medial/
5. basal/

#118. subrostrum <hermaphrodite or female, presence>/
1. present/
2. absent/

#119. subrostrum <hermaphrodite or female, form>/
1. triangular/
2. not triangular <stipulate shape>/

#120. <subrostrum (if triangular)> apex <hermaphrodite or female, form>/
1. curved inwards/
2. erect/

#121. peduncle <mature hermaphrodite or female, maximum height>/ mm high/

#122. peduncle <mature hermaphrodite or female, maximum width>/ mm wide/

#123. peduncle <hermaphrodite or female, length>/
1. shorter than capitulum/
2. as long as capitulum/
3. longer than capitulum/

#124. chitinous or calcareous peduncular scales or spines <hermaphrodite or female, presence>/
1. present/
2. absent <naked>/

#125. chitinous or calcareous peduncular scales or spines <hermaphrodite or female, form>/

#126. chitinous or calcareous peduncular scales or spines <hermaphrodite or female, arrangement>/
1. contiguous/
2. closely spaced/
3. distantly spaced/

#127. chitinous or calcareous peduncular scales or spines <hermaphrodite or female, form>/
1. overlapping/
2. not overlapping/

#128. chitinous or calcareous peduncular scales or spines <hermaphrodite or female, arrangement>/

#129. chitinous or calcareous peduncular scales or spines <hermaphrodite or female, arrangement>/
1. in well-developed growth zone immediately below the capitulo-peduncular junction, but whorls not well developed in this area/
2. not in well-developed growth zone immediately below the capitulo-peduncular junction/

#130. chitinous or calcareous peduncular scales or spines <hermaphrodite or female, form>/
1. decreasing in size towards base/
2. retaining their size towards base/

#131. cirrus I (juvenile form) anterior ramus with <number of segments>/ segments/

#132. cirrus I (juvenile form) posterior ramus with <number of segments>/ segments/

#133. cirrus II (juvenile form) anterior ramus with <number of segments>/ segments/

#134. cirrus II (juvenile form) posterior ramus with <number of segments>/ segments/

#135. cirrus III (juvenile form) anterior ramus with <number of segments>/ segments/

#136. cirrus III (juvenile form) posterior ramus with <number of segments>/ segments/

#137. cirrus IV (juvenile form) anterior ramus with <number of segments>/ segments/

#138. cirrus IV (juvenile form) posterior ramus with <number of segments>/ segments/

#139. cirrus V (juvenile form) anterior ramus with <number of segments>/ segments/

#140. cirrus V (juvenile form) posterior ramus with <number of segments>/ segments/

#141. cirrus VI (juvenile form) anterior ramus with <number of segments>/ segments/

#142. cirrus VI (juvenile form) posterior ramus with <number of segments>/ segments/

#143. cirrus I (adult form) anterior ramus with <number of segments>/ segments/

#144. cirrus I (adult form) posterior ramus with <number of segments>/ segments/

#145. cirrus II (adult form) anterior ramus with <number of segments>/ segments/

#146. cirrus II (adult form) posterior ramus with <number of segments>/ segments/

#147. cirrus III (adult form) anterior ramus with <number of segments>/ segments/

#148. cirrus III (adult form) posterior ramus with <number of segments>/ segments/

#149. cirrus IV (adult form) anterior ramus with <number of segments>/ segments/

#150. cirrus IV (adult form) posterior ramus with <number of segments>/ segments/

#151. cirrus V (adult form) anterior ramus with <number of segments>/ segments/

#152. cirrus V (adult form) posterior ramus with <number of segments>/ segments/

#153. cirrus VI (adult form) anterior ramus with <number of segments>/ segments/

#154. cirrus VI (adult form) posterior ramus with <number of segments>/ segments/

#155. chaetotaxy <type of arrangement of cirri>/
1. ctenopod/
2. lasiopod/
3. acanthopod/

#156. cirrus I <position>/
1. set apart from CII/
2. set close to CII/

#157. cirrus I <form>/
1. similar to cirrus II/
2. dissimilar to cirrus II/

#158. cirri I and II <form>/
1. similar to cirrus III/
2. dissimilar to cirrus III/

#159. rami of cirrus I <size>/
1. equal/
2. subequal/
3. unequal/

#160. rami of cirrus II <size>/
1. equal/
2. subequal/
3. unequal/

#161. rami of cirri III to VI <size>/
1. equal/
2. subequal/
3. unequal/

#162. median segments of cirri III to VI <form>/
1. with bunch of long setae postero-distally/
2. with 1-3 spines postero-distally/
3. without bunch of long setae or 1-3 spines postero-distally/

#163. long setae on anterior margins of segments of cirrus VI in <range>/ pairs/

#164. dense scales on posterior margins of segments of cirri V and VI <presence>/
1. present/
2. absent/

#165. maxilla <form>/
1. quadrangular/
2. triangular/
3. elongated/

#166. maxilla <form>/
1. setose/
2. naked/

#167. maxillule <form>/
1. entire/
2. slightly notched/
3. distinctly notched/

#168. <maxillule> cutting edge <form>/
1. straight/
2. sinuous/
3. at an angle/

#169. <maxillule> cutting edge <form>/
1. extending most of length of margin/
2. extending 1/2 length of margin/
3. extending 1/3 length of margin/
4. extending less than 1/3 length of margin/

#170. <maxillule> cutting edge <form>/
1. with large spines/
2. with medium-sized spines/
3. with small spines/

#171. <maxillule> upper angle <form>/
1. with strong spines/
2. without strong spines/

#172. <maxillule> lower angle <form>/
1. with small strong spines/
2. without small strong spines/

#173. <maxillule> surface <form>/
1. with fine hairs/
2. without fine hairs/

#174. mandible <range>/-toothed/

#175. <mandible> first main tooth <form>/
1. well separated from other teeth/
2. not well separated from other teeth/

#176. <mandible> second main tooth <form>/
1. smallest/
2. not smallest/

#177. <mandible> denticles <form>/
1. present/
2. absent/

#178. <mandible> denticles <form>/
1. between teeth 1 and 2/
2. between teeth 2 and 3/
3. between main teeth/
4. along upper margins of teeth 2-4/

#179. <mandible> upper margin of tooth 3 <form>/
1. with small spines/
2. without small spines/

#180. <mandible> enlarged spine on upper margin of tooth 2 <form>/
1. present/
2. absent/

#181. <mandible> lower angle <form>/
1. pectinate/
2. bifid/
3. entire <undivided>/

#182. <mandible> lower angle <form>/
1. forming an acute point <molariform>/
2. expanded <blunt>/

#183. <mandible> lower part <form>/
1. setose/
2. naked/

#184. labrum <form>/
1. flattened <not bullate>/
2. slightly bullate/
3. bullate/
4. strongly bullate/

#185. labrum/
1. notched/
2. entire/

#186. labrum <dentition>/
1. without small denticles or teeth/
2. with small denticles/
3. with small teeth/

#187. mandibular palp <form>/

#188. caudal appendages <presence>/
1. present/
2. absent/

#189. caudal appendages <form>/
1. unsegmented <uni-articulate>/
2. segmented <multi-articulate>/

#190. caudal appendages <unsegmented form>/
1. leaf-like/
2. elongated/
3. broad and blunt/
4. club-shaped/

#191. caudal appendages <length>/
1. minute <less than $\frac{1}{4}$ length of basal segment of pedicel of CVI>/
2. short <$\frac{1}{4}$ to almost total length of basal segment of pedicel of CVI>/
3. moderate <equal to length of pedicel of CVI>/
4. long <greater than length of pedicel of CVI>/

#192. caudal appendages <form>/
1. densely setose/
2. naked/

#193. <caudal appendages> segments <range>/

#194. <caudal appendages> segments <form>/
1. indistinct/
2. distinct/

#195. <caudal appendages> distal tuft of setae <presence>/
1. present/
2. absent/

#196. <caudal appendages> distal circlet of setae <presence>/
1. present/
2. absent/

#197. penis <form>/
1. short/
2. long/

#198. penis <form>/
1. thin/
2. stout/

#199. penis <form>/
1. compressed laterally/
2. cylindrical/

#200. penis <form>/
1. not annulated <smooth>/
2. faintly annulated/
3. distinctly annulated/

#201. penis <form>/
1. setose <hirsute>/
2. naked/

#202. filamentary appendages <presence>/
1. present/
2. absent/

#203. filamentary appendages <number>/ pairs/

#204. ovigerous frenae <form>/

#205. ova <size>/ μ/

#206. ova <type>/
1. lecithotrophic/
2. planktotrophic/

#207. nauplii <form>/
1. with gnathic spines/
2. without gnathic spines/

#208. nauplii <form>/
1. with flotation setae on appendages/
2. without flotation setae on appendages/

#209. complemental or dwarf male <position>/
1. in scutal pouch at base of occludent margin and base of orifice, just above adductor muscle <normal position>/
2. in deep pits in scutal membrane at occludent margin, just above adductor muscle/
3. in depression just below adductor muscle between scuta/
4. in shallow pocket in sub-rostral region/

#210. complemental or dwarf male <capitulum and peduncle differentiation>/
1. with capitulum and peduncle differentiated/
2. with capitulum and peduncle not differentiated/

#211. complemental or dwarf male <form>/
1. sac-like/
2. globulo-ovoid/
3. elongate-ovoid/

#212. complemental or dwarf male <total height>/ mm high/

#213. complemental male or dwarf <total width>/ mm wide/

#214. <complemental or dwarf male> capitulum <height>/ mm high/

#215. <complemental or dwarf male> capitulum <width>/ mm wide/

#216. <complemental or dwarf male> peduncle <height>/ mm high/

#217. <complemental or dwarf male> peduncle <height>/ mm wide/

#218. <complemental or dwarf male> capitular plates <presence>/
1. present/
2. absent/

#219. <complemental or dwarf male> capitular plates <number>/

#220. <complemental or dwarf male> capitular plates <form>/
1. small <reduced>/
2. well developed <large>/

#221. <complemental or dwarf male> penis <presence>/
1. present/
2. absent/

#222. <complemental or dwarf male> cirri <presence>/
1. present/
2. absent/

#223. <complemental or dwarf male> mouthparts <presence>/
1. present/
2. absent/

#224. <outstanding diagnostic feature; diagnostic character for taxon>/

#225. <references>/

#226. <notes>/

#227. <specimens examined>/

APPENDIX 2

*SHOW Barnacle species. Items file. 22 June 1994 Revised Fri Jul 22 16:35:24 1994

*ITEM DESCRIPTIONS

Scillaelepas arnaudi Newman, 1980/ 1<[*IScillaelepas* (*Aurivillialepas*) *arnaudi*] Newman, 1980: 389, fig. 9 A-F, fig. 10 A-I> 2,4<Walters Shoals, S of Madagascar, SW Indian Ocean> 3,3 4,3<with a single complemental male situated in the cavity between the rostrum and subrostrum> 14,14<including 1 subrostrum> 16,1<upper 2T+2S+1C, lower 1SR+1R+2RL+2ML+2CL+1SC; those of the lower whorl with the apexes curved inwards and their lateral margins, where overlapped by adjacent plates, marked by articular furrows; those of the upper whorl lightly sculpted, primarily by transverse growth lines on the scutum and carina, and 2 sets of chevron-shaped criss-crossed growth lines on the tergum; those of the lower whorl less distinctly sulptured by transverse and/or longitudinal growth lines, and only the medianlaterals and carinolaterals are marked by an articular ridge running from the apex to the base> 19<for the most part>,1<and one another> 21,2 26,1<lightly sculpted, primarily by transverse growth lines> 31<free of carinal margin of tergum; wide in lateral aspect> 32<reaching apex of median latus> 33<free of carinal margin of tergum> 37,1 39<traversed by a narrow median ridge running from the apex to the base; lightly sculpted with 2 sets of chevron-shaped, criss-crossed growth lines> 44,1 47,1 51<lightly sculpted, primarily by transverse growth lines> 60,2 68,1 70<less distinctly sculptured by growth lines (than carina)> 72,1 76,1 78<marked by an articular ridge running from the apex to the base, formed by the overlap between the carinolatus and median latus, and the subcarina and the carinolatus> 83<incurved> 88,1 91<marked by an articular ridge running from the apex to the base, formed by the overlap between the carinolatus and median latus, and the subcarina and the carinolatus> 96,1 98<acute; apical angle 60-70°; less distinctly sculptured by growth lines (than carina)> 105,2 109,1 111<less distinctly sculptured by growth lines (than carina)> 118,1 119<less distinctly sculptured by growth lines (than carina)>,1 124,1 126<imbricating> 127,1<imbricating> 129,1<ie"imbricating scales fitting closely around the bases of the lower whorl of capitular plates, including subrostrum"> 130<relatively large (to what?)> 143,10-11 144,8-9 145,11-12 146,13-14 147,13-14 148,14 149,15 150,13-15 151,16 152,14-16 153,16 154,15-16 155,2<?, from sketch> 156<separated from the posterior folds by an intercalary fold> 157,1<segments about as high as wide and clad with thin spine-like simple setae> 158,1<ie with segments about as high as wide> 159,2 160,2 161, ½ 166,1<supports relatively stout spines above and below the medial notch, and the papilla of the maxillary gland is long and conspicuous> 170,1<long spines of upper cutting edge reach medially as far as those of the lower edge> 174, 3/2 <abnormal, observed on holotype only> 179,1<1, stout> 180,2 183,1<with a comb of stout spines in a single row along the superior and apical margins; about ⅔ of the inferior margin supports a comb of closely spaced setae> 186,2<20 fine, low teeth, arranged more or less in pairs, with individuals of a pair being separted by approximately the diameter of a tooth.> 187<short; lower surfaces and anterior margins support short, simple setae> 188,1 189,1 191,2< ½ the length of the proximal article of CVI> 197<about ½ length of the first article of the pedicel of CVI and, as apparently in all [*IScillaelepas*] examined by Newman (1960), it is apparently immature> 201,1<sparely clothed with fine, short setae> 202,1 203,3<along the median dorsal surface of the prosoma; the first opposite the oral cone, is shaped like the symbol for Aires (first sign of the Zodiac); the second, situated somewhat short of the posterior margin of the prosoma, is in the form of a V that does not quite meet at the base; the third, on the intercalary fold between CI and CII, consists of well-separated dorsally directed, nipple-like projections> 209,4<in the cavity between rostrum and subrostrum; 1 each in 2 of the 3 hermaphrodites examined; lying on their sides> 212<about ½ the size of the males of [*IS.* (*A.*) *bocquetae*]> 218,1 219,6 220<plus a median latus on 1 side; carinolatus absent> 224<Distinguished from [*IS.* (*A.*) *bocquetae*] by the overlap between the rostrolatus and the carinolatus, and from both [*IS.* (*A.*) *bocquetae*] and [*IS.* (*A.*) *calycula*] by acute (60-70°) rather than obtuse (90°) rostrolateral apexes. Distinguished from [IS. (A.) falcata] by the wider carina meeting the apex of the medianlatus rather than that of the carinolaterals, and the carina being apically free of, rather

than closely applied to the carinal margin of the tergum. Distinguished from [*IS. (A.) bocquetae*] and [*IS. (A.) calycula*] by differences in the filamentary appendages and, presumably, from [*IS. (A.) falcata*] in this regard, although its appendages have not yet been described.>

227<[IM/S Marion-Dufresne] Cruise MD. 08 (1976): 6, 33°08.7'S 43°59.7'E, 600 m, Walters Shoals, SW Indian Ocean, 1 6. 3.76, Charcot dredge sample 46, fine sand, 1 specimen; 6, 33°1 1.4'S 44°00.4'E, 620-535 m, Walters Shoals, S of Madagascar, SW Indian Ocean, 1 6. 3.76, beam trawl 47, 2 specimens, each with complemental male; H (6, sample 47) Laboratoire des Crustacés, Museum National d'Histoire Naturelle, Paris, no. entrée 7910; P (6, sample 47) U.S.N.M. Cat. No. 173134; P (6, sample 46) British Museum (Nat. Hist.) Reg. No. 1979-38 2.>

225<Newman, W.A. (1980). A review of extant [*IScillaelepas*] including recognition of new species from the North Atlantic, western Indian Ocean and New Zealand. [ITethys] 9: 379-398.>

APPENDIX 3

*SHOW Barnacle species. Specifications file. 22 June 1994 Revised Fri Jul 22 16:35:22 1994

*NUMBER OF CHARACTERS 227
*MAXIMUM NUMBER OF STATES 6
*MAXIMUM NUMBER OF ITEMS 14
*DATA BUFFER SIZE 12000

*CHARACTER TYPES 1,TE 3,OM 5-6,RN 8,OM 11-12,OM 14,IN 15,TE 16-17,OM 20-21,OM 27-30,OM 32,OM 39-40,OM 45-46,OM 48-50,OM 52,OM 57,OM 79,OM 86,OM 90-91,OM 99-100,OM 104,OM 106,IN 108,TE 110-111,OM 115-116,OM 121-122,RN 123,OM 125,TE 128,TE 131-154,RN 155,OM 159-162,OM 163,IN 167-170,OM 174,IN 184,OM 186,OM 187,TE 191,OM 193,IN 200,OM 203,IN 204,TE 205,RN 212-217,RN 219,IN 224-227,TE

*NUMBERS OF STATES 2,6 3,5 4,3 7,4 8,3 11-12,3 13,5 16-17,3 20-21,3 22,5 27-30,3 32,3 39,4 40,3 43,5 45-46,3 48-50,3 51,4 52,3 53,5 57,3 58,6 62,4 63,5 70,4 75,5 78,4 79,3 82,5 85-86,3 90,3 91,5 95,5 98,4 99-100,3 101,5 104,3 110-111,3 113,3 115-116,3 117,5 123,3 126,3 155,3 159-162,3 165,3 167-168,3 169,4 170,3 178,4 181,3 184,4 186,3 190-191,4 200,3 209,4 211,3

*INAPPLICABLE CHARACTERS 4,1:209-223 10,2:11 23,2:24-25 26,2:27-36 28,1-2:29 37,2:38-46:59 47,2:48-59 60,2:61-67 68,2:69-75 73,2:74 76,2:77-87 88,2:89-95 96,2:97-104 105,2:106-108 109,2:110-117 118,2:119 119,2:120 124,2:125-130 177,2:178 188,2:189-196 189,1:193-194 189,2:190 202,2:203 218,2:219-220

*SPECIAL STORAGE

APPENDIX 4

Barnacles 16:01 20-DEC-94

Scillaelepas arnaudi Newman, 1980

SYNONYMY. *Scillaelepas (Aurivillialepas) arnaudi* Newman, 1980: 389, fig. 9 A-F, fig. 10 A-I.

DISTRIBUTION/DEPTH. Distribution temperate; deep.

SECONDARY SEXUAL DIMORPHIC TYPE. Androdioecious.

CAPITULAR PLATES. Capitular plates 14; in 2 whorls; overlapping bases of upper whorl; with fine furrows.

TERGUM. Apex recurved towards carina.

SUBCARINA. Subcarina overlapping basal portion of carinolatus.

SUBROSTRUM. Subrostrum triangular.

PEDUNCLE. Chitinous or calcareous peduncular scales or spines present; overlapping; in well-developed growth zone immediately below the capitulo-peduncular junction, but whorls not well developed in this area.

CIRRI. Cirrus I (adult form) anterior ramus with 10-11 segments; posterior ramus with 8-9 segments; II (adult form) anterior ramus with 11-12 segments; II (adult form) posterior ramus with 13-14 segments; III (adult form) anterior ramus with 13-14 segments; III (adult form) posterior ramus with 14 segments; IV (adult form) anterior ramus with 15 segments; IV (adult form) posterior ramus with 13-15 segments; V (adult form) anterior ramus with 16 segments; V (adult form) posterior ramus with 14-16 segments; VI (adult form) anterior ramus with 16 segments; VI (adult form) posterior ramus with 15-16 segments.

CHAETOTAXY. Chaetotaxy lasiopod. Cirrus I similar to cirrus II; cirri I and II similar to cirrus III. Rami of cirrus I subequal; II subequal; cirri III to VI equal, or subequal.

MOUTHPARTS. Maxilla setose. Cutting edge with large spines. Mandible 3-toothed, or 2-toothed; upper margin of tooth 3 with small spines; enlarged spine on upper margin of tooth 2 absent; lower part setose. Labrum with small denticles. Mandibular palp short; lower surfaces and anterior margins support short, simple setae.

CAUDAL APPENDAGES. Caudal appendages present; unsegmented; short.

PENIS. Penis setose.

FILAMENTARY APPENDAGES. Filamentary appendages present; 3 pairs.

COMPLEMENTARY MALES Complemental or dwarf male in shallow pocket in sub-rostral region; capitular plates present; capitular plates 6.

OUTSANDING DIAGNOSTIC FEATURES. Distinguished from *S.* (*A.*) *bocquetae* by the overlap between the rostrolatus and the carinolatus, and from both *S.* (*A.*) *bocquetae* and *S.* (*A.*) *calycula* by acute (60-70°) rather than obtuse (90°) rostrolateral apexes. Distinguished from *S.* (*A.*) *falcata* by the wider carina meeting the apex of the medianlatus rather than that of the carinolaterals, and the carina being apically free of, rather than closely applied to the carinal margin of the tergum. Distinguished from *S.* (*A.*) *bocquetae* and *S.* (*A.*) *calycula* by differences in the filamentary appendages and, presumably, from *S.* (*A.*) *falcata* in this regard, although its appendages have not yet been described.

REFERENCES. Newman, W.A. (1980). A review of extant *Scillaelepas* including recognition of new species from the North Atlantic, western Indian Ocean and New Zealand. Tethys 9: 379-398.

BSPECIMENS EXAMINED. M/S Marion-Dufresne Cruise MD. 08 (1976): 6, 33°08.7'S 43°59.7'E, 600 m, Walters Shoals, SW Indian Ocean, 1 6. 3.76, Charcot dredge sample 46, fine sand, 1 specimen; 6, 33°1 1.4'S 44°00.4'E, 620-535 m, Walters Shoals, S of Madagascar, SW Indian Ocean, 1 6. 3.76, beam trawl 47, 2 specimens, each with complemental male; H (6, sample 47) Laboratoire des Crustacés, Museum National d'Histoire Naturelle, Paris, no. entrée 7910; P (6, sample 47) U.S.N.M. Cat. No. 173134; P (6, sample 46) British Museum (Nat. Hist.) Reg. No. 1979-38 2.

APPENDIX 5

Barnacles 16:00 20-DEC-94

Scillaelepas arnaudi Newman, 1980

SYNONYMY. *Scillaelepas (Aurivillialepas) arnaudi* Newman, 1980: 389, fig. 9 A-F, fig. 10 A-I.

DISTRIBUTION/DEPTH. Distribution temperate (Walters Shoals, S of Madagascar, SW Indian Ocean); deep.

SECONDARY SEXUAL DIMORPHIC TYPE. Androdioecious (with a single complemental male situated in the cavity between the rostrum and subrostrum).

CAPITULAR PLATES. Capitular plates 14 (including 1 subrostrum); in 2 whorls (upper 2T+2S+1C, lower 1SR+1R+2RL+2ML+2CL+1SC; those of the lower whorl with the apexes curved inwards and their lateral margins, where overlapped by adjacent plates, marked by articular furrows; those of the upper whorl lightly sculpted, primarily by transverse growth lines on the scutum and carina, and 2 sets of chevron-shaped criss-crossed growth lines on the tergum; those of the lower whorl less distinctly sulptured by transverse and/or longitudinal growth lines, and only the medianlaterals and carinolaterals are marked by an articular ridge running from the apex to the base); for the most part overlapping bases of upper whorl (and one another); with fine furrows.

CARINA. Apex free of carinal margin of tergum; wide in lateral aspect; reaching apex of median latus; free of carinal margin of tergum.

TERGUM. Tergum traversed by a narrow median ridge running from the apex to the base; lightly sculpted with 2 sets of chevron-shaped, criss-crossed growth lines; apex recurved towards carina.

SCUTUM. Scutum lightly sculpted, primarily by transverse growth lines.

SUBCARINA. Subcarina less distinctly sculptured by growth lines (than carina); overlapping basal portion of carinolatus.

CARINOLATUS. Carinolatus marked by an articular ridge running from the apex to the base, formed by the overlap between the carinolatus and median latus, and the subcarina and the carinolatus; apex incurved.

INFRAMEDIAN LATUS. Inframedian latus marked by an articular ridge running from the apex to the base, formed by the overlap between the carinolatus and median latus, and the subcarina and the carinolatus.

ROSTROLATUS. Rostrolatus acute; apical angle 60-70°; less distinctly sculptured by growth lines (than carina).

ROSTRUM. Rostrum less distinctly sculptured by growth lines (than carina).

SUBROSTRUM. Subrostrum less distinctly sculptured by growth lines (than carina), triangular.

PEDUNCLE. Chitinous or calcareous peduncular scales or spines present; imbricating; overlapping (imbricating); in well-developed growth zone immediately below the capitulo-peduncular junction, but whorls not well developed in this area (ie"imbricating scales fitting closely around the bases of the lower whorl of capitular plates, including subrostrum"); relatively large (to what?).

CIRRI. Cirrus I (adult form) anterior ramus with 10-11 segments; posterior ramus with 8-9 segments; II (adult form) anterior ramus with 11-12 segments; II (adult form) posterior ramus with 13-14 segments; III (adult form) anterior ramus with 13-14 segments; III (adult form) posterior ramus with 14 segments; IV (adult form) anterior ramus with 15 segments; IV (adult form) posterior ramus with 13-15 segments; V (adult form) anterior ramus with 16 segments; V (adult form) posterior ramus with 14-16 segments; VI (adult form) anterior ramus with 16 segments; VI (adult form) posterior ramus with 15-16 segments.

CHAETOTAXY. Chaetotaxy lasiopod (?, from sketch). Cirrus I separated from the posterior folds by an intercalary fold; similar to cirrus II (segments about as high as wide and clad with thin spine-like simple setae); cirri I and II similar to cirrus III (ie with segments about as high as wide). Rami of cirrus I subequal; II subequal; cirri III to VI equal, or subequal.

MOUTHPARTS. Maxilla setose (supports relatively stout spines above and below the medial notch, and the papilla of the maxillary gland is long and conspicuous). Cutting edge with large spines (long spines of upper cutting edge reach medially as far as those of the lower edge). Mandible 3-toothed, or 2-toothed (abnormal, observed on holotype only); upper margin of tooth 3 with small spines (1, stout); enlarged spine on upper margin of tooth 2 absent; lower part setose (with a comb of stout spines in a single row along the superior and apical margins; about ⅔ of the inferior margin supports a comb of closely spaced setae). Labrum with small denticles (20 fine, low teeth, arranged more or less in pairs, with individuals of a pair being separted by approximately the diameter of a tooth.). Mandibular palp short; lower surfaces and anterior margins support short, simple setae.

CAUDAL APPENDAGES. Caudal appendages present; unsegmented; short (1/2 the length of the proximal article of CVI).

PENIS. Penis about 1/2 length of the first article of the pedicel of CVI and, as apparently in all Scillaelpas examined by Newman (1960), it is apparently immature; setose (sparely clothed with fine, short setae).

FILAMENTARY APPENDAGES. Filamentary appendages present; 3 pairs (along the median dorsal surface of the prosoma; the first opposite the oral cone, is shaped like the symbol for Aires (first sign of the Zodiac); the second, situated somewhat short of the posterior margin of the prosoma, is in the form of a V that does not quite meet at the base; the third, on the intercalary fold between CI and CII, consists of well-separated dorsally directed, nipple-like projections).

COMPLEMENTARY MALES Complemental or dwarf male in shallow pocket in sub-rostral region (in the cavity between rostrum and subrostrum; 1 each in 2 of the 3 hermaphrodites examined; lying on their sides); about 1/2 the size of the males of *S.* (*A.*) *bocquetae*; capitular plates present; capitular plates 6; capitular plates plus a median latus on 1 side; carinolatus absent.

OUTSANDING DIAGNOSTIC FEATURES. Distinguished from S. (A.) bocquetae by the overlap between the rostrolatus and the carinolatus, and from both *S.* (*A.*) *bocquetae* and *S.* (*A.*) *calycula* by acute (60-70°) rather than obtuse (90°) rostrolateral apexes. Distinguished from *S.* (*A.*) *falcata* by the wider carina meeting the apex of the medianlatus rather than that of the carinolaterals, and the carina being apically free of, rather than closely applied to the carinal margin of the tergum. Distinguished from *S.* (*A.*) *bocquetae* and *S.* (*A.*) calycula by differences in the filamentary appendages and, presumably, from *S.* (*A.*) *falcata* in this regard, although its appendages have not yet been described.

REFERENCES. Newman, W.A. (1980). A review of extant *Scillaelepas* including recognition of new species from the North Atlantic, western Indian Ocean and New Zealand. Tethys 9: 379-398.

BSPECIMENS EXAMINED. M/S Marion-Dufresne Cruise MD. 08 (1976): 6, 33°08.7′S 43°59.7′E, 600 m, Walters Shoals, SW Indian Ocean, 16.3.76, Charcot dredge sample 46, fine sand, 1 specimen; 6, 33°1 1.4′S 44°00.4′E, 620-535 m, Walters Shoals, S of Madagascar, SW Indian Ocean, 1 6.3.76, beam trawl 47, 2 specimens, each with complemental male; H (6, sample 47) Laboratoire des Crustacés, Museum National d'Histoire Naturelle, Paris, no. entrée 7910; P (6, sample 47) U.S.N.M. Cat. No. 173134; P (6, sample 46) British Museum (Nat. Hist.) Reg. No. 1979-382.

Systematic cirripedeology in the 21st century: Multimedia biodiversity information systems

Ronald Sluys & Peter H. Schalk
Expert-center for Taxonomic Identification, Amsterdam, Netherlands

Frederick R. Schram
Institute for Systematics and Population Biology, University of Amsterdam, Amsterdam, Netherlands

ABSTRACT

An overview is provided of the computer program LINNAEUS II for multimedia documentation of biodiversity. Because of its low-learning threshold the program is eminently suited both for scientists entering their specialized knowledge into the database, e.g. on thoracican cirripedes, and for users of the database when consulting published CD-ROMs.

1 INTRODUCTION

The growing world-wide awareness about the present biodiversity crisis has increased the demand for pertinent information about life on Earth. It is only on the basis of a good documentation and understanding of the species of organisms that inhabit our world, such as barnacles, that adequate measures can be taken for the conservation of biodiversity and the sustainable development of living resources. Consequently, there is presently a great deal of pressure on systematic biologists to provide nature conservationists, ecologists, biomonitoring agencies, industry, agriculture, and fisheries with the required information because it is the science of systematics that documents biodiversity and seeks an understanding of what that diversity means.

The systematics of barnacles offers an excellent example of what is needed. The particular kind of expertise requested by society concerns synoptic knowledge of taxonomic groups and ready identification of species. This kind of information has been generated by barnacle systematists but has to be retrieved from both specialists and the usually scattered literature.

In a popular publication, Gould (1993: 427-428) observed that biology is a rather unique science in that it incorporates and depends on so much illustrative and graphic information. As far as cirripedes are concerned a good picture of a barnacle or cirripede structural plan can tell a non-expert user more than an extensive textual description.

Mission 3 of Systematics Agenda 2000 (1994) mentions the need to construct electronic data bases for storing and organizing knowledge about biodiversity. The

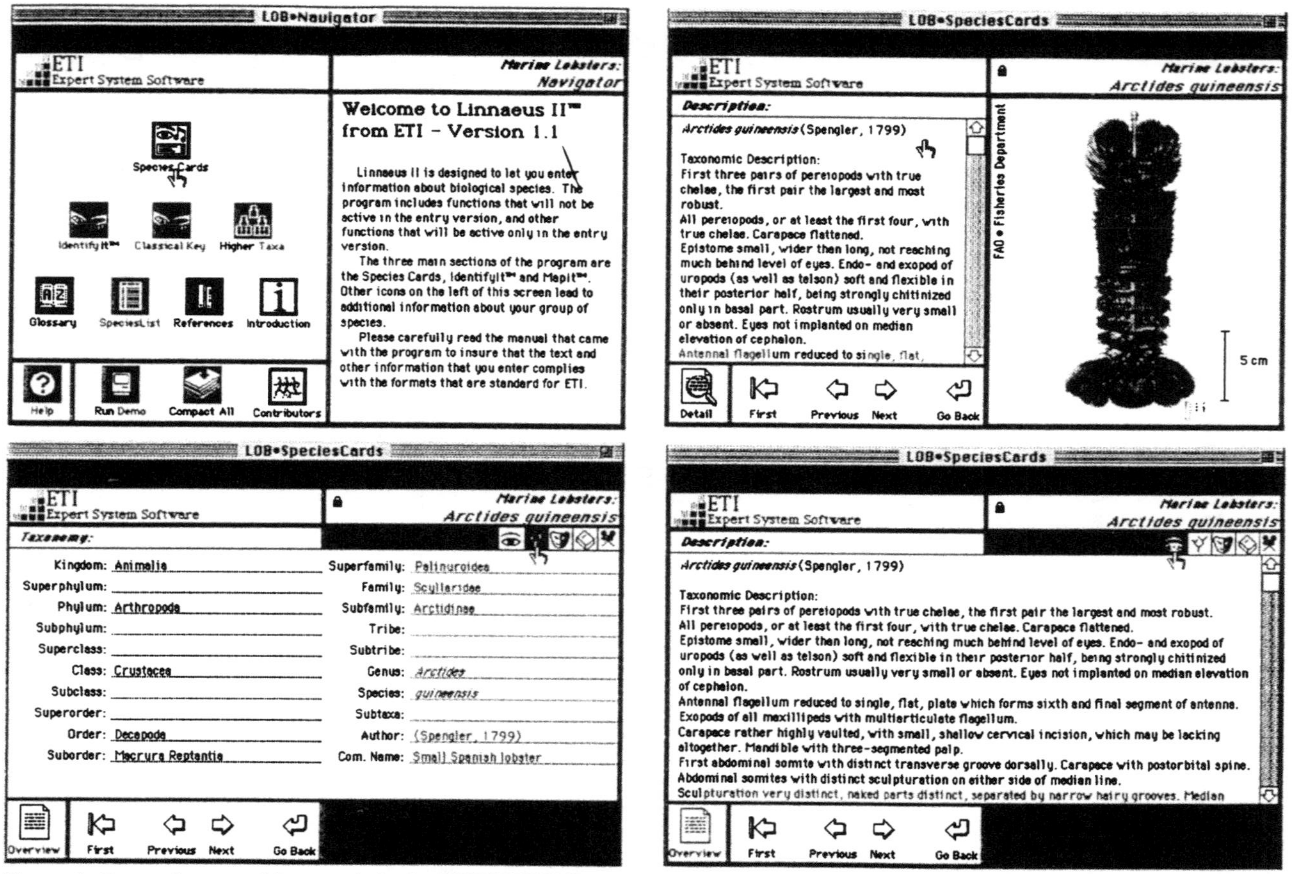

Figure 1. Screen images of four cards in the LINNAEUS II program, taken from a published program on commercial lobsters. The Navigator card (top left) is the central place from where other sections of the program can be reached. Species information is presented on an Overview Card (top right) and a Detail Card (lower right). On the Detail Card several fields can be activated, e.g. Taxonomy (lower left), and Description (lower right).

report addresses the need to implement an information system that can be accessed by a broad international user community. In view of Gould's apt observation above, this electronic information system should be able to incorporate various kinds of graphic data (e.g. color pictures, videos) and other illustrative material such as sound recordings (cf. Estep et al. 1993). To these ends, the Expert-center for Taxonomic Identification (ETI), was founded several years ago with the goal of creating a multimedia World Biodiversity Data Base. Sections of this data base are published as separate CD-ROM titles on various subjects. Many of these CD-ROMs have converted monographic studies into the form of interactive identification and information systems (see below). The advantages of making such a system available for barnacles are obvious

ETI has developed the software shell LINNAEUS II that would enable barnacle scientists to store information on taxa of cirripedes in the form of text, pictures, videos, and (even) sounds. Specialists can use LINNAEUS II for two purposes, viz., creation of their private barnacle research data base, and data transfer to ETI's World Biodiversity Data Base. Furthermore, LINNAEUS II can provide barnacle specialists and non-specialists with interactive tools for identifcation of species and biogeographic data storage.

2 LINNAEUS II: A MULTIMEDIA BIODIVERSITY INFORMATION SYSTEM

The LINNAEUS II program (available both for MacIntosh and WINDOWS platforms) is distributed on four floppies accompanied by a manual. It consists of seven main sections: data bases for information about 1) species as well as 2) higher taxa (the Species and Higher Taxa Cards), 3) the *IdentifyIt* program used for needed identification of specimens, 4) a reference database, 5) a glossary section, 6) an introduction section, and 7) the *MapIt* program for biogeographic information. The program looks like a computerized filing card system and opens with a central card, the Navigator, that gives quick access to all other parts of the software.

Clicking on the Species Card icon on the navigator card (Fig. 1) leads to the section of the program containing data related to the species taxon. The example illustrated here utilizes a card from a data base on lobsters, but the same could easily apply to one on barnacles. Data on species is structured on an Overview Card and a Detail Card. The Overview Card displays a single picture, usually showing the entire organism, and the first section of the description field. Clicking on the appropriate icon on this card makes the program jump to the Detail Card from which the following data fields can be activated: a) Description, b) Taxonomy, c) Synonyms, d) Literature, and e) Media Clips.

In the Description field, the barnacle specialist would type information regarding the taxonomic description, geographic distribution, detailed morphological data, facts concerning reproduction, ecology data, and molecular and biochemical data. Specialists can enter as much text as they want, the only restriction being that longer texts are ordered in chapters and paragraphs, as in a scientific paper.

The Taxonomy field enables the cirripedologist to enter the higher taxa that are relevant for each species in the program. The higher taxa database is similar to the species cards. They function in the same way but store information on higher taxa.

The Synonyms field is used for entering the junior synonyms of the species described (according to standards commonly used for the various groups of barnacles and as specified by the code of zoological nomenclature).

The Media Clips field is designed to hold multimedia information of various types: pictures, even sound recordings, videos, and text. The individual media clips are activated by clicking on the name of the clip, after which the multimedia data is displayed in a separate window.

IdentifyIt allows the specialist user to build a multimedia identification system, based on a matrix of taxa and characters. Individual characters and their states, such as barnacle plate morphology or detailed larval structures, are linked to descriptive texts and pictures (including even videos as well as the potential for sound recordings) when the use of such is deemed necessary by the specialist.

The system is designed to work equally well in the other direction. A non-specialist user of an *IdentifyIt* program when faced with trying to identify some specimen at hand may select any character from the list to begin with, say for example asymmetrical opercular plates. Once the user has made several additional choices as offered by the program, *IdentifyIt* then lists the individual species or series of species in order of probability. In this example, *IdentifyIt* might come up with a species of *Verruca*, shown with a 'hit percentage' (in this case possibly 98% confidence). A click on the species name, e.g. *Verruca stroemia*, leads into the species data base for a quick check of the description or other information, after which one can return to the identification program. It is also possible to compare species in the hitlist for both common and exclusive characters.

MapIt enables cirripedologists to enter and display geographic information on grid-based maps. The user of the *MapIt* program is offered the facility for 'zooming in' on maps, thus obtaining more detailed information on specific barnacle distributional records.

There are several support programs in LINNAEUS II. The glossary portion defines and describes specific technical terms used in the program, e.g. ovigerous frenae, thus increasing the ease of use for people who are not experienced systematists in barnacle groups. All text in LINNAEUS II is hyperlinked to the glossary: A mouse click on a word induces a search in the glossary section. In addition to the glossary, the introduction section also has a supporting function. This section provides a general account of the particular taxonomic group and deals with specific collecting and examination techniques.

It is important to note that the LINNAEUS II user interface is the same for both specialists entering data and for non-experts consulting the published CD-ROMs. This implies that the user interface has a very transparent lay out and does not require any special knowledge about computers and programming. The user interface for specialist's entering data has to have the same low learning threshold as the one for users of the data. Systematists are not computer scientists and should not be obstructed in their work by unfriendly and opaque interfaces. The LINNAEUS II package actually functions as a knowledge engineer, guiding the specialist painlessly in the construction of an expert system in his/her particular group (Kaesler 1993).

3 IMPORTING AND EXPORTING TO OTHER FORMATS

All data entered into LINNAEUS II can be easily exported to other data bases or used for instance for developing printed publications. Export facilities are built into all sections of the LINNAEUS II program. LINNAEUS II, originally available for Macintosh, has been exported also to the WINDOWS platform. This means that the LINNAEUS II software runs on the two currently most used computer platforms and allows scientists to start on one platform and to continue on the other. The datasets used for identifications can also be exported. For instance, this will allow the user to employ the information in a system like DELTA or in cladistic programs, such as PAUP. Special support is offered by the ETI staff to those who want to port back and forth to any of the other currently popular data bases.

The data entered into the World Biodiversity Data Base by ETI partners working together in a network of cirripede taxonomists shall be on-line accessible to that network. This facility shall be functional when a network of specialists becomes active and needs a platform for ready communication and data exchange. This allows each partner to work on his/her own machine and to down load and exchange information to ETI's ftp site.

4 LINNAEUS II AS PUBLICATION FORMAT

ETI has published already several CD-ROMs in various taxonomic groups with multimedia information on biodiversity. Although CD-ROM titles may be customized according to specific needs, most of the more recently published titles have the LINNAEUS II format as basic user interface. Examples of such published programs are 'North Australian Sea Cucumbers,' 'Marine Planarians of the World,' and 'Fishes of the North-Eastern Atlantic and the Mediterranean.' Customization of programs for barnacles could consist of incorporation of interactive dichotomous identification keys, pictorial keys, or a quiz. Other CD-ROM titles available or about to be available from ETI are the following: 'Birds of Europe,' 'Marine Mammals of the World,' 'Pelagic Molluscs of the World,' 'Turbellarians of the World – a guide to families and genera,' 'Lobsters of the World,' 'Five Kingdoms: Life on Earth,' and 'Protoctist Glossary.'

5 CONCLUSION

Presently, software and hardware have evolved to such an extent that tools are available, such as the LINNAEUS II program, enabling scientists to make their results available to science and society in a user-friendly, cheap, and an easy-to-update format. The availability of these techniques does not obviate the need for skilled taxonomists (who else is going to provide and check the data that is fed into these systems?) and cannot permanently solve the problem of the lack of human resources in systematic biology. However, modern electronic databases can 1) provide ready access to systematic knowledge, 2) enhance better communication between scientists and society, and 3) make the utilization and retrieval of information more efficient.

REFERENCES

Estep, K., R. Sluys & E. E. Syvertsen 1993. 'Linnaeus' and beyond: workshop report on multimedia tools for the identification and database storage of biodiversity. *Hydrobiologia* 269/270: 519-525.

Gould, S. J. 1993. *Eight little piggies*. London: Penguin Books Ltd.

Kaesler, R. L. 1993. A window of opportunity: peering into a new century of paleontology. *J. Paleont.* 67: 329-333.

Systematics Agenda 2000 1994. *Systematics Agenda 200: Charting the Biosphere*. Technical report. 34 pp.

Chemical signals in barnacles: Old problems, new approaches

Anthony S. Clare
Marine Biological Association of the UK, The Laboratory, Citadel Hill, Plymouth, UK

ABSTRACT

In keeping with most marine animals, barnacles rely primarily on their chemical senses to gather information from the external environment. Chemical signals, for their part, are used by barnacles to control and coordinate fundamental processes such as growth, reproduction, and development. Yet few areas of research on barnacle chemical communication have attracted special attention. Here, recent progress on the pheromonal control of larval settlement and egg hatching, and the regulation of moulting and metamorphosis by hormones is reviewed. Possible avenues for future research are also suggested in light of the advances made in our understanding of chemical signalling in other organisms. Barnacles will continue to serve as useful models for the study of chemical signalling, especially if new techniques are adopted. For example, the tools of molecular biology, electrophysiology, and image analysis have been applied to barnacle research only relatively recently. Coupled to the study of more tractable species than has formerly been the case, rapid progress should be possible regarding the role that chemical signals play in the life of barnacles.

1 INTRODUCTION

The title of this chapter reflects the mandate that I have been given: not merely to review past research, but to 'go out on a limb' and 'blaze some new trails'. Considering the current pace of change and discovery in the biological sciences, this is indeed a difficult task. In attempting to fulfil this charge, I shall confine myself to a discussion of research on chemical communication in barnacles (reflecting my own interests), illustrating possible new approaches by reference to research on chemical communication in other organisms. In the context of barnacle research, this is a relatively new field, having its roots in the pioneering experiments on barnacle settlement by the British researchers, E.W. Knight-Jones and D.J. Crisp. In fact, the now dynamic field of marine chemical ecology was founded by these and another British researcher, D.P Wilson, who also worked on larval settlement (see Hadfield 1986).

Before proceeding we should first consider some definitions. For the purposes of our discussion I shall use the term chemical communication in its widest sense, embracing both internal, e.g., endocrine, and external (pheromonal) communication. Accordingly, by chemical signals we mean hormones, neurotransmitters, neuromodulators, pheromones, allelochemicals, and so forth.

My last review of research on barnacle chemical communication (Clare 1987), which covered research up to 1983, discussed only endocrinology (*sensu* Yates 1981). I compared the then state of knowledge to advances made relating to the endocrinology of other arthropods, principally decapod crustaceans and insects, in order to provide a framework for future research. Seven years later, our understanding of the hormonal control of barnacle reproduction and development has advanced little, in marked contrast to the situation for higher arthropods (see Loeb 1993). With the advent of modern molecular and electrophysiological techniques, the oft cited anatomical impedement to endocrine research – the calcareous shell – should no longer be a major hurdle. Advances made in muscle and photoreceptor physiology using barnacle tissues as model sytems (Stuart 1983; Hoyle 1987; Horn 1989; Flatman 1991; Huang & Bittar 1991; Werner et al. 1992) are testimony to what can be achieved if there is sufficient interest and will.

I believe that over the next 5-10 years the primary areas of research relating to chemical communication in barnacles will be concerned with the larval stages. A major driving force to these studies is the need to prevent biofouling of man-made structures that are immersed into the marine environment; barnacles being a major component of the hard fouling (Christie & Dalley 1987). Having essentially solved the problem of macrofouling through the introduction of organotin-containing coatings, the environmental problems posed by their use (Bryan et al. 1986; Alzieu et al. 1986) have resulted in restrictive legislation (Kjaer 1992) and a search for alternatives to toxic antifoulants (Clare et al. 1992b). A corollary to studies on the prevention of barnacle fouling has been the re-kindling of interest in the fundamentals of barnacle settlement (defined as attachment and metamorphosis). Two further areas of research that continue to attract attention, and are likely to do so into the foreseeable future, are the hormonal control of moulting and the pheromonal induction of egg hatching in barnacles. It is with these three areas of barnacle biology – egg hatching, moulting and settlement – that this chapter will primarily be concerned. Attention is focused on the Balanomorpha that have received most attention in the past. However, renewed interest in other barnacle groups, notably the Rhizocephala (see, for example, Høeg 1992) is likely to lead to other 'model systems' for studies of barnacle chemical communication.

2 PHEROMONAL CONTROL OF EGG HATCHING

The embryos of thoracican barnacles are brooded in the mantle cavity of the adult. The cavity is confined by the wall plates, opercular plates and basis (whether calcareous or membranous), and is lined by a cuticle and underlying hypodermis (= epidermis). Even when the opercular plates are closed, the mantle cavity can remain confluent with the external seawater *via* a micropyle. In the context of the present discussion, therefore, it is important to note that the embryos are considered to be brooded externally; a compound that is released by the adult into the mantle cavity to effect egg hatching (see below) is, strictly speaking, not as was originally thought, a hormone (cf. Clare 1987), but a pheromone (Gerhart et al. 1990).

The boreo-arctic barnacle, *Semibalanus balanoides* (Clare & Rittschof 1989), has received most attention with respect to the control of egg hatching. The eggs of this species are laid in November in the United Kingdom and are then brooded in the mantle cavity until the following Spring. In a now classic paper, Barnes (1957) drew attention to the close synchrony between egg hatching in *S. balanoides* and the Spring phytoplankton bloom. A direct chemical linkage between the two processes was demonstrated independently by Barnes (1957) and Crisp (1956). The chemical inducer was found to be released by the adult barnacle, following feeding, to effect egg hatching (Crisp & Spencer 1958; Barnes & Barnes 1982).

In a series of papers, Clare and co-workers (Clare et al. 1982, 1985; Clare & Walker 1986) reported that the egg hatching pheromone of *S. balanoides* is an eicosanoid, the physiological action of which may be mediated by the release of embryonic dopamine. The *in vitro* assay developed by Crisp & Spencer (1958) and adopted, in a slightly modified form, by Clare et al. (1982), was instrumental in these studies. Subsequent analyses (gas chromatography/mass spectometry) by researchers at Imperial Chemical Industries identified the hatching pheromone as the novel eicosanoid, 10, 11, 12 – trihydroxy – 5, 8, 14, 17 – eicosatetraenoic acid (Holland et al. 1985). More recent findings (Song et al. 1990) have highlighted the danger of assuming that biological activity is associated with the major peak in a gas chromatograph. Moreover, it is highly unlikely that all components will be resolved as pure peaks by gas chromatography (see Wilkins 1994 and references cited therein). The true identity of the hatching pheromone of *S. balanoides*, therefore, remains in some doubt. Indeed, hatching pheromone may comprise a mixture of hydroxy fatty acids, which would be analogous to the situation found for the sex pheromones of insects (eg. Bjostad et al. 1984). The chemical synthesis and bioassay of putative pheromones will be required before the issue of structural identity can be resolved unequivocally.

Modest progress has thus been made towards identifying the nature of *S. balanoides* hatching pheromone. We know much less concerning the larger biological question: how does feeding induce egg hatching? Crisp (1956) was the first to demonstrate that the egg hatching pheromone was a product of the barnacle's own metabolism. Barnes (1957), on the other hand, suggested that diatoms might be the source of the inductive cue. Indeed, a concentrated culture of *Skeletonema costatum* was reported to significantly increase hatching of egg lamellae *in vitro* over that in seawater. Barnes (1957) further suggested that the active substance might be concentrated by the barnacle and act on the eggs in the form of an excretory metabolite. The hypotheses of Barnes and Crisp may be reconciled in that both considered that hatching pheromone is a metabolite of the adult barnacle. More recently, Starr et al. (1991) have obtained evidence in support of the importance of physical contact with plankton during feeding activity. Furthermore, the size of the plankton was suggested to be of importance: larger plankton were more effective inducers of egg hatching than smaller plankton. Although the liberation of nauplii can occur within 1 hour of exposure to algae, there is no evidence to preclude that the intake of food is requisite to naupliar liberation (contra Anderson 1994).

A possible causal link between feeding and egg hatching that has yet to be examined is that hatching pheromone is an excretory metabolite of eicosapentaenoic acid (EPA); the putative precursor of hatching pheromone. This theory is similar to that propounded by Barnes (1957). Significantly, marine plankton are rich in EPA. For example, the diatom, *Skeletonema costatum* is an effective inducer of egg hatching when fed to gravid barnacles (Crisp & Spencer 1958; Barnes & Barnes 1982; Starr et al. 1991; Clare unpublished) and the EPA content of this diatom is reported to comprise between 5 and 30% of the total free fatty acids (Groth-Nard & Robert 1993 and references cited therein). An alternative working hypothesis for feeding-induced egg hatching suggests itself from this information. Following the winter anecdysis and commensurate with a resumption in feeding, the excretion of dietary metabolites into the adult's mantle cavity will rise. A threshold concentration of a particular metabolite(s) of EPA (egg hatching pheromone) might then be attained to induce egg hatching. Certainly, pheromone-induced egg hatching *in vitro* is concentration dependent (Crisp & Spencer 1958; Hughes & Clare, unpublished). This hypothesis could be tested with controlled feeding experiments employing artificial diets. For example, gravid barnacles could be fed with liposomes, or a microencapsulated diet, in which the lipid composition has been tailored. The importance of EPA to egg hatching might then be tested *in vivo*.

Since there is some evidence of a disparity in the biosynthetic pathways of eicosanoids in certain marine invertebrates compared to vertebrates (Corey et al. 1975), it will be of interest to determine how hatching pheromone is produced from the putative precursor, EPA. The results of early pharmacological experiments on the possible inhibition of hatching pheromone biosynthesis *in vitro* by aspirin and indomethacin (Clare et al. 1985) can now be reconciled with current knowledge that hatching pheromone is not synthesised via a cyclo-oxygenase pathway. Since the structures of hatching pheromones elucidated thus far do not contain a cyclopentane ring, they are produced via a linear pathway of eicosanoid synthesis. Although this is likely to involve lipoxygenases, the involvement of cytochrome P450 (see Takahashi et al. 1990; Piomelli 1993) cannot be ruled out at this stage.

The significance of barnacle egg hatching extends beyond the environmental control of a reproductive event. In only one other aquatic 'system' has it been demonstrated that eicosanoids act as pheromones. Female goldfish release metabolites of $PGF_{2\alpha}$ into the water which then act as sex pheromones stimulating male sexual behaviour (Sorensen et al. 1988). Radiotracer studies have indicated that the pheromone is a mixture of several novel PGF metabolites (Sorensen 1992). More recently, evidence has been obtained using electro-olfactograms that F prostaglandins may be widely used as sex pheromones in Cypriniformes (Irvine & Sorensen 1993, Kitamura et al. 1994). Interestingly, $PGF_{2\alpha}$ also has a hormonal function, modulating ovulation (Stacey & Goetz 1982) and a paracrine function, stimulating female spawning behaviour (Stacey 1987) in many fish species (reviewed by Sorensen 1992). Consequently, the term 'hormonal-pheromone' has been applied to the prostaglandin metabolites, as well as to certain steroids, that have a demonstrable hormonal and pheromonal function.

In the process of studying barnacle egg hatching pheromone, it has become clear that barnacles have the capacity to synthesise a multitude of eicosanoids *in vitro* (Clare 1984). Whether all these compounds are synthesised *in vivo* is questionable. However, since eicosanoids have diverse physiological roles in invertebrates

(Ruggeri & Thoroughgood 1985; Stanley-Samuelson 1987, 1991; Stanley-Samuelson & Loher 1990), as well as vertebrates (Shimizu & Wolfe 1990), we can predict that functions other than egg hatching will be established for barnacle eicosanoids. Recent developments in the assay of eicosanoids, in particular radio- (RIA) and enzyme immunoassay (EIA), will greatly facilitate these studies. Perhaps 'hormonal-pheromones' will be discovered. Indeed, since hatching pheromone is a stable end-product of barnacle metabolism, intermediate metabolites of the putative precursor, EPA, may be found to have physiological functions, perhaps serving as autocoids or local hormones.

Recently, an investigation has been instigated to attempt to unravel the signal transduction pathway for barnacle egg hatching pheromone. Like hormonal receptors, the receptors for eicosanoids are thought to be G protein-linked (Sonnenberg et al. 1990; Hirata et al. 1991). Moreover, since hatching pheromone is detected in solution, it would not be surprising if its receptor is related to the superfamily of G protein-linked receptors that are expressed in vertebrate olfactory epithelium (Buck & Axel 1991). Indeed the lysine receptor which is present in the cilia of *Haliotis rufescens* larvae is G protein-linked, though it does not appear to be closely related to G_{olf} (Wodicka & Morse 1991). Lysine facilitates the response of the larvae to metamorphic inducers present on the surfaces of crustose coralline red algae (Baxter & Morse 1992; Morse 1993). However, contrary to our expectations, preliminary pharmacological evidence suggests that the egg hatching pheromone receptor is not G protein-linked (Clare & Hughes, unpublished). Moreover, we have been unable to obtain evidence for the involvement of inositol trisphosphate or diacylglycerol. The only intracellular messenger which we have obtained consistent evidence for is cyclic AMP. All pharmacological tests that are commonly applied to assess the involvement of this cyclic nucleotide, namely, inhibition of phosphodiesterase, stimulation of adenylate cyclase and application of cAMP analogues (cf. Clare et al. 1995), were positive in the *in vitro* egg hatching assay. Unfortunately, since these experiments employed the whole organism (the embryo), they are non-specific and we are unable to determine whether cAMP is involved in the transduction of hatching pheromone; it could equally well be acting downstream from, or in parallel to, the transduction of the pheromone. In order to make real progress, it will be necessary to determine the site of reception of the pheromone. Having obtained structural information for putative pheromones (Hill et al. 1988; Song et al. 1990) it should be possible to acquire radio-labelled pheromone for autoradiographic studies.

Barnacle egg hatching pheromone remains one of the few marine eicosanoids with an established physiological function. As such it is, and will remain for the foreseeable future, an important model system for studies of eicosanoids, particularly as the biosynthesis, functions and mechanisms of action of hydroxy fatty acids are relatively poorly understood.

3 ENDOCRINE CONTROL OF MOULTING AND METAMORPHOSIS

As mentioned earlier, our knowledge of barnacle endocrinology has advanced little over the past 10 years or so (cf. Clare 1987). Fingerman (1988) has recently reviewed the subject from the standpoint of antifouling. Suggestions for the control of barnacle fouling included manipulating gonad-stimulating hormone to inhibit reproduction. While plausible, the control of fouling requires that larval settlement of fouling organisms is prevented (cf. Clare et al. 1992b). The point of attack is not, therefore, the adult and its reproduction but the settlement stage (Visscher 1927), which in barnacles is the cypris larva. Effective control measures are more likely to involve repelling larvae or preventing their adhesion to the substratum. This strategy forms the basis of the new technology of nontoxic antifouling (Clare et al. 1992b). Nevertheless, a credible approach for barnacle control, in some applications, may be to interfere with metamorphosis of the cyprid to the juvenile barnacle. For this approach to be realised, fundamental studies on the endocrine control of moulting and metamorphosis in barnacles will be needed.

To date, the majority of studies that have tried to elucidate the hormonal regimes operating to control moulting and metamorphosis of barnacles have been pharmacological in nature (Clare 1987; Fingerman 1988; Clare et al. 1992a). Exceptions include a recent study of barnacle neuropeptides by Webster (personal communication) who tested a range of antisera raised against decapod neuropeptides on the central nervous systems (CNS's) of 3 barnacle species: *Balanus balanus*, *B. perforatus* and *Chirona hameri*. Only 2 antisera, anti-pigment-dispersing hormone (PDH) and anti-crustacean cardioactive peptide (CCAP), were reported to give convincing patterns of immunoreactivity. Since there is a high degree of similarity between insect and crustacean PDH, it is not surprising that immunoreactivity could be detected in barnacles. However, barnacles do not possess chromatophores (Clare 1987; Fingerman 1988), so a role for PDH is problematic. Nevertheless, a factor with crustacean PDH-like activity has been extracted from barnacle CNS (Sandeen & Costlow 1961; Costlow 1963). Given that there is evidence for the inhibition of methyl farnesoate (see below) sythesis by PDH (Loeb 1993), perhaps PDH has a role in barnacle reproduction and/or moulting.

In the context of the present discussion it is of interest that anti-moult inhibiting hormone (anti-MIH) gave negative results (Webster personal communication) in spite of evidence for the existence of MIH in barnacles (Davis & Costlow 1974). One possible explanation for this discrepancy is that the amino acid sequences of barnacle and crab MIH's may be sufficiently unrelated for the MIH epitope to be recognised by the antiserum. This lack of recognition would not be surprising since the amino acid sequence identity between crab (*Carcinus maenas*) and lobster (*Homarus americanus*) MIH is only 29% (Chang et al. 1993). The small size of barnacle nervous systems and the paucity of neurosecretory material that they contain (see Clare & Walker 1989; Walker 1992), means that it is likely to be technologically impractical at present to obtain sufficient purified barnacle MIH (assuming that this hormone exists in barnacles) for sequencing. Of course, separation and sequencing technologies continue to improve, but in the interim, alternatives to the conventional approach of isolating and sequencing neuropeptides can be attempted. It may be possible, for example, to clone and sequence barnacle MIH. This would involve PCR amplification of barnacle cDNA using degenerate oligonucleotide probes based on the amino acid sequence of decapod MIH (see Klein et al. 1993a,b). Another approach might be subtractive hybridisation (see Palazzolo et al. 1989, 1990) wherein a spatial and temporal pattern of mRNA expression is assumed. For example, a reasonable assumption might be that MIH mRNA is expressed in barnacle CNS, but not

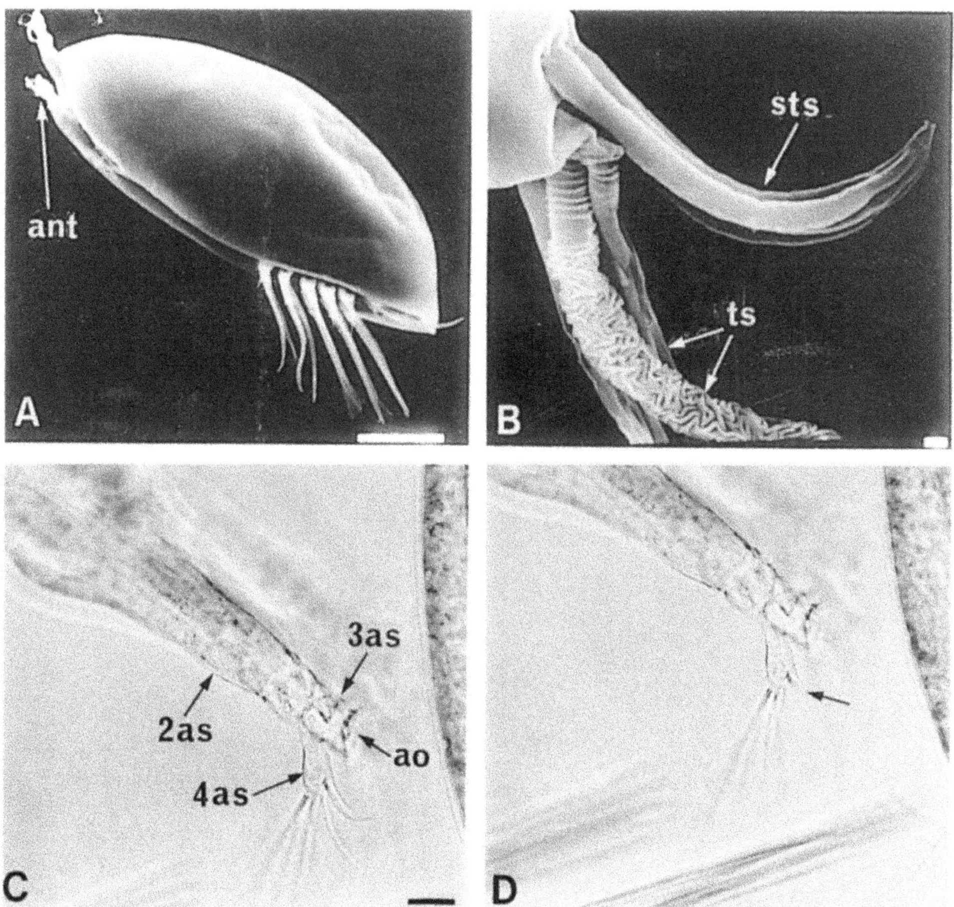

Plate 1. *Balanus amphitrite*. (A) Scanning electron micrograph of the cypris larva. Scale bar, 100 µm. ant = antennule; (B) Scanning electron micrograph of the fourth antennular segment. Scale bar, 1 µm. sts = subterminal setae; ts = terminal setae; (C) Light micrograph of an antennule prior to laser surgery. Scale bar, 30 µm. 2as, 3as and 4as = second, third and fourth antennular segments; ao = attachment organ; (D) Light micrograph of the same antennule shown in C after laser ablation of the subterminal setae and one of the terminal setae. Scale same as C. From the 1993 Annual Report of the Marine Biological Association.

muscle, and that the level of expression increases during intermoult; the action of MIH being to inhibit ecdysteroid (moulting hormone) biosynthesis and thus moulting (Skinner 1985; Chang et al. 1993). Effectively, the number of potential clones for screening is reduced.

Evidence for the involvement of ecdysteroids in barnacle moulting is fragmentary and has been reviewed recently (Clare 1987; Fingerman 1988). Only one study has sought to isolate and identify barnacle ecdysteroids (Bebbington & Morgan 1977). Although 20-hydroxyecdysone (20-HE) and ecdysone were identified (the first report for Crustacea), it is unfortunate that evidently no account was taken of dipteran contamination and so the source of the ecdysteroids is equivocal (Clare 1987). Techniques for the isolation and identification of ecdysteroids have advanced considerably since Bebbington & Morgan's study (see Morgan & Marco 1990 for a review) and it should now be possible, using for example high performance liquid chromatography in combination with RIA or EIA, to measure the titre of the putative moulting hormone of barnacles – 20-HE – in relation to the moult cycle, both of larvae and the adult. Other outstanding issues which can now be addressed in barnacles include the biosynthesis and metabolism of ecdysteroids.

Ecdysteroid receptors have also been characterised in insects (Koelle et al. 1991; Riddiford & Truman 1993). By using *Drosophila melanogaster* ecdysteroid receptor (EcR) cDNA, Palli and Riddiford (unpublished, cited Riddiford & Truman 1993) were able to isolate the EcR gene from *Manduca sexta*. Subsequent studies on EcR expression have indicated that the EcR may have an important part to play in developmental responses to changes in the ecdysteroid titre. There have been no comparable investigations published for Crustacea.

Evidence is emerging in support of a 'molecular interplay' between ecdysteroids and insect juvenile hormone (JH) (cf. Riddiford & Truman 1993). In insects, JH prevents metamorphosis, but the cellular and molecular events underlying this action are unclear. Recent evidence of a putative 29 kD receptor promises to shed some light in this area (Riddiford & Truman 1993). The equivalent hormone in crustaceans is the unepoxidated precursor of JHIII – methyl farnesoate (MF) (Chang et al. 1993). This terpenoid hormone is produced by the mandibular organs of crabs (Laufer & Borst 1988) and apparently acts to increase the ecdysteroid titre. Presumably MF acts on the Y organ (the moulting gland), but how the ecdysteroid increase is achieved remains to be elucidated (Chang et al. 1993). The result of this 'interplay' in malacostracan crustaceans is the reverse of the situation in insects, where moulting is stimulated (Yudin et al. 1980; Borst et al. 1987).

It was through the study of barnacle cyprid metamorphosis, however, that the first evidence was obtained for a role for 'JH' in crustaceans. Details of these early experiments and their findings have been reviewed by Clare (1987) and Fingerman (1988). Essentially, JH and its analogues induced premature, and abnormal, metamorphosis. In other words, metamorphosis occurred prior to attachment of the cypris larva to a substratum. As pointed out by Clare (1987), neither JH, nor a hormone with JH-like activity, has been isolated from cypris larvae. In view of the evidence that MF is the crustacean JH, coupled to the availability of a sensitive bioassay (Tobe et al. 1989) and RIA (Goodman et al. 1990) for MF and insect JH respectively, the time is now opportune to examine whether a juvenile hormone is present in cyprids. Analysis by gas chromatography-mass spectrometry will, of course, be required to confirm the identity of 'barnacle JH'.

4 CHEMICAL CUES TO SETTLEMENT

The subject of barnacle larval settlement (defined as attachment and metamorphosis) has been reviewed (Gabbott & Larman 1987; Anderson 1994) and is discussed in the present volume (Walker 1995). There is no need, therefore, to reiterate the background to this area of research in depth, other than to emphasise some key issues that point the way forward to future research.

The settlement stage larva of barnacles, the cyprid, is commonly regarded as having the sole purpose of finding a suitable place to attach and metamorphose into a juvenile barnacle (Crisp 1976). Settlement is obviously a crucial, if not the most important, stage in the life cycle of barnacles (Bourget 1988). While on a comparatively large scale the site of settlement may be dictated by physical factors, e.g., hydrodynamics (see Roughgarden et al. 1987; LeFèvre & Bourget 1992; Pineda 1994), at a fine scale, at least, some species are able to exercise 'choice' over the place of settlement. This is especially true of species that have stringent, specific requirements for a settlement site (see below). Behavioural mechanisms are also likely to operate for barnacles that are gregarious. There is good laboratory evidence to support the importance of cypris behaviour in gregariousness, but it has only been relatively recently that field evidence has been obtained (Barnett et al. 1979; Bourget 1988; Chabot & Bourget 1988; Raimondi 1988). Much still remains to be done concerning the factors determining the site of settlement in the field. Given the multitude of potential factors that operate on the cypris larva, it will be a difficult task to ascertain their relative importance. This task may be rendered somewhat more tractable by the use of flume experimentation, where some approximation to the real world can be obtained while retaining the ability to exercise control over environmental variables (see Mullineaux & Butman 1991).

That there is a chemical basis to gregariousness has long been accepted. The first work on the nature of the chemical cue was done over 30 years ago. Actually, the cue may be regarded as a pheromone, if the term is used in its widest sense (see Burke 1986). The main points that have been established since then are that the settlement pheromone is a component of the barnacle cuticle; it is proteinaceous in character; and it is probably perceived by the antennules of the cyprid (see Gabbott & Larman 1987; Walker 1992 for reviews). Crisp & Meadows (1963) claimed that the cyprid recognises the chemical cue (which they termed arthropodin) via a tactile chemical sense, because the cyprid must come into contact with the substratum for settlement to be induced. This theory draws an analogy with the antibody-antigen reaction (Crisp & Meadows 1962) and takes account of the importance of the molecular configuration of the settlement pheromone, whether in its natural form as a component of the cuticle, or when adsorbed to the substratum. Conceivably, molecular forces between the settlement pheromone and the antennular adhesive (Walker & Yule 1984; Clare et al. 1994a) can be sensed by the cyprid as increased adhesion or friction (Crisp & Meadows 1963). Indeed, evidence has been presented of increased adhesion of cyprids on arthropodin-treated versus untreated surfaces (Yule & Crisp 1983; Yule & Walker 1984, 1987). If adhesion is important to settlement pheromone recognition, then its effect on settlement may be mediated by cell deformation. Binding of chemotactic compounds to cells can alter cell shape, which in turn can effect changes in cell biochemistry; a process termed mechanotransduction (Watson 1991).

The most obvious point of contact between the cyprid and the substratum during searching behaviour is the attachment disc of the antennule (Nott 1969; Nott & Foster 1969). However, the settlement pheromone could equally well be detected in the boundary layer over the substratum by the fourth antennular segment (Clare & Nott 1994). In fact, evidence that cypris larvae can discriminate conspecific from allospecific arthropodin does not tally with the adhesion theory of pheromone recognition (Pawlik 1992). Settlement pheromone is in fact a soluble protein (Crisp & Meadows 1962) so it is likely to emanate from the cuticle of recently moulted barnacles and newly settled spat. While Crisp & Meadows (1963) categorically state that settlement pheromone cannot be detected in solution, they present evidence that searching behaviour and attachment are promoted by dilute solutions of barnacle extracts: '... fairly dilute solutions caused some cyprids to change from normal positive to a negative reaction to light' (Crisp & Meadows 1963: 369). However, in a separate experiment it was shown that cyprids could discriminate clean from pheromone-treated slate panels, even when immersed in seawater containing pheromone at the same concentration that was used to treat the slate; a result that supports the concept of a tactile chemical sense, because, in the words of Crisp & Meadows (1963: 370): '... the cyprids could not experience any change in concentration of the settling factor in solutions as they approached the treated panels.' Nevertheless, a difference in concentration might exist between the solution above, and within the diffusion boundary layer over the treated panels that would be sufficient to facilitate discrimination by the cyprids; molecular diffusion dominates within the diffusion boundary (Dodds 1990). The thickness of the diffusion boundary in the experiments of Crisp & Meadows (1962, 1963) is not known, although, generally speaking, they are less than 1 mm thick (Nowell & Jumars 1984).

It will be difficult to resolve the issue of the mode of detection of settlement pheromone in experiments that do not control for the adsorption of pheromone onto substrata from solution. One possible approach would be to tether cyprids in seawater, such that they could not make contact with a substratum. Electrophysiological recordings, by conventional techniques, could then be made before and after the introduction of settlement pheromone into the seawater. Although small in size, the problems associated with recording from larvae are not unsurmountable as evidenced by recent progress in this area (Arkett et al. 1989; Barlow 1990).

If settlement pheromone is detected in solution, and this idea is not a new one (Knight-Jones 1953; Nott & Foster 1969), then attention will undoubtedly focus on the fourth antennular segment of the cyprid (as opposed to the third segment with its attachment disc) as the putative site of pheromone reception. The fourth segment bears an impressive array of sensory setae (Gibson & Nott 1971; Clare & Nott 1994). Clare & Nott (1994) have drawn attention to the outward resemblance of some of these setae to crustacean aesthetascs; the putative olfactory receptors (Hallberg et al. 1992). The distinction between olfaction and gustation is blurred in aquatic environments (Ache 1991), but some features of cypris settlement suggest that a mechanism akin to olfaction may operate in the perception of settlement pheromone. First is the observation that the fourth antennular segment is flicked through the water column during cyprid searching behaviour (Gibson & Nott 1971; Clare & Nott 1994) in a manner reminiscent of decapod antennular flicking (Clare & Nott 1994). The latter action has been found to facilitate stimulus access to olfactory receptors, and is akin to mammalian sniffing (Schmitt & Ache 1979; Moore et al. 1991). Secondly, evidence has recently been obtained in support of a role for cyclic AMP in cyprid settlement (Clare et al. 1995). This result is in accord with our knowledge of mammalian olfaction for which cAMP is a component of the signal transduction pathway (Anholt 1991). Likewise, cAMP is involved in the signal transduction pathway of some odorants in the lobster (Michel & Ache 1992). Finally, settlement of other marine invertebrate larvae is modulated by mechanisms that bear a strong resemblance to those operating in mammalian olfaction.

Perhaps the best characterised example is that of the abalone, *Haliotis rufescens* (Morse 1990, 1992), although significant progress has been made towards elucidating the signal transduction pathway involved in settlement of larval hydroids, *Hydractinia echinata* (Leitz 1993). For the abalone, two pathways operate in parallel. The inductive cue, a GABA mimetic peptide, is associated with the surface of crustose coralline red algae; the settlement site favoured by the larvae. Signal transduction of the cue involves cAMP and calcium. The parallel pathway potentiates the settlement induction by the cue. In this pathway, lysine apparently acts via a G protein-linked receptor, diacylglycerol- and calcium-stimulated protein kinase C. Although the mechanism of potentiation is not known, it is perhaps significant that separate olfactory signal transduction pathways interact to enhance odour discrimination in vertebrates (Anholt 1991; Berridge 1993) and the lobster (Schmiedel-Jakob et al. 1990). Could it be that separate pathways mediate chemosensory and physical (tenacity of adhesion) stimulation of cypris settlement and that there is a molecular interplay between these pathways? Such a scenario would account for the apparent incongruity between the adhesion theory of pheromone recognition and evidence that the efficacy of settlement pheromone induction is related to systematic affinity (see above).

Two outstanding problems, the resolution of which is fundamental to an appreciation of how cypris larvae are able to perceive and discriminate between potential settlement sites are: a) The molecular structures of settlement pheromones and b) their site(s) of reception. Some progress has been made towards the first of these problems. Gabbott & Larman (1987) reviewed research up to 1984, most of it their own, on the purification of the settlement pheromone of *Semibalanus balanoides*. This research established that the pheromone consists of a polymorphic system of closely related proteins, the most stable sub-unit of which has a molecular weight of 23,500-25,000 daltons. This conclusion has since been questioned by Naldrett (1992) who claimed that the 'settlement factor proteins' were in fact ferritins. Instead another 'adhesive protein' ($IX_{(a)}$) was suggested to be a candidate settlement pheromone of *S. balanoides*. Unfortunately, neither this protein, nor ferritins, have been examined for settlement inducing properties.

Studies on the settlement pheromone of *S. balanoides* are hampered by the seasonal availability of larvae of this species (the effective window for research is only about 1-2 months/year) and the fact that the cypris larva requires flowing water in order to set. In contrast, adults of *B. amphitrite amphitrite* can be manipulated in the laboratory to produce larvae throughout the year which are then easily cultured to the cypris stage (Rittschof et al. 1992; Clare et al. 1994b). Moreover, the cypris larva of this species will set in the absence of flow. The ability to obtain larvae of known age, coupled to the absence of a requirement for flow, has enabled a simple settlement assay to be developed. This assay has recently been used in a number of studies that have addressed the inhibition (e.g. Rittschof et al. 1992) and stimulation (e.g. Rittschof et al. 1984; Tegtmeyer & Rittschof 1989; Clare et al. 1992a, 1994, 1995) of settlement by various compounds and substrata.

There is some controversy in the literature regarding the settlement pheromone of *B. amphitrite* (see Crisp 1990). Rittschof (1985) extracted 'settlement pheromone' from barnacle-conditioned seawater. Partially purified pheromone induced larval settlement at µg/L concentrations in static assays. This is a much lower concentration than is required for pheromone extracted from *B. amphitrite* tissues and tested in the presence (Crisp 1990) or absence (Clare et al. 1994a) of flow. As pointed out by Crisp (1990), it is not clear whether these two lines of research are dealing with the same phenomenon, or indeed the same compound(s). In order to resolve this issue unequivocally, it will be necessary to obtain sequence information for the proteins in question. It is likely that such information will also contribute substantially to our understanding of how cypris larvae are able to distinguish between conspecific and allospecific pheromones. Since a DNA sequence can be deduced from that of the amino acids, oligonucleotide probes can be designed that may enable the gene(s) for settlement pheromone to be cloned and characterised (see below).

Two alternative, and indirect, approaches to elucidating the nature of barnacle settlement pheromone are in progress. The first is based on the finding that synthetic peptides with a carboxy-terminal arginine or lysine and a neutral or basic amino-terminal amino acid will stimulate settlement of *B. amphitrite* cyprids (Tegtmeyer & Rittschof 1989; Rittschof et al. 1991; Pettis 1991). Whether these peptides are acting as pheromone mimics or are acting downstream or in parallel to settlement pheromone is unclear. Again sequence information for the native pheromone would help to resolve this problem. It is perhaps significant, however, that settlement of oyster larvae is also induced by peptides (Zimmer-Faust & Tamburri 1994). Of the synthetic peptides assayed, glycyl-glycyl-L-arginine was the most potent inducer of settlement, being active at picomolar concentrations; this tripeptide is also a potent inducer of barnacle settlement (Tegtmeyer & Rittschof 1989). The second approach is based on a modelling study of *S. balanoides* settlement pheromone that has utilised amino acid compositional data of Larman et al. (1982) to generate synthetic proteins (Yamamoto & Nagai 1992). While the rationale of Yamamoto's approach is to develop novel bioadhesives, it will be of interest to assay these model proteins as settlement inducers. Such a study is now in progress (Clare & Yamamoto, unpublished).

While the availability of a rapid, reproducible assay (see above) will greatly facilitate the purification of barnacle settlement pheromone, a more straightforward strategy might be to choose a species for study that has a stringent, specific requirement for a settlement site. Examples of barnacles that meet this criterion include Rhizocephala (especially the male cyprid) that parasitise crustaceans (cf. Clare et al. 1993), *Octolasmis mulleri* that settles on the gills of blue crabs (Walker 1974; Gannon 1990), and *Conopea galeata* that colonises the octocoral, *Leptogorgia virgulata* and arguably settles on the egg masses of the gastropod, *Simnia* (Crisp 1990) that feeds on this octocoral (Patton 1972). Members of the Rhizocephala may prove the most amenable to study since their larvae are lecithotrophic and are thus readily cultured to the cypris stage.

The proposition that the settlement inducer that is present in seawater extracts of barnacles is derived from barnacle tissues is firmly established in the literature (Gabbott & Larman 1987; Pawlik 1992) and has not been questioned. However, barnacle cyprids, in common with most other larvae, prefer to settle on substrata that possess a well developed biofilm (Crisp 1984; Clare et al. 1992b). There is no evidence for the involvement of diatomaceous fouling in barnacle settlement induction. However, bacterial films have been shown to induce larval settlement in a wide range of marine invertebrates (Zobell & Allen 1935; Wilson 1955; Cameron & Hindgardner 1974; Brancato & Woollacott 1982; Kirchman et al. 1982; Mitchell & Kirchman 1984; Weiner et al. 1985; Bonar et al. 1986; Maki & Mitchell 1985; Fitt et al. 1989; Johnson et al. 1991; Tamburri et al. 1992; Davis 1994), including barnacles (Maki et al. 1990). The situation for barnacles, and perhaps other larvae, is complicated by the fact that some bacterial isolates stimulate barnacle settlement, whereas others inhibit this process (Maki et al. 1988). The ecological relevance of this observation is, however, debatable since a barnacle cyprid may only rarely encounter a biofilm that is composed of a pure bacterial strain in the field (see Paerl 1985). Nevertheless, it is clearly important, when the isolation of the inductive cue is being considered, to know the source of that cue.

The solution to the second question – the site of pheromone reception – will be greatly facilitated by isolating and characterising a settlement pheromone (see above). Ligand binding studies will then be possible; a requisite if a molecular cloning of the pheromone receptor is to be attempted (e.g. by PCR amplification of degenerate oligonucleotides based on conserved sequences in G protein-linked receptors) because the ligand will be needed to confirm the identity of the receptor. Assuming that the pheromone receptor can be cloned and a sequence obtained, *in situ* hybridisation studies can be carried out to locate sites of expression in the cypris larva.

An alternative approach that is currently being examined is to ablate the sensory setae of the cypris antennule and then to examine the ability of cyprids that have been operated on to detect settlement pheromone in choice settlement assays (Clare et al. 1994a). The ablation technique involves using a pulsed nitrogen laser with the cyprid immobilised on the cold stage of an inverted microscope. A focused spot diameter of 0.3 μm can be achieved (Taylor & Brownlee 1993) and consequently individual setae can be ablated (Pl. 1). The choice settlement assays can then be used to determine whether cyprids that have had one or more of their setae ablated can perceive settlement pheromone and, moreover, distinguish conspecific from allospecific pheromone.

5 CONCLUDING REMARKS

It will be apparent from this brief account of barnacle chemical communication that although barnacles have attracted comparatively little attention in this field they can potentially serve as useful model systems. For example, the study of the chemical induction of egg hatching led to the first demonstration that eicosanoids can serve as pheromones. Evidence for crustacean ecdysones and 'juvenile hormone' was also first obtained for barnacles; and the study of cyprid metamorphosis is likely to make important contributions to our knowledge of crustacean juvenile hormone function.

New approaches to traditional problems in barnacle biology, including the use of molecular, electrophysiological and laser microsurgical techniques, promise new insights, the significance of which is unlikely to be confined to the Crustacea.

NOTE ADDED IN PROOF

Recent evidence obtained in this laboratory does not support the hypothesis of egg hatching pheromone of *S. balanoides* is an excretory metabolite of EPA (Clare 1994).

ACKNOWLEDGEMENTS

I am indebted to Drs. Graham Walker, Alan Southward and two anonymous reviewers for their comments on the manuscript. Financial support for the authors' studies on barnacle chemical communication has been provided by Duke University Research Council, the US Office of Naval Research, the Natural Environment Research Council and a grant-in-aid from the Marine Biological Association.

REFERENCES

Ache, B.W. 1991. Phylogeny of smell and taste. In T.V. Getchell (ed.), *Smell and Taste in Health and Disease*, pp. 3-18. New York: Raven Press.

Alzieu, Cl., J. Sanjuan, J.P. Deltreil, & M. Borel 1986. Tin contamination in Arcachon Bay: Effects on oyster shell anomalies. *Mar. Poll. Bull.* 17: 494-498.

Anderson, D.T. 1994. *Barnacles. Structure, Function, Development and Evolution.* London: Chapman & Hall.

Anholt, R.R.H. 1991. Odor recognition and olfactory transduction: the new frontier. *Chemical Senses* 16: 421-427.

Arkett, S.A., F.-S. Chia, J.I. Goldberg, & R. Koss 1989. Identified settlement receptor cells in a nudibranch veliger respond to specific cue. *Biol. Bull.* 176: 155-160.

Barlow, L.A. 1990. Electrophysiological and behavioral responses of larvae of the red abalone (*Haliotis rufescens*) to settlement-inducing substances. *Bull. Mar. Sci.* 46: 537-554.

Barnes, H. 1957. Processes of restoration and synchronization in marine ecology. The spring diatom increase and the 'spawning' of the common barnacle, *Balanus balanoides* (L.). *Année Biol.* 33: 68-85.

Barnes, M. & H. Barnes 1982. Effect of turbulence on the feeding and moulting of the cirripede *Balanus balanoides* (L.) given an algal diet. *J. Exp. Mar. Biol. Ecol.* 65: 163-172.

Barnett, B.E., S.C. Edwards, & D.J. Crisp 1979. A field study of settlement behaviour in *Balanus balanoides* and *Elminius modestus* (Cirripedia: Crustacea) in relation to competition between them. *J. Mar. Biol. Ass. UK.* 59: 575-580.

Baxter, G.R. & D.E. Morse 1992. Cilia from abalone larvae contain a receptor-dependent G protein transduction system similar to that in mammals. *Biol. Bull.* 183:147-154.

Bebbington, P.M. & E.D. Morgan 1977. Detection and identification of molting hormone (ecdysones) in the barnacle *Balanus balanoides*. *Comp. Biochem. Physiol.* 56(B): 77-79.

Berridge, M.J. 1993. Inositol trisphosphate and calcium signalling. *Nature* 361: 315-325.

Bjostad, L.B., C.E. Linn, J.W. Du, & E.L. Roelofs 1984. Identification of new sex pheromone components in *Trichoplusia ni* predicted from biosynthetic precursors. *J. Chem. Ecol.* 10: 1309-1323.

Bonar, D.B., R.M. Weiner, & R.R. Colwell 1986. Microbial-invertebrate interactions and potential for biotechnology. *Microb. Ecol.* 12: 101-110.

Borst, D.W., H. Laufer, M. Landau, E.S. Chang, W.A. Hertz, F.C. Baker, & D.A. Schooley 1987. Methyl farnesoate and its role in crustacean reproduction and development. *Insect Biochem.* 17: 1123-1127.

Bourget, E. 1988. Barnacle larval settlement: the perception of cues at different spatial scales. In G. Chelazzi & M. Vannini (eds.), *Behavioral Adaptation to Intertidal Life*, pp. 153-172. New York: Plenum Press.

Brancato, M.S. & R.M. Woollacott 1982. Effect of microbial films on settlement of bryozoan larvae (*Bugula simplex, B. stolonifera, B. turrita*). *Mar. Biol.* 71: 51-56.

Bryan, G.W., P.E. Gibbs, G.R. Burt, & L.G. Hummerstone 1987. The effects of tributyltin (TBT) accumulation on adult dog-whelks, *Nucella lapillus*: long-term field and laboratory experiments. *J. Mar. Biol. Ass. UK* 67: 525-544.

Buck, L. & R. Axel 1991. A novel multigene family may encode odorant receptors: a molecular basis for odor recognition. *Cell* 65: 175-187.

Burke, R.D. 1986. Pheromones and the gregarious settlement of marine invertebrate larvae. *Bull. Mar. Sci.* 39: 323-331.

Cameron, R.A. & R.T. Hinegardner 1974. Initiation of metamorphosis in laboratory cultured sea urchins. *Biol. Bull.* 146: 335-342.

Chabot, R. & E. Bourget 1988. Influence of substratum heterogeneity and settled barnacle density on the settlement of cypris larvae. *Mar. Biol.* 97: 45-56.

Chang, E.S., M.J. Bruce & S.L. Tamone 1993. Regulation of crustacean molting: a multi-hormonal system. *Amer. Zool.* 33: 324-329.

Christie, A.O. & R. Dalley 1987. Barnacle fouling and its prevention. *Crustacean Issues* 5: 419-433.

Clare, A.S. 1984. *Hormones in cirripedes*. Ph.D. Thesis, Univ. Wales.

Clare, A.S. 1987. Endocrinology of cirripedes. *Crustacean Issues* 5: 249-266. Rotterdam: Balkema.

Clare, A.S. 1994. *The Physiologist* 37: A-82.

Clare, A.S., R.K. Freet, & M.J. McClary 1994. On the antennular secretion of the cyprid of *Balanus amphitrite amphitrite*, and its role as a settlement pheromone. *J. Mar. Biol. Ass. UK* 74: 243-250.

Clare, A.S. & J. A. Nott 1994. Scanning electron microscopy of the fourth antennular segment of *Balanus amphitrite amphitrite* (Crustacea: Cirripedia). *J. Mar. Biol. Ass. UK* 74: 967-970.

Clare, A.S. & D. Rittschof 1989. What's in a name? *Nature* 338: 627.

Clare, A.S., D. Rittschof, & J.D. Costlow 1992a. Effects of the nonsteroidal ecdysone mimic RH 5849 on larval crustaceans. *J. Exp. Zool.* 262: 436-440.

Clare, A.S., D. Rittschof, D.J. Gerhart, & J.S. Maki 1992b. Molecular approaches to nontoxic antifouling. *Invert. Reprod. Devel.* 22: 67-76.

Clare, A.S., R.F. Thomas, & D. Rittschof 1995. Evidence for the involvement of cyclic AMP in the pheromonal modulation of barnacle settlement. *J. Exp. Biol.* 198: 655-664.

Clare, A.S. & G. Walker 1986. Further studies on the hatching process in *Balanus balanoides* (L.). *J. Exp. Mar. Biol. Ecol.* 97: 295-304.

Clare, A.S. & G. Walker 1989. Morphology of the nervous system of barnacles: the median ocellus of *Balanus hameri* (= *Chirona hameri*) (Crustacea: Cirripedia). *J. Mar. Biol. Ass. UK* 69: 769-784.

Clare, A.S., G. Walker & J.T. Høeg 1993. The Rhizocephala: a model system for studies on the chemical basis of barnacle settlement. *Environs* 14: 3-5.

Clare, A.S., G. Walker, D.L. Holland, & D.J. Crisp 1982. Barnacle egg hatching: a novel rôle for a prostaglandin-like compound. *Mar. Biol. Lett.* 3: 113-120.

Clare, A.S., G. Walker, D.L. Holland, & D.J. Crisp 1985. The hatching substance of the barnacle, *Balanus balanoides* (L.). *Proc. R. Soc. Lond.* 224B: 131-147.

Corey, E.J., H.E. Ensley, M. Hamberg, & B. Samuelsson 1975. Disparate pathways of prostaglandin biosynthesis in coral and mammalian systems. *J. Chem. Soc. Chem. Commun.* 1176: 277-278.

Costlow, J.D. 1963. Molting and cyclic activity in chromatophorotropins of the central nervous system of the barnacle, *Balanus eburneus*. *Biol. Bull.* 124: 254-261.
Crisp, D.J. 1956. A substance promoting hatching and liberation of young in cirripedes. *Nature* 178: 263.
Crisp, D.J. 1976. The role of the pelagic larva. In P.S. Davies (ed.), *Perspectives in Experimental Biology*: 145-155. Oxford: Pergamon Press.
Crisp, D.J. 1984. Overview of research on marine invertebrate larvae. In J.D. Costlow & R.C. Tipper (eds.), *Marine Biodeterioration: an Interdisciplinary Study*: 103-126. Annapolis: Naval Institute Press.
Crisp, D.J. 1990. Gregariousness and systematic affinity in some North Carolinian barnacles. *Bull. Mar. Sci.* 47: 516-525.
Crisp, D.J. & P.S. Meadows 1962. The chemical basis of gregariousness in cirripedes. *Proc. R. Soc. Lond.* B156: 500-520.
Crisp, D.J. & C.P. Meadows 1963. Adsorbed layers: the stimulus to settlement in barnacles. *Proc. R. Soc. Lond.* B158: 364-387.
Crisp, D.J. & C.P. Spencer 1958. The control of the hatching process in barnacles. *Proc. R. Soc. Lond.* B148: 278-299.
Davis, C.W. & J.D. Costlow 1974. Evidence for molt inhibiting hormone in the barnacle *Balanus improvisus* (Crustacea, Cirripedia). *J. Comp. Physiol.* 93: 85-91.
Dodds, W.K. 1990. Hydrodynamic constraints on evolution of chemically mediated interactions between aquatic organisms in unidirectional flows. *J. Chem. Ecol.* 16: 1417-1430.
Fingerman, M. 1988. Application of endocrine manipulations to the control of marine fouling crustaceans. In M.-F. Thompson, R. Sarajini & R. Nagabhushanam (eds.), *Marine Biodeterioration*, pp. 81-91. Rotterdam: Balkema.
Fitt, W.K., M.P. Labare, W.C. Fuqua, M. Walch, S.L. Coon, D.B. Bonar, R.R. Colwell, & R.M. Weiner 1989. Factors influencing bacterial production of inducers of settlement behaviour of larvae of the oyster *Crassostrea gigas*. *Microb. Ecol.* 17: 287-298.
Flatman, P.W. 1991. Mechanisms of magnesium transport. *Annu. Rev. Physiol.* 53: 259-271.
Gabbott, P.A. & V.N. Larman 1987. The chemical basis of gregariousness in cirripedes: A review (1953-1984). *Crustacean Issues* 5: 377-388.
Gannon, A.T. 1990. Distribution of *Octolasmis muelleri*, an ectocommensal gill barnacle, on the blue crab. *Bull. Mar. Sci.* 46: 55-61.
Gerhart, D.J., A.S. Clare, K. Eisenman, D. Rittschof, & R.B. Forward 1990. Eicosanoids in corals and crustaceans: primary metabolites that function as allelochemicals and pheromones. In A. Epple, C.G. Scanes & M.H. Stetson (eds.), *Progress in Comparative Endocrinology*, pp. 598-602. New York: Wiley.
Gibson, P.H. & J.A. Nott. 1971. Concerning the fourth antennular segment of the cypris larva of *Balanus balanoides*. In D.J. Crisp (ed.), *4th European Marine Biology Symposium*: 227-236. Cambridge: University Press.
Goodman, W.G., D.C. Coy, F.C. Baker, L. Xu, & Y.C. Toong 1990. Development and application of a radioimmunoassay for the juvenile hormones. *Insect Biochem.* 20: 357-364.
Groth-Nard, C. & J.-M. Robert 1993. Les lipides des diatomees. *Diatom Res.* 8: 281-308.
Hadfield, M.G. 1986. Settlement and recruitment of marine invertebrates: a perspective and some proposals. *Bull. Mar. Sci.* 39: 418-425.
Hallberg, E., K.U.I. Johansson, & R. Elofsson 1992. The aesthetasc concept: structural variations of putative olfactory receptor cell complexes in Crustacea. *Microsc. Res. Tech.* 22: 325-335.
Hill, E.M., D.L. Holland, K.H. Gibson, E. Clayton & A. Oldfield 1988. Identification and hatching factor activity of monohydroxyeicosapentaenoic acid in homogenates of the barnacle *Elminius modestus*. *Proc. R. Soc. Lond.* B234: 455-461.
Hirata, M., Y. Hayashi, F. Ushikubi, S. Nakanishi, & S. Narumiya 1991. Cloning and expression of cDNA for a human thromboxane A2 receptor. *Nature* 349: 617-620.
Høeg, J.T. 1992. Rhizocephala. In *Microscopic Anatomy of Invertebrates, Vol. 9: Crustacea*, pp. 313-345. New York: Wiley-Liss.
Holland, D.L., J. East, K.H. Gibson, E. Clayton, & A. Oldfield 1985. Identification of the hatching

factor of the barnacle *Balanus balanoides* as the novel eicosanoid 10, 11, 12-trihydroxy-5, 8, 14, 17-eicosatetraenoic acid. *Prostaglandins* 29: 1021-1029.

Horn, L.W. 1989. L-Glutamate transport in internally dialyzed barnacle muscle fibers. *Amer. J. Physiol.* 257: C442-C450.

Hoyle, G. 1987. The giant muscle cells of barnacles. *Crustacean Issues* 5: 213-225.

Huang, Y.P. & E.D. Bittar 1991. Protection by GTP from the effects of aluminum on the sodium efflux in barnacle muscle fibers. *Biochim. Biophys. Acta* 1062: 255-263.

Irvine, I.A.S. & P.W. Sorensen 1993. Acute olfactory sensitivity of wild common carp, *Cyprinus carpio*, to goldfish hormonal sex pheromones is influenced by gonadal maturity. *Can. J. Zool.* 71: 2199-2210.

Johnson, C.R., D.G. Muir, & A.L. Reysenbach 1991. Characteristic bacteria associated with surfaces of coralline algae: a hypothesis for bacterial induction of marine invertebrate larvae. *Mar. Ecol. Prog. Ser.* 74: 281-294.

Kirchman, D., S. Graham, D. Reish, & R. Mitchell 1982. Bacteria induce settlement and metamorphosis of *Janua (Dexiospira) brasiliensis* Grube (Polychaeta: Spirorbidae). *J. Exp. Mar. Biol. Ecol.* 56: 153-163.

Kitamura, S., H. Ogata, & F. Takashima 1994. Olfactory responses of several species of teleost to F-prostaglandins. *Comp. Biochem. Physiol.* 107A: 463-467.

Kjaer, E.B. 1992. Bioactive materials for antifouling coatings. *Progr. Org. Coat.* 20: 339-352.

Klein, J.M., D.P.V. de Kleijn, G. Hünemeyer, R. Keller, & W.M. Weidemann 1993a. Demonstration of the cellular expression of genes encoding molt-inhibiting hormone and crustacean hyperglycemic hormone in the eyestalk of the shore crab *Carcinus maenas*. *Cell Tiss. Res.* 274: 515-519.

Klein, J.M., S. Mangerich, D.P.V. de Kleihn, R. Keller, & W.M. Weidemann 1993b. Molecular cloning of crustacean putative molt-inhibiting hormone (MIH) precursor. *FEBS* 334: 139-142.

Knight-Jones, E.W. 1953. Laboratory experiments on gregariousness during setting in *Balanus balanoides* and other barnacles. *J. Exp. Biol.* 30: 584-598.

Koelle, M.R., W.S. Talbot, W.A. Segraves, M.T. Bender, P. Cherbas, & D.S. Hogness 1991. The *Drosophila* EcR gene encodes an ecdysone receptor, a new member of the steroid receptor superfamily. *Cell* 67: 59-77.

Larman, V.N., P.A. Gabbott & J. East 1982. Physico-chemical properties of the settlement factor proteins from the barnacle *Balanus balanoides*. *Comp. Biochem. Physiol.* 72B: 329-338.

Laufer, H. & D.W. Borst. 1988. Juvenile hormone in Crustacea. In H. Laufer & R.G.H. Downer (eds.), *Invertebrate Endocrinology*, pp. 305-313. New York: Liss.

Le Fèvre, J. & E. Bourget 1992. Hydrodynamics and behaviour: transport processes in marine invertebrate larvae. *Trends Evol. Ecol.* 7: 288-289.

Leitz, T. 1993. Biochemical and cytological bases of metamorphosis in *Hydractinia echinata*. *Mar. Biol.* 116: 559-564.

Loeb, M.J. 1993. Hormonal control of growth and reproduction in the arthropods: introduction to the symposium. *Amer. Zool.* 33: 303-307.

Maki, J.S. & R. Mitchell 1985. Involvement of lectins in the settlement and metamorphosis of marine invertebrate larvae. *Bull. Mar. Sci.* 37: 675-683.

Maki, J.S., D. Rittschof, J.D. Costlow, & R. Mitchell 1988. Inhibition of attachment of larval barnacles, *Balanus amphitrite*, by bacterial surface films. *Mar. Biol.* 97: 199-206.

Maki, J.S., D. Rittschof, M.-O. Samuelsson, U. Szewzyk, A.B. Yule, S. Kjelleberg, J.D. Costlow, & R. Mitchell 1990. Effect of marine bacteria and their exopolymers on the attachment of barnacle cypris larvae. *Bull. Mar. Sci.* 46: 499-511.

Michel, W.C. & B.W. Ache 1992. Cyclic nucleotides mediate an odor-evoked potassium conductance in lobster olfactory receptor cells. *J. Neurosci.* 12: 3979-3984.

Mitchell, R. & D. Kirchman. 1984. The microbial ecology of marine surfaces. In J.D. Costlow & R.C. Tipper (eds.), *Marine Biodeterioration: an Interdisciplinary Study*, pp. 49-58. London: Spon.

Moore, P.A., J. Atema, & G.A. Gerhardt 1991. Fluid dynamics and microscale chemical movement in the chemosensory appendages of the lobster *Homarus americanus*. Chem. Senses 16: 663-674.

Morgan, E.D. & M.P. Marco 1990. Advances in techniques for ecdysteroid analysis. *Invert. Reprod. Devel.* 18: 55-66.

Morse, D.A. 1990. Recent progress in larval settlement and metamorphosis: closing the gaps between molecular biology and ecology. *Bull. Mar. Sci.* 46: 465-483.

Morse, D.E. 1992. Molecular mechanisms controlling metamorphosis and recruitment in abalone larvae. In S.A. Sheperd, M.J. Tegner & S.A. Guzenan (eds.), *Abalones of the World*, pp. 107-119. Oxford: Blackwell.

Morse, D.E. 1993. Signalling in planktonic larvae. *Nature* 363: 406.

Mullineaux, L.S. & C.A. Butman 1991. Initial contact, exploration and attachment of barnacle (*Balanus amphitrite*) cyprids settling in flow. *Mar. Biol.* 110: 93-103.

Naldrett, M.J. 1992. *Cement and other adhesives in the barnacle.* Ph.D. Thesis, Univ. Reading.

Nott, J.A. 1969. Settlement of barnacle larvae: surface structure of the antennular attachment disc by scanning electron microscopy. *Mar. Biol.* 2: 248-251.

Nott, J.A. & B.A. Foster 1969. On the structure of the antennular attachment organ of the cypris larva of *Balanus balanoides* (L.). *Phil. Trans. R. Soc. Lond.* 256: 115-133.

Nowell, A.R.M. & P.A. Jumars 1984. Flow environments of aquatic benthos. *Annu. Rev. Ecol. Systemat.* 15: 303-328.

Paerl, H.W. 1985. Influence of attachment on microbial metabolism and growth in aquatic ecosystems. In D.C. Savage & M. Fletcher (eds.), *Bacterial Adhesion: Mechanisms and Physiological Significance*: 363-400. New York: Plenum Press.

Palazzolo, M.J., D.R. Hyde, K.V. Raghavan, K. Mecklenburg, S. Benzer, & E. Meyerowitz 1989. Use of a new strategy to isolate and characterize 436 *Drosophila* cDNA clones corresponding to RNAs detected in adult heads but not in early embryos. *Neurons* 3: 527-539.

Palazzolo, M.J., B.A. Hamilton, D. Ding, C.H. Martin, D.A. Mead, M.C. Mierendorf, K.V. Raghavan, E.M. Meyerowitz, & H.D. Lipshitz 1990. Phage lambda cDNA cloning vectors for substractive hybridization, fusion-protein synthesis and Cre-IoxP automatic plasmid subcloning. *Gene* 88: 25-36.

Patton, W.K. 1972. Studies on the animal symbionts of the gorgonian coral *Leptogorgia virgulata* (Larmark). *Bull. Mar. Sci.* 22: 419-431.

Pawlik, J.R. 1992. Induction of marine invertebrate larval settlement: evidence for chemical cues. In V.J. Paul (ed.), *Ecological Roles of Marine Natural Products*, pp. 189-236. Ithaca: Comstock.

Pettis, R.J. 1991. *Biologically active arginine-terminal peptides*. Ph. D. Thesis: UNC Chapel Hill.

Pineda, J. 1994. Spatial and temporal patterns in barnacle settlement rate along a southern California rocky shore. *Mar. Ecol. Prog. Ser.* 107: 125-138.

Piomelli, D. 1993. Arachidonic acid in cell signalling. *Curr. Opinion Cell Biol.* 5: 274-280.

Raimondi, P.T. 1988. Settlement cues and determination of the vertical limit of an intertidal barnacle. *Ecology* 69: 400-407.

Riddiford, L.M. & J.W. Truman 1993. Hormone receptors and the regulation of insect metamorphosis. *Amer. Zool.* 33: 340-347.

Rittschof, D. 1985. Oyster drills and the frontiers of chemical ecology: unsettling ideas. *Amer. Malacol. Bull.* Special Ed. 1: 111-116.

Rittschof, D., E.S. Branscomb, & J.D. Costlow 1984. Settlement and behavior in relation to flow and surface in larval barnacles, *Balanus amphitrite* Darwin. *J. Exp. Mar. Biol. Ecol.* 82: 131-146.

Rittschof, D., A.S. Clare, D.J. Gerhart, Sr. A. Mary & J. Bonaventura 1992. Barnacle *in vitro* assays for biologically active substances: toxicity and settlement inhibition assays using mass cultured *Balanus amphitrite amphitrite* Darwin. *Biofouling* 6: 115-122.

Rittschof, D., A.R. Schmidt, I.R. Hooper, D.J. Gerhart, D. Gunster & J. Bonaventura. 1991. Molecular mediation of settlement of selected invertebrate larvae. In M.-F. Thompson, R. Sarajini & R. Nagabhushanam (eds.), *Bioactive Compounds from Marine Organisms with Emphasis on the Indian Ocean*, pp. 317-330. New Delhi: Oxford & IBH.

Roughgarden, J.S., S.D. Gaines & S. Pacala. 1987. Supply side ecology: the role of physical transport processes. In *Orginization of Communities: Past and Present. Proc. Brit. Ecol. Soc.*, pp. 459-489. London: Blackwell.

Ruggeri, B. & C.A. Thoroughgood 1985. Prostaglandins in aquatic fauna: a comprehensive review. *Mar. Ecol. Prog. Ser.* 23: 301-306.

Sandeen, M.L. & J.D. Costlow 1961. The presence of decapod-pigment-activating substances in the central nervous system of representative Cirripedia. *Biol. Bull.* 120: 192-205.

Schmiedel-Jakob, I., W.C. Michel, P.A.V. Anderson, & B.W. Ache 1990. Whole cell recording from lobster olfactory receptor cells: multiple ionic bases for the receptor potential. *Chem. Senses* 15: 397-405.

Schmitt, B.C. & B.W. Ache 1979. Olfaction: responses of a decapod crustacean are enhanced by flicking. *Science* 205: 204-206.

Shimizu, T. & L.S. Wolfe 1990. Arachidonic acid cascade and signal transduction. *J. Neurochem.* 55: 1-15.

Skinner, D.M. 1985. Interacting factors in the control of the crustacean molt cycle. *Amer. Zool.* 25: 275-284.

Song, W.-C., D.L. Holland, K.H. Gibson, E. Clayton, & A. Oldfield 1990. Identification of novel hydroxy fatty acids in the barnacle *Balanus balanoides*. *Biochim. Biophys. Acta* 1047: 239-246.

Sonnenburg, W.K., J. Zhu, & W.L. Smith 1990. A prostaglandin E receptor coupled to a pertussis toxin-sensitive guanine nucleotide regulatory protein in rabbit cortical collecting tubule cells. *J. Biol. Chem.* 265: 8479-8483.

Sorensen, P.W. 1992. Hormonally derived sex pheromones in goldfish: a model for understanding the evolution of sex pheromone systems in fish. *Biol. Bull.* 183: 173-177.

Sorensen, P.W., T.J. Hara, N.E. Stacey, & F.W. Goetz. 1988. F prostaglandins function as potent olfactory stimulants comprising the postovulatory female sex pheromone in goldfish. *Biol. Reprod.* 39: 1039-1050.

Stacey, N.E. 1987. Roles of hormones and pheromones in fish reproductive behavior. In D. Crews (ed.), *Psychobiology of Reproductive Behavior*, pp. 28-69. New York: Prentice-Hall.

Stacey, N.E. & F.W. Goetz 1982. Role of prostaglandins in fish reproduction. *Can. J. Fish. Aquat. Sci.* 39: 92-98.

Stanley-Samuelson, D.W. 1987. Physiological roles of prostaglandins and other eicosanoids in invertebrates. *Biol. Bull.* 173: 92-109.

Stanley-Samuelson, D.W. 1991. Comparative eicosanoid physiology in invertebrate animals. *Amer. J. Physiol.* 260: 849-853.

Stanley-Samuelson, D.W. & W. Loher 1990. Evolutionary aspects of prostaglandins and other eicosanoids in invertebrates. In A. Epple, C.G. Scanes & M.H. Stetson, *Progress in Comparative Endocrinology*, pp. 614-619. New York: Wiley.

Starr, M., J.H. Himmelman, & J. Therriault 1991. Coupling of nauplii release in barnacles with phytoplankton blooms: a parallel strategy to that of spawning in urchins and mussels. *J. Plankton Res.* 13: 561-571.

Stuart, A.E. 1983. Vision in barnacles. *Trends Neurosci.* 6: 137-140.

Takahashi, K., J. Capdevila, A. Karara, J.R. Falck, H.R. Jacobson, & K.F. Badr 1990. Cytochrome P-450 arachidonate metabolites in rat kidney: characterization and hemodynamic responses. *Amer. J. Physiol.* 258: 781-789.

Tamburri, M.N., R.K. Zimmer-Faust, & M.L. Tamplin 1992. Natural sources and properties of chemical inducers mediating settlement of oyster larvae: a re-examination. *Biol. Bull.* 183: 327-338.

Taylor, A.R. & C. Brownlee 1993. Patch clamping plant cells: use of ultraviolet laser microsurgery to remove cell wall. *Axobits* 12: 12-14.

Tegtmeyer, K. & D. Rittschof 1989. Synthetic peptide analogs to barnacle settlement pheromone. *Peptides* 9: 1403-1406.

Tobe, S.S., D.A. Young, H.W. Khoo, & F.C. Baker 1989. Farnesoic acid as a major product of release from crustacean mandibular organs in vitro. *J. Exp. Zool.* 249: 165-171.

Visscher, P. 1927. Nature and extent of fouling of ships' bottoms. *Bull. Bur. Fish.* 43: 193-252.

Walker, G. 1974. The occurrence, distribution, and attachment of the pedunculate barnacle *Octolasmis mülleri* (Coker) on the gills of crabs, particularly the blue crab, *Callinectes sapidus* Rathbun. *Biol. Bull.* 147: 678-689.

Walker, G. 1992. Cirripedia. In, *Microscopic Anatomy of Invertebrates Vol. 9: Crustacea*, pp. 249-311. New York: Wiley-Liss.

Walker, G. & A.B. Yule 1984. Temporary adhesion of the barnacle cyprid: the existence of an antennular adhesive secretion. *J. Mar. Biol. Ass. UK* 64: 679-686.

Watson, P.A. 1991. Function follows form: generation of intracellular signals by cell deformation. *FASEB* 5: 2013-2019.

Werner, U., E. Suss-Toby, A. Rom & B. Minke 1992. Calcium is necessary for light excitation in barnacle photoreceptors. *J. Comp. Physiol.* A170: 427-434.

Weiner, R.M., A.M. Segall, & R.R. Colwell 1985. Characterization of a marine bacterium associated with *Crassostrea virginica* (the eastern oyster). *Appl. Environ. Microbiol.* 49: 83-90.

Wilkins, C.L. 1994. Multidimensional GC for qualitative IR and MS of mixtures. *Analyt. Chem.* 66: 295-301.

Wilson, D.P. 1955. The role of microorganisms in the settlement of *Ophelia bicornis* Savigny. *J. Mar. Biol. Assoc. UK.* 34: 531-543.

Wodicka, L.M. & D.E. Morse 1991. cDNA sequences reveal mRNAs for two G_α signal transducting proteins from larval cilia. *Biol. Bull.* 180: 318-327.

Yamamoto, H. & A. Nagai 1992. Polypeptide models of the arthropodin protein of the barnacle *Balanus balanoides. Mar. Chem.* 37: 131-143.

Yates, F.E. 1981. Analysis of endocrine signals: the engineering and physics of biochemical communication systems. *Biol. Reprod.* 24: 73-94.

Yudin, A.I., R.A. Diener, W.H. Clark, & E.S. Chang 1980. Mandibular gland of the blue crab, *Callinectes sapidus. Biol. Bull.* 159: 760-772.

Yule, A.B. & D.J. Crisp 1983. Adhesion of cypris larvae of the barnacle, *Balanus balanoides*, to clean and arthropodin treated surfaces. *J. Mar. Biol. Ass. U. K.* 63: 261-271.

Yule, A.B. & G. Walker 1984. The adhesion of the barnacle, *Balanus balanoides*, to slate surfaces. *J. Mar. Biol. Ass. UK.* 64: 147-156.

Yule, A.B. & G. Walker 1987. Adhesion in barnacles. *Crustacean Issues* 5: 389-402. Rotterdam: Balkema.

Zimmer-Faust, R.K. & M.N. Tamburri 1994. Chemical identity and ecological implications of a waterborne, larval settlement cue. *Limnol. Oceanogr.* 39: 1075-1087.

Zobell, C.E. & E.C. Allen 1935. The significance of marine bacteria in the fouling of submerged surfaces. *J. Bacteriol.* 29: 230-251.

Larval settlement: Historical and future perspectives

Graham Walker
University of Wales, Bangor, School of Ocean Sciences, Menai Bridge, Gwynedd, UK

ABSTRACT

The barnacle cypris larva is initially a pelagic organism. During its planktonic phase the cyprid positions itself in the water column to maximise survival, dispersal and the chance of contacting the adult habitat; it responds to environmental cues – principally light and pressure – by active upward swimming and passive sinking. When the competent cyprid contacts a surface it moves on to the settlement phase. Contact with a surface is maintained by a temporary adhesive on the antennules allowing exploration of the surface when various factors are being sensed at different temporal and spatial scales. Two factors only, chemical cues and microbial films, have been selected for detailed discussion. The consummatory act of settlement is permanent fixation, when the cyprid cements itself down onto a surface. When the fixed cyprid has metamorphosed to the juvenile it is then considered to be recruited to the surface.

1 INTRODUCTION

Settlement in barnacles involves a specific settlement-stage larva, the cyprid. This conservative larval form is instantly recognisable by the shape of its bivalved carapace, the anterior antennules and posterior thoracic swimming appendages when extended out ventrally from the carapace (Walley 1969; Walker et al. 1987; Anderson 1994). The cyprid metamorphoses from the final pelagic nauplius stage (both planktotrophic and lecithotrophic forms). In abysso-benthic thoracicans particularly, certain rhizocephalans and acrothoracicans the cypris larvae are released from the parent, the product of direct development.

The cyprid is a lecithotrophic larva endowed with a finite amount of energy to carry out all its tasks – swimming, temporary attachment whilst exploring surfaces, permanent fixation to a surface and finally a further metamorphosis (Lucas et al. 1979). As all barnacle cyprids carry out these various tasks, each will need to be considered and discussed. However, before embarking on this exercise it is perhaps pertinent to define the phases of a cyprid's existence.

A cyprid is initially pelagic and so there is the *planktonic phase* in which active swimming and passive sinking in response to stimuli will serve to position the cyprid at the level in the water column (larval navigation?) to maximise survival, dispersal and eventually the chance of contacting target surfaces. During the planktonic phase the larva acquires the 'competence for settlement', i.e., physiological readiness including fully matured secretions. Exceptions to this planktonic phase include the cyprid of an akentrogonid rhizocephalan, which lacks a thorax and hence swimming appendages and can only disperse by antennular walking on surfaces (Glenner et al. 1989) and certain acrothoracican cyprids, which also disperse by walking – their thoracic appendages are rudimentary and have no locomotory function (Turquier 1985).

The *settlement phase* follows when the antennules make contact with surfaces. Temporary attachment enables exploration of surfaces and, if suitably stimulated by an array of 'releasers', permanent fixation then follows. If a location fails to stimulate sufficiently, then as a free agent the cyprid can detach itself and revert to being a pelagic larva.

The *metamorphic phase* following permanent fixation usually sees the cyprid transform into the juvenile form [for rhizocephalans it is more complex – the metamorphosis is to an inoculation stage (female) or another larval form (trichogon) if male]. Once metamorphosed to the juvenile form the individual is then considered to have been recruited to the surface (Keough & Downes 1982).

In a recent review, Pawlik (1992) preferred to take the view that for invertebrate larvae generally settlement involves the entire transition from planktonic larva to benthic juvenile. This review prefers to keep the phases separate (following Crisp 1984) and so will concentrate on the events within the settlement phase as defined above. For barnacle cyprids the three phases are discrete, but for the larvae of other benthic invertebrates the metamorphic phase may not be so easily recognised.

2 THE PLANKTONIC PHASE

The majority of free-living barnacles release planktotrophic nauplius larvae, which moult and grow through six stages before the metamorphosis to the cyprid. Our knowledge of this pelagic part of the life history in nature is restricted. What proportion of the nauplii perish through lack of food, what proportion are predated, what proportion of the surviving cyprids reach suitable habitats? – these are some of the questions that cannot yet be answered. However, some progress is being made, albeit slowly. Methods are emerging for tagging and tracking larvae (Levin 1990; Levin et al. 1993; Dixon et al. 1994). An understanding is being gained of nearshore water flows not only crucial for dispersing larvae, but also for returning them to the littoral zone (Denny 1987; Pineda 1994a). Advective transport is possible with direction dependent on the depth in the water column and hence the positional control of the larvae. Turbulent mixing (eddy diffusion) can both disperse and aggregate larvae depending on conditions (De Wolf 1973). Tidally-generated internal (density) waves create circulating cells near the water surface and larvae can become concentrated in the slicks between cells and carried onshore (Shanks 1986; Shanks & Wright 1987). Modelling such water flows for particular coastlines with a view to predicting larval distance travelled and larval patchiness in space and time must become increasingly important. Already emerging are models concerned with larval settlement in turbulent bottom boundary layers (Eckman 1990; Gross et al. 1992).

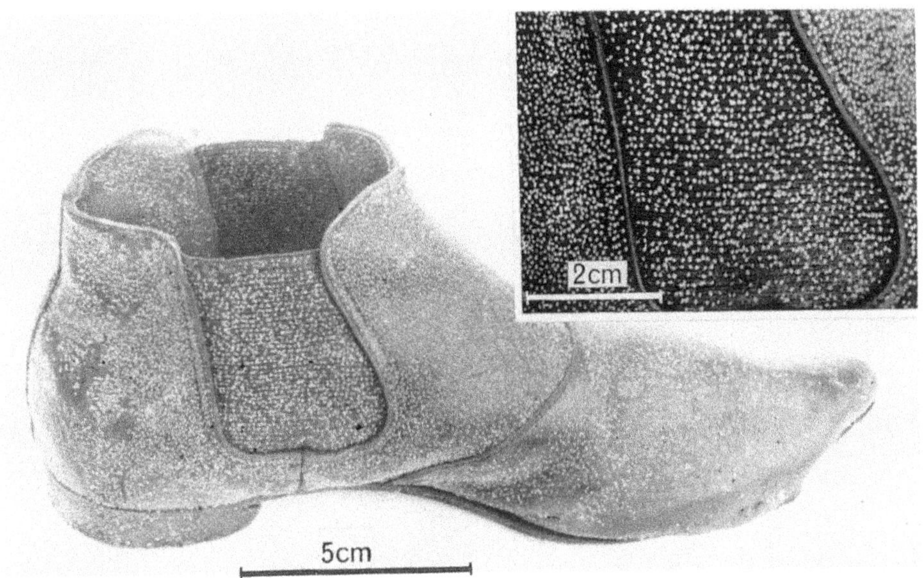

Plate 1. Shoe taken from the shore of The Menai Strait, North Wales, UK showing a heavy settlement (and recruitment) of the intertidal barnacle, *Semibalanus balanoides*. Note (i) the gregarious settlement, (ii) absence of adult barnacles, (iii) settlement along surface contours (inset), (iv) the spacing out or 'territoriality' effect, and (v) progressive reduction of settlement on inside of shoe due to restricted water flow.

Anderson (1994) listed the lengths of thoracican larvae, including the cyprids of thirty species. Most of these cyprids lie within the range 375-940 µm with an approximate mean length of 600 µm. Barnes (1953) highlighted the variation in cyprid carapace length for a single species, which can be quite marked. Species with large cyprids include *Lepas anatifera* (ca. 1500 µm) and *Chirona hameri* (ca. 1450 µm). Such larvae have to remain opportunistic over an extended period as their target surfaces are ephemeral; their size in part presumably reflects the increased energy stores needed to remain viable for longer. Conversely, rhizocephalan cyprids are small, 60-400 µm (Glenner et al. 1989), carry relatively less fuel and therefore need to locate their targets more quickly. In the Kentrogonida, the female cyprid is smaller than the male cyprid, again a condition reflecting the increased energy stores needed by the male cyprid to search for the more ephemeral virgin externa target (Høeg 1991).

The fusiform cyprid can swim quite rapidly in short bursts (Yule 1982). *Semibalanus balanoides* cyprids, for instance, are capable of attaining speeds up to 95 body lengths per second (ca. 9.5 cm s^{-1}). The hydrodynamic body shape and the combined effort of six pairs of thoracic appendages in bursts of 10-40 limb beats ensures both controlled lift and forward progression. Little feathering occurs on the recovery stroke, but the speed of recovery is roughly one-third that of the propulsive stroke, ensuring somewhat jerky movement. This discontinuous motion occurs so rapidly as to produce apparent smooth motion when viewed at normal speed. Nevertheless Yule (1982) concluded that the swimming stroke of the cyprid is relatively inefficient in comparison to the gliding swimming motion of copepods. Following a limb beat sequence the cyprid becomes inactive, and being negatively buoyant, passively sinks. For *Balanus crenatus* cyprids the maximum sinking rate is 45 cm min^{-1} (De Wolf 1973) but for the larger *S. balanoides* cyprid it is 26 cm min^{-1}. If the thoracic appendages remain extended outside of the carapace then sinking rate is reduced significantly (by 7 cm min^{-1} for *S. balanoides*) and the sinking path prescribed is a spiral. Cyprids need strong stimulation to remain in swimming mode – in the confined space of a container in the laboratory those of certain species tend to lie inactive on the bottom after a short time in still water conditions (Crisp 1955).

The energy budget of the boreo-arctic *Semibalanus balanoides* cyprid has been thoroughly worked out (Lucas et al. 1979). Because of its lowered metabolic rate (0.6 ml O_2h^{-1} g^{-1} dry wt. at 10°C) compared with the previous nauplius stage (1.9 ml O_2h^{-1} g^{-1} dry wt.) (Lucas 1980) and principal storage material being lipid, this cyprid is capable of remaining viable for settlement and subsequent metamorphosis for up to several weeks in the plankton. The classical energy budget study of Lucas et al. (1979) has not been repeated, in the same way, for any other barnacle species, although metabolic data are available for cyprids of several species (Jorgensen & Vernberg 1982; Harms 1987; Collis 1991). All these studies show that the cyprid consumes less oxygen than the final nauplius stage, underlining the intermittent activity of the cyprid and the need to use its energy reserves economically. Work aimed at comparing the energy budget of the cyprid of a tropical or sub-tropical species with that of *S. balanoides* would be especially informative. The raised metabolic rate, a consequence of higher temperature, would probably limit such a cyprid's existence in the plankton to days. Time pressure to settle and metamorphose successfully therefore will be more acute.

During the planktonic phase the cyprid is likely to respond to light, gravity(?) (see

Sulkin 1990) and pressure, but possibly water movement and pheromones are also important cues. The cyprids of estuarine species will also be sensitive to seawater ionic composition (Rittschof et al. 1986b; Dineen & Hines 1992, 1994a,b). These stimuli ensure the larva maintains itself at the optimum depth for dispersal and then for contacting the adult habitat (Grosberg 1982; Gaines et al. 1985; Shanks 1986; Le Fèvre & Bourget 1991). Cyprids have obvious sense organs – most have eyes (nauplius eye + two compound eyes), frontal filaments (Walker 1974), carapace setae (Walker & Lee 1976), lattice organs (Elfimov 1986; Jensen et al. 1994) and antennular setae (Nott & Foster 1969), but except for the eyes we can only speculate at present on functions of the other sensors. The sensitivity of cyprids to various stimuli *over time* and the means of detecting such stimuli is a wide open field for research (see Clare this volume). The eyes of *Balanus amphitrite* cyprids are most sensitive in the blue-green range (530 nm) (Visscher & Luce 1928), but Yule & Walker (1984) showed that *S. balanoides* cyprids have a much broader spectral response. Walker (1974) considered that the frontal filament vesicles might be the region for pressure perception. The hypothesis he put forward was of pressure acting on the prolific ciliary membranes within the vesicles causing differential ion fluxes across them initiating nerve transduction. Digby (1972) has proposed an alternative and more general mechanism for pressure detection in crustaceans. His hypothesis suggests that pressure change will alter the volume of an extremely thin layer of hydrogen gas at or near the cuticle surface. These volume changes will alter the electrode or semiconductor properties of the cuticle, which will be appreciable to the organism. Neither hypothesis has been investigated experimentally in the interim, which is unfortunate because pressure is undoubtedly the most reliable and constant indicator of depth and cyprids are extremely sensitive to pressure change (Knight-Jones & Qasim 1967). A cyprid's depth-regulating capability, which links pressure change with light as environmental cues, is well illustrated by *Lepas* cyprids. They are very rarely sampled at 10 m depth by the Continuous Plankton Recorder (Roskell 1975), but are commonly taken in neuston nets in the top 10 cm (Weikert 1972) where contact with flotsam is more assured. However, the trade-off is exposure near the sea surface to predation by small hovering seabirds (see Moyse 1987). A more recent study by Conway et al. (1990) on *L. pectinata* larvae has revealed that the nauplii occur in the upper 150 m, but the cyprids were only ever found between 300-400 m! *L. pectinata* attaches to floating *Sargassum* weed, so these cyprids will have to ascend in the water column to be successful settlers. The deep distribution may enable the cyprids to survive longer (lower temperature at this depth), so forming a pool from which to repopulate the surface waters over an extended period.

Even when the pelagic cyprid has attained competence to settle, surfaces may not be immediately available. Such a larva merely survives on its reserves, but must remain opportunistic, able to react when a surface is contacted or close by. It is not known whether cyprids are brought to the vicinity of target surfaces passively or by certain stimuli, i.e., pheromones from existing conspecifics, pheromones from host (if epizoic), pressure waves from surfaces etc., causing them to change behaviour and actively seek contact with surfaces. Denny & Shibata (1989) concluded by modelling that in surf-zone areas larvae can be effectively transported to solid surfaces by the turbulent mixing of the water rather than by their own sinking or directed swimming.

3 THE SETTLEMENT PHASE

When contact is made with a surface an adhesive on the attachment discs of the antennules maintains the contact (see Yule & Walker 1987). The morphology of the antennule, including the fourth segment, has been fully described for the *Semibalanus balanoides* cyprid (Nott 1969; Nott & Foster 1969; Gibson & Nott 1971) and more recently for the *Balanus amphitrite* cyprid (Clare & Nott 1994; Glenner & Høeg 1995). The appearance of the attachment disc with encircling skirt of cuticle, the velum, gives an overall impression of a sucker and the term 'antennulary sucker' has been used in the past (Crisp 1955). Some workers have perpetuated the sucker function (see Lindner 1984), but suction was disproved by Yule & Crisp (1983) when they measured the force to remove temporarily attached *S. balanoides* cyprids from a slate surface. The measured tenacities (force to remove per unit area of attachment disc) were $2\text{-}3 \times 10^5$ Nm^{-2}, equivalent to 2-3 atmospheres) so precluding suction. Temporary adhesion to surfaces with differing physico-chemical properties has also been measured for *S. balanoides* cyprids (Yule & Walker 1987) and the conclusion reached was that the cyprid can assess the nature of a surface (roughness and critical surface tension) from the strength of adhesion achieved. More recently Neal & Yule (1992) have shown that increased temporary adhesion is indicative of a propensity to settle by *S. balanoides* cyprids, whereas Maki et al. (1994) show a far less clear correlation between settlement and temporary adhesion for *Balanus amphitrite* cyprids.

Walker & Yule (1984) discovered the proteinaceous adhesive covering the attachment discs of competent cyprids, some of which is left behind (footprints) on the surface where an attachment disc is applied (see Walker 1987). This adhesive is produced from unicellular antennulary glands first described by Nott & Foster (1969). Similar antennular adhesive has also now been reported for *B. amphitrite* cyprids (Clare et al. 1994).

Temporary adhesion allows exploration of surfaces as antennular discs are alternately attached, detached and reattached in the familiar cyprid 'walk'. Exploration by *Semibalanus balanoides* cyprids is said to take place in three spatial phases (Crisp 1976, 1984), the first termed 'wide searching' (<1 m) is when antennular walking takes place with few changes in direction. If the cyprid receives favourable stimuli, or remains beyond a certain time on the surface (self stimulation?), the behaviour changes. Movement is slowed because the cyprid pauses longer at each step, apparently testing the surface more thoroughly. This second behaviour phase is termed 'close searching' (<5 mm) and involves many changes in direction. With continued positive stimulation the behaviour pattern moves on to the third phase – 'inspection' (<1 mm), which involves the cyprid stepping to and fro in a confined area, such that any translatory movement has virtually ceased. Although this exploration sequence may be the same for other cyprids (more observations are needed), it should be remembered that 'short-circuiting' within the sequence can and does occur when optimum conditions prevail. During all phases the antennular fourth segments are flicked regularly (Clare & Nott 1994) and the caudal appendages at the rear of the larva are swept across the surface, or in some chthamalid species raised up away from the surface. In addition, the thoracic appendages beat in short bursts so pulling on the attached antennules giving the subjective impression that the strength of the bond between the temporary adhesive and the surface is being tested. The carapace is also drawn down onto the surface when the antennules are reflexed. The anterior carapace rubs onto the surface and may be part of surface preparation needed to ensure efficient bonding of the adhesive discharged for permanent fixation.

During the exploration phases the cyprid is exposed to various environmental factors which indicate the settlement suitability of the location. The combination of factors will have a powerful influence on where and when final fixation will take place (Pl. 1). Such factors include light intensity, water flow, surface texture and topography, microbial/microalgal films and various chemical cues (stimulatory and inhibitory). These factors have been reviewed previously (Lewis 1978; Crisp 1984; Bourget 1988; Pawlik 1992), but two are worthy of further discussion here – chemical cues and microbial films.

Investigations into the chemical basis for gregarious settlement by cyprids go back forty years or so. Gregarious settlement was first described for *Elminius modestus* by Knight-Jones & Stevenson (1950), with later laboratory experiments concluding that such a settlement pattern might be due to the cyprids recognising the quinone-tanned protein in the cuticle covering the established adult shell (Knight-Jones 1953; Knight-Jones & Crisp 1953). The discovery of such a strong stimulus led onto work to determine how the proteins (arthropodins) in arthropod cuticles were involved, resulting in the now classical studies of Crisp & Meadows (1962, 1963). Subsequent work, admirably reviewed by Gabbott & Larman (1987), culminated in the identification of the 'settlement factor' in *S. balanoides* as being a mixture of polymorphic acidic proteins having amino acid compositions similar to that of actin. These proteins are only an active stimulant to fixation when adsorbed onto a surface and were thought to be recognised by antennular sense organs. Exactly how this sensory interaction occurs still remains controversial. Nott & Foster (1969) considered that the antennular sense organs might detect particular amino acid sequences (= peptides?) in settlement factor proteins following release of enzyme from the antennules, a theory originally proposed by Knight-Jones (1953). Support for such a mechanism comes from the work of Tegtmeyer & Rittschof (1989), who were able to stimulate the settlement of *B. amphitrite* cyprids using synthetic peptides.

Since *S. balanoides* settlement factor is chemically related to actin (Larman 1984), a 'sticky' protein, the possibility exists that the cyprid recognises the increased adhesion afforded by its 'stickiness', a remarkable interaction if proved (see Yule & Crisp 1983). The magnitude of temporary adhesion is clearly increased when settlement factor is adsorbed to a surface (Yule & Crisp 1983; Yule & Walker 1984), although bovine serum albumin on a surface does not enhance temporary adhesion over that measured on control surfaces, confirming that not all adsorbed proteins cause an increase in temporary adhesion. This 'tactile chemical sense' theory was first proposed by Crisp & Meadows (1962), without them knowing of the existence of the antennulary gland secretion on the attachment discs. Because cyprids can discriminate between settlement factors derived from different species (Crisp 1990; Whillis et al. 1990; Dineen & Hines 1992, 1994a,b) and inconceivable resolution of a tactile sense would be required to differentiate 'stickiness' between very similar proteins, the present consensus is that settlement factor recognition should be a chemical rather than a physical phenomenon.

In laboratory experiments crude settlement factor is prepared by homogenising whole barnacles in seawater or distilled water. In nature, such water-soluble molecules are likely to be short-lived because they will only be present on newly-formed cuticle and will diffuse into the water to be quickly diluted outside of the boundary layer. Wethey (1984) therefore considered a chemical cue to have minimal effect in the field and placed greater emphasis on physical factors such as surface contour and water flow. This objection to the involvement of chemistry in the patterns of settlement was overcome with the discovery that the temporary adhesive of cyprids was left behind on surfaces where cyprids had walked (footprints) (Walker & Yule 1984). This proteinaceous material, derived from modified hypodermal cells, when deposited onto surfaces undoubtedly stimulates the gregarious response (Yule & Walker 1984; Clare et al. 1994). The most 'attractive' surfaces will acquire more footprints per unit area over time as the larvae will remain longer on these surfaces. Protrusions or pits are surface features which cyprids readily explore, so footprints will become concentrated around and within these areas. The presence of footprints will enhance the 'attractiveness' of a surface and gregarious settlement will result, even in the absence of conspecific adults. This larva-larva interaction demands much more research, particularly in understanding whether the recognition of footprints is a chemical or physical phenomenon. If chemical, how specific is it, i.e., conspecific footprints vs. those of other species, and is there a minimal concentration which will stimulate settlement? Footprints are important for gregarious settlement because they are produced at the appropriate time and their pattern on a surface, coupled with the degree of aggregation, is a major stimulus to an alighting conspecific cyprid. The presence of adults and spat undoubtedly stimulates settlement close by, but as Yule & Walker (1987) argue, it may be the contour effect of the overall adult shape, particularly the basal shell margin contour that stimulates cyprid exploration. Settlement in response to the pattern of footprints would then follow gradually moving away from the adult, so there is no need to consider that each cyprid must necessarily contact an already settled barnacle before settling gregariously. Cyprid footprints might increase the strength of temporary adhesion when antennules are applied on them and as a consequence the exploratory behaviour might be short-circuited directly to the inspection phase. This more physical or mechanical hypothesis has yet to be tested.

A further way in which a chemical cue might act is that of water-soluble chemicals emanating from adults and recently settled spat into the water column. Such chemicals, if detected, will not stimulate settlement per se, but may alter the behaviour of the pelagic cyprid, so promoting contact with surfaces (Crisp & Meadows 1963; Rittschof, pers. comm.). At the same time such chemicals may also be continuously adsorbed onto near surfaces and in this form they will stimulate settlement.

Holland et al. (1984) observed that extracts of oil shales promoted settlement of *S. balanoides* cyprids. Hill & Holland (1985) fractionated the oil shale and reported enhanced settlement in response to the fraction containing metalloporphyrins – the hydrocarbon and asphaltene fractions inhibited settlement. Metalloporphyrins are found naturally bound to proteins, so it is hypothesised that the metalloporphyrins effectively bind to the temporary adhesive on the cypris antennules when contact is made, increasing the strength of temporary adhesion and so promoting settlement. Work to test this hypothesis (see Gabbott & Larman 1987) is still awaited.

All the known factors have now been considered for chemical cues acting in gregarious settlement, but the exact way(s) in which they work still remains to be unequivocally proved (see also Hui & Moyse 1987).

There are other non-gregarious situations, epizoic and parasitic associations particularly, in which settlement must be stimulated strongly by a chemical cue(s) emanating from the 'host'. In some cases, the association is highly specific (single host species), whilst for others it is less exact in that several host species may be settled on (Lewis 1978; Foster 1987; Anderson, 1994). It is work on the specific associations which is likely to begin to clarify how chemical cues induce settlement. An example is the male cyprid of kentrogonid rhizocephalans, which will only settle on virgin externae. It is predicted that a highly specific pheromone is released from the externa to attract these cyprids to such an ephemeral site. Rearing of rhizocephalan larvae is straightforward as they are lecithotrophic and male cyprids are easily recognised; it is only the low settlement rate which may prove an obstacle to what otherwise promises to be highly relevant future settlement research.

Competition for space on hard substrata is intense and established organisms may deter settlement of larvae on them in different ways. Of relevance here is chemical inhibition to settlement. Standing et al. (1984) isolated both inhibitor and inducer substances from octocorals when tested with *Balanus amphitrite* cyprids. The inhibitors were low molecular weight compounds (< 20 kilodaltons). Rittschof and co-workers have taken these initial findings further (Rittschof et al. 1985, 1986a; Gerhart et al. 1988). The inhibitors worked when adsorbed to a surface and were not toxic to the cyprids. They were identified as diterpenoid hydrocarbons – pukalide and epoxypukalide. There is no direct evidence that these artificially-produced extracts actually work under natural conditions, but the circumstantial evidence is strong. This sort of work is part of the wider screening programme searching for natural products (or their analogues) with a view to their incorporation into antifouling coatings (Clare et al. 1992). Such compounds are more acceptable environmentally over the more traditional toxins presently in use, but their success hinges on having an impact on the full spectrum of settling larvae of foulers, as well as a vehicle allowing continuous release; for a commercial coating both factors remain insuperable currently, but a recent short term study with an experimental coating incorporating octocoral extract or proprietory analogue did show promise (Price et al. 1992).

Other environmental chemical cues are varied in their reported effect on barnacle settlement. Mucus deposited on a surface may be stimulatory (limpets, nudibranchs, whelks) (Raimondi 1988) or inhibitory (cnidarians, limpets, whelks) (Johnson & Strathmann 1989) as are the microbial/microalgal films that occur on surfaces. Bacteria can change the nature of the surface by altering surface wettability as well as creating an interface (exopolymers) capable of producing adsorbed or soluble chemical cues which are stimulatory or inhibitory to barnacle settlement (Maki et al. 1990, 1992) or to cyprid attachment (Maki et al. 1988; Holmström et al. 1992). In the screening study of Maki et al. (1988) where 18 different bacterial films were tested – 7 were inhibitory, 10 showed no effect and only 1 was stimulatory to *Balanus amphitrite* settlement. Similarly, Mary et al. (1993) showed only 4 stimulatory bacterial strains out of 16 tested, indicating that inhibitory cues are common on surfaces. However, it must be emphasised that these laboratory experiments only tested single bacterial strains. Nevertheless such a wide inhibitory phenomenon forms the basis of a further strategy for biological antifouling coatings (Gatenholm et al. 1995).

In a recent study Neal & Yule (1994) measured the temporary adhesion of cyprids of two barnacles on *multispecies* bacterial films allowed to develop under contrasting water shear regimes. The results clearly showed both sets of cyprids to have reduced tenacity on low shear films compared with high shear films. However, there were significant differences in tenacity between the two cyprids on the low shear films. *Elminius modestus* cyprids gave similar tenacities on both experimental and control surfaces, whilst for *Balanus perforatus* cyprids tenacity was significantly reduced on the experimental surface over that on the control. This species-specific difference indicates that cyprids can detect different properties of bacterial films and highlights the exciting prospect that bacterial biofilms may help dictate the pattern of settlement for certain barnacles. It is already known that microalgal films cue-guide settlement of certain intertidal barnacle cyprids to their natural level within the intertidal zone (Strathmann & Branscomb 1979; Strathmann et al. 1981; Le Tourneux & Bourget 1988).

Further research on biofilm-antennule interactions should solve the more physical problem, often assumed, that when permanent fixation takes place either much of the biofilm has to be cleared from the surface or the cyprid actively seeks out a natural break in the biofilm.

Studies have emphasised that settlement is a key determinant of adult barnacle population structure (Grosberg 1982; Underwood & Denley 1984; Connell 1985; Gaines & Roughgarden 1985; Roughgarden et al. 1987; Underwood & Fairweather 1988; Raimondi 1990, 1991; Minchinton & Scheibling 1991, 1993). The discriminatory behaviour of cyprids during site selection is usually inferred from the positions in which the recently settled spat are found. It is extremely tedious to observe the actual movements of cyprids during the process of fixation site selection, even using video-recording equipment, as the field of view is too restricted at the scale needed and the cryptic cyprids are hard to detect unless moving. It is much more convenient to count the spat in the field (directly or from still photographs) or to set up settlement assays in the laboratory and compare counts of settled cyprids on experimental surfaces with those on controls. Both approaches have inherent shortcomings, not always appreciated and acknowledged.

As pointed out earlier, once a cyprid has permanently fixed itself to a surface (settled) and metamorphosed to the juvenile it is then regarded as recruited to that surface and open to the conditions operating there. Usually in field counts it is recruitment which is being measured because the impact of immediate post-settlement mortality and mortality of newly-metamorphosed juveniles cannot be assessed if counts only take place every few days (Denley & Underwood 1979) or every 30 days (Caffey 1985). It may prove possible to use density of recruits, which are easier to count, as an indicator of density of settlers but only if mortality between the two stages acts in a density-independent way (Connell 1985). Counts every day must give a better indication of settlement rate (Raimondi 1990; Pineda 1994b) and if intertidal sites are being considered then counts at each low water period should be aimed for (Wethey 1984). De Wolf (1973) carried out the ultimate sampling – settlement counts every hour for (*B. crenatus* and *B. improvisus*) cyprids on panels suspended from a raft. Such frequent sampling has also been carried out more recently by Roberts et al. (1991). Gotelli (1990) has highlighted the fact that gregarious settlement and more particularly the larva-larva interaction (see earlier) has been largely ignored in laboratory settlement assays and recommends the use of different larval densities (see Clare et al. 1994) or even single larvae as being more appropriate in the future.

The original settlement assay apparatus (see Crisp 1976) was developed for *Semibalanus balanoides* cyprids which need water movement to stimulate settlement. The circular holding trough is rotated to negate any light variability and to ensure all surfaces under test are exposed to the cyprids. Because of the variability in numbers settling, even on identical replicate surfaces, it is imperative that large numbers of replicates are used, hence the large holding trough. Furthermore, each batch of cyprids used in an experiment will differ in number and in rate of settlement. The settlement rate of freshly caught *S. balanoides* cyprids is usually well below 10% and even for cyprids maintained in the laboratory for 7 days the rate only reaches 50% or so on settlement factor-treated slate (Lucas et al. 1979). Many thousands of cyprids are used in a settlement experiment and time allowed for settlement is 4 hours at ambient air temperature (see Larman & Gabbott 1975).

Rittschof et al. (1984) working with *Balanus amphitrite* cyprids developed an assay for the quantitative study of their responses to surfaces. It is rapid (2-10 mins) and simple in design – a glass tube through which a constant water flow runs. They also developed a settlement assay using petri dishes. Low numbers of cyprids (20-250) are needed for each container and settlement time allowed is 22h at 28°C (Rittschof et al. 1992).

Crisp and co-workers relied on the short seasonal abundance of *S. balanoides* cyprids (April-June) in North Wales. These larvae were collected in plankton net hauls, maintained for at least a day in the laboratory, then used in the settlement assay. Their age (time from the nauplius-cyprid metamorphosis) is unknown and variable, whereas Rittschof and co-workers rear *B. amphitrite* to the cypris stage and can experiment with synchronised batches of larvae of known history (see Clare et al. 1995) or even single larvae. Settlement rates for *B. amphitrite* cyprids can also be very high, reaching 90% on a settlement factor-treated surface in still water conditions.

Minchinton & Scheibling (1993) are bold enough to state that extrapolating the findings of laboratory settlement studies to the field is unlikely to be successful. I am not so pessimistic. More realistic conditions are now being introduced into laboratory settlement work, such as controlled water flow over surfaces in flumes (Mullineaux & Butman 1991), and testing of two variables concurrently has also begun (Dineen & Hines 1992, 1994a,b). Studies in the laboratory are artificial and contrived since they are specifically designed to test hypotheses. However, it should never be overlooked that the factors inducing settlement act in combination, with their degree of influence variable in both time and space (Hudon et al. 1983). Combined laboratory and field studies can be achieved (Wethey 1986) and should be the aim in the future.

At the end of inspection the cyprid orients itself to surface contours, light and water flow (Crisp 1976). Contour has the greatest influence and water flow the least. The cyprids will also space themselves out from each other or from spat or adults (Hui & Moyse 1987). Such territoriality allows enough space for the post-metamorphosed juvenile to grow during a very vulnerable time in the life cycle. *Semibalanus balanoides* cyprids also show specific avoidance behaviour for the shells of conspecifics, but this behaviour is not universal (Hui & Moyse 1987). The consummatory act of settlement, permanent fixation, then finally takes place. The cyprid cement is released as a fluid to flow into and around all surface irregularities and embed the antennular attachment organs, including the fourth segments in some species (Yule & Walker 1987). Measurement of the force of adhesion of *S. balanoides* cyprid cement over time shows an asymptotic increase reaching a maximum after 1-3 hours (mean maximum tenacity = 9×10^5 Nm^{-2} on a slate surface). The assumption is that this time is needed for the cement to become effectively bonded to the surface, which probably involves the polymerisation of the cement proteins. This polymerisation process also makes the cement chemically inert and therefore non-biodegradable, an essential property for an organic glue in the marine environment. The tenacity of cyprid cement will vary dramatically depending on the surface characteristics of the substrata settled on (Yule & Walker 1987). Such work confirms the rationale of developing coatings with low surface free energy that will have inherent antifouling potential because any adhesion (temporary or permanent) of the organisms is impaired.

4 THE METAMORPHIC PHASE

The fixed cyprid undergoes metamorphosis to the juvenile. This reorganisational change takes place over many hours and has been studied morphologically for *S. balanoides* cyprids by Walley (1969), with the concomitant metabolic cost measured by Lucas et al. (1979). A more recent study of Glenner & Høeg (1993) follows the metamorphosis of several species using the SEM.

5 CONCLUSIONS

Barnacle settlement has become a diverse subject, now demanding more of a co-ordinated approach from larval biologists, biochemists, chemical ecologists, micro-biologists, hydrodynamicists and ecologists. How the cyprid perceives many of the stimuli operating in the water column and on a surface remains problematic. It is still generally believed that these stimuli act as 'releasers' for instinctive behavioural responses by the cyprid which culminates in the final settlement act, permanent fixation. The sequences in settlement behaviour undoubtedly progress along reducing spatial scales.

Field studies of barnacle settlement have largely been centered on specific rocky coasts around the world – west and east coasts of N. America (east coast of Canada), Great Britain and S. E. Australia. These coasts are all different in terms of larval supply (short term vs. more continuous availability) and biotic and abiotic factors, so it is hardly surprising that ideas differ on how settlement affects adult population structure; consistent settlement differences can occur among sites in a restricted coastal region, suggesting that even the local coastal morphology and geology have important influences.

I am confident that barnacle settlement research on all fronts will continue to make as significant a contribution to the further understanding of the settlement processes of sessile invertebrates in the future as it has over the past half century.

ACKNOWLEDGEMENTS

Dr J. A. Nott kindly supplied the photographs in Plate 1. and Drs A. S. Clare and A. B. Yule read through and made comments on a manuscript draft.

REFERENCES

Anderson, D.T. 1994. Barnacles. Structure, function, development and evolution. London: Chapman Hall.

Barnes, H. 1953. Size variations in the cyprids of some common barnacles. *J. Mar. Biol. Ass. UK* 32: 297-304.

Bourget, E. 1988. Barnacle larval settlement: the perception of cues at different spatial scales. In G. Chelazzi & M. Vannini (eds.), *Behavioural adaptation of intertidal life*: 153-172. New York & London: Plenum Press.

Caffey, H.M. 1985. Spatial and temporal variation in settlement and recruitment of intertidal barnacles. *Ecol. Monogr.* 55: 313-332.

Clare, A.S., D. Rittschof, D.J. Gerhart & J.S. Maki 1992. Molecular approaches to nontoxic antifouling. *Invert. Reprod. Develop.* 22: 67-76.

Clare, A.S. & J.A. Nott 1994. Scanning electron microscopy of the fourth antennular segment of *Balanus amphitrite amphitrite. J. Mar. Biol. Ass. UK* 74: 967-970.

Clare, A.S., R.K. Freet & M. Jr. McClary 1994. On the antennular secretion of the cyprid of *Balanus amphitrite amphitrite*, and its role as a settlement pheromone. *J. Mar. Biol. Ass. UK* 74: 243-250.

Clare, A.S., R.F. Thomas & D. Rittschof 1995. Evidence for the involvement of cyclic AMP in the pheromonal modulation of barnacle settlement. *J. Exp. Biol.* 198: 655-664.

Collis, S.A. 1991. Aspects of the biology of *Sacculina carcini* (Crustacea: Cirripedia: Rhizocephala), with particular emphasis on the larval energy budget. PhD Thesis, University of Wales, Bangor.

Connell, J.H. 1985. The consequence of variation in initial settlement vs. post-settlement mortality in rocky intertidal communities. *J. Exp. Mar. Biol. Ecol.* 93: 11-45.

Conway, D.V.P., C.J. Ellis & I.G. Humpheryes 1990. Deep distributions of oceanic cirripede larvae in the Sargasso Sea and surrounding North Atlantic Ocean. *Mar. Biol.* 105: 419-428.

Crisp. D.J. 1955. The behaviour of barnacle cyprids in relation to water movement over a surface. *J. Exp. Biol.* 32: 569-590.

Crisp, D.J. 1976. Settlement responses in marine organisms. In R.C. Newell (ed.), *Adaptations to the environment: essays on the physiology of marine animals*: 83-124. London: Butterworths.

Crisp, D.J. 1984. Overview of research on marine invertebrate larvae, 1940-1980. In J.D. Costlow & R.C. Tipper (eds.), *Marine biodeterioration: an interdisciplinary study*: 103-126. Annapolis, Maryland: Naval Institute Press.

Crisp, D.J. 1990. Gregariousness and systematic affinity in some North Carolinian barnacles. *Bull. Mar. Sci.* 47: 516-525.

Crisp, D.J. & P.S. Meadows 1962. The chemical basis of gregariousness in cirripedes. *Proc. R. Soc.* (B)156: 500-520.

Crisp, D.J. & P.S. Meadows 1963. Adsorbed layers: the stimulus to settlement in barnacles. *Proc. R. Soc.* (B)158: 364-387.

Denley, E.J. & A.J. Underwood 1979. Experiments on factors influencing settlement, survival, and growth of two species of barnacles in New South Wales. *J. Exp. Mar. Biol. Ecol.* 36: 269-293.

Denny, M.W. 1987. Life in the maelstrom: the biomechanics of wave-swept rocky shores. *Trends in Ecol. & Evolut.* 2: 61-66.

Denny, M.W. & M.F. Shibata 1989. Consequences of surf-zone turbulence for settlement and external fertilization. *Amer. Nat.* 134: 859-889.

De Wolf, P. 1973. Ecological observations on the mechanisms of dispersal of barnacle larvae dur-

ing planktonic life and settling. *Neth. J. Sea Res.* 6: 1-129.
Digby, P.S. 1972. Detection of small changes in hydrostatic pressure by crustacea and its relation to electrode action in the cuticle. In M.A. Sleigh & A.G. MacDonald (eds.), *The effects of pressure on living organisms. Symp. Soc. Exp. Biol.* 16: 445-472: New York: Academic Press.
Dineen, J.F.Jr. & A.H. Hines 1992. Interactive effects of salinity and adult extract upon settlement of the estuarine barnacle *Balanus improvisus* (Darwin, 1854). *J. Exp. Mar. Biol. Ecol.* 156: 239-252.
Dineen, J.F.Jr. & A.H. Hines 1994a. Effects of salinity and adult extract on settlement of the oligohaline barnacle *Balanus subalbidus. Mar. Biol.* 119: 423-430.
Dineen, J.F.Jr. & A. H. Hines 1994b. Larval settlement of the polyhaline barnacle *Balanus eburneus* (Gould): cue interactions and comparisons with two estuarine congeners. *J. Exp. Mar. Biol. Ecol.* 179: 223-234.
Dixon, D.R., H.B.S.M. Corte-Real, D. Jollivet, L.R.J. Dixon & P.W.H. Holland 1994. Molecular tracking of planktonic larvae. *J. Mar. Biol. Ass. UK* 74: 710-711.
Eckman, J.E. 1990. A model of passive settlement by planktonic larvae onto bottoms of differing roughness. *Limnol. Oceanog.* 35: 887-901.
Elfimov, A.S. 1986. Morphology of the carapace of cypris larva of the barnacle *Heterolepas mystacophora. The Soviet J. Mar. Biol.* 12: 152-156.
Foster, B.A. 1987. Barnacle ecology and adaptation. *Crustacean Issues* 5: 113-133.
Gabbott, P.A. & V.N. Larman 1987. The chemical basis of gregariousness in cirripedes: a review (1953-1984). *Crustacean Issues* 5: 377-388.
Gaines, S., S. Brown & J. Roughgarden 1985. Spatial variation in larval concentration as a cause of spatial variation in settlement for the barnacle, *Balanus glandula. Oecologia* 67: 267-272.
Gaines, S. & J. Roughgarden 1985. Larval settlement rate: a leading determinant of structure in an ecological community of the marine intertidal zone. *Proc. Natl. Acad. Sci. USA* 82: 3707-3711.
Gatenholm, P., C. Holmström, J.S. Maki & S. Kjelleberg 1995. Toward biological antifouling surface coatings: marine bacteria immobilised in hydrogel prevent settlement of larvae. *Biofouling* (in press).
Gerhart, D.J., D. Rittschof & S.W. Mayo 1988. Chemical ecology and the search for marine antifoulants. Studies of predator-prey symbiosis. *J. Chem. Ecol.* 14: 1905-1917.
Gibson, P. & J.A. Nott 1971. Concerning the fourth antennular segment of the cypris larvae of *Balanus balanoides*. In D.J. Crisp (ed.), *4th European Marine Biology Symposium*: 227-236. Cambridge: University Press.
Glenner, H. & J.T. Høeg 1993. Scanning electron microscopy of metamorphosis in four species of barnacles (Cirripedia: Thoracica: Balanomorpha). *Mar. Biol.* 117: 431-439.
Glenner, H. & J.T. Høeg 1995. Scanning electron microscopy of cypris larvae of *Balanus amphitrite* (Cirripedia: Thoracica: Balanomorpha). *J. Crust. Biol.* (in press).
Glenner, H., J.T. Høeg, A. Klysner & B. Brodin Larsen 1989. Cypris ultrastructure, metamorphosis and sex in seven families of rhizocephalan barnacles (Crustacea: Cirripedia: Rhizocephala). *Acta Zool. (Stockh.)* 70: 229-242.
Gotelli, N.J. 1990. Stochastic models of gregarious larval settlement. *Ophelia* 32: 95-108.
Grosberg, R.K. 1982. Intertidal zonation of barnacles: the influence of planktonic zonation of larvae on vertical distribution of adults. *Ecology* 53: 894-899.
Gross, T.F., F.E. Werner & J.E. Eckman 1992. Numerical modelling of larval settlement in turbulent bottom boundary layers. *J. Mar. Res.* 50: 611-642.
Harms, J. 1987. Energy budget for the larval development of *Elminius modestus* (Crustacea: Cirripedia). *Helgol. Wiss. Meersunters* 41: 45-67.
Hill, E.M. & D.L. Holland 1985. Influence of oil shale on intertidal organisms: isolation and characterisation of metalloporphyrins that induce the settlement of *Balanus balanoides* and *Elminius modestus. Proc. R. Soc.* (B)225: 107-120.
Høeg, J.T. 1991. Functional and evolutionary aspects of the sexual system in the Rhizocephala (Crustacea: Thecostraca: Cirripedia). In R. Bauer & J. Martin (eds.), *Crustacean sexual biology*: 208-227. New York: Columbia University Press.
Holland, D.L., D.J. Crisp, R. Huxley & J. Sisson 1984. Influence of oil shale on intertidal organ-

isms: effect of oil shale extract on settlement of the barnacle *Balanus balanoides* (L.). *J. Exp. Mar. Biol. Ecol.* 75: 245-255.

Holmström, C., D. Rittschof & S. Kjelleberg 1992. Inhibition of settlement by larvae of *Balanus amphitrite* and *Ciona intestinalis* by a surface-colonising marine bacterium. *Appl. Environ. Microbiol.* 58: 2111-2115.

Hudon, C., E. Bourget & P. Legendre 1983. An integrated study of the factors influencing the choice of the settling site of *Balanus crenatus* cyprid larvae. *Can. J. Fish. Aquatc. Sci.* 40: 1186-1194.

Hui, E. & J. Moyse 1987. Settlement patterns and competition for space. *Crustacean Issues* 5: 363-376.

Jensen, P.G., J. Moyse, J.T. Høeg & H. Al-Yahya 1994. Comparative SEM studies of lattice organs: putative sensory structures on the carapace of larvae from Ascothoracica and Cirripedia (Crustacea Maxillopoda Thecostraca). *Acta Zool. (Stockh.)* 75: 125-142.

Johnson, L.E. & R.R. Strathmann 1989. Settling barnacle larvae avoid substrata previously occupied by a mobile predator. *J. Exp. Mar. Biol. Ecol.* 128: 87-103.

Jorgensen, D.D. & W.B. Vernberg 1982. Oxygen uptake in a barnacle: scaling to body size from nauplius to adult. *Can. J. Zool.* 60: 1231-1235.

Keough, M.J. & B.J. Downes 1982. Recruitment of marine invertebrates: the role of active larval choices and early mortality. *Oecologia* 54: 348-352.

Knight-Jones, E.W. 1953. Laboratory experiments on gregariousness during setting in *Balanus balanoides* and other barnacles. *J. Exp. Biol.* 30: 584-598.

Knight-Jones, E.W. & D.J. Crisp 1953. Gregariousness in barnacles in relation to the fouling of ships and to anti-fouling research. *Nature* 171: 1109.

Knight-Jones, E.W. & S.Z. Qasim 1967. Responses of Crustacea to changes in hydrostatic pressure. *Proceedings of the Symposium on Crustacea held at Ernakulum* 3: 1132-1150. Bangalore: Marine Biological Association of India.

Knight-Jones, E.W. & J.P. Stevenson 1950. Gregariousness during settlement in the barnacle *Elminius modestus* Darwin. *J. Mar. Biol. Ass. UK* 29: 281-297.

Larman, V.N. 1984. Protein extracts from some marine animals which promote barnacle settlement: possible relationship between a protein component of arthropod cuticle and actin. *Comp. Biochem. Physiol.* 77B: 73-81.

Larman, V.N. & P.A. Gabbott 1975. Settlement of cyprid larvae of *Balanus balanoides* and *Elminius modestus* induced by extracts of adult barnacles and other marine animals. *J. Mar. Biol. Ass. UK* 55: 183-190.

Le Fèvre, J. & E. Bourget 1991. Neustonic niche for cirripede larvae as a possible adaptation to long range dispersal. *Mar. Ecol. Prog. Ser.* 74: 185-194.

Le Tourneux, F. & E. Bourget 1988. Importance of physical and biological settlement cues used at different spatial scales by the larvae of *Semibalanus balanoides*. *Mar. Biol.* 97: 57-66.

Levin, L.A. 1990. A review of methods for labeling and tracking marine invertebrate larvae. *Ophelia* 32: 115-144.

Levin, L.A., D. Huggett, P. Myers, T. Bridges & J. Weaver 1993. Rare earth tagging methods for the study of larval dispersal by marine invertebrates. *Limnol. Oceanog.* 38: 346-360.

Lewis, C.A. 1978. A review of substratum selection in freeliving and symbiotic cirripedes. In F.-S. Chia & M.E. Rice (eds.), *Settlement and metamorphosis of marine invertebrate larvae*: 207-218. New York: Elsevier.

Lindner, E. 1984. The attachment of macrofouling invertebrates. In J.D. Costlow & R.C. Tipper (eds.), *Marine biodeterioration: an interdisciplinary study*: 183-202. Annapolis, Maryland: Naval Institute Press.

Lucas, M.I. 1980. Studies on energy flow in a barnacle population. PhD Thesis, University of Wales, Bangor.

Lucas, M.I., G. Walker, D.L. Holland & D.J. Crisp 1979. An energy budget for the free-swimming and metamorphosing larvae of *Balanus balanoides* (Crustacea: Cirripedia). *Mar. Biol.* 55: 221-229.

Maki, J.S., D. Rittschof, J.D. Costlow & R. Mitchell 1988. Inhibition of attachment of larval barnacles, *Balanus amphitrite*, by bacterial surface films. *Mar. Biol.* 97: 199-206.

Maki, J.S., D. Rittschof, M.-O. Samuelsson, U. Szeuzyk, A.B. Yule, S. Kjelleberg, J.D. Costlow & R. Mitchell 1990. Effect of marine bacteria and their exopolymers on the attachment of barnacle cypris larvae. *Bull. Mar. Sci.* 46: 499-511.

Maki, J.S., D. Rittschof & R. Mitchell 1992. Inhibition of larval barnacle attachment to bacterial films - an investigation of physical properties. *Microb. Ecol.* 23: 99-106.

Maki, J.S., A.B. Yule, D. Rittschof & R. Mitchell 1994. The effect of bacterial films on the temporary adhesion and permanent fixation of cypris larvae, *Balanus amphitrite* Darwin. *Biofouling* 8: 121-131.

Mary, A.Sr., V. Mary, D. Rittschof & R. Nagabhushanam 1993. Bacterial-barnacle interaction: potential using juncellins and antibiotica to alter structure of bacterial communites. *J. Chem. Ecol.* 19: 2155-2167.

Minchinton, T.E. & R.S. Scheibling 1991. The influence of larval supply and settlement on the population structure of barnacles. *Ecology* 72: 1867-1879.

Minchinton, T.E. & R.S. Scheibling 1993. Free space availability and larval substratum selection as determinants of barnacle population structure in a developing rocky intertidal community. *Mar. Ecol. Prog. Ser.* 95: 233-244.

Moyse, J. 1987. Larvae of lepadomorph barnacles. In A.J. Southward (ed.), *Barnacle biology*, Crustacean Issues 5: 329-362, Rotterdam: A.A. Balkema.

Mullineaux, L.S. & C.A. Butman 1991. Initial contact, exploration and attachment of barnacle (*Balanus amphitrite*) cyprids settling in flow. *Mar. Biol.* 110: 93-103.

Neal, A.L. & A.B. Yule 1992. The link between cypris temporary adhesion and settlement of *Balanus balanoides* (L.). *Biofouling* 6: 33-38.

Neal, A.L. & A.B. Yule 1994. The tenacity of *Elminius modestus* and *Balanus perforatus* cyprids to bacterial films grown under different shear regimes. *J. Mar. Biol. Ass. UK* 74: 251-257.

Nott, J.A. 1969. Settlement of barnacle larvae: surface structure of the antennular attachment disc by scanning electron microscopy. *Mar. Biol.* 2: 248-251.

Nott, J.A. & B.A. Foster 1969. On the structure of the antennular attachment organ of the cypris larva of *Balanus balanoides*. *Phil. Trans. R. Soc.* (B)256: 115-134.

Pawlik, J.R. 1992. Chemical ecology of the settlement of benthic marine invertebrates. *Oceanog. Mar. Biol. Ann. Rev.* 30: 273-335.

Pineda, J. 1994a. Internal tidal bores in the nearshore: warm-water fronts, seaward gravity currents and the onshore transport of neustonic larvae. *J. Mar. Res.* 52: 427-458.

Pineda, J. 1994b. Spatial and temporal patterns in barnacle settlement rate along a southern California rocky shore. *Mar. Ecol. Prog. Ser.* 107: 125-138.

Price, R.R., M. Patchan, A.S. Clare, D. Rittschof & J. Bonaventura 1992. Performance enhancement of natural antifouling compounds and their analogs through microencapsulation and controlled release. *Biofouling* 6: 207-216.

Raimondi, P.T. 1988. Settlement cues and determination of the vertical limit of an intertidal barnacle. *Ecology* 69: 400-407.

Raimondi, P.T. 1990. Patterns, mechanisms, consequences of variability in settlement and recruitment of an intertidal barnacle. *Ecol. Monogr.* 60: 283-309.

Raimondi, P.T. 1991. Settlement behavior of *Chthamalus anisopoma* larvae largely determines the adult distribution. *Oecologia* 85: 349-360.

Rittschof, D., E.S. Branscomb & J.D. Costlow 1984. Settlement and behavior in relation to flow and surface in larval barnacles, *Balanus amphitrite* Darwin. *J. Exp. Mar. Biol. Ecol.* 82: 131-146.

Rittschof, D., I.R. Hooper, E.S. Branscomb & J.D. Costlow 1985. Inhibition of barnacle settlement and behavior by natural products from whip corals, *Leptogorgia virgulata* (Lamarck, 1815). *J. Chem. Ecol.* 11: 551-563.

Rittschof, D., I.R. Hooper & J.D. Costlow 1986a. Barnacle settlement inhibitors from sea pansies, *Renilla reniformis*. *Bull. Mar. Sci.* 39: 376-382.

Rittschof, D., J. Maki, R. Mitchell & J.D. Costlow 1986b. Ion and neuropharmacological studies of barnacle settlement. *Neth. J. Sea Res.* 20: 269-275.

Rittschof, D., A.S. Clare, D.J. Gerhart, Sr.A. Mary & J. Bonaventura 1992. Barnacle *in vitro* assays for biologically active substances: toxicity and settlement inhibition assays using mass cultured *Balanus amphitrite amphitrite* Darwin. *Biofouling* 6: 115-122.

Roberts, D., D. Rittschof, E. Holm & A.R. Schmidt 1991. Factors influencing initial larval settlement: temporal, spatial and surface molecular components. *J. Exp. Mar. Biol. Ecol.* 150: 203-211.

Roskell, J. 1975. Continuous plankton records: plankton atlas of the North Atlantic and the North Sea. Supplement 2. The oceanic cirripede larvae, 1955-1972. *Bull. Mar. Ecol.* 8: 185-199.

Roughgarden, J., S.D. Gaines & S.W. Pacala 1987. Supply side ecology: the role of physical transport processes. In P. Giller & J. Gee (eds.), *Organization of communities: past and present*: 459-486, London: Blackwell Scientific.

Shanks, A.L. 1986. Tidal periodicity in the daily settlement of intertidal barnacle larvae and an hypothesized mechanism for the cross-shelf transport of cyprids. *Biol. Bull.* 170: 429-440.

Shanks, A.L. & W.G. Wright 1987. Internal-wave-mediated shoreward transport of cyprids, megalopae, and gammarids and correlated longshore differences in the settling rate of intertidal barnacles. *J. Exp. Mar. Biol. Ecol.* 114: 1-13.

Standing, J.D., I.R. Hooper & J.D. Costlow 1984. Inhibition and induction of barnacle settlement by natural products present in octocorals. *J. Chem. Ecol.* 6: 823-834.

Strathmann, R.R. & E.S. Branscomb 1979. Adequacy of cues to favorable sites used by settling larvae of two intertidal barnacles. In S.E. Stancyk (ed.), *Reproductive ecology of marine invertebrates*: 77-89, Columbia S.C.: University of South Carolina Press.

Strathmann, E.R., E.S. Branscomb & K. Vedder 1981. Fatal error in set as a cost of dispersal and the influence of intertidal flora on set of barnacles. *Oecologia* 48: 13-18.

Sulkin, S.D. 1990. Larval orientation mechanisms: the power of controlled experiments. *Ophelia* 32: 49-62.

Tegtmeyer, K. & D. Rittschof 1989. Synthetic peptide analogs to barnacle settlement pheromone. *Peptides* 9: 1403-1406.

Turquier, Y. 1985. Cirripèdes acrothoraciques des côtes occidentales de la Mediterranée et de l'Afrique du Nord. I. Cryptophialidae. *Bull. Soc. Zool.* France 110: 151-168.

Underwood, A.J. & E.J. Denley 1984. Paradigms, explanations, and generalizations in models for the structure of intertidal communities on rocky shores. In D.R. Strong, D. Simberloff, L.G. Abele & A. Thistle (eds.), *Ecological communities: conceptual issues and the evidence*: 151-180, Princeton, New Jersey: Princeton University Press.

Underwood, A.J. & P.G. Fairweather 1988. Supply side ecology and benthic marine assemblages. *Trends in Ecol. & Evolut.* 4: 16-20.

Visscher, J.P. & R.M. Luce 1928. Reactions of the cyprid larvae of barnacles to light with special reference to spectral colours. *Biol. Bull.* 54: 336-350.

Walker, G. 1974. The fine structure of the frontal filament complex of barnacle larvae (Crustacea: Cirripedia). *Cell Tiss. Res.* 152: 449-465.

Walker, G. 1987. Marine organisms and their adhesion. In W.C. Wake (ed.), *Synthetic adhesives and sealants*. Critical Reports on Applied Chemistry, Vol. 16: 112-135, Chichester, New York: Wiley & Sons.

Walker, G. & V.E. Lee 1976. Surface structures and sense organs of the cypris larva of *Balanus balanoides* as seen by scanning and transmission electron microscopy. *J. Zool., Lond.* 178: 161-172.

Walker, G. & A.B. Yule 1984. Temporary adhesion of the barnacle cyprid: the existence of an antennular adhesive secretion. *J. Mar. Biol. Ass. UK* 64: 679-686.

Walker, G., A.B. Yule & J.A. Nott 1987. Structure and function in balanomorph larvae. In A.J. Southward (ed.), *Barnacle biology*, Crustacean Issues 5: 307-328, Rotterdam: A.A. Balkema.

Walley, L.J. 1969. Studies on the larval structure and metamorphosis of *Balanus balanoides* (L.). *Phil. Trans. R. Soc.* (B)256: 237-280.

Weikert, H. 1972. Verteilung und Tagesperiodik der Evertebraten-neuston im subtropischen Nordostatlantik wahrend der 'Atlantischen Kuppenfahrten 1967' von F.S. 'Meteor'. *'Meteor' Forsch-Ergebnisse* (D), 11: 29-87.

Wethey, D.S. 1984. Spatial pattern in barnacle settlement: day to day changes during the settlement season. *J. Mar. Biol. Ass. UK* 64: 687-698.

Wethey, D.S. 1986. Ranking of settlement cues by barnacle larvae: influence of surface contour. *Bull. Mar. Sci.* 39: 393-400.

Whillis, J.A., A.B. Yule & D.J. Crisp 1990. Settlement of *Chthamalus montagui* Southward cyprids on barnacle arthropodin. *Biofouling* 2: 95-99.

Yule, A.B. 1982. The application of new techniques to the study of planktonic organisms. PhD Thesis, University of Wales, Bangor.

Yule, A.B. & D.J. Crisp 1983. Adhesion of cypris larvae of the barnacle, *Balanus balanoides*, to clean and arthropodin treated surfaces. *J. Mar. Biol. Ass. UK* 63: 261-271.

Yule, A.B. & G. Walker 1984. The temporary adhesion of barnacle cyprids: effects of some differing surface characteristics. *J. Mar. Biol. Ass. UK* 64: 429-439.

Yule, A.B. & G. Walker 1987. Adhesion in barnacles. *Crustacean Issues* 5: 389-402.

Naupliar evidence for cirripede taxonomy and phylogeny

Olga M. Korn
Institute of Marine Biology, Russian Academy of Sciences, Vladivostok, Russia

ABSTRACT

The comparative larval morphology of balanoid and chthamaloid barnacles is illustrated by six barnacle species (*Balanus crenatus, Balanus improvisus, Balanus rostratus, Chthamalus dalli, Semibalanus cariosus* and *Solidobalanus hesperius*) from the Russian waters of the Sea of Japan. The development of all the balanoid barnacles investigated follows the pattern of temperate water species. The nauplii possess an elongated, pear-shaped cephalic shield with posterior spines, a trilobed labrum without teeth, and antennae with cuspidate and plumodenticulate setae. The sequence of emergence of the abdominal spines is typical of this group. The development of chthamaloid barnacles follows a different pattern. Their nauplii possess a rounded, convex cephalic shield without posterior spines and a unilobed labrum with teeth. The antennae have feathered and hispid setae, and the abdominal process has more spines than that in balanoid larvae. All six barnacle species have different areas of distribution and follow peculiar reproductive patterns. Cirripede larvae occur in the plankton from March to January, creating an almost permanent danger of barnacle fouling. Larval evidence agrees in general with the current taxonomy of the order Thoracica. Naupliar morphology seems to confirm the sistergroup relationship between the Iblidae and the remaining thoracicans. With regard to a number of larval characters, scalpelloid forms are more primitive than lepadoid forms. The resemblance between pollicipedid and verrucomorph, and especially between pollicipedid and chthamaloid nauplii mainly rests on plesiomorphic traits. The following naupliar characters traced from ibloids through pollicipedids, verrucomorphs and chthamaloids to coronulids and pyrgomatids are presumed to be primitive: rounded, convex cephalic shield with marginal cuticular ridges and without any spines, unilobed, tongued labrum, short dorsal thoracic spine and abdominal process, and feathered setae in the antennal setation. Studies on larvae provide some evidence for the polyphyletic origin of balanomorph barnacles. The characters of chthamaloid nauplii clearly separate them from larvae of other balanomorphs. Most of features shared by larvae of higher balanomorphs (Coronulidae, Tetraclitidae, Pyrgomatidae, Balanidae and Archaeobalanidae) are probably synapomorphies. In terms of larval morphology, the Balanidae and Archaeobalanidae are heterogeneous groups.

1 INTRODUCTION

Barnacle larvae make up an essential component of meroplankton assemblages, especially in high latitudes; however, their specific identification involves difficulties. Information on the morphology of larvae and their seasonal occurrence is of great importance for planktology, larval ecology, reproductive biology, and also for numerous applied investigations related, as a rule, to the problem of fouling. Studies in comparative larval morphology provide additional, often important material for the taxonomy and phylogeny of crustaceans.

The nauplius as a larva of the sessile barnacles was discovered in plankton samples in 1823 by John Tompson (Winsor 1969). The sequence of developmental stages (six naupliar and one cyprid) was described for the first time based on *Dosima fascicularis* (Willemöes-Suhm 1876). Many papers dealing with the complete or partial larval development of barnacles belonging to different families have been published during recent decades. The patterns of larval development of balanoid barnacles from temperate waters (Norris & Crisp 1953; Jones & Crisp 1954; Moyse 1961; Lang 1979) and chthamaloid barnacles (Sandison 1967; Korn & Ovsyannikova 1979; Lang 1979; Egan & Anderson 1989) are established today. Moyse (1987) summarized all available information on pedunculate cirripede larvae. Egan & Anderson (1988) outlined the naupliar development of coronuloid barnacles. To describe the larval limbs, Bassindale (1936) proposed using a setation formula analogous to that accepted for copepods. It noted only the number of setae and their disposition. Jones & Crisp (1954) employed a graphical method. Newman (1965) combined the preceding variants and made an alphabetical list that is now in universal use. Subsequent authors modified Newman's formula, adding supplementary notes (Sandison 1967; Lang 1979; Stewart et al. 1989). Grygier (1987a, 1994) showed that the patterns of development of antennular setation and segmentation in the Cirripedia also have taxonomic value.

Most of these investigations have been devoted to the description of the larval development of one or more barnacle species, occasionally with keys to naupliar and cypris stages (Hirano 1953; Moyse 1961; Barker 1976; Lang 1979, 1980; Standing 1980; Kado 1982; Geraci & Romairone 1986; Elfimov 1987; Korn 1988a). In the last decade comparative larval morphology has found application in the taxonomy of adult barnacles (Lang 1979; Egan & Anderson 1985, 1987, 1989; Moyse 1987; Korn 1988b; Miller et al. 1989; Lee & Kim 1991; Choi et al. 1992; Kado & Hirano 1994). In spite of the ever-increasing interest in barnacle larvae, the complete naupliar development has been described for less than a tenth of the presently known species of the order Thoracica.

2 LARVAE OF BARNACLES INHABITING RUSSIAN WATERS OF THE SEA OF JAPAN

Six barnacle species belonging to three families are common in the Russian waters of the Sea of Japan: *Balanus rostratus* Hoek, *Balanus crenatus* Bruguière and *Balanus improvisus* Darwin (Balanidae), *Semibalanus cariosus* (Pallas) and *Solidobalanus hesperius* Pilsbry (Archaeobalanidae), and *Chthamalus dalli* Pilsbry (Chthamalidae).

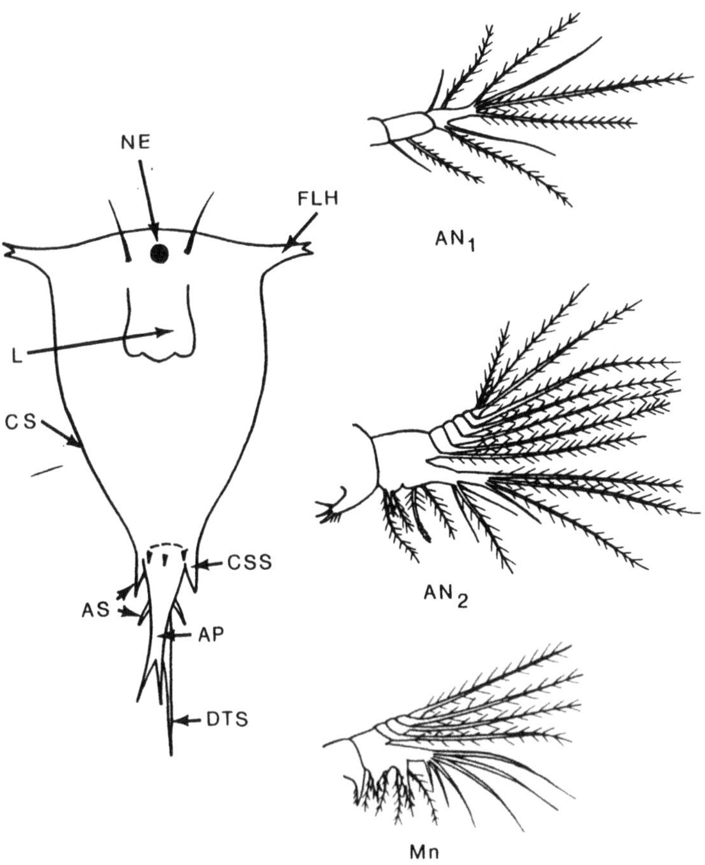

Figure 1. Body outline and limbs of barnacle nauplius. NE-naupliar eye; FLH-frontolateral horn; L-labrum; CS-cephalic shield; CSS-cephalic shield spines; AP-abdominal process; AS-abdominal spines; DTS-dorsal thoracic spine, An1-antennule, An2-antenna; Mn-mandible; C-cuspidate, D-plumodenticulate setae.

Their naupliar stages have been described in detail for this locality (Korn 1989, 1991; Korn & Ovsyannikova 1979, 1981; Ovsyannikova & Korn 1981; 1984). Some other species, such as *Balanus eburneus, Balanus trigonus, Balanus amphitrite*, and *Lepas anatifera*, noted in fouling communities of this region (Zevina 1976), are introduced via ships in summer, but evidently do not survive in winter.

2.1 *Brief description of naupliar diagnostic characters*

A typical nauplius hatches from the barnacle egg (Fig. 1). Its dorsal surface comprises by the cephalic shield (carapace) with frontolateral horns anteriorly and the dorsal thoracic spine posteriorly. In stages 1-3, the cephalic shield tapers directly into the dorsal thoracic spine. In stage 4, a distinct posterior shield border is formed, which may bear two spines. The shield may have also dorsal and/or marginal spines in some taxonomic groups. The presence of the various cephalic shield spines and their lengths are taxonomic features. Frontolateral horns are found only in cirripede

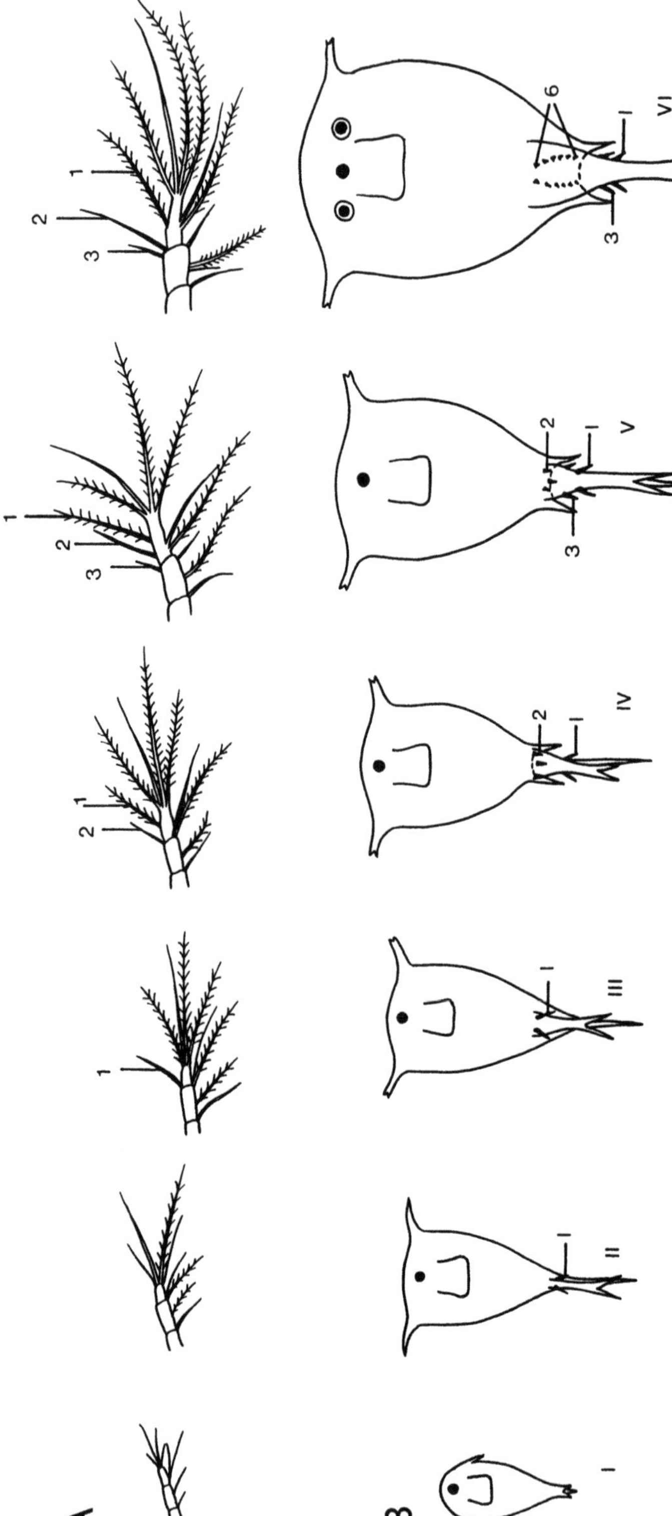

Figure 2. The six naupliar stages of balanoid barnacles. A-antennules, 1-3-preaxial setae; B-body outlines of nauplii I-VI, 1-6-abdominal spines.

nauplii. In stage 1, they are usually directed posteriorly, but from stage 2 they are deflected anteriolaterally. Their length and direction are traits used in taxonomy.

The abdominal process is placed ventral to the dorsal thoracic spine, is provided with numerous spines, and terminates in a caudal furca. The number of abdominal spines and their arrangement are taxonomic features. The length of abdomen relative to the dorsal thoracic spine and the length of the both caudal processes relative to the cephalic shield are also used in taxonomy.

The upper lip (labrum) covers the mouth. Its form and armament with teeth are systematic characters. All barnacle larval stages have one unpaired naupliar eye. In stage 6 compound eyes appear, the pigmentation of which is usually seen a few days after the moult.

The nauplius has three pairs of limbs: uniramous antennules, biramous antennae and mandibles. The number and types of setae, and also the number of limb segments (articles) and the precise placement of setae (i.e. their individual identity) are the important morphological characters.

2.2 *Naupliar development of balanoid barnacles*

The development of all the balanoid barnacles investigated follows a typical pattern observed in temperate balanoid species (Norris & Crisp 1953; Jones & Crisp 1954; Barnes & Barnes 1959a,b; Barnes & Costlow 1961; Crisp 1962a,b; Lang 1979, 1980; Branscomb & Vedder 1982; Brown & Roughgarden 1985) (Fig. 2). The nauplii have an elongated (the length greater than the width) and often pear-shaped cephalic

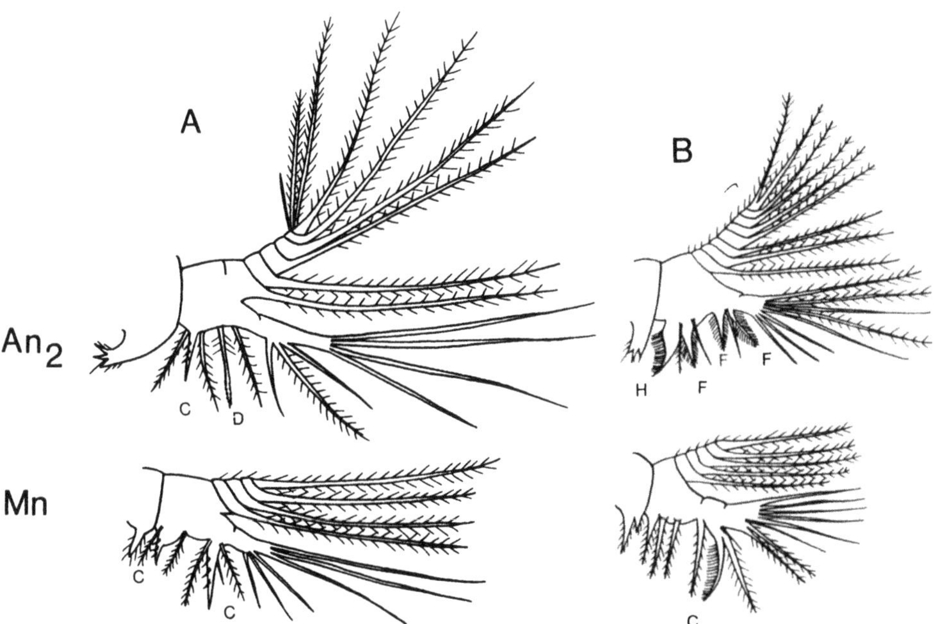

Figure 3. Antennae (An2) and mandibles (Mn) of balanoid (A) and chthamaloid (B) nauplii. C-cuspidate, D-plumodenticulate, H-hispid, F-feathered setae.

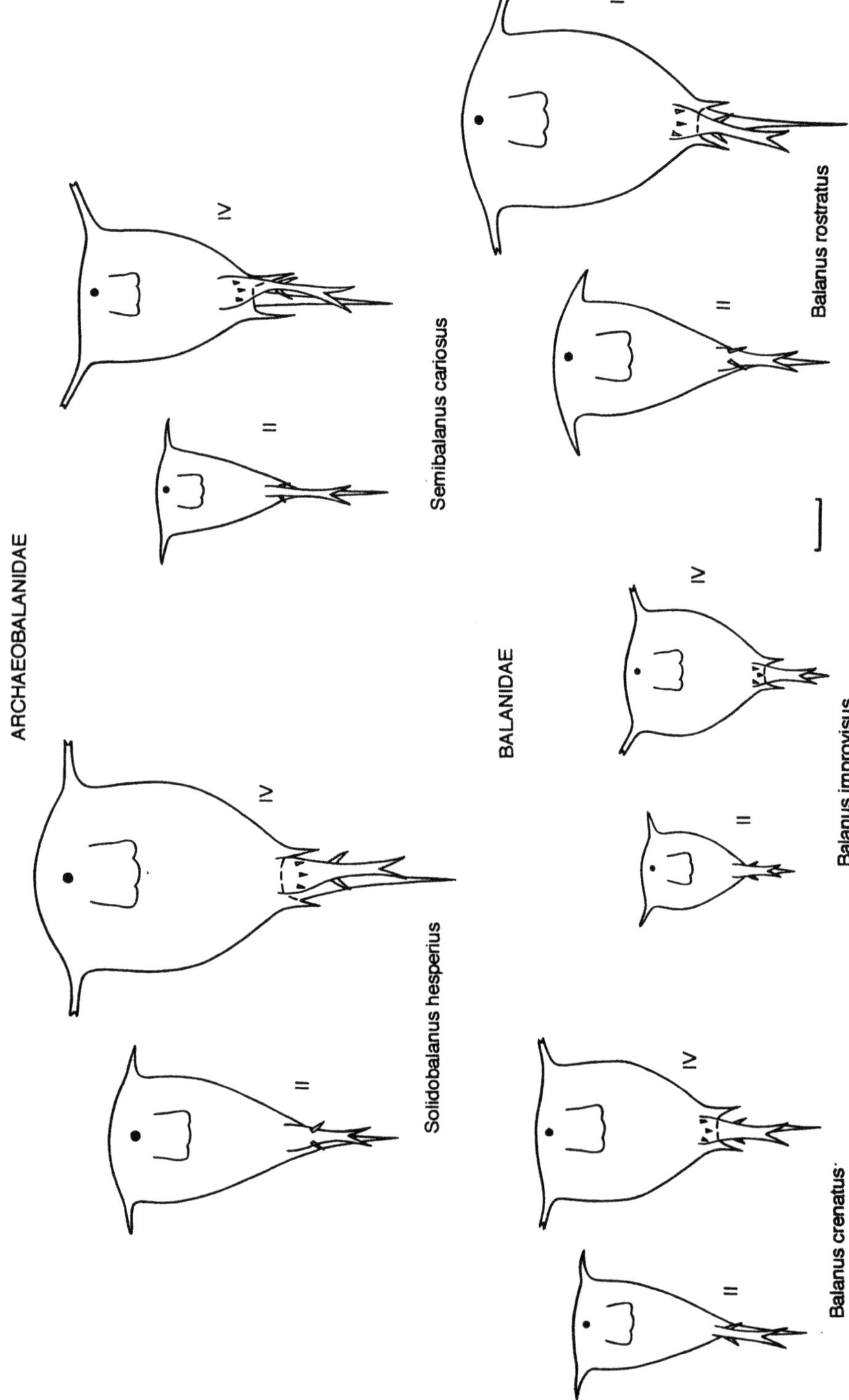

Figure 4. Body outlines of balanoid larvae from Peter the Great Bay. Scale 100 μm.

shield (carapace) with a pair of posterior spines. The frontolateral horns are directed anteriorly or laterally. The trilobed labrum generally has no teeth. The both caudal processes are some shorter than the cephalic shield itself. The sequence of emergence of the abdominal spines is typical of this group . In addition to common simple (S) and plumose (P) setae, the antennae bear cuspidate (C) and plumodenticulate (D) setae and the mandibles possess two cuspidate setae (Fig. 3A). The mandibular D-setation noted by Lang (1979) and some other authors mainly in larvae of warm-water barnacle species (Egan & Anderson 1986, 1987; Lee & Kim 1991a,b) are ill-defined in our material. The total number of setae in the antenna and mandible of stage 6 is 21:17 (*Balanus crenatus*), 26:21 (*Balanus improvisus*), 22:19 (*Balanus rostratus*), 24:22 (*Semibalanus cariosus*), and 20:18 (*Solidobalanus hesperius*).

Nauplius 1. The frontolateral horns are closely appressed to the sides of the cephalic shield and directed posteriorly. The dorsal thoracic spine and the abdominal process are short and poorly developed.

Nauplius 2. The horns are directed laterally. The dorsal thoracic spine and the abdominal process are well developed. The abdominal process has the first (distal) pair of spines.

Nauplius 3. The horns are directed frontolaterally. The antennule bears the first preaxial seta.

Nauplius 4. The cephalic shield has a posterior border with a pair of spines; dorsal and marginal cephalic spines are absent. The abdominal process has the second (proximal) pair of spines with a medial spine between them.

Nauplius 5. The abdominal process has the third pair of spines in a position between the first and second pairs.

Nauplius 6. The abdominal process has the first and the third pairs of spines. The proximal (second) pair is replaced with six pairs of small spines. The lateral compound eyes acquire pigmentation 1-3 days after moulting.

2.3 *Comparative naupliar morphology of balanoid barnacles*

Larvae of *Solidobalanus hesperius* can be easily distinguished from the other balanoid nauplii described (Fig. 4). Nauplii 2-3 are slender, elongated, and the cephalic shield is almost triangular. The frontolateral horns are long in all larval stages. The horns of nauplius 2 are directed laterally and have a slight curvature; from stage 3, they are deflected forwards. The cephalic shield of nauplii 4-6 is trapeziform with a straight anterior margin and slightly convex lateral edges. The posterior shield spines are long. Setae on segments 1 and 2 of the mandibular endopods are without setules. An interesting peculiarity is the absence of the naupliar cuspidate (C) setae on the antennal endopods where they typically occur in most balanoids; they are replaced with small, rudimentary s-setae.

Larvae of *Balanus rostratus* and *Semibalanus cariosus* belong to the different morphological group. They have a large pear-shaped body with comparatively short horns directed laterally, and only in late stages (5 or 6) slightly deflected forward. The posterior shield spines are rather short. The anterior shield margin is always convex; it becomes more convex in late stages, while in *Solidobalanus hesperius* its convexity decreases. Some setae on segments 1 and 2 of the mandibular endopods are with setules. The antennal endopods bear C-setae. The larvae of *Balanus rostra-*

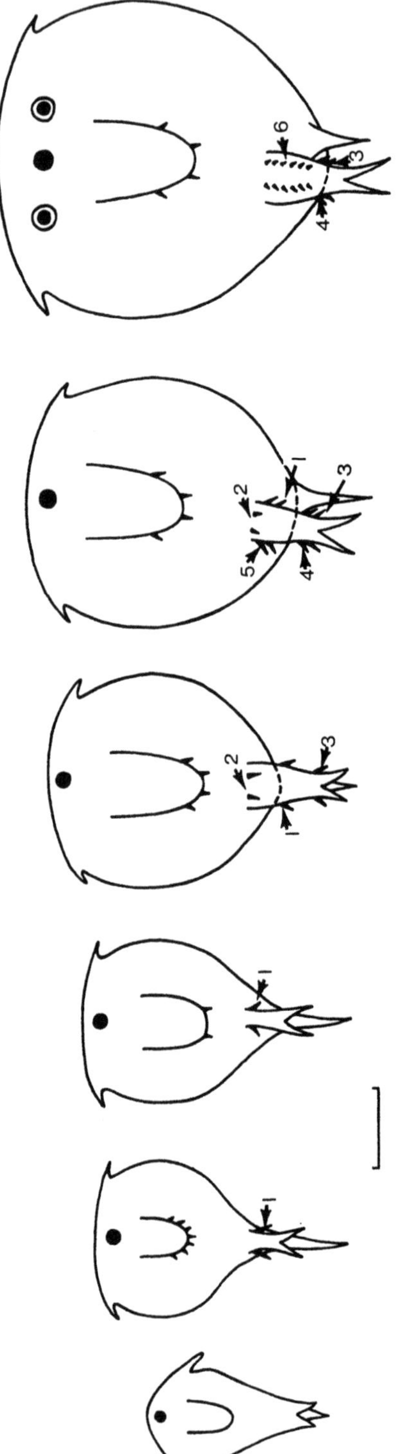

Figure 5. The six naupliar stages of *Chthamalus dalli* (I-VI). 1-6-abdominal spines. Scale 100 μm.

tus and *Semibalanus cariosus* have the closest morphological resemblance, but their co-occurrence in plankton samples is impossible, since *B. rostratus* releases nauplii in autumn and *S. cariosus* does so in early spring. Besides, the larvae of *S. cariosus* larvae are somewhat larger (they are the largest in the Sea of Japan); they also have some longer posterior shield spines, and longer frontolateral horns in late stages. The anterior margin of *B. rostratus* nauplii is less convex and is in line with the horns. The abdomen/dorsal thoracic spine ratio in early larval stages is about 1/2 in *S. cariosus*, 2/3 in *B. rostratus* and *Solidobalanus hesperius*, and 3/4 in *Balanus crenatus*.

The nauplii of *Balanus crenatus* are rather small. They occupy an intermediate position between the two foregoing groups. Early stages of *B. crenatus* larvae bear a strong resemblance to nauplii of *Solidobalanus hesperius*. The larvae of these two species concurrently occur in the plankton from early spring to late autumn, and difficulties with their identification are inevitable. Early stages of *B. crenatus* are only slightly smaller than those of *S. hesperius*. From stage 4, the cephalic shield of *B. crenatus* nauplii becomes roundish and posseses a convex anterior margin until the last stage, which makes it very similar to nauplii of *Semibalanus cariosus* and *Balanus rostratus*. At the same time, the frontolateral horns of *B. crenatus* are deflected forward from stage 3. This is also characteristic of *S. hesperius* larvae. In stage 6, the abdomen of *B. crenatus* nauplii reaches the dorsal thoracic spine, which is not the case in the above-mentioned species. Instead of typical cuspidate (C) setae, the antennal endopodites of *B. crenatus* nauplii have rudimentary (s) setae, as in *S. hesperius*.

The nauplii of *Balanus improvisus* are easy to distinguish in plankton samples. They are roundish and among the smallest of the balanoid larvae. There is a small protuberance on the anterior margin of nauplius 1; such a feature is not typical for the larvae of other species. The anterior margin remains convex in all larval stages. The frontolateral horns are of medium length, and from stage 3 are directed forward and ventrally. On the contrary, the posterior shield spines are deflected dorsally. In comparison with the nauplii of other balanoid barnacles, those of *B. improvisus* have very short caudal processes less than half as long as the cephalic shield. The abdominal process matches the dorsal thoracic spine already in stage 4, and becomes longer in stages 5 and 6. On the basis of this feature, *B. improvisus* nauplii, in contrast to the foregoing species, are referred to the third group of species in Sandison›s (1967) classification. The teeth on the labrum of *B. improvisus* nauplii noted by Jones & Crisp (1954) are ill-defined in specimens from Peter the Great Bay.

2.4 *Morphology and development of the nauplius of* Chthamalus dalli

Chthamalus larvae differ strikingly from balanoid nauplii (Fig. 5). Their identification in plankton samples is not difficult. Nauplii of *C. dalli* have a convex, globular cephalic shield without posterior spines. The frontolateral horns are directed ventrally and it is difficult to define their direction relative to the longitudinal body axis. The unilobed labrum bears numerous teeth. The both caudal processes are considerably shorter than the cephalic shield itself, especially in the late stages. The abdominal process has considerably more spines than that of balanoid larvae. In addition to common simple (S) and plumose (P) setae, the antennae bear one hispid (H) and several feathered (F) setae and the mandibles possess cuspidate (C) setae (Fig. 3B). The total number of setae on the naupliar limbs in *Chthamalus* is greater than in balanoid larvae: at stage 6, there are 31 setae on the antenna and 22 setae on the mandible.

Nauplius 1. The frontolateral horns are not appressed to the sides of the cephalic shield but directed posteriorly. Two small protuberances are disposed on each side of the body.

Nauplius 2. The labrum is equiped with 5-7 medial teeth. The abdominal process has the first (distal) pair of spines.

Nauplius 3. The labrum has 2 medial teeth. The antennule develops the first preaxial seta.

Nauplius 4. The cephalic shield is without posterior spines. From stage 4 the labrum has 2 medial and 2 lateral teeth. The abdominal process has the second (proximal) pair of spines without a medial spine and the third pair located above the caudal furca, which is typically absent in balanoid nauplii.

Nauplius 5. The fourth and fifth pairs of spines appear on the abdominal process. The abdomen becomes longer than the dorsal thoracic spine.

Nauplius 6. The abdominal process has two pairs of spines above the caudal furca. The others spines are replaced with six pairs of small spines. The lateral compound eyes acquire pigmentation 1-3 days after moulting.

2.5 *Seasonal occurrence of barnacle larvae*

All the barnacle species that are of frequent occurrence in Peter the Great Bay near Vladivostok have different areas of distribution and show different patterns of reproduction, and therefore the breeding season of barnacles in this locality continues almost throughout the year (Table 1). Data on their reproductive biology are valuable for predicting the time and intensity of their settlement on artificial substrata.

The reproductive cycle of a high-boreal barnacle, *Semibalanus cariosus*, is similar to the annual life cycle of boreo-arctic cirripedes with one brood per year, as in the well-studied *Semibalanus balanoides* and *Balanus balanus* (Crisp 1954; Barnes & Barnes 1954; Barnes 1959; Crisp & Patel 1969). Active gametogenesis occurs in summer. Spawning takes place once a year, in November. The incubation period lasts about four months over winter. Nauplii are released in March at a water temperature of about 0°C. Larval settlement probably occurs in April-May (Korn 1989).

A wide-boreal barnacle, *Balanus rostratus*, also spawns once a year in September, and produces a single large larval generation. Active gametogenesis also occurs in

Table 1. Seasonal occurrence of barnacle larvae in Peter the Great Bay.

Species/Months	I	II	III	IV	V	VI	VII	VIII	IX	X	XI	XII
Balanus crenatus		----	----	----	----	----	----	----	----	----		
Balanus improvisus					----	----	----	----	----	----	----	
Balanus rostratus	--								----	----	----	----
Chthamalus dalli				----	----	----	----					
Semibalanus cariosus		----	----									
Solidobalanus hesperius		----	----	----	----	----	----	----	----	----	----	

summer, but the incubation period lasts only about two weeks. Larvae are found in the plankton from the end of September to January. Hatching of nauplii is associated with the autumn diatom outburst. The larvae settle onto the substrata at the beginning of winter. The reproductive cycle of this species resembles that of boreo-arctic cirripedes (Korn 1985). Both this and the previous species belong to the first group of barnacles with one spawning each year (Hines 1978).

Balanus crenatus and *Solidobalanus hesperius* produce several small broods during the year. Their larvae occur in plankton during the greater part of the year, from March to December. Mass hatching of nauplii of both species is observed in late spring and early autumn. In contrast to the long-term breeding cycles in *Semibalanus cariosus* and *Balanus rostratus*, breeding of these species appears to be possible in any season if environmental conditions are favorable (Crisp & Patel 1969).

Balanus improvisus is an immigrant that was first recorded in Peter the Great Bay in 1969 (Zevina & Gorin 1971). In recent times it has spread rapidly, forcing the common native species *Balanus crenatus* and *Solidobalanus hesperius* out of shallow depth. At present this warm water species is common in Russian waters of the Sea of Japan in spite of the harsh temperature regime of this region. Its larvae are encountered from June to December at a wide range of temperatures (from 22-23°C to -1.7°C), and 2-3 peaks of abundance occur from August to October. At the surface their number reaches 800 ind.m^{-3}, which markedly exceeds the abundance of other barnacle species and is an indication of successful acclimation to the new region (Korn 1991). According to Hines's classification (1978), *Balanus improvisus* belong to group 5, as probably do *Balanus crenatus* and *Solidobalanus hesperius*.

A majority of chthamalid barnacles produces numerous small broods during the year and fits into group 3 according to Hines's classification (Crisp 1950; Iwaki 1975; Karande & Thomas 1976; Hines 1978; Egan & Anderson 1989) while *Chthamalus dalli* spawns only twice a year in spring, with a low fecundity. Gonadal growth in this species continues from October to February, and the breeding season lasts about three months, from March to June. The larvae hatch for the first time in May and are encountered in the plankton until the end of summer (Korn & Kolotukhina 1983). Hence, the reproductive pattern of this species is intermediate between the two groups of balanoid barnacles.

Barnacle larvae occur in Peter the Great Bay from March to January, thus creating an almost permanent danger of barnacle fouling (Table1).

3 NAUPLIUS LARVAE AND TAXONOMY OF BARNACLES

At present, side by side with methods of classic zoological systematics, other kinds of evidence have come into use in cirripede taxonomy and phylogeny. These unclude sexuality (Yamaguchi 1986), comparative morphology of dwarf males (Klepal 1987), genetic identity (Dando 1987), feeding mechanisms (Anderson & Southward 1987), sperm ultrastructure (Healy & Anderson 1990), and 18 S ribosomal DNA sequences (Applegate et al. 1991). In studies of barnacle larval development, one question formulated by Dineen (1987) is inevitable – are naupliar characterictics reliable indicators of phylogenetic affinities among the Cirripedia? This is relevant since the major changes in the evolution of thoracican adults from pedunculate to

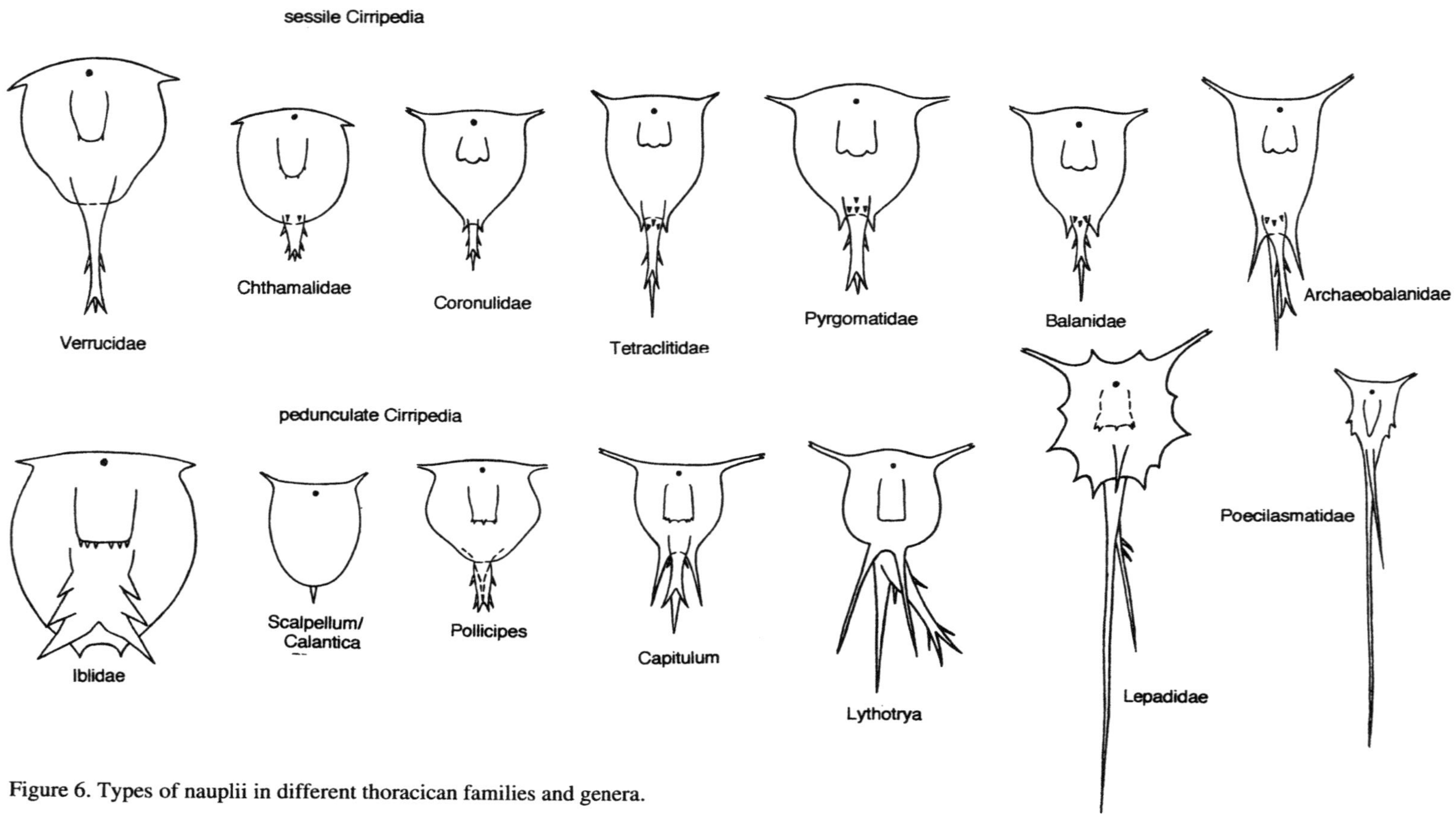

Figure 6. Types of nauplii in different thoracican families and genera.

sessile forms were not accompanied by similar significant changes in larvae (Moyse 1987). However, ontogenetic evidence in various marine invertebrates indicates that relationships at all levels of the taxonomic hierarchy can be judged from the similarity or differences among larval stages. Crustacean larvae may obviously also indicate phylogenetic lineages. For example, if adult morphology is useless in discussing the monophyly of the Cirripedia since rhizocephalan and acrothoracican adults are too reduced, numerous detailed similarities in the nauplii and cyprids of the Thoracica, Acrothoracica, and Rhizocephala show that these three orders form a monophylum (Høeg 1992).

It is essential to synthesize all available information on the larval morphology of the Thoracica in order to compare the larval types in different taxonomic groups and attempt to use these data while discussing some contentious problems of cirripede taxonomy and phylogeny.

At present, the complete larval development is known for about 70 species of Thoracica belonging mainly to the shallow-water suborder Balanomorpha. This information is definitely insufficient to allow any profound conclusions. Nevertheless, careful examination of these results shows that some characters such as the shape of the cephalic shield and the labrum, the length of the dorsal thoracic spine and abdominal process, the number of the abdominal spines, and peculiarities of limb setation, may be used primarily to distinguish larvae at the family level. The shape of the

Table 2. The proposed list of diagnostic characters of thoracican nauplii, with preliminary evaluations of their cladistic status. 0 = plesiomorphy, 1-3 = apomorphies.

1. Naupliar stage number: six naupliar stages (0); abbreviated development (1).
2. Mode of development: planktotrophic (0); lecithotrophic (1).
3. Nauplius size increase: 2-4 times (0); insignificant (1); about 10 times (2).
4. Planktonic life duration: about 2-3 weeks (0); not more than 1 week (1); about 1-2 months (2).
5. Cephalic shield surface: with cuticular ridges (0); without cuticular ridges (1).
6. Posterior cephalic shield border: defined in stage 4 (0); defined in stage 5 (1).
7. Cephalic shield shape: rounded, convex (0); pear-shaped, flattened (1); ornate (2); angular (3).
8. Posterior shield spines: absent (0); short (1); long (2).
9. Shield marginal spines: absent (0); present (1).
10. Shield dorsal spines: absent (0); present (1).
11. Frontolateral horns: medium-sized (0); short (1); long (2).
12. Labrum shape: unilobed with narrowed distal part (0); unilobed with enlarged distal part (1); trilobed (2); reduced (3).
13. Labral teeth: present (0); absent (1).
14. Dorsal thoracic spine: well-developed (0); reduced (1).
15. Abdominal process: short (0); medium-sized (1); long (2); reduced (3).
16. Abdominal process/dorsal thoracic spine ratio: abdomen is subequal to or longer than dorsal thoracic spine (0); abdomen is shorter than dorsal thoracic spine (1).
17. Abdominal spines: present (0); absent (1).
18. Abdominal furca: present (0); absent (1).
19. Antennal gnathobase: well-developed (0); reduced (1).
20. Limbs: well-developed (0); reduced (1).
21. Antennae: with hispid setae (0); with cuspidate setae (1); with neither (2).
22. Antennae: with feathered setae (0); with plumodenticulate setae (1); with neither (2).

Table 3. Character state matrix for the nauplii of thoracican genera. Numerals correspond to characters in Table 1. Superscripts indicate variability within a genus; ? - unknown.

OTU	Characters 1-22																					
Ibla	0^{1}	1^{0}	1^{0}	1^{0}	0	0	0	1	0	0	1	0	1^{0}	1	0	0	0^{1}	0	1^{0}	1^{0}	?	?
Scalpellum	0^{1}	1	1	1	1	0	0	0	0	0	0	3	1	0	3	1	1	0	1	1	2	2
Calantica	0	1	1	0	1	0	0	0	0	0	1	3	1	1	3	?	1	0	1	1	2	2
Pollicipes	0	0	0	0	1	0	0	0	0	0	1	0	0	0	0	0	0	0	0	0	0	0
Capitulum	0	0	0	0	1	0	3	2	0	0	2	0	0	0	1	1	0	0	0	0	?	?
Lithotrya	0	0	0	0	1	0	3	2	1	0	2	0	0	0	1	1	0	0	0	0	1	0
Lepas	0	0	2	2	1	0	2	0	1	1	2	1	0	0	2	1	0	0	0	0	1	0
Dosima	0	0	2	2	1	0	2	0	1	1	2	1	0	0	1	1	0	0	0	0	1	0
Conchoderma	0	0	2	2	1	0	2	0	1	1	2	1	0	0	2	1	0	0	0	0	1	0
Octolasmis	0	0	2	0	1	1	2	0	1	0	2	0	1	0	2	1	1	1	1	1	1	2
Verruca	0	0	0	0	1	0	0	0	0	0	0	0	0	0	1	0	0	0	0	0	?	0
Catomerus	0	0	0	0	1	0	3	2	1	0	0	0	0	0	1	0	0	0	0	0	0	0
Octomeris	0	0	0	0	1	0	0	1	0	0	0	0	0	0	0	1	0	0	0	0	0	0
Chamaesipho	0	0	0	0	1	0	3	2^{0}	0	0	0	0	0	0	0	0	0	0	0	0	0	0
Euraphia	0	0	0	0	1	0	0	0	0	0	0	0	0	0	0	0	0	0	0	0	0	0
Chthamalus	0	0	0	0	1	0	0	0	0	0	0	0	0	0	0	0	0	0	0	0	0	0
Chelonibia	0	0	0	0	1	0	0	1	0	0	1	2	1	0	1	0	0	0	0	0	1	2
Tetraclita	0	0^{1}	0^{1}	0^{1}	1	0	1	1	0	0	1	2^{3}	1	0	1	1	0	0	0^{1}	0^{1}	1^{2}	2
Tetraclitella	0^{1}	0^{1}	0^{1}	0^{1}	1	0	1	1^{0}	0^{1}	0	1	2^{3}	1	0	1	1	0	0	0^{1}	0^{1}	1^{2}	1^{2}
Tesseropora	0	0	0	0	1	0	1	1	1	0	1	2	1	0	1	1	0	0	0	0	1	1
Epopella	0	0	0	0	1	0	1	1	0	0	0	2	1	0	1	1	0	0	0	0	?	?
Austrobalanus	0	0	0	0	1	0	1	1	1	0	1	2	1	0	1	0	0	0	0	0	1	1
Megatrema	0	0	0	0	1	0	0	1	0	0	1	2	1	0	1	0	0	0	0	0	?	?
Savignium	0	0	0	0	1	0	0	1	0	0	1	2	1	0	1	0	0	0	0	0	1	2
Semibalanus	0	0	0	0	1	0	1	1	0	0	1	2	1	0	1	1	0	0	0	0	1	1
Solidobalanus	0	0	0	0	1	0	3	2	0	0	2	2	1	0	1	1	0	0	0	0	1	1
Elminius	0	0	0	0	1	0	1^{3}	1^{2}	0^{1}	0	1^{0}	2	1	0	1	0^{1}	0	0	0	0	1	1
Hexaminius	0	0	0	0	1	0	2	1	1	1	0	2	0	0	1	0	0	0	0	0	1	1
Conopea	0	0	0	0	1	0	3	2	1	1	0	2	0	0	1	1	0	0	0	0	1	1
Acasta	0	0	0	0	1	0	3	2	1	0	0	2	1	0	1	1	0	0	0	0	1	1
Chirona	0	0	0	0	1	0	3	1	0	0	0	2	1	0	1	1	0	0	0	0	1	1
Balanus	0	0	0	0	1	0	1^{3}	1	0^{1}	0^{1}	1^{0}	2	1^{0}	0	1	1^{0}	0	0	0	0	1	1
Megabalanus	0	0^{1}	0^{1}	0^{1}	1	0	1^{3}	2	0^{1}	1^{0}	0	2	0^{1}	0	1	1	0	0	0	0	1	1
Austromegabalanus	0	0	0	0	1	0	3	1	1	1	0	2	1	0	1	0	0	0	0	0	1	1
Notomegabalanus	0	0	0	?	1	0	1	1	0	0	1	2	1	0	1	1	0	0	0	0	1	?

labrum and presence or absence of labral teeth are two of the main diagnostic characters for early larval stages because it would be difficult to measure the length of the cephalic shield (it lacks a distinct posterior margin), and, as a rule, there is only one pair of abdominal spines. Such characters as the length and direction of the frontolateral horns, length of the shield spines, and individual peculiarities of body shape may serve as distinguishing features at generic and specific levels. Modern techniques such as SEM and TEM will assist greatly to improve the level of detail available from larval characters.

Figure 6 represents the typical larval forms in different cirripede families and genera. Table 2 lists the proposed diagnostic characters of thoracican larvae, with preliminary evaluations of their cladistic status. Table 3 is the character state matrix for the nauplii of thoracican genera.

3.1 *Larvae of pedunculate Cirripedia*

All authors, since and including Darwin, agree that among the Thoracica the stalked barnacles are more primitive than the sessile barnacles (Newman 1987). The most ancient fossil remains are known for the pedunculate barnacles. In addition to the paleontological record, the presence of well-developed males and the simpler structure of nervous system are the primitive characters (Zevina 1982).

Known larvae of pedunculate barnacles constitute a very heterogeneous group. Moyse (1987) reviewed all available descriptions of nauplii of Pedunculata and divided them into three distinct groups: 1) generalized nauplii of intertidal scalpellomorphs with brief planktonic life (*Pollicipes* and *Capitulum*); 2) highly specialized, oceanic lepadomorph larvae with an extended planktonic life (*Lepas, Dosima, Conchoderma*, and *Octolasmis*); and 3) lecithotrophic nauplii of a variety of pedunculate species, from abyssobenthic to intertidal, and parasitic forms (*Calantica, Scalpellum, Arcoscalpellum, Ibla*).

3.1.1 *Iblid nauplii*

Within the Pedunculata, the family Iblidae has been considered to be the most primitive one. Iblid males are not situated within special pockets in the scutum, as in lepadomorphs and scalpellomorphs, but just at the bottom of the mantle cavity (Zevina 1982). The morphological features of these animals, such as the plesiomorphic location of the adductor muscle in a post-esophagal position, in contrast to other thoracicans that have a pre-esophagal adductor muscle (Klepal 1985), and a rather unique mineral composition of the shell plates (Whyte 1988), have no equivalents in other thoracican groups. The most recent phylogenetic investigation emphasized the isolated position of ibloids, classifying them as superorder Prothoracica, superfamily Ibloidea (Anderson 1993). A cladistic analysis shows that the Iblidae is the sister group to all remaining Thoracica (Glenner et al. 1995).

Among the five known species of Ibloidea, the larvae of three species have already been described (Batham 1946a, Anderson 1965, 1987, Karande 1974a). The only planktotrophic nauplii, those of *Ibla cumingi*, possess a rounded, convex cephalic shield with two small posterior spines, short, ventrally directed frontolateral horns, a very short abdominal process with numerous spines, and a unilobed labrum with numerous teeth. Unfortunatelly, the setation formulae of this species is unknown. The most significant distinguishing characters of *I. cumingi* are the total disappearance of the dorsal thoracic spine in nauplii 4-6 (reduction?) and the characteristic sculpture of the upper surface of cephalic shield (Karande 1974a). The last feature is not described in detail but it is probably similar to the tesselated dorsal shield of Hansen's y-larvae (Grygier 1987b; Itô 1990) and represents a plesiomorphic condition.

Most of the peculiarities of the nauplii of *Ibla idiotica* and *I. quadrivalvis* are connected with being lecithotrophic (Batham 1946a; Anderson 1965, 1987). The free-swimming but not-feeding nauplius of *I. quadrivalvis* does not grow in size, but moults regularly, and its limb setation is simple. The duration of larval development is abbreviated and the development of cypris features begins precociously; the cyprids themselves appear in 7 days (Anderson 1965, 1987). Scanning electron microscopy (Anderson 1987) showed the presence of marginal cuticular ridges on the cephalic shield similar to ascothoracidan larvae (Grygier 1990, 1992; Itô & Grygier

1990). In *I. idiotica* the direct development to the cypris stage is completed in the parental mantle cavity. The body of the larva is full of yolk, the cephalic shield is hemisherical, the frontolateral horns are reduced, and the nauplius is almost motionless. Antennae and mandibles are absent, whereas the cypris thoracic limbs are precociously developed (Batham 1946a).

Among the related groups, cirripede nauplii have their closest resemblance to ascothoracidan larvae. In spite of the presence of frontolateral horns, rhizocephalan nauplii are not suitable for outgroup comparison, being always lecithotrophic and often having abbreviated development (Høeg 1992a). Acrothoracican nauplii are also lecithotrophic and their features connected with feeding are reduced (Nilsson-Cantell 1978). Many ascothoracidan nauplii are also lecithotrophic, containing yolk and lacking substantial feeding armament on the appendages. However, some members of two families, the Lauridae and Petrarcidae, have nauplii which seem adapted to very early release and planktotrophy (Grygier 1990). In spite of the study of ascothoracidan nauplii being at its initial stage, various larval types have already described in this group (Boxshall & Böttger-Schnack 1988; Grygier 1983, 1987a, 1990, 1992, 1993; Itô & Grygier 1990). Grygier (1987a) believed that early ascothoracidan nauplii are most similar to early cirripede nauplii but that features of late cirripede stages, many of which are associated with feeding, are delayed or entirely suppressed in many ascothoracidans due to brooding.

Ascothoracidan nauplii share many characters with ibloid larvae. The six lecithotrophic naupliar stages of *Baccalaureus falsiramus* have an oval bowl-shaped dorsal shield ornamented with concentric cuticular ridges as in *I. quadrivalvis*. The labrum is small and unilobed. The short caudal region consists of a terminal spine and 2-4 furcal setae (Itô & Grygier 1990; Grygier 1992). The apparently planktotrophic nauplii of an other *Baccalaureus* sp. has a round dorsal shield with a pair of short posterior projections as in *I. cumingi*, a large labrum with a narrowly rounded rear margin, and a short caudal region (Grygier 1990).

The following features of *Ibla* larvae which are traced in ascothoracidan nauplii are assumed to be plesiomorphic: a rounded, convex cephalic shield with cuticular ridges on its surface, a short abdominal process, and a unilobed labrum with a narrowed distal part. The presence of the cuticular ridges, which have not yet been demonstrated in nauplii of the other cirripede taxa, and also the reduced dorsal thoracic spine could be considered as primitive to all other Thoracica and confirms the results of a cladistic analysis (Glenner et al. 1995). The presence of cuticular ridges on the naupliar cephalic shield is probably a primitive character in the same way as the reticulated or cellular surface pattern of the cyprid carapace (Elfimov 1989).

3.1.2 *Lepadoid nauplii*

Among lepadoid barnacles only the nauplii of Lepadidae (pelagic stalked barnacles) and Poecilasmatidae (commensals of decapods) are described at present (Bainbridge & Roskell 1966; Lang 1976, 1979; Colón-Urban et al. 1979; Dalley 1984; Moyse 1987). The larval morphology and also the ecology of this group were thoroughly analyzed by Moyse (1987); therefore, we shall restrict ourselves to a brief account of its peculiarities.

Unusual nauplii in the closely related families Lepadidae and Poecilasmatidae show an apparent resemblance. They possess polygonal, 'ornate' cephalic shield with numerous marginal gland spines and an extraordinary long dorsal thoracic spine and abdominal process that exceed several times the length of the cephalic shield. They have very long frontolateral horns and B-setae in the antennular setation. All lepadomorph late nauplii also have five or six antennular segments and two rather than three preaxial setae (Grygier 1994).

Both groups also exhibit clear differences at the family level: the different forms of the labrum (with an enlarged distal part and teeth in the Lepadidae, but tapered without teeth in the Poecilasmatidae); the presence of the dorsal spine in most lepadid larvae and its absence in poecilasmatid nauplii; the absence in the latter of the numerous lepadid abdominal spines and the furca; and some degeneration of limbs in poecilasmatids (reduction of mandibles, the absence of feathered setae in antennal setation, etc.). The pelagic period of *Lepas, Dosima,* and *Conchoderma* lasts about 1-2 months, whereas *Octolasmis* nauplii develop in the plankton only for the 2-3 weeks that is usual for cirripede larvae, and the growth increments between its stages are somewhat less than in the Lepadidae.

Undoubtedly, the peculiar body shape in lepadid larvae, the remarkably long appendages, and also the large size increments between stages (nauplii increase more than 10 times in length from stage 1 to stage 6) represent adaptations of these pelagic species to an extended planktonic life when the larvae are searching for a suitable substratum for settlement. These adaptative features, e.g., marginal spines, are not so well expressed in poecilasmatid larvae, and this is apparently connected with the residence of adult poecilasmatids on decapod gills. Lang (1976) assumed that the reduction of the mandibular gnathobases in *Octolasmis* larvae suggests a modified feeding pattern. The significance of the tongued labrum without teeth similar to that of lecithotrophic larvae of *Tetraclita squamosa* (M. Barnes & Achituv 1981) and *Ibla quadrivalvis* (Anderson 1965, 1987) is yet unknown.

The similarities between the Lepadidae and Poecilasmatidae seem to rest mainly in synapomorphies. Hence, larval characters indicate that these two taxa form a monophylum, and the Poecilasmatidae is probably the sister group of all the Lepadidae. The separation of the pelagic barnacles into two independent families, Lepadidae and Dosimidae, of which the former had a common ancestor with Poecilasmatidae, and the latter with the Oxynaspididae (Memmi 1983), is not confirmed by the morphology of the larvae. Nauplii of *Dosima fascicularis* (Bainbridge & Roskell 1966) share so many synapomorphies with those of *Lepas* and *Conchoderma* – e.g. a polygonal 'ornate' cephalic shield with numerous marginal spines and a long dorsal spine, exeedingly long caudal processes, very long frontolateral horns, a unilobed labrum with an enlarged distal part, and bristled (B) setae in the antennular setation (Lang 1976, 1979; Dalley 1984; Moyse 1987) – that they could hardly have evolved convergently. Larval characters seems to confirm a monophyletic Lepadidae. Unfortunately, larvae of the family Oxynaspididae have not yet been described.

It is interesting that lepadid nauplii have a great facies resemblance with recently recorded planktotrophic, ascothoracidan nauplii of *Zibrowia* sp. *Zibrowia* nauplii have an octagonal dorsal shield with four pairs of large marginal papillae, an elongated caudal region, as in lepadid larvae, and a tapered labrum as in octolasmid larvae (Grygier 1990). Late planktonic petrarcid nauplii have six pairs of blunt marginal processes, an elongated caudal region, a tapered labrum, and feathered setae in the antennal setation (Grygier 1993).

3.1.3 *Scalpelloid nauplii*

Among scalpelloid barnacles, a great number of species probably have lecithotrophic nauplii. The lecithotrophic larvae exhibit particular features connected with their mode of development, as in the case with iblids. Generally, they retain swimming structures and lose feeding ones. The newly-hatched nauplius of *Calantica spinosa* is unusually large and full of yolk, but its growth at successive moults is slight. The larvae have no intestine and possess reduced caudal processes and a reduced number of limb setae (Batham 1946b). Larvae of *Scalpellum scalpellum* are similar to those mentioned above. They are also lecithotrophic and develop in the plankton. Their labrum, dorsal thoracic spine, and abdominal process are reduced, gnathobases are absent (Kaufmann 1965; Nilsson-Cantell 1978). The development of some deep-water scalpelloid forms (*Ornatoscalpellum stroemii, Tarasovium cornutum*) is abbreviated and proceeds in the mantle cavity of the female until the cypris stage (Nilsson-Cantell 1978).

Three scalpelloids, *Pollicipes polymerus, P. pollicipes*, and the closely related *Capitulum mitella* (originally classified as *Pollicipes mitella*) (Pollicipedinae), and a burrowing barnacle *Lithotrya dorsalis* (Lithotryinae) have planktotrophic nauplii in common (Lewis 1975; Yasugi 1937; Dineen 1987; Molares et al. 1994). In spite of the available description of *C. mitella* nauplii being insufficient, it is still evident that larvae of the foregoing pollicipedid species possess some distinctive features usually appropriate to the family level. The carefully described nauplii of *P. polymerus* have a rounded, convex cephalic shield, a unilobed labrum with numerous teeth, short dorsal thoracic spine and abdominal process, numerous abdominal spines, and F-and H-setae in the antennal setation. Most of these characters are also found in iblid larvae (outgroup) and likely represent the plesiomorphic condition. Larvae of *C. mitella* have fairly long dorsal thoracic spine and abdominal process, fewer abdominal spines than *P. polymerus*, unusually long posterior cephalic spines, which are entirely absent in *P. polymerus*, and very long frontolateral horns, whereas the latter in *P. polymerus* are rather short. The abdominal process in *C. mitella* remains shorter than the dorsal thoracic spine until the late stages, whereas in *P. polymerus* it becomes gradually longer. The recently described nauplii of *P. pollicipes* similar the *P. polymerus* larvae considerably more than those of *C. mitella*. They have rounded, convex cephalic shield without any spines, unilobed labrum with teeth, frontolateral horns of medium length, and numerous abdominal spines in late naupliar stages. Judge by figures, their dorsal thoracic spine and abdominal process are some longer than in *P. polymerus*. The setation formulae of naupliar limbs of *P. pollicipes* are unknown.

Nauplii of *Lithotrya dorsalis* bear a great resemblance to those of *C. mitella* (Dineen 1987). They also have very long frontolateral horns and posterior cephalic shield spines, a unilobed labrum with teeth, and a similar shield outline. At the same time, the dorsal thoracic spine and abdominal process of *L. dorsalis* nauplii, like those of *Lepas*, are considerably longer than those in *C. mitella*. Most of the similarities between the nauplii of *Capitulum* and *Lithotrya* are probably synapomorphic indicating the close relationship between these two taxa.

It is worthy of note that late nauplii of all known scalpellomorphs (as well as sessilians) obviously have four antennal segments and three preaxial setae (Grygier 1994).

3.1.4 *Relationships between larval forms of pedunculates*

For many years there was no consensus of opinion as to which forms of Pedunculata are the more ancient, those possessing numerous capitular plates (scalpellomorphs) or those with few capitular plates (lepadomorphs). The presence of males seems to testify that scalpelloids are more primitive, whereas paleontological records and the structure of the nervous system confirm the more ancient origin of lepadomorphs (Zevina 1980). Newman (1987) believed that the general trend in lepadomorphan evolution progressed from cyprilepadoid-praelepadoid to lepadoid, and hence to scalpelloid forms.

In terms of naupliar morphology, lecithotrophy represents an apomorphic and irreversible condition (Moyse 1987). Therefore, lecithotrophic scalpellomorph larvae (*Scalpellum, Calantica*), as well as lecithotrophic iblid nauplii, cannot be used as models for the ancestral form of thoracican larvae. The highly specialized lepadid and poecilasmatid nauplii are apparently also apomorphic in most characters, although five- or six-segmented antennules of lepadomorphan late nauplii are probably plesiomorphic compared with four-segmented antennules of scalpellomorphans/sessilians, since planktotrophic ascothoracidan late nauplii also have six articles (Grygier 1994). There are also generalized planktotrophic nauplii that possess a majority of plesiomorphic characters, but such naupliar forms have so far been described only in scalpelloids, namely pollicipedids.

Moyse (1987) believed that the larva of *C. mitella* is the closest to the form of the original cirripede nauplius. Using outgroup comparison to ibloid and also to ascothoracidan nauplii, the *Pollicipes* nauplius would seem to be closer to an ancestral model than that of *Capitulum* because the former has the greatest suite of plesiomorphic characters. However, comparative SEM studies of cyprids showed that the lattice organs of *C. mitella* most closely resemble those of the Ascothoracida and are therefore concluded to represent the relatively plesiomorphic condition (Jensen et al. 1994).

Thus, in most larval characters scalpellomorph forms are more primitive than lepadomorph forms. The lepadid larval type can be derived from the compact pollicipedid larva, namely the *Capitulum* nauplius, by a lengthening of the dorsal thoracic spine and the abdominal process and the addition of numerous marginal cephalic spines. The nauplii of *Lithotrya dorsalis* (Dineen 1987) can possibly serve as an intermediate model between those of *Capitulum* and *Lepas*. The possible future discovery of non-pelagic planktotrophic nauplii in any shallow-water lepadomorph would be evidence in support of the comparatively primitive status of this group.

It is likely that forms with numerous capitular plates and forms with few both evolved early from unknown ancestors. The evolutionary pathways in both groups were different, and evolutionary rates were not equal among morphological structures (Zevina 1982).

3.2 *Larvae of sessile Cirripedia*

The adaptive radiation in pedunculate barnacles has given rise to the more advanced sessile barnacles. It is generally accepted that the Balanomorpha, the Verrucomorpha, and the extinct Brachylepadomorpha evolved from one or more scalpelloid-like ancestors (Yamaguchi 1977; Anderson 1993).

3.2.1 *Verrucomorph nauplii*

In spite of the absence of the peduncle, adults of Verrucomorpha more closely resemble the Lepadomorpha than the Balanomorpha. They are similar to the former in the presence of caudal appendages and in the morphology of the mouth apparatus (Zevina 1982).

Among verrucomorphs, the complete larval development has been described only for one species, *Verruca stroemia* (Bassindale 1936; Pyefinch 1948). This information is insufficient; however, verrucomorph nauplii exhibit features similar to those of *Pollicipes polymerus* among scalpellomorph forms. Both are characterized by a rounded cephalic shield devoid of posterior spines and bearing horns of medium length, a similar shield outline, and F-setae in the antennal setation. The abdominal process is elongated in comparison with the dorsal thoracic spine and exceeds the latter in the late stages of development. Therefore, it is easy to derive the *Verruca* larval type from the *Pollicipes* nauplius by some lengthening of the dorsal thoracic spine and abdominal process. The resemblance between these two taxa probably rests on plesiomorphic traits.

3.2.2 *Chthamaloid nauplii*

Darwin and subsequent authors considered the Chthamaloidea to be primitive in comparison with the more advanced groups Coronuloidea and Balanoidea (Newman & Ross 1976).

Chthamalid and pollicipedid adults share a great number of plesiomorphic fea-

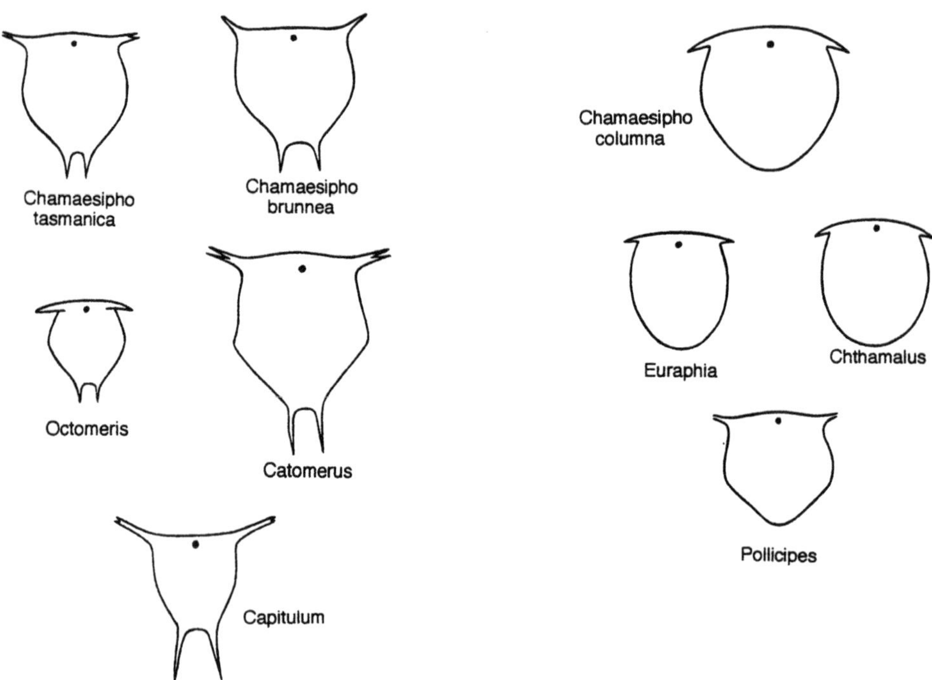

Figure 7. Cephalic shield outlines in chthamaloid and pollicipedid nauplii.

tures. The trophi of chthamaloids are similar to those of scalpellomorphs. The labrum is thick and bullate, and this is basically a scalpellomorphan character. The tridentoid mandibles and the multiarticulate caudal appendages of the primitive chthamaloids are also typical of pollicipedid scalpellomorphans (Newman & Ross 1976).

Many morphological features of chthamaloid nauplii clearly separate them from larvae of other balanomorph barnacles. Among the former, larvae of the genera *Chthamalus* and *Euraphia* show the greatest resemblance. Nauplii of seven species of the genus *Chthamalus* have been described so far. These are: *C. stellatus* (Bassindale 1936; Daniel 1958); *C. malayensis* (Karande & Thomas 1976); *C. fragilis* (Lang 1979); *C. dalli* (Korn & Ovsyannikova 1979; Miller et al. 1989); *C. dentatus* (Achituv 1986); *C. antennatus* (Egan & Anderson 1989); and *C. fissus* (Miller et al. 1989). Within the genus *Euraphia*, the larvae of only two species are known: *E. aestuarii*, originally classified as *C. aestuarii* (Sandison 1967) and *E. depressa*, originally classified as *C. depressus* (Le Reste 1965). Some features of *E. withersi* are outlined by Karande & Thomas (1976). This information is not always sufficient. Comparative analyses of chthamaloid larvae have been made by Korn & Ovsyannikova (1979) and Egan & Anderson (1989).

The larvae of all the foregoing species have a broad, convex, globular cephalic shield without posterior or any other spines. The frontolateral horns are medium-sized, directed ventrally and slightly curved under the cephalic shield. Therefore, it is difficult to specify their direction relative to the longitudinal axis of the body, which is a taxonomic character for many balanoid species (Barnes & Barnes 1959a,b). All chthamalid nauplii also possess a unilobed labrum with numerous teeth in late stages which are partially reduced in late stages.

The dorsal thoracic spine and abdominal process of all *Chthamalus* are considerably shorter than the cephalic shield. In stages 5-6 the abdominal process become longer than the dorsal thoracic spine, which is not typical of most balanoid larvae. The number of abdominal spines in *Chthamalus* larvae is greater than in balanoid larvae (see Section 2.4). The number of setae on the limbs of *Chthamalus* nauplii is always larger than in balanoid larvae. Cuspidate (C) and plumodenticulate (D) setae are typical of balanoid nauplii, whereas chthamalid larvae have no D-setae, and only one C-seta differing from those of the former found on the mandibular endopod. *Chthamalus* nauplii possess two setal types that are never found in balanoid larvae: hispid (H) setae and 3-4 feathered (F) setae located on the antennal endopods. The structure of the antennal gnathobase in chthamalid larvae in comparison with balanoid nauplii is described by Egan & Anderson (1989). The former is angular, bears no more than three major teeth on the dorsal edge, and bends backward against the body of the antenna.

Peculiarities of limb setation of chthamaloid nauplii are undoubtedly connected with their feeding mechanism. Stone (1989) believes that nauplii of verrucids and chthamalids, like lepadids, possess a plesiomorphic fine-meshed filter capable of filtering very small cells (see Section 3.2.3.5), whereas in the balanoids this fine-meshed filter has been lost.

The typical *Chthamalus* larvae exhibit considerable morphological similarity to some scalpellomorph and verrucomorph nauplii (Lang 1979; Korn 1988b; Egan & Anderson 1989; Miller et al. 1989). Like those of verrucomorphs, the chthamalid nauplii bear the greatest resemblance to larvae of *Pollicipes polymerus*. Both possess a rounded, convex cephalic shield devoid of spines, a short dorsal thoracic spine and an abdominal process with numerous spines and which in stage 6 becomes longer than the dorsal thoracic spine, a unilobed labrum with teeth, and F- and H-setae in the antennal setation. The most remarkable feature is the presence of H-setae, described so far only for the larvae of these two groups. When comparing *Chthamalus fissus* and *C. dalli* with *Pollicipes polymerus* in Monterey Bay, it is difficult to discriminate the chthamalid and pollicipedid nauplii. The similarity between the larvae of pollicipedids and chthamalids accords with the similarity of the adults (Miller et al. 1989). The similar features are obviously plesiomorphies, as in the case with verrucomorph larvae.

The above characteristics are manifested mainly in the nauplii of *Chthamalus* and *Euraphia*. Other genera of chthamaloid barnacles differ from the typical pattern of development to some extent. For instance, the naupliar cephalic shield of *Octomeris angulosa* bears a pair of posterior spines (Sandison 1954). Nauplii of *Chamaesipho tasmanica* have two spines on the cephalic shield (Egan & Anderson 1989). So does *C. brunnea*, while *C. columna* has none (Barker 1976). It should be noted that the body shape of nauplii of *C. columna* is similar to that of typical chthamalid larvae rather than to *C. brunnea* and *C. tasmanica*. The presence of only two teeth on the labrum is considered to be a characteristic feature of *Chamaesipho* nauplii (Foster 1967). Nauplii of *Catomerus polymerus* (Catophragmidae) appreciably differ from the typical chthamalid larvae. They have an angular body shape and a flat cephalic shield with long posterior spines. The frontolateral horns are directed anteriorly. The dorsal thoracic spine and the abdominal process are slender and elongated (Egan & Anderson, 1989). *Catomerus polymerus* is a primitive catophragmid chthamaloid barnacle. According to Darwin's concept, it represent the basic form among balanomorphs and had evolved from a scalpellid lepadomorph ancestor (Anderson 1983). *Catomerus* larvae (Egan & Anderson 1989) resemble most closely those of *Chamaesipho*. The nauplii of *Catomerus* and *Chamaesipho* resemble *Capitulum* larvae to a greater degree, whereas the larvae of *Chthamalus* and *Euraphia* are similar to those of *Pollicipes*.

According to larval morphology, all the chthamaloid barnacles may be divided into two groups: (1) *Catomerus*, *Octomeris*, and *Chamaesipho*; (2) *Euraphia* and *Chthamalus* (Fig. 7). Sperm ultrastructure data suggest an independent origin of *Chamaesipho* from the Catophragmidae, however *Chthamalus* and *Octomeris* represent a single evolutionary line (Healy & Anderson 1990).

It is worthy of note that chthamalid larvae resemble the planktotrophic nauplii of iblids, which of course are quite distant phyletically from chthamalids. The nauplii in both families have a rounded, convex cephalic shield (posterior spines are present in iblids and lacking in chthamalids), a unilobed labrum provided with numerous teeth, a very short dorsal thoracic spine and abdominal process, and the same sequence of emergence of the abdominal spines. Therefore, the larvae of *Ibla cumingi* and *Chthamalus malayensis* from the Indian Ocean are strikingly similar (Karande & Thomas 1976). Nevertheless, these groups are unlikely to represent relatives. Similarities between these two taxa may represent symplesiomorphic traits and may also reflect convergencies that appeared in the Iblidae and Chthamalidae due to the same mode of life in the intertidal zone. Such as a compact shape of the larvae is often found in intertidal barnacles.

3.2.3 *The other balanomorph families*

The other balanomorph families represent an evolutionary young group whose radiation did not much affect larval development. The nauplii of Coronulidae, Tetraclitidae, Pyrgomatidae, Balanidae, and Archaeobalanidae differ to a lesser degree from each other than the larvae of other cirripede families. They have a typical, trilobed labrum usually devoid of teeth. The cephalic shield provided with two posterior spines is pear-shaped and relatively flat in balanids, archaeobalanids, and tetraclitids, and to a greater degree rounded, broad, and dorsoventrally convex in coronulids and pyrgomatids. The dorsal thoracic spine and the abdominal process are elongated, and abdomen is usually shorter than the former. The nauplii possess cuspidate (C) and plumodenticulate (D) setae in the antennal and mandibular setation, and, although some deviations exist, the balanoid type of abdominal spine arrangement is usual. Most of these characters (trilobed labrum, abdomen with 3 spines in series 2, D-setae) are obviously synapomorphies. The rounded, convex cephalic shield and the sometimes noted labral teeth are probably primitive characters.

3.2.3.1 *Coronulid nauplii.* The nauplii of only two coronulid species, both of them obligatory commensals of marine animals, have been described – complete larval development of Chelonibia patula (Lang 1979) and stages 1 and 2 of C. testudinaria (Pillai 1958). Chelonibia larvae have a broad nearly circular, highly arched cephalic shield. The trilobed labrum is of the same form as Balanus sp. (see Section 3.2.3.4), and the longer median lobe lacks denticulation in all stages. The abdominal process is distinctly shorter than the dorsal thoracic spine only in stage 2, while in later stages it is subequal or longer. The abdominal spination is not typical: stages 2 and 3 have an additional pair of spines. Stages 4 and 5 have a complex of small spines instead of the usual medial spine in series 2. Some deviations from the typical balanoid setation formulae are also observed (Lang 1979).

Thus, coronulid larvae have some primitive features similar to pollicipedid nauplii: the rounded, convex cephalic shield, the abdominal processes elongated in comparison with the dorsal thoracic spine. The availability of only a single complete larval description prevents us from proposing a diagnosis for the larvae of this family. No larval peculiarities related to the commensalism of coronulid adults, resembling for example those of poecilasmatid larvae, are evident.

3.2.3.2 *Tetraclitid nauplii.* All members of the Tetraclitidae are intertidal or restricted to very shallow waters (Newman & Ross 1976).

Among tetraclitids, a number of species have the usual kind of planktotrophic nauplii: *Tetraclitella karandei* (Karande 1974b, 1982); *Tetraclitella depressa* and *Epopella plicatus* (Barker 1976); *Tetraclita serrata* (Griffiths 1979); *Tetraclitella purpurascens*, *Austrobalanus imperator* and *Tesseropora rosa* (Egan & Anderson 1988). Some species have lecithotrophic nauplii that show particular features similar to characters of ibloid and scalpelloid larvae. The nauplii of *Tetraclita squamosa* develop in the plankton, but their increase in size during development is small compared with many other species. The lack of much setation coupled with a reduced development of labrum is evidently a consequence of the unnecessity for this species to feed during its planktonic life (Barnes & Achituv 1981). The larvae of *Tetraclitella divisa* pass through four naupliar stages to the cyprids within the mantle cavity. These nauplii show a vestigial limb setation and reduced posterior shield spines (Anderson 1986).

Tetraclitid naupliar characters such as the body shape, labral shape, limb setation, and form of the gnathobase are similar to those of balanoid larvae. Australian tetraclitids (*Austrobalanus imperator*, *Tetraclitella purpurascens* and *Tesseropora rosa*) have small lateral shield spines in some naupliar stages (Egan & Anderson 1988). A greater number of small abdominal spines compared with those of balanoid species is typical of some coronuloid species (Lang 1979; Karande 1974b; Griffiths 1979; Egan & Anderson 1988). Some peculiarities of the setation formulae (the SSPS setation on the terminal segment of the antennules and PPPSPP on the terminal segment of the antennal endopodites) are common to most coronuloid species (Egan & Anderson 1988).

3.2.3.3 *Pyrgomatid nauplii.* The family Pyrgomatidae includes coral-inhabiting barnacles. Larvae of only two pyrgomatid species are known: *Megatrema anglica* (Moyse 1961) and *Savignium milleporae* (Stewart et al. 1989). Pyrgomatid nauplii differ from typical balanoid larvae and are similar to foregoing coronulid nauplii by the rounded, convex outline of cephalic shield, in consequence of which the frontolateral horns are deflected ventrally. In later stages the abdominal process becomes gradually longer than the dorsal thoracic spine. The middle labral lobe projects beyond the lateral ones (Moyse 1961). These features are obviously primitive. The remaining traits (trilobed labrum, abdominal spination, etc) are shared by pyrgomatids with typical balanoid larvae and are probably derived.

Some deviations of setation formulae from the typical balanoid pattern (for instance, the appearance of a serrate (R) seta instead of a plumodenticulate (D) seta on the antennal endopod) are reported for *Savignium milleporae* (Stewart et al. 1989). A subapical placement of the fourth apical seta on the antennule is clearly figured in *S. milleporae,* although the authors believe that it is a first preaxial seta appearing in stage 2. The presence of three apical setae in comparison with four apical setae in most balanomorph larvae is regarded as a plesiomorphic condition (Grygier 1994). No naupliar features determined by habitat conditions, namely their symbiotic relationships, have been established which would distinguish them from larvae of species inhabiting non-living substrata.

3.2.3.4 *Balanid and archaeobalanid nauplii.* In most cirripede families, the larval development has been described for only few species, whereas in the Balanidae and Archaeobalanidae a comparatively wide range of larval forms is known (Figs 8 and 9). An analysis of their morphological features was published earlier (Korn 1988b). In an unsuccessful attempt to propose a diagnosis for the larvae of both families, I found no differences at the family level between the nauplii of Balanidae and Archaeobalanidae. The larvae in both groups possess medium-long dorsal thoracic spine and abdominal process (somewhat shorter than the cephalic shield), a pair of posterior spines, a trilobed labrum usually devoid of teeth, the same arrangement of abdominal spines, and C- and D-setae in the limb setation.

The family Archaeobalanidae comprises three subfamilies: Archaeobalaninae, Semibalaninae, and Elminiinae. Let us assume that the typical characters for all archaeobalanids are those exemplified by the larvae of *Solidobalanus hesperius*, viz.,

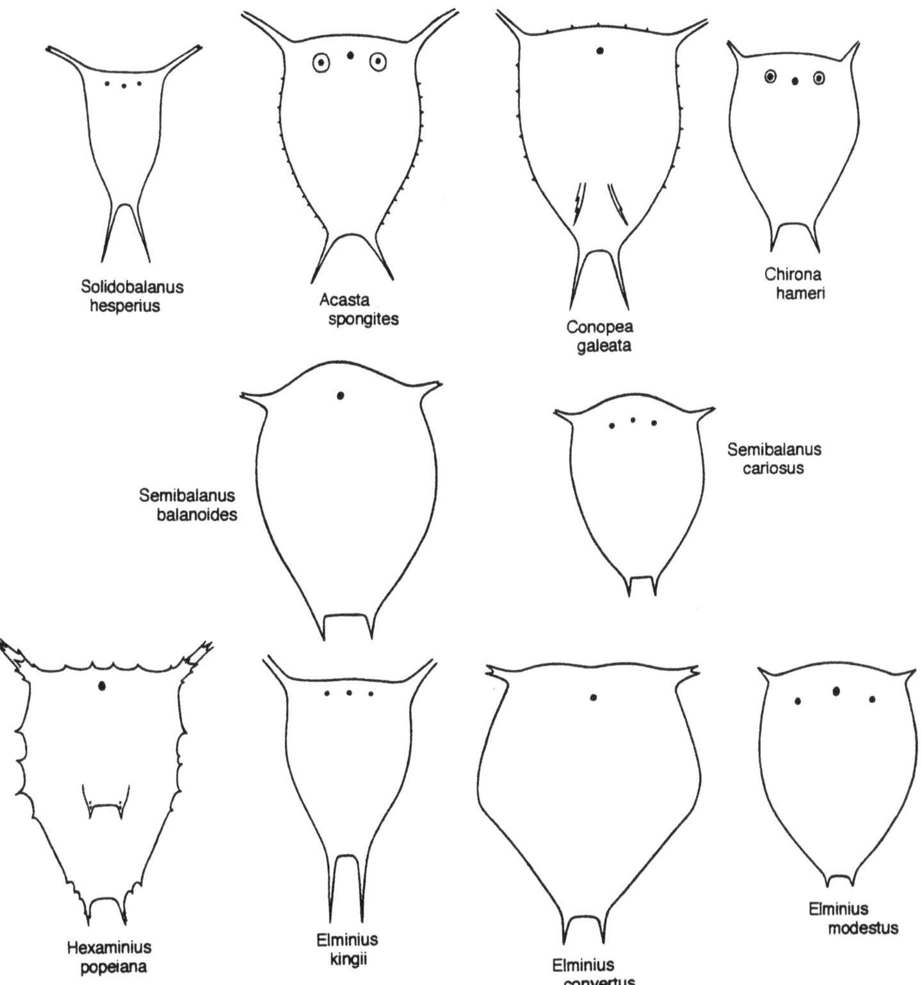

Figure 8. Cephalic shield outlines in archaeobalanid nauplii.

long, forwardly directed horns, a straight anterior margin of the larva, long posterior spines, and a trapeziform body shape (Barnes & Barnes 1959a; Korn & Ovsyannikova 1981). These features are traced in most of the archaeobalanine species (Moyse 1961; Molenock & Gomez 1972; Crisp 1962b), whereas the larvae of most species in the family Balanidae exhibit the opposite set of characters: fairly short horns primarily directed laterally, a convex anterior margin of the larvae, short posterior spines, and a pear-shaped body outline (Figs 8 and 9).

The known larvae in the Semibalaninae, *Semibalanus balanoides* and *S. cariosus* (Bassindale 1936; Crisp 1962a; Branscomb & Vedder 1982; Korn 1989) are very similar to each other and show characters peculiar to balanids, namely the nauplii of *Balanus rostratus* and *B. balanus* (Barnes & Costlow 1961; Ovsyannikova & Korn 1981). In the subfamily Elminiinae, the nauplii of *Elminius covertus* and *E. modestus* are of the balanid type (Knight-Jones & Waugh 1949; Barker 1976; Egan & Anderson 1985), while those of *Hexaminius popeianus* and *E. kingii* are more similar to the

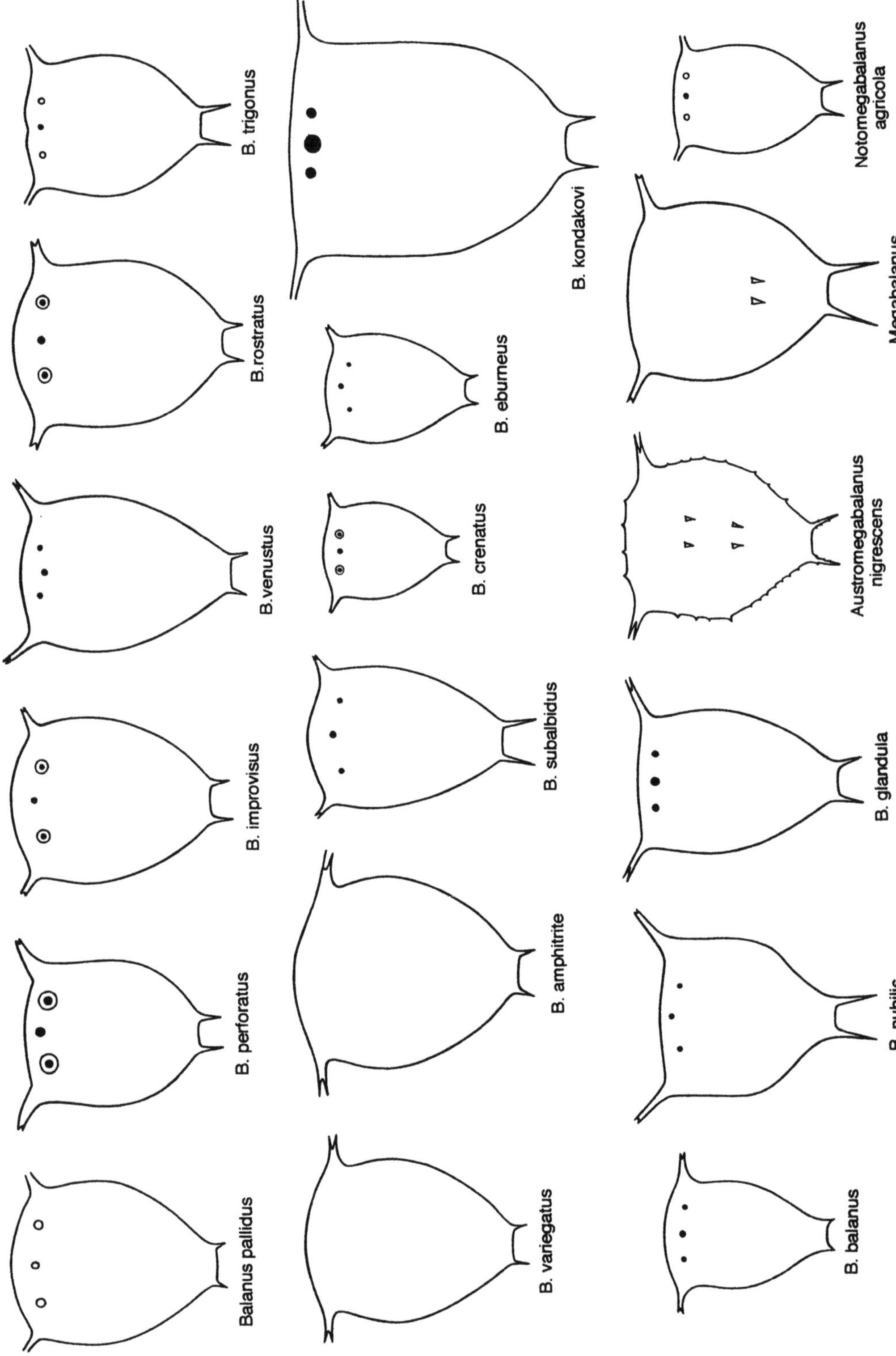

Figure 9. Cephalic shield outlines in balanid nauplii.

archaeobalanid type (Arenas 1982; Egan & Anderson 1985).

On the other hand, the nauplii of *Balanus glandula* (Brown & Roughgarden 1985) and *B. nubilis* (Barnes & Barnes 1959b), belonging to the family Balanidae, show some characters typical of archaeobalanids. The larvae of some other species, *B. venustus* and *B. subalbidus*, possess an intermediate set of characters (Lang 1979). Thus, it seems impossible to separate the Balanidae and Archaeobalanidae on the basis of their larval morphology.

It should be noted that within the vast family Balanidae deviations from the typical larval development occur very often. All these variations concern warm-water species, while species from arctic and temperate waters exhibit more constant morphological characters and always follow the pattern described in the earlier papers (Norris & Crisp 1953; Jones & Crisp 1954; Crisp 1962a). Teeth emerge on the labrum in larvae of *Balanus improvisus* (Jones & Crisp 1954), *B. pallidus* (Sandison 1967), *B. venustus* (Lang 1979), *Elminius covertus* (Egan & Anderson 1985), and *B. albicostatus* (Lee & Kim 1991). Very small marginal spines on the cephalic shield and even dorsal 'lepadid' spines in stages 2-5 are reported for nauplii of *B. venustus* (Lang 1979). The larvae of *Megabalanus rosa* have two dorsal spines (Choi et al. 1992), *M. volcano* has 2-3 dorsal and numerous small marginal spines (Kado & Hirano 1994), and *Austromegabalanus nigrescens* nauplii possess 1-4 dorsal and marginal spines (Egan & Anderson 1987). Some species also show an aberrant pattern of abdominal spines. In larvae of *B. amphitrite* the second pair of spines appears in stage 3, whereas in *B. cirratus* in stage 2. The third pair of spines of these species is absent in stage 5 (Karande 1973, 1974c). Many of the species listed above belong to the '*Balanus amphitrite*' complex (Henry & McLaughlin 1975).

Some warm-water archaeobalanid species have additional spines on the cephalic shield. Larvae of *Conopea galeata* and *Hexaminius popeianus* possess marginal spines and two dorsal spines (Molenock & Gomez 1972; Egan & Anderson 1985). Nauplii of *Acasta spongites* have lateral spines (Moyse 1961).

The functional significance of the dorsal shield spines remains unknown and cannot be correlated with larval habitat. Nauplii with these spines develop in such disparate environments as the open sea (*Lepas, Conchoderma*), estuarine waters (*Hexaminius*) and coastal waters (*Austromegabalanus nigrescens, Balanus venustus*) (Egan & Anderson 1989). Zevina (1981) proposed that the spines of the cephalic shield serve as an antipredator device and, possibly, for increasing the body surface area, which is necessary in planktonic forms. Dorsal shield spines seem to be convergent features in view of the occurrence of these spines in unrelated groups (Choi et al. 1992).

3.2.3.5 *Generic and specific relationships among balanomorph larvae*. Larval characters can be useful in establishing the taxonomic status of barnacle adults at the generic or specific levels. Barker (1976) pointed out that the similarity between the larvae of *Elminius plicatus* and *Tetraclitella depressa* supports the classification of *E. plicatus* as tetraclitid. Later studies on adult barnacles indicated that this species is rightly classified as *Epopella plicatus* (Tetraclitidae) (Foster 1978). Naupliar peculiarities of *Chamaesipho columna* corroborate the recent separation of this species from *C. tasmanica* by Foster and Anderson (1986) (Fig. 6). The larval differences between *Austromegabalanus*, *Notomegabalanus*, and *Megabalanus* support the classification of the megabalanines into three genera (Egan & Anderson 1987). Recent data on sperm ultrastructure (Healy & Anderson 1990) raised the possibility that *Megabalanus* and *Austromegabalanus* had separate origins. The contrasting larval development of the two genera adds further weight to this possibility (Choi et al. 1992).

Lee & Kim (1991a) showed that the numerical setation of the second antennae and mandibles of 11 *Balanus* species agree with the division of adult barnacles into four groups (Newman & Ross 1976). The progressive increase in setal complexity in nauplii of the group of *B. amphitrite* compared with the group of *B. balanus* is apparently caused by inhabiting warm waters. It was repeatedly mentioned that there is a correlation between the geographical distribution of a cirripede species and the diet of its larvae: the larvae of high latitude barnacles need diatoms as food, but the larvae of oceanic and low latitude species need flagellates (Moyse 1963; Moyse & Knight-Jones 1967). Moyse (1984, 1987) also emphasised taxonomic distinctions and suggested that the feathered (F) setae in the nauplii of *Chthamalus* and *Lepas* form a fine-meshed filtering mechanism. Stone (1989) also noted that the nauplii of balanoids have widely spaced setules, well suited to harvest diatoms typical of high latitude waters, while the chthamaloids have a fine-meshed filter capable of filtering small flagellates which are predominant in warm waters.

3.2.4 *The origin of the sessile barnacles*

Until recently the question of a mono- or polyphyletic origin of the sessile barnacles was debatable. The majority of authors supported the 'Darwinian' hypothesis that the Balanomorpha, Verrucomorpha, and Brachylepadomorpha evolved from the Lepadomorpha as independent lineages. *Pollicipes*, *Capitulum*, and *Scillaelepas* had a central position as models for the pedunculate ancestor of sessile Cirripedia (Newman & Ross 1976; Newman 1982). Data on the adult morphology and postlarval ontogeny of the primitive sessile barnacles *Chionelasmus* and *Brachylepas* showed that Brachylepadomorpha, Verrucomorpha and Balanomorpha (three sessile suborders) form a monophyletic taxon and evolved from a sessile, brachylepadomorphan-like ancestor (Newman 1987). The discovery of the most primitive living sessile barnacle, *Neoverruca brachylepadoformis*, from abyssal hydrothermal vents, constituting a 'missing link' between the Brachylepadomorpha and Verrucomorpha, confirmed this concept (Newman 1989; Newman & Hessler 1989). On the contrary, polyphyletic scheme of the origin of sessile barnacles from pollicipedine-like ancestors have been recently presented by Anderson (1993).

Larval features provide some evidence for a polyphyletic origin of the Balanomorpha. As has been noted earlier (Moyse 1987; Korn 1988b), pollicipedid nauplii are the most plausible model for the ancestral form of the larvae of sessile barnacles. Verrucomorph and chthamaloid nauplii resemble the larvae of *Pollicipes polymerus*, while remain balanomorph nauplii are similar in some traits to the larvae of *Capitulum mitella*. Both balanomorphs and *Capitulum* have two posterior cephalic spines, laterally directed horns, and a fairly long dorsal thoracic spine and abdominal process. Moyse (1987) noticed that the form of the caudal furca in the nauplius of *C. mitella* and the gnathobase of its antennal basis resemble those of balanoid larvae. However, Kado et al. (1994) believed that the antennal gnathobases of *C. mitella* are very similar to those of chthamaloid species and differ in shape from those of coronuloids and balanoids. Nauplii of *C. mitella* nauplii differ from balanomorph larvae in the presence of a unilobed labrum with numerous teeth, which may be considered as plesiomorphic character. Unfortunately, the detailed setation formulae of this species is unknown. It is worthy of note that the spines armouring the abdominal process of the related species *Lithotrya dorsalis* (Dineen 1987) are identical to those of balanoid larvae (2 pairs with medial one between the proximal pair in stage 4).

Among balanomorph nauplii, larvae of *C. mitella* bear the closest resemblance to archaeobalanid larvae, for instance, those of *Solidibalanus hesperius* (Barnes & Barnes 1959a; Korn & Ovsyannikova 1981). Both have a straight anterior margin, long frontolateral horns, and posterior shield spines.

Thus, the nauplius of *Capitulim* shares some synapomorpies with certain balanomorph larvae that are absent in chthamaloid nauplii. This may indicate a sistergroup relation and argues against a monophyletic Balanomorpha. However, the alternative hypothesis also is possible: the synapomorphic characters shared by the larvae of *Capitulum* and some Balanomorpha, e.g., the posterior cephalic shield spines, are simply convergences observed in distantly related thoracican groups. Moreover, the description of the larval development of *C. mitella*, which is so important in barnacle phylogeny, is until now insufficiant, so definitive conclusions are premature. In any case, *Pollicipes* and *Capitulum* are not very close relatives and according to larval characters likely represent different barnacle families.

4 CONCLUSIONS

Comparative larval development along with data of classic zoological taxonomy can be of great use in clarifying the relationships among the Cirripedia. Careful examination of larval types in different barnacle groups shows that larval evidence agree in general with the current taxonomy of the adults. As with other ontogenetic stages, nauplii can obviously indicate phylogenetic lineages, though this depends upon the correct identification of primitive and derived characters. At the same time, some difficulties are inevitable: a small number of completely described larval forms in comparison with the abundance of adult barnacles, the lack of paleontological records of the larvae, and the most essential, a possibly high level of convergent characters determined by lecithotrophy, pelagic habitat and feeding mechanism. Therefore, larval characters should not predominate in phylogenetic considerations. Naupliar morphology can be used as additional material for the achievement of a consensus in phylogeny.

ACKNOWLEDGMENTS

This work was supported by the 'Biological Diversity' program of The George Soros International Science Foundation. I thank Dr. J. T. Høeg, Prof. F. R. Schram, and two annonymous referees for careful and very useful comments on the manuscript.

REFERENCES

Achituv, Y. 1986. The larval development of *Chthamalus dentatus* Krauss (Cirripedia) from South Africa. *Crustaceana* 51: 259-269.

Anderson D.T. 1965. Embryonic and larval development and segment formation in *Ibla quadrivalvis* Cuv. (Cirripedia). *Austral. J. Zool.* 13: 1-15.

Anderson, D.T. 1983. *Catomerus polymerus* and the evolution of the balanomorph form in barnacles (Cirripedia). *Mem. Austral. Mus.* 18: 7-20.

Anderson, D.T. 1986. The circumtropical barnacle *Tetraclitella divisa* (Nilsson-Cantell) (Balanomorpha, Tetraclitidae): cirral activity and larval development. *Proc. Linn. Soc. N.S.W.* 109: 107-116.

Anderson, D.T. 1987. The larval musculature of the barnacle *Ibla quadrivalvis* Cuvier (Cirripedia, Lepadomorpha). *Proc. R. Soc. Lond.* (B) 231: 313-338.

Anderson, D.T. 1993. *Barnacles - Structure, Function, Development and Evolution.* London: Chapman & Hall.

Anderson, D.T. & A.J. Southward 1987. Cirral activity of barnacles. *Crustacean issues* 5: 135-174.

Applegate, M., L.G. Abele & T. Spears 1991. A phylogenetic study of the Cirripedia based on 18 S ribosomal DNA sequences. *Amer. Zool.* 31(5): 101A.

Arenas, J.N. 1982. Estadios larvales de *Elminius kingii* Gray, Cirripedio del Sur de Chile. *Stud. Neotrop. Fauna Environ.* 17: 61-67.

Bainbridge, V. & J. Roskell 1966. A re-description of the larvae of *Lepas fascicularis* Ellis and Solander with observations on the distribution of *Lepas* nauplii in the north-eastern Atlantic. In H. Barnes (ed.), *Some Contemporary Studies in Marine Science*: pp. 67-81. London: George Allen & Unwin Ltd.

Barker, M.F. 1976. Culture and morphology of some New Zealand barnacles (Crustacea: Cirripedia). *N.Z.J. Mar. Freshw. Res.* 10: 139-158.

Barnes, H. 1959. Temperature and the life cycle of *Balanus balanoides* (L.). In D.L. Ray (ed.), *Marine Boring and Fouling Organisms*: pp.234-245. Seattle: Univ.Washington Press.

Barnes, H. & M. Barnes 1954. The general biology of *Balanus balanus* (L.) Da Costa. *Oikos* 5: 63-76.

Barnes, H. & M. Barnes 1959a. The naupliar stages of *Balanus nubilis* Darwin. *Can. J. Zool.* 37: 15-23.

Barnes, H. & M. Barnes 1959b. The naupliar stages of *Balanus hesperius* Pilsbry. *Can. J. Zool.* 37: 237-244.

Barnes, H. & J.D. Costlow 1961. The larval stages of *Balanus balanus* (L.) Da Costa. *J. Mar. Biol. Ass. UK.* 41: 59-68.

Barnes, M. & Y. Achituv 1981. The nauplius stages of the cirripede *Tetraclita squamosa rufotincta* Pilsbry. *J. Exp. Mar. Biol. Ecol.* 54: 149-165.

Bassindale, R. 1936. The developmental stages of three English barnacles, *Balanus balanoides* (Linn.), *Chthamalus stellatus* (Poli) and *Verruca stroemia* (O.F.Müller). *Proc. Zool. Soc. Lond.* 106: 57-74.

Batham, E.J. 1946a. Description of female, male and larval forms of a tiny stalked barnacle, *Ibla idiotica* n. sp. *Trans. R. Soc. N.Z.* 75: 347-356.

Batham, E.J. 1946b. *Pollicipes spinosus* Quoy and Gaimard. II. Embryonic and larval development. *Trans. R. Soc. N.Z.* 75: 405-418, pls. 36-43.

Boxshall, G.A. & R. Böttger-Schnack 1988. Unusual ascothoracid nauplii from the Red Sea. *Bull. Brit. Mus. (Nat. Hist.), Zool.* 54: 275-283.

Branscomb, E.S. & K. Vedder 1982. A description of the naupliar stages of the barnacles *Balanus glandula* Darwin, *Balanus cariosus* Pallas, and *Balanus crenatus* Bruguière (Cirripedia, Thoracica). *Crustaceana* 42: 83-95.

Brown, S.K. & J. Roughgarden 1985. Growth, morphology, and laboratory culture of larvae of *Balanus glandula* (Cirripedia: Thoracica). *J. Crust. Biol.* 5: 574-590.

Choi, K.H., D.T. Anderson & C.H. Kim 1992. Larval development of the megabalanine balanomorph *Megabalanus rosa* (Pilsbry) (Cirripedia, Balanidae). *Proc. Linn. Soc. N.S.W.* 113: 175-184.

Colón-Urban, R., P.J. Cheung, G.D. Ruggieri & R.F. Nigrelli 1979. Observations on the development and maintenance of the deep sea barnacle, *Octolasmis aymonini geryonophila* (Pilsbry). *Int. J. Invert. Reprod.* 1: 245-252.

Crisp, D.J. 1950. Breeding and distribution of *Chthamalus stellatus*. *Nature* 166: 311-312.

Crisp, D.J. 1954. The breeding of *Balanus porcatus* (Da Costa) in the Irish Sea. *J. Mar. Biol. Ass. UK.* 33: 473-496.

Crisp, D.J. 1962a. The planktonic stages of the Cirripedia *Balanus balanoides* (L.) and *Balanus balanus* (L.) from north temperate waters. *Crustaceana* 3: 207-221.

Crisp, D.J. 1962b. The larval stages of *Balanus hameri* (Ascanius, 1767). *Crustaceana* 4: 123-130.

Crisp, D.J. & B. Patel 1969. Environmental control of the breeding of three boreo-arctic cirripedes. *Mar. Biol.* 2: 283-295.

Dalley, R. 1984. The larval stages of the oceanic, pedunculate barnacle *Conchoderma auritum* (L.) (Cirripedia, Thoracica). *Crustaceana* 46: 39-54.

Dando, P.R. 1987. Biochemical genetics of barnacles and their taxonomy. *Crustacean Issues* 5: 73-87.

Daniel, A. 1958. The development and metamorphosis of three species of sessile barnacles. *J. Madras Univ.* (B) 28: 23-47.

Dineen, J.F.Jr. 1987. The larval stages of *Lithotrya dorsalis* (Ellis & Solander, 1786): a burrowing thoracican barnacle. *Biol. Bull.* 172: 284-298.

Egan, E.A. & D.T. Anderson 1985. Larval development of *Elminius covertus* Foster and *Hexaminius popeiana* Foster (Cirripedia: Archaeobalanidae: Elminiinae) reared in laboratory. *Austal. J. Mar. Freshw. Res.* 36: 383-404.

Egan, E.A. & D.T. Anderson 1986. Larval development of *Balanus amphitrite* Darwin and *Balanus variegatus* Darwin (Cirripedia, Balanidae) from New South Wales, Australia. *Crustaceana* 51: 188-207.

Egan, E.A. & D.T. Anderson 1987. Larval development of the megabalanine balanomorph *Austromegabalanus nigrescens* (Lamarck) (Cirripedia, Balanidae). *Austral. J. Mar. Freshw. Res.* 38: 511-522.

Egan, E.A. & D.T. Anderson 1988. Larval development of the coronuloid barnacles *Austrobalanus imperator* (Darwin), *Tetraclitella purpurascens* (Wood) and *Tesseropora rosea* (Krauss) (Cirripedia, Tetraclitidae). *J. Nat. Hist.* 22: 1379-1405.

Egan, E.A. & D.T. Anderson 1989. Larval development of the chthamaloid barnacles *Catomerus polymerus* Darwin, *Chamaesipho tasmanica* Foster & Anderson and *Chthamalus antennatus* Darwin (Crustacea: Cirripedia). *Zool. J. Linn. Soc.* 95: 1-28.

Elfimov, A.S. 1987. Comparative morphological study of the cypris larvae of the genus *Lepas* (Cirripedia, Lepadomorpha). *Proc. 18th Conf. Young Sci., Biol. Fac. Moscow Univ.* 3: 2-6 (in Russian).

Elfimov, A.S. 1989. *The cypris larvae of cirripedes and their significance in the formation of fouling*. Ph. D. thesis. Moscow: Moscow University (in Russian).

Foster, B.A. 1967. The early stages of some New Zealand shore barnacles. *Tane* 13: 33-42.

Foster, B.A. 1978. The marine fauna of New Zealand; barnacles (Cirripedia). *Mem. N.Z. Ocean. Inst.* 69: 1-160.

Foster, B.A. & D.T. Anderson 1986. New names for two well known shore barnacles (Cirripedia, Thoracica) from Australia and New Zealand. *J. Roy . Soc. N.Z.* 16: 57-69.

Geraci, S. & V. Romairone 1986. Larval stages and *Balanus* (Cirripedia) settlement in a port environment with a key to naupliar stages of Tyrrhenian species. *P.S.Z.N.I.: Marine Ecology* 7: 151-164.

Glenner, H., M.J. Grygier, J.T. Høeg, P.G. Jensen & F.R. Schram. 1995. Cladistic analysis of the Cirripedia. *Zool. J. Linn. Soc.,London.* 114: 365-404.

Griffiths, R.J.I. 1979. The reproductive season and larval development of the barnacle *Tetraclita serrata* Darwin. *Trans. Roy. Soc. S. Afr.* 44: 97-111.

Grygier, M.J. 1983. *Ascothorax*, a review with descriptions of new species and remarks on larval development, biogeography, and ecology (Crustacea: Ascothoracida). *Sarsia* 68: 103-126.

Grygier, M.J. 1987a. Nauplii, antennular ontogeny, and the position of the Ascothoracida within the Maxillopoda. *J. Crust. Biol.* 7: 87-104.

Grygier, M.J. 1987b. New records, external and internal anatomy, and systematic position of Hansen's y-larvae (Crustacea: Maxillopoda: Facetotecta). *Sarsia* 72: 261-278.

Grygier, M.J. 1990. Early planktotrophic nauplii of *Baccalaureus* and *Zibrowia* (Crustacea: Ascothoracida) from Okinawa, Japan. *Galaxea* 8: 321-337.

Grygier, M.J. 1992. Laboratory rearing of ascothoracidan nauplii (Crustacea: Maxillopoda) from plankton at Okinawa, Japan. *Publ. Seto Mar. Biol. Lab.* 35: 235-251.

Grygier, M.J. 1993. Late planktonic naupliar development of an ascothoracidan crustacean (?Petrarcidae) in the Red Sea and a comparison to the Cirripedia. *L. A. Co. Mus. Contr. Sci.* 437: 1-14.

Grygier, M.J. 1994. Developmental patterns and hypothesis of homology in the antennules of thecostracan nauplius larvae (Crustacea). *Acta Zoologica*, Stockh. 75: 219-234.

Healy, J.M. & D.T. Anderson 1990. Sperm ultrastructure in the Cirripedia and its phylogenetic significance. *Rec. Austral. Mus.* 42: 1-26.

Henry, D.P. & P.A. McLaughlin 1975. The barnacles of the *Balanus amphitrite* complex (Cirripedia, Thoracica). *Zool. Verh.* 141: 1-254.

Hines, A.H. 1978. Reproduction of three species of intertidal barnacles from central California. *Biol. Bull.* 154: 262-281.

Hirano, R. 1952. On the rearings and metamorphoses of four important barnacles in Japan. *J. Oceanogr. Soc. Japan* 8: 139-143.

Høeg, J.T. 1992a. The phylogenetic position of the Rhizocephala: are they truly barnacles? *Acta Zool.* Stockh. 73: 323-326.

Høeg, J.T. 1992b. Rhizocephala. In F.Harrison (ed.), *Microsc. Anat. Invertebr.* (9): pp. 313-345. New York: John Wiley.

Itô, T. 1990. Naupliar development of *Hansenocaris furcifera* Itô (Crustacea: Maxillopoda: Facetotecta) from Tanabe Bay, Japan. *Publ. Seto Mar. Biol. Lab.* 34: 201-224.

Itô, T. & M.J. Grygier 1990. Description and complete larval development of a new species of *Baccalaureus* (Crustacea: Ascothoracida) parasitic in a zoanthid from Tanabe Bay, Honshu, Japan. Zool. Sci. 7: 485-515.

Iwaki, T. 1975. Breeding and settlement of *Chthamalus challengeri* Hoek on the southern coast of Hokkaido. *Bull. Fac. Fish. Hokkaido Univ.* 26: 1-10.

Jensen, P.G., J.Moyse, J.Høeg & H.Al-Yahya 1994. Comparative SEM studies of lattice organs: putative sensory structures on the carapace of larvae from Ascothoracida and Cirripedia (Crustacea Maxillopoda Thecostraca). *Acta Zool.* Stockh. 75: 125-142.

Jones, L.W.G. & D.J. Crisp 1954. The larval stages of the barnacle *Balanus improvisus* Darwin. *Proc. Zool. Soc. Lond.* 123: 765-780.

Kado, R. 1982. Ecological and taxonomical studies on the free-living nauplius of barnacles (Crustacea, Cirripedia). Ph. D. thesis, Tokyo: University of Tokyo (in Japanese).

Kado, R. & R. Hirano 1994. Larval development of two Japanese megabalanine barnacles, *Megabalanus volcano* (Pilsbry) and *Megabalanus rosa* (Pilsbry) (Cirripedia, Balanidae), reared in the laboratory *J. Exp. Mar. Biol. Ecol.* 175: 17-41.

Karande, A.A. 1973. Larval development of *Balanus amphitrite amphitrite*.D., reared in the laboratory. *Proc. Ind. Acad. Sci.* B 77: 56-63.

Karande, A.A. 1974a. Development of the pedunculate barnacle *Ibla cumingi* Darwin. *Ind. J. Mar. Sci.* 3: 173-177.

Karande, A.A. 1974b. Larval development of the barnacle *Tetraclitella karandei* reared in the laboratory. *Biol. Bull.* 146: 249-257.

Karande, A.A. 1974c. *Balanus variegatus* Darwin: the laboratory reared larvae compared with *Balanus amphitrite amphitrite* Darwin (Cirripedia). *Crustaceana* 26: 229-232.

Karande, A.A. 1982. The nauplius appendages of the cirripede *Tetraclitella karandei* Ross. *J. Exp. Mar. Biol. Ecol.* 62: 87-92.

Karande, A.A. & M.K. Thomas 1976. The larvae of the inter-tidal barnacle *Chthamalus malayensis* Pilsbry. *Proc. Ind. Acad. Sci.* (B) 83: 210-219.

Kaufmann, R. 1965. Zur Embryonal- und Larvalentwicklung von *Scalpellum scalpellum* L. (Crust. Cirr.) mit einem Beitrag zur Autökologie dieser Art. *Z. Morph. Ökol. Tiere* 55: 161-232.

Klepal, W. 1985. *Ibla cumingi* (Crustacea, Cirripedia) - a gonochoristic species (anatomy, dwarfing and systematic implications). *P.S.Z.N.I.: Marine Ecology* 6: 47-119.

Klepal, W. 1987. A review of the comparative anatomy of the males of cirripedes. *Oceanogr. Mar. Biol. Ann. Rev.* 25: 285-351.

Knight-Jones, E.W. & G.D. Waugh 1949. On the larval development of *Elminius modestus* Darwin. *J. Mar. Biol. Ass. UK.* 28: 413-428.

Korn, O.M. 1985. The reproductive cycle of the barnacle *Balanus rostratus* in Peter the Great Bay, Sea of Japan. *Biol. Morya* 3: 36-43 (in Russian).

Korn, O.M. 1988a. *Key to larvae of common barnacle species (Cirripedia, Thoracica) of Peter the Great Bay, Sea of Japan.* Preprint No 23. Vladivostok: Far East Branch, USSR Academy of Sciences (in Russian).

Korn, O.M. 1988b. Larvae morphology and phylogenetic system of barnacles of the order Thoracica. *Zool. Zh.* 67: 1644-1651 (in Russian).

Korn, O.M. 1989. Reproduction of the barnacle *Semibalanus cariosus* in the Sea of Japan. *Biol. Morya* 5: 40-48 (in Russian).

Korn, O.M. 1991. Larvae of the barnacle *Balanus improvisus* in the Sea of Japan. *Biol. Morya* 1: 52-62 (in Russian).

Korn, O.M. & I.I. Ovsyannikova 1979. Larval development of the barnacle *Chthamalus dalli. Biol. Morya* 5: 60-69 (in Russian).

Korn, O.M. & I.I. Ovsyannikova 1981. Larval development of the barnacle *Solidobalanus hesperius hesperius* (Cirripedia, Thoracica) in laboratory conditions. *Zool. Zh.* 60: 1472-1479 (in Russian).

Korn, O.M. & N.K. Kolotukhina 1983. Reproduction of the barnacle *Chthamalus dalli* in the Sea of Japan. *Biol. Morya* 2: 31-38 (in Russian).

Lang, W.H. 1976. The larval development and metamorphosis of the pedunculate barnacle *Octolasmis mülleri* (Coker, 1902) reared in the laboratory. *Biol. Bull.* 150: 255-267.

Lang, W.H. 1979. Larval development of shallow water barnacles of the Carolinas (Cirripedia: Thoracica) with keys to naupliar stages. *NOAA Tech. Rep. NMFS* Circ. 421: 1-39.

Lang, W.H. 1980. Crustacea. Cirripedia: Balanomorph nauplii of the NW Atlantic shores. *Fiche Ident. Zooplankton, Cons. Int. Explor. Mer.* 163: 1-6.

Lee, C. & C.H. Kim 1991a. Larval development of *Balanus albicostatus* Pilsbry (Cirripedia, Thoracica) reared in the laboratory. *J. Exp. Mar. Biol. Ecol.* 147: 231-244.

Lee, C. & C.H. Kim 1991b. The larval development of a fouling organism *Balanus kondakovi* Tarasov & Zevina (Cirripedia, Thoracica). *Korean J. Zool.* 34: 81-92.

Le Reste, L. 1965. Contribution a l›étude des larves de Cirripèdes dans le Golfe de Marseille. *Recueil des Travaux de la Station Marine d›Endoume, Bulletin* 38: 33-121.

Lewis, C.A. 1975. Development of the gooseneck barnacle *Pollicipes polymerus* (Cirripedia: Lepadomorpha); fertilization through settlement. *Mar. Biol.* 32: 141-153.

Memmi, M. 1983. A new point of view on the origin and position of the pelagic barnacles in the system of suborder Lepadomorpha (Cirripedia, Thoracica). *Dokl. Akad. Nauk SSSR* 273: 1271-1275 (in Russian).

Miller, K.M., S.M. Blower, D. Hedgecock & J. Roughgarden 1989. Comparison of larval and adult stages of *Chthamalus dalli* and *Chthamalus fissus* (Cirripedia: Thoracica). *J. Crust. Biol.* 9: 242-256.

Molares, J., F. Tilves & C. Pascual 1994. Larval development of the pedunculate barnacle *Pollicipes cornucopia* (Cirripedia: Scalpellomorpha) reared in the laboratory. *Mar. Biol.* 120: 261-264.

Molenock, J. & E.D. Gomez 1972. Larval stages and settlement of the barnacle *Balanus (Conopea) galeatus* (L.) (Cirripedia Thoracica). *Crustaceana* 23: 100-108.

Moyse, J. 1961. The larval stages of *Acasta spongites* and *Pyrgoma anglicum* (Cirripedia). *Proc. Zool. Soc. London* 137: 371-392.

Moyse, J. 1963. A comparison of the value of various flagellates and diatoms as food for barnacle larvae. *J. Cons. Int. Explor. Mer.* 28: 175-187.

Moyse, J. 1984. Some observations on the swimming and feeding of the nauplius larvae of *Lepas pectinata* (Cirripedia: Crustacea). *Zool. J. Linn. Soc. Lond.* 80: 323-336.

Moyse, J. 1987. Larvae of lepadomorph barnacles. *Crustacean Issues* 5: 329-362.

Moyse, J. & E.W. Knight-Jones 1967. Biology of cirripede larvae. *Proc. Symp. Crustacea* M.B.A. India 2: 595-611. Bangalore: Marine Biological Association of India.

Newman, W.H. 1965. Prospectus on larval cirriped setation formulae. *Crustaceana* 9: 51-56.

Newman, W.H. 1982. Cirripedia. In L.G.Abele (ed.), *The Biology of Crustacea* 1: 197-221. New York: Academic Press.

Newman, W.H. 1987. Evolution of cirripedes and their major groups. *Crustacean issues* 5: 3-42.

Newman, W.H. 1989. Juvenile ontogeny and metamorphosis in the most primitive living sessile barnacle, *Neoverruca*, from abyssal hydrothermal springs. *Bull. Mar. Sci.* 45: 467-477.

Newman, W.H. & R.R. Hessler 1989. A new abyssal hydrothermal verrucomorphan (Cirripedia; Sessilia): The most primitive living sessile barnacle. *Trans. San Diego Soc. Nat. Hist.* 21: 259-273.

Newman, W.A. & A. Ross 1976. Revision of the balanomorph barnacles, including a catalog of the species. *Mem. San Diego Soc. Nat. Hist.* 9: 1-108.

Nilsson-Cantell, C.A. 1978. *Cirripedia Thoracica and Acrothoracica. Mar. Invert. Scand.* 5. Oslo: Universitetsforlaget.

Norris, E. & D.J. Crisp 1953. The distribution and planktonic stages of the cirripede *Balanus perforatus* Bruguière. *Proc. Zool. Soc. Lond.* 123: 393-409.

Ovsyannikova, I.I. & O.M. Korn 1981. Naupliar stages of the acorn barnacle *Balanus rostratus eurostratus* Broch. In A.I. Kafanov (ed.), *Systematics and Chorology of Benthic Invertebrates of Far Eastern Seas*: pp. 36-42. Vladivostok: Academy of Sciences of the USSR, Far East Scientific Centre (in Russian).

Ovsyannikova, I.I. & O.M. Korn 1984. Naupliar development of the barnacle *Balanus crenatus* in Peter the Great Bay (Sea of Japan). *Biol. Morya* 5: 34-40 (in Russian).

Pillai, N.K. 1958. Development of *Balanus amphitrite*, with a note on the early development of *Chelonibia testudinaria*. *Bull. Cent. Res. Inst. Univ. Kerala.* C 6: 117-130.

Pyefinch, K.A. 1948. Methods of identification of the larvae of *Balanus balanoides* (L.), *Balanus crenatus* Brug. and *Verruca stroemia* O.F. Müller. *J. Mar. Biol. Ass. UK.* 27: 451-463.

Sandison, E.E. 1954. The identification of the nauplii of some South African barnacles with notes on their life histories. *Trans. Roy. Soc. S. Afr.* 34: 69-101.

Sandison, E.E. 1967. The naupliar stages of *Balanus pallidus stutsburi* Darwin and *Chthamalus aestuarii* Stubbings (Cirripedia Thoracica). *Crustaceana* 13: 161-174.

Standing, J.D. 1980. Common inshore barnacle cyprids of the Oregonian faunal province (Crustacea: Cirripedia). *Proc. Biol. Soc. Wash.* 93: 1184-1203.

Stewart, B.A., P.A. Cook & Y. Achituv 1989. Naupliar stages of the coral inhabiting barnacle *Savignium milleporum* (Darwin) (Cirripedia: Pyrgomatidae) from the Gulf of Eilat, Red Sea. *Bull. Mar. Sci.* 45: 164-173.

Stone, C.J. 1989. A comparison of algal diets for cirripede nauplii. *J. Exp. Mar. Biol. Ecol.* 132: 17-40.

Whyte, M.A. 1988. The mineral composition of the valves and peduncle scales of *Ibla quadrivalvis* (Cuvier) (Cirripedia. Thoracica). *Crustaceana* 55: 219-224.

Willemöes-Suhm, R. 1876. On the development of *Lepas fascicularis* and the 'archizoea' of Cirripedia. *Phil. Trans. Roy. Soc. London* (B) 166: 131-154.

Winsor, M.P. 1969. Barnacle larvae in the nineteenth century. A case study in taxonomic theory. *J. Hist. Med. Allied Sci.* 24: 294-309.

Yamaguchi, T. 1977. A review of phylogenetic studies on Cirripedia. *J. Geogr.* 86: 285-304 (in Japanese).

Yamaguchi, T. 1986. Sexual polymorphism in Cirripedia. *J. Geogr.* 95: 126-142 (in Japanese).

Yasugi, R. 1937. On the swimming larvae of *Mitella mitella* L. *Bot. Zool. Tokyo* 5: 792-796 (in Japanese).

Zevina, G.B. 1976. The order Cirripedia. In A.V. Zhirmunsky (ed.), *The Animals and Plants of Peter the Great Bay*: pp. 42-46. Leningrad: Nauka (in Russian).

Zevina, G.B. 1980. A new classification of Lepadomorpha (Cirripedia). *Zool. Zh.* 59: 689-698 (in Russian).

Zevina, G.B. 1981. Barnacles of the suborder Lepadomorpha (Cirripedia, Thoracica) of the World Ocean. I. Family Scalpellidae. Opredeliteli po Faune SSSR (Guides to the fauna of the USSR) 127. Leningrad: Nauka (in Russian).

Zevina, G.B. 1982. Barnacles of the suborder Lepadomorpha (Cirripedia, Thoracica) of the World Ocean. II. Opredeliteli po Faune SSSR (Guides to the fauna of the USSR) 133. Leningrad: Nauka (in Russian).

Zevina, G.B. & A.N. Gorin 1971. The invasion of *Balanus improvisus* and *B. eburneus* into the Sea of Japan. *Zool. Zh.* 50: 771-773 (in Russian).

An unusual barnacle nauplius illustrating several hitherto unappreciated features useful in cirripede systematics

Mark J. Grygier
Silver Spring, Maryland, USA

ABSTRACT

An extremely elongate, sixth-stage cirripede nauplius from plankton in Osaka Bay, Japan, is similar in some respects to previously described nauplii of both the Lepadidae and Poecilasmatidae, but is distinguished from each of them by many features. Characters which have not been described in detail before for any cirripede nauplius, or for at most one or two species, include the setation, pores, cuticular ornamentation, and muscle attachment sites on the cephalic shield, the glands associated with the shield's marginal gland spines, the interruptions in the cuticular hoops that make up the antennal exopod, and evidence for four segments in the antennal endopod. The significance and potential utility of these features in maxillopodan systematics is discussed. The distribution of setae, pores, and muscle attachment sites on the cephalic shield of a typical facetotectan nauplius ('nauplius y') is illustrated for comparison. The presence in the present nauplius of an array of strong setae on the basis and endopod of the mandible, despite the absence of a fine filter apparatus on the antenna, suggests that earlier concepts of the functional morphology of feeding in the nauplii of pedunculate cirripedes have been over-simplified.

1 INTRODUCTION

Moyse (1987) reviewed the current state of knowledge of the nauplii of stalked barnacles or Pedunculata sensu Newman (1987), who recognized four extant orders: Iblomorpha, Lepadomorpha, Heteralepadomorpha, and Scalpellomorpha. One of the three major morphological categories of nauplii that Moyse (1987) recognized are those adapted to a pelagic, planktotrophic existance by the possession of long spines and elongate appendages. Several species of Lepadidae and Poecilasmatidae in the Lepadomorpha have such nauplii. In the Lepadidae, *Lepas anatifera* L., *L. pectinata* L., *Dosima fascicularis* (Ellis & Solander), and *Conchoderma auritum* (L.) have been studied in detail (Bainbridge & Roskell 1966; Dalley 1984; Moyse 1987). An assortment of lepadid-type nauplii whose precise identities are unknown has also been described (cited by Dalley 1984); many of these have been assigned to the collective groups *Archizoea* and *Nauplius*. In the Poecilasmatidae, knowledge of nauplii is effectively limited to several species of *Octolasmis* (Lang 1976, 1979; Colón-Urban et al. 1979), although some unidentified but similar forms have also been described (Rose 1929; Sandison 1950). In the Iblomorpha, Scalpellomorpha, and Anelasmatidae of the Heteralepadomorpha, the known nauplii are lecithotrophic or, if planktotrophic, do not exhibit adaptations for pelagic life (Moyse 1987). The nauplii of the many other families of Lepadomorpha and Heteralepadomorpha are completely unknown; therefore, it should not be surprising to find unfamiliar kinds of cirripede nauplii in the plankton. The present report concerns one very striking and distinctive form from Japan, the antennules of which have already been discussed by Grygier (1994).

2 MATERIALS AND METHODS

One last-stage nauplius (NVI) with a developing cyprid larva inside was caught in plankton in Osaka Bay, Japan ('Toyoshio-maru' stn. 4-b (A-26), 28 June 1985, 34°25.5'N, 135°13.0'E, depth not recorded). It was examined whole in dorsal, ventral, and side views in glycerine at about 400X using normal and phase contrast optics, and detailed drawings were made with the aid of a camera lucida. The labrum, left appendages, and left frontolateral horn were dissected and mounted in glycerine jelly for examination at higher power; unfortunately, many setae were broken in the process, and few additional details compared to the whole mount were noted. An attempt to match the setae and aesthetascs of the developing cyprid's antennules to the naupliar setae (setal identity by formation; see Grygier 1994) failed.

3 DESCRIPTION

3.1 *Dimensions*

Total length 6.64 mm (Fig. 1A). Cephalic shield 0.64 mm long along midline, 0.49 mm wide behind frontolateral horns. Distance across tips of horns, including apical hairs, 0.84 mm. Dorsal thoracic spine 5.89 mm long, 77 µm wide at base; abdominal process broken but reaching at least to 40% length of dorsal thoracic spine.

3.2 *Cephalic shield*

Shape of shield a square-fronted hexagon (Fig. 1B). Frontolateral horns extending outward and about 30° forward from front corners. Six pairs of marginal gland spines: one pair anterior, four lateral, and one posterior. Third lateral gland spines longest, then fourth pair, then posterior pair, then second lateral pair; anterior pair and first lateral pair shortest. Each gland spine with apical pore serving as exit for one gland cell in first two lateral pairs and posterior pair (gland cells of latter collapsed), and for two gland cells in rear two lateral pairs; glands of anterior pair not observed. Four additional minute spines on right margin, five on left, and two on anterior margin, unassociated with pores or glands.

Dorsal ornament consisting of bumps, hair-like setae, and pores (Fig. 1B). Seven

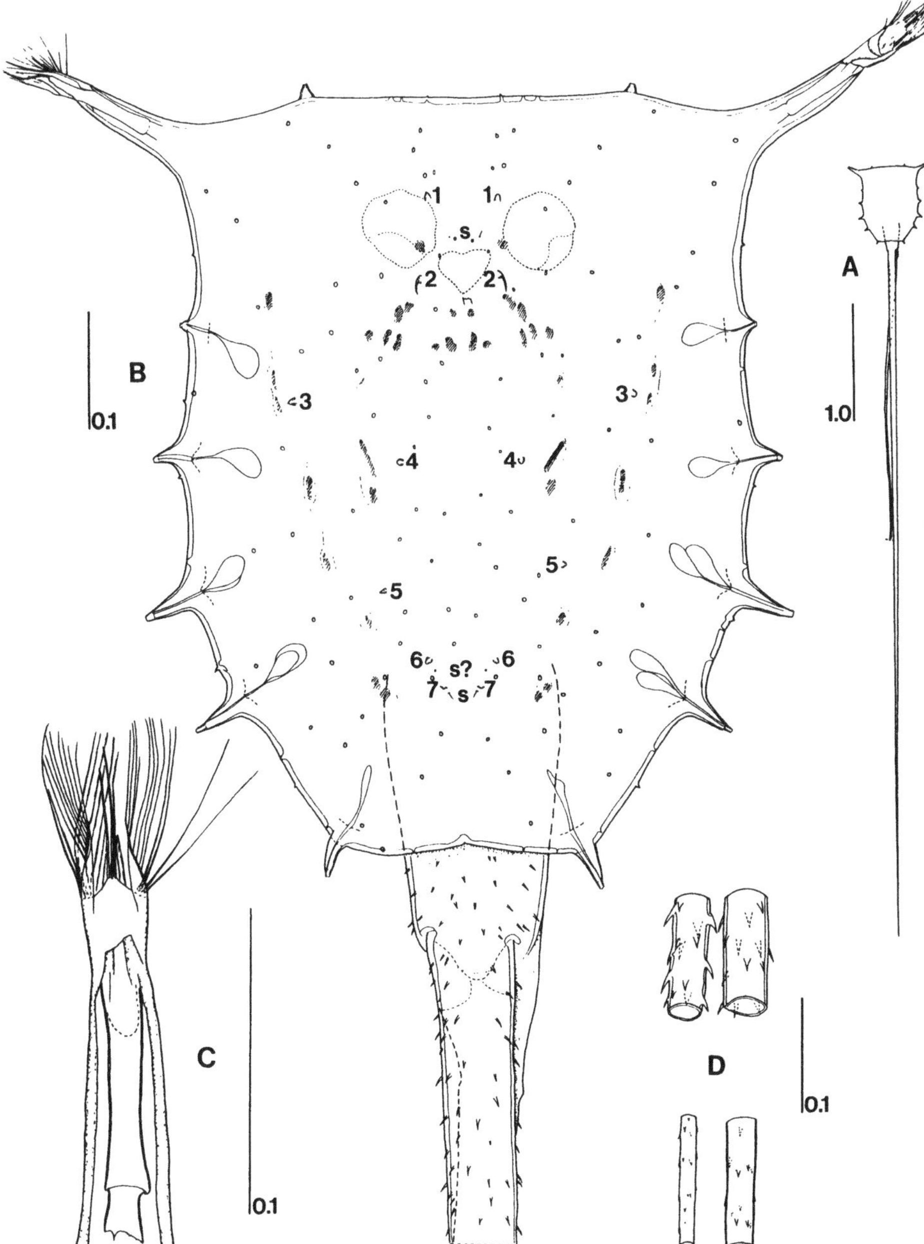

Figure 1. Unidentified stage VI pedunculate cirripede nauplius from Osaka Bay. (A) Habitus, dorsal view, appendages omitted; (B) Cephalic shield, trunk region, and proximal parts of dorsal thoracic spine and abdominal process, dorsal view, showing nauplius and compound eyes (dashed), pores (small circles), muscle attachment sites (hatched ovals), small bumps (1-7), hair-like setae (s) and supposed setal sockets (s?), and position of retracted hypodermis in relation to marginal gland spines (dashed lines intersecting gland ducts); (C) Distal two-thirds of left frontolateral horn, ventral view; (D) Same-level sections of successively more posterior regions of dorsal thoracic spine (right) and abdominal process (left, including broken tip), dorsal view. Scale bars in mm.

pairs of small bumps generally close to midline, but third pair much farther lateral and fifth pair somewhat farther lateral than others. Most bumps hemispherical and similar in size, but second pair elaborated as backwardly curved, hooklike spines, fifth pair conical, and seventh pair smaller than rest. Small, asymmetrical plate with spinulose anterior margin on midline just behind hooklike spines. Apparently three pairs of minute, hair-like setae originally present, but some now lost (Fig. 1B): anterior pair between first and second pairs of bumps (right seta and socket of left one seen); questionable posterior pair between sixth and seventh pairs of bumps (only supposed sockets seen); rearmost pair behind seventh pair of bumps.

About 110 pores scattered over entire dorsal surface (Fig. 1B), each appearing conical in optical section. Few pores in outer 30% of shield, remainder distributed to some degree in paired radiating rows. Four pairs of pores on ventral side of cephalic shield (Fig. 2A): two anterior pairs between frontal filaments and bases of frontolateral horns, and two posterolateral pairs flanking bases of posterior marginal gland spines.

Twenty pairs of dorsal muscle attachments visible as crater-like formations in retracted hypodermis (Fig. 1B). One pair between first and second pairs of bumps, three pairs close to midline behind second pair of bumps, and cluster of five pairs slightly farther laterally. Single pair behind most lateral of these five, followed by two pairs just outside fourth pair of bumps. Arcs of eight pairs much farther laterally; of these, second and third close together just outside third pair of bumps, and seventh and eighth close together lateral to seventh pair of bumps. In addition, some large muscles apparently attached close to midline between fourth and fifth pairs of bumps, but without forming craters (not illustrated).

Frontolateral horns containing cuticle-lined duct, seemingly ending in two pores (Fig. 1C). Apex partly encircled by delicate flange with short spine on edge and with sides developed as furled wings bearing many long, delicate, marginal hairs.

3.3 *Caudal armament*

Dorsal side of trunk region armed with small, posteriorly directed spinules (Fig. 1B), ventral side with six pairs of large spines (series-2 spines) representing thoracopods (Fig. 2: ser2). Dorsal thoracic spine and abdominal process both armed for whole length with posteriorly directed spinules (Figs 1B, D and 2A). Spinules of dorsal thoracic spine similar in size dorsally and ventrally, at first increasing in size posteriorly, then decreasing greatly (Fig. 1D). Spinules of abdominal process about twice as long on ventral side compared to dorsal side for part of length (Fig. 1D), then also becoming reduced in size posteriorly; ventral midline of anterior one-third of preserved length armed with pavement of minute spinules except for extreme anterior part (Fig. 2A).

3.4 *Frontal filaments and eyes*

Frontal filaments 0.14-0.15 mm long, basal 22-25% of length thicker than distal part and possessing much thicker cuticle (Fig. 2A-ff, B). A small nauplius eye and a pair of larger compound eyes of the cyprid are present (Fig. 1B).

3.5 *Labrum*

Total length 0.50 mm, width of basal part 0.16 mm, width of distal part 0.12 mm. Tip extending beyond rear margin of dorsal shield, to level of fourth pair of ventral thoracic spines (Fig. 2A-lb). Basal part with parallel lateral margins, distal part with narrow, biconcave section followed by distal expansion and narrow, conical apex ending in pore. Right side of distal expansion bearing thickened knob, left bearing broken spine, each side also bearing about a dozen vibrissa-like setae. On dorsal side (Fig. 2C), pair of pores at midlength of biconcave region and medial pore at mouth. Biconcave margins lined with two rows of dorsally slanting setae, those of inner row finer and more numerous. Basal part with pair of diagonal rows of short setae on dorsal side bounding sides of atrium oris; latter bounded posteriorly by two transverse rows of setae on transverse swelling of ventral body surface (Fig. 2A), but paragnaths apparently absent.

3.6 *Antennule*

Six-segmented (Fig. 2A-an). First segment short, circular, unarmed except for row of spinules along anteromedial, distal margin. Second segment longer than wide, with slight constriction at base; armed with row of spinules along anteromedial, distal margin, another row at midlength posteromedially, and several isolated spinules. Third segment somewhat longer and narrower than second, also unarmed except for spinule row on distal margin and several isolated spinules. Fourth segment longer than third and becoming somewhat thicker distally, with sensillum or pore on medial side of postaxial angle and bearing four long setae: postaxial b- and c-setae plumose with setules in two medial and two lateral rows, medial setules long and widely spaced, lateral ones short and closely spaced; simple seta arising from middle of lateral margin; preaxial seta arising lateral to fifth segment, biserially plumose with long, widely spaced setules. Fifth segment small, narrow, cylindrical, with four long setae: postaxial e-seta simple, arising lateral to thicker d-seta, latter armed like b- and c-setae; preaxial seta armed like that on preceding segment. Sixth, apical segment about half size of fifth, bearing few distal spinules and one short and three long setae, posterolateral long one armed like b-, c-, and d-setae, others simple.

3.7 *Antenna*

Twice as long as antennule, reaching far beyond margin of cephalic shield (Fig. 2A-at). Coxa divided into large, thin- walled proximal region and short, thick-walled distal annulus, these regions being separated laterally by a constriction. Patch of spinules on anterior face of coxa. Medial margin proximal to annulus bearing anteriorly-pointing gnathobase with tip reaching mouth opening, a few spinules proximal to gnathobase, and claw-like spine ventral to gnathobase, serving as stop against labrum to limit antennal adduction (Fig. 2D). Gnathobase not seen perfectly, but bearing setiform structure anterolaterally, distal prong with tooth, and shorter, medial prong; row of bristles proximal to medial prong and a few more bristles at base of distal prong.

Basis divided laterally into at least two and probably three partial annuli (distal one may actually belong to exopod, Fig. 2E), distal two each armed with row of

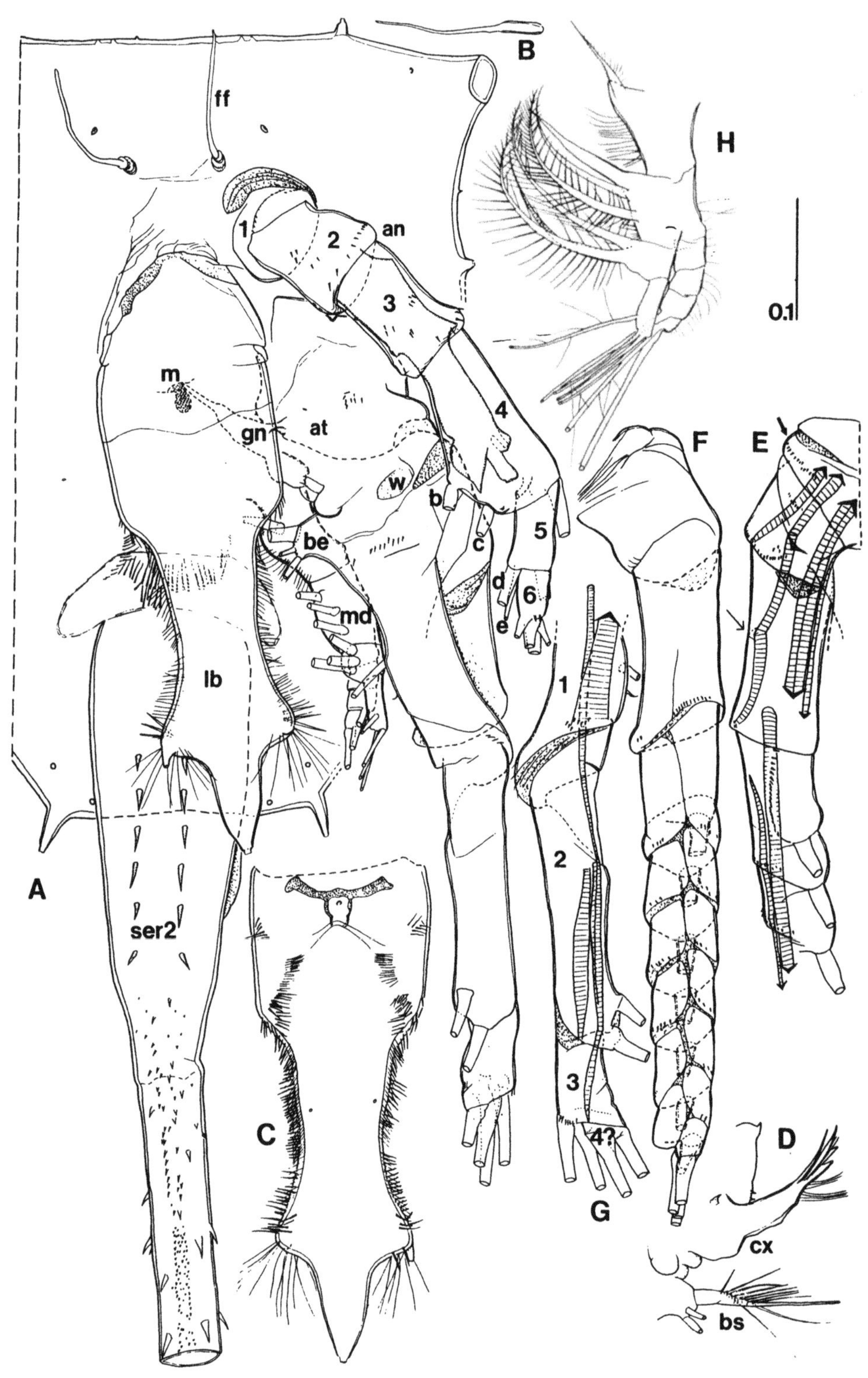

A
B
C
D
E
F
G
H
0.1
ff
an
1
2
3
4
5
6
m
gn
at
w
b
c
d
e
be
md
lb
ser2
4?
cx
bs

spinules. Medial margin of basis undivided, but produced into blunt process with three setae and few spinules (Fig. 2D). Most proximal seta hispid, i.e. spinelike, with 15-17 long, stiff bristles in two closely-spaced rows near base (rows of 6 and 9 or 7 and 10 bristles, respectively) and distal spinules (Fig. 2D). Other two setae simple. Proximal 'window' of thin cuticle on anterior face of basis (Fig. 2A-w). Medial part of basis more or less continuous with first endopodal segment, with some fine sutures perhaps marking boundary anteriorly.

Exopod nine-segmented, or 10-segmented if supposed distal part of basis actually belongs to exopod (Fig. 2E, F). First segment of nine short with distal marginal row of spinules; such spinules found on each segment except apical one, two subapical segments additionally armed with patches of spinules on dorsal side. Second segment twice as long as first, with longitudinal seam along at least distal half of dorsal side; site of muscle insertion and origin found laterally at midlength, suggesting derivation from two fused segments (Fig. 2E). Third segment a little shorter than second, with longitudinal row of spinules along proximal two-thirds of anterior side and longitudinal suture along dorsal side (Fig. 2F); this segment, formed by partial fusion of two ventrally separate segments, bearing two setae. Next five segments (4-8) each bearing one seta, their cuticular rings incomplete with seam along dorsal margin (Fig. 2F). Ninth, apical segment very small, with four long setae. Proximal seven setae long with very few (rarely none) long, laterally directed setules; perhaps most setules lost. Of apical setae, medial two simple, lateral two with sparse double row of long setules.

Endopod at least three-segmented, possibly four-segmented (Fig. 2A, G). First segment largely fused with basis, elongate, with long, simple seta and minute, simple seta on medial margin distal to midlength; patch of spinules on distal part of anterior face and row of spinules along ventral part of distal margin. Second segment longer than first, with two biserially plumose, distal setae. Articulation between these first two segments very peculiar, helical with a branch; proximal part of second segment thereby showing two secondary pseudo-articulations. Third segment shorter than others, with two long setae on lateral part of apex, outer one uniserially plumose, inner one biserially plumose. Medial part of apex forming minute, articulated region with muscle inserted at base and bearing two long, sparsely biserially plumose setae with long setules. This medial region possibly interpretable as fourth segment (Fig. 2G; see Discussion, Section 4.2.2).

Figure 2. Unidentified stage VI pedunculate cirripede nauplius from Osaka Bay. (A) Partial ventral view, emphasizing medial and left-hand structures with all setae cut short for clarity (full descriptions of all appendage setae provided in text), antennular segments numbered (1-6), some antennular setae labeled (b-e) and outline of cyprid antennule shown within naupliar antennule; (B) Frontal filament; (C) Labrum, dorsal view; (D) Medial armament of right antennal coxa and basis, ventral view; (E) Proximal half of right antennal exopod, lateral view showing musculature (thin arrow showing muscle insertion and origin indicative of segment fusion), assignment of arrowed partial annulus to basis or exopod uncertain; (F) Entire right antennal exopod, dorsal view, showing breaks in annuli; (G) Right antennal endopod with musculature, lateral view, segments numbered (1-4?); (H) Left mandible, lateral view, setulation of some elements omitted for clarity. Abbreviations: an, antennule; at, antenna; ba, basis; be, antennal basal endite; bs, basis; cx, coxa; ff, frontal filament; gn, antennal gnathobase; lb, labrum; m, mouth; md, mandible; ser2, series-2 spines (six pairs); w, 'window' of thin cuticle. Scale bar in mm.

3.8 *Mandible*

Small but not vestigial (Fig. 2A-md, H). Coxa with proximal knob, short, plumose seta, and medial longitudinal row of about 14 hairs. Basis with plumose spine (so-called hispid seta), two curved, plumose setae slightly longer than spine, and rows of hairs along medial and lateral margins. Exopod four-segmented with lateral row of hairs continuous with that on basis. First three exopodal segments equally short, fourth one smaller; one long, plumose seta on third segment, two setae on fourth, at least one of them plumose. Endopod two-segmented. First segment bearing two plumose setae like those on basis and, posterior to distal seta, a spine (so-called hispid seta) similar in length and appearance to the setae, but bearing longer, stiffer bristles. Second segment bearing one medium-long, plumose seta subapically and three long, simple, apical setae.

3.9 *Maxillule*

Maxillules absent, but pit or pore observed on left side and four fine hairs on right at expected site of maxillular rudiments, on anterior part of lateral side of trunk (not illustrated).

4 DISCUSSION

4.1 *Affinities*

The present nauplius can only be compared to previously described lepadid, lepadid-type, and poecilasmatid nauplii (see Moyse 1987), which are all characterized in later instars by a very long abdominal process and dorsal thoracic spine, gland spines on the margin of the cephalic shield, an elongated labrum, five- or six-segmented antennules, and very long antennae.

Two specifically lepadid features exhibited by the present form include hooks on the outer distal corners of the labrum (on one side only in the present specimen) and a multi-segmented mandibular exopod (four- or five-segmented in various lepadids (Dalley 1984; Moyse 1987), four-segmented here). At the level of Cirripedia, with the Ascothoracida employed as the outgroup (see detailed descriptions of ascothoracidan nauplii: Grygier 1990, 1993; Itô & Grygier 1990) and the application of an oligomerization criterion of polarity (see Huys & Boxshall 1991), both of these features are apomorphies in the present nauplius. Namely, ascothoracidan nauplii have a simple, tapered labrum and mandibular exopods with up to eight segments. However, the vestigial mandibular exopods in poecilasmatid nauplii are even more apomorphic than those of the present nauplius.

Despite the similarity of the labrum, a list of the differences from lepadid nauplii would be very long. Some of the more striking differences between the present nauplius and identified lepadids (Dalley 1984; Moyse 1987) include: lack of a dorsal spine on the cephalic shield and presence of a large, distal, conical protrusion of the labrum [although both states are found in the lepadid-type *'Nauplius hastatus'* (Chun 1896)], four rather than three pairs of lateral gland spines and many fewer small spines interspersed among the gland spines, no maxillular rudiments, a very restricted ventral 'setose region', and no series-1 or series-3 spines on the abdominal process. Differences in the appendages of the present specimen include: the a-seta and one b-seta missing from the antennules, no feathered setae and a near absence of fringes of fine hairs on the antenna, lower setal counts on the antennal basis and endopod, a two- rather than three-segmented mandibular endopod with many fewer setae, and setae only on the two distal segments rather than all segments of the mandibular exopod.

Specifically poecilasmatid features that the present nauplius shares include the lack of a dorsal spine on the cephalic shield (plesiomorphic; ascothoracidan nauplii have no such spine - see references cited two paragraphs above), a very restricted ventral 'setose region' (apomorphic; ventral setal arrays are well-developed in ascothoracidans), the lack of feathered setae on the antenna (polarity status unclear; feathered setae are present or not in different ascothoracidans), and similar setal armature of the proximal part of the antennal endopod (apomorphic; smaller number of elements than in planktotrophic ascothoracidans). However, the present larva also shows several differences from known poecilasmatids (Lang 1976, 1979; Colón-Urban et al. 1979): larger gland spines and, specifically, four lateral pairs of them; no large ventral spine partway along the abdominal process, the wide distal part of the labrum with hooked corners instead of a simple, tapered cone; the presence of the e-seta on the antennule; and a much less reduced mandible with a well-developed rather than vestigial exopod and a clearly two-segmented endopod.

In sum, the present nauplius cannot be assigned surely to either the Lepadidae or the Poecilasmatidae, although it has features in common with each and perhaps bears a closer superficial similarity to the latter. To identify it, it will be necessary either to capture more specimens and induce the cyprid larvae that molt from them to metamorphose, or to rear nauplius larvae from brooded eggs of the many other as yet unstudied pedunculate families. The first-stage nauplii of the hitherto unstudied Microlepadidae (Grygier & van den Spiegel, in prep.) shed no light on the present problem, however.

4.2 *Little-appreciated naupliar features of systematic value*

In recent studies of the planktonic nauplii of other groups of Thecostraca, such as the Ascothoracida (e.g. Grygier 1990, 1992; Itô & Grygier 1990) and Facetotecta (e.g. Grygier 1987b; Itô 1987, 1990), other characters than those usually mentioned in the literature on cirripede larvae have received much attention. For example, in the former two groups, the distribution of pores and setae and the pattern outlined by cuticular ridges on the dorsal shield have been described in detail. Indeed, in the Facetotecta, the ridge pattern has received the most attention of any character (see Fig. 3). In contrast, there has been almost no mention of dorsal ornamentation in cirripede nauplii, despite considerable attention to the marginal spines of the cephalic shield. While the present plankton-caught nauplius of unknown pedunculate affinities represents little more than a curiosity in its own right, it has proven to be a perfect specimen to examine for overlooked morphological features comparable to those in other thecostracans. Its use in this way can help to establish a better paradigm for descriptions of laboratory-reared nauplii in the Cirripedia.

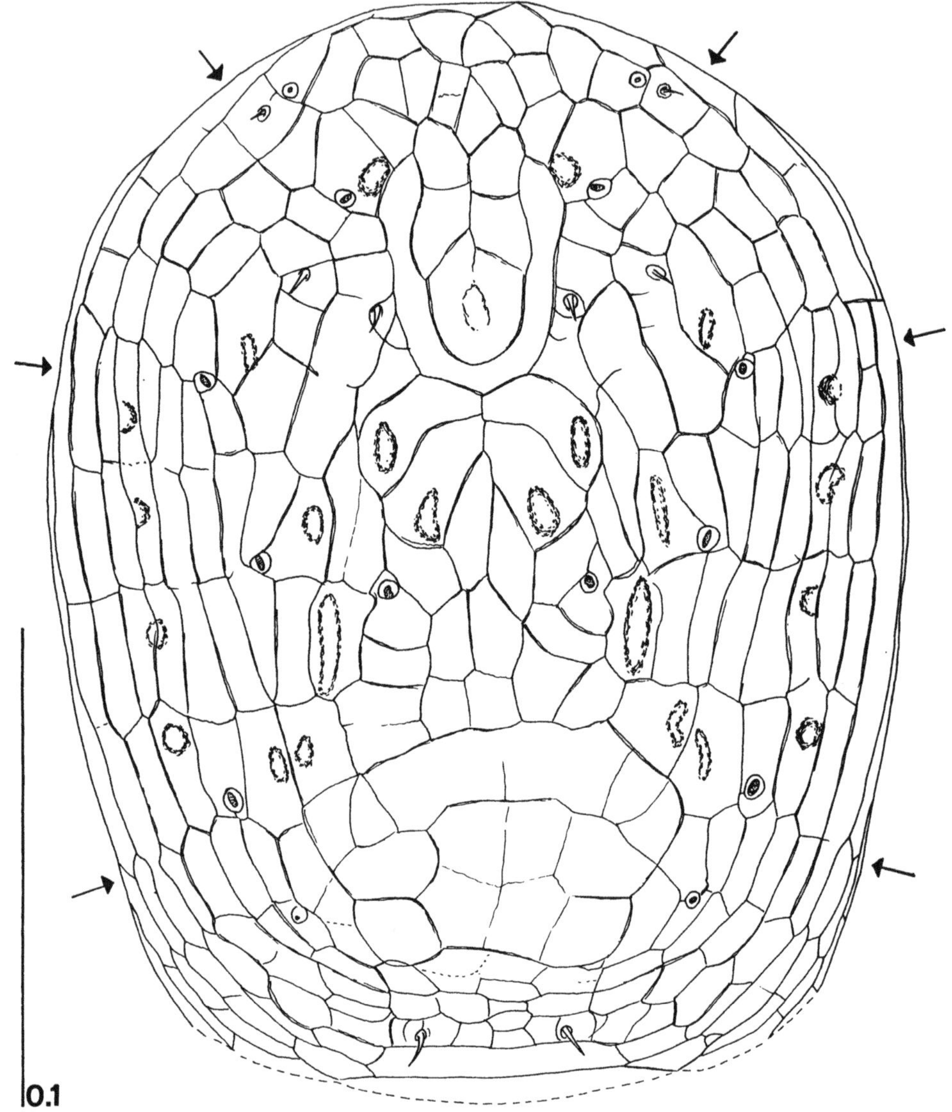

Figure 3. Exuvia of cephalic shield of fourth-stage nauplius of undescribed facetotectan from Tanabe Bay, Honshu, Japan, anterior end at top, showing cuticular ridges (meshwork between them omitted), setae (four pairs), dorsal pores (arrows point to locations of marginal pores on downturned brim of shield), and muscle attachment sites. Specimen housed at Seto Marine Biological Laboratory, Itô slide 31-X ≈ 4-XI-86(10)4Nceph. Scale bar in mm.

4.2.1 *Ornamentation of dorsal surface of cephalic shield*

Very little comparative information exists concerning the fine-scale dorsal ornamentation of the cephalic shield in cirriped nauplii. In nauplii of *Semibalanus balanoides* (L.), Mauchline (1977) observed about ten pores scattered evenly over the shield, but Walker & Lee (1976: Fig. 1a) used scanning electron microscopy to map 32 dorsal pores in sixth stage nauplii of that species (mostly near the margins, unlike the present case), and also two pairs of minute, mid-dorsal setae (rather than one pair). The present nauplius has many more pores than that, as do ascothoracidan nauplii, especially in later stages (e.g. Itô & Grygier 1990; Grygier 1992, 1993). In contrast, facetotectan nauplii have rather few pores, arranged in a stereotypical pattern (Fig. 3).

Anderson (1987) provided scanning electron micrographs showing concentric ridges near the shield margin in nauplii of *Ibla quadrivalvis* Cuvier, similar to those in some ascothoracidans (e.g. Itô & Grygier 1990; Grygier 1992), but the present nauplius has no such ridges. Such dorsal cuticular excresences have only been used once as a taxonomic feature: Kado & Hirano (1994) observed dorsal spinules, many more than the few bumps on the present nauplius, on NIV-VI nauplii of *Megabalanus volcano* (Pilsbry), but not in *M. rosa* (Pilsbry).

The maximum of two gland cells per marginal gland spine in the present specimen contrasts with the gland spines of petrarcid ascothoracidan metanauplii from the Red Sea, which bear up to six pores and associated gland cells each (Grygier 1993). Depictions of these glands in other cirripedes are rare; Chun (1896) showed one cell per gland spine in '*Nauplius hastatus.*' Accurate information on this point is essential in order to evaluate the hypothesis (Grygier 1990, 1993) that cirripede frontolateral horns are merely hypertrophied gland spines and homologous with an anterior pair of marginal gland spines in petrarcid ascothoracidan nauplii.

The pattern of dorsal muscle attachment sites in the nauplii has never been utilized in cirripede systematics in the manner of muscle scars on ostracod valves, for example. The dorsal attachment sites for muscles have been shown only in *Semibalanus balanoides* by Walley (1969), semidiagrammatically. There are more muscle attachment sites in a long row parallel to the midline, and fewer of them farther laterally, than in the present nauplius; also, there are only four rather than the present eight muscles in the outermost arc on each side. The naupliar musculature has also been depicted in some detail by Chun (1896) in '*Nauplius hastatus*' and by Anderson (1987) in *Ibla quadrivalvis*, but only in side view, so a direct comparison with Figure 1B is not possible except in terms of the number of muscles present. Dorsal muscle attachment sites have not been recorded in ascothoracidan nauplii. Those of facetotectan nauplii are rather few in number and generally fixed in position throughout the group (eight dorsal and four dorsolateral pairs shown in Fig. 3). Itô (1990) showed them in the successive stages of *Hansenocaris furcifera* Itô as oval marks that are clear of mesh-like ornamentation, but he did not recognize them as muscle attachments. Again, in the Facetotecta the muscles associated with these sites have not been identified.

4.2.2 *Segmentation of antennal rami*

It has been shown here for the first time in a cirripede nauplius that the segments of the antennal exopod are not simple, chitinous hoops. Rather, they are incomplete hoops with slightly offset ends, and the narrow zones of thin cuticle that separate the ends of each hoop are in line down the length of the ramus opposite the setal insertions. A similar arrangement exists in lecithotrophic nauplii of *Dendrogaster* in the Ascothoracida (Grygier, unpublished) and in the Upper Cambrian so-called branchiopod *Rehbachiella* (see Walossek 1993: Pl. 11:1). Thus, this kind of exopodal segmentation must be regarded as plesiomorphic within the Cirripedia, if it is not in fact universal therein.

The possibility of a tiny, fourth, apical segment bearing two setae on the antennal endopodite has never been suggested in any thecostracan nauplius. At most three segments have been reported. The structure in question in the present specimen has a muscle inserted at its base, as one would expect if it truly is the vestige of a once larger, fully articulated segment. Argument by muscle insertion is not foolproof, however. For example, the antennular claw of ascothoracidans can be retracted by muscular contraction, but that claw represents an embellished spine of the naupliar antennule, not a reduced segment (Grygier 1987a); in ascothoracidans, it is unknown whether the relevant muscles are already present in the naupliar antennule and what their function there is.

Nonetheless, Dahms (1991) suggested that a four-segmented antennal endopod may have been the ancestral condition in the nauplii of the Copepoda, even though the endopod is one- or two-segmented in extant forms. The one-segmented endopod of nauplius VI of *Longipedia minor* T. & A. Scott (Harpacticoida, Polyarthra) has three transverse spinule rows and a cluster of setae associated with the second row, all of which suggests a fusion of four ancestral segments. This would seem to agree with the four-segmented interpretation in the present cirripede nauplius. However, Dahms's (1991) analysis was flawed by his not taking into account the actual first endopodal segment (as expressed in the first copepodid larva), which, with its marker setae, is fused to the basis in copepod nauplii (Izawa 1987), including *Longipedia.* Therefore, Dahms's (1991) model actually implies a five-segmented ancestral endopod, for which there is no published example in any crustacean nauplius.

4.3 *Functional morphology*

In comparing the functional morphology of feeding in nauplii of *Lepas* and *Poecilasma*, Moyse (1987) noted a correlation between the antennal and mandibular armature in the two genera and ascribed a functional significance to this. *Lepas* has an antenna that serves in part as a fine filter owing to feathered setae and fringes of hairs, and the basis and endopod of the mandible have six strong feeding setae. Moyse (1987) supposed that the fine filter is used to capture small, flagellated phytoplankton, which are combed out and passed to the mouth by the mandibular setae. The coarse filter formed by the setae of the antennal exopod, on the other hand, captures larger food items (diatoms), which are transferred to the labrum by the antennular setae. In contrast, *Poecilasma* lacks the feathered setae and fringes on the antennae, and, correspondingly, has a vestigial mandible; it possesses only the machinery for capturing larger food items. The present nauplius, like *Poecilasma*, lacks the fine-meshed antennal filter, but yet has a *Lepas*-like mandibular armature. Therefore, one can assume that the strong setae of the basis and endopod in the mandible have another function, in *Lepas* as well as the present nauplius, besides the transfer of food from an antennal fine filter. Perhaps they assist in shoving food under the distal half of the labrum and in preventing the loss of unswallowed food items before they reach the rather distant mouth.

ACKNOWLEDGMENTS

This work was carried out at the Seto Marine Biological Laboratory, Kyoto University, at the invitation of its director, Dr. Eiji Harada, with the support of a postdoctoral fellowship from the Japan Society for the Promotion of Science under the auspices of the U.S. National Science Foundation Japan Program. I thank Dr. Susumu Ohtsuka for providing the specimen and for hosting me later at the Hiroshima University Fisheries Laboratory, where the production of the manuscript was supported by an international fellowship from the Dan Charitable Trust Fund for Research in the Biological Sciences. I also thank Drs. John Moyse, Ryusuke Kado, and William A. Newman for their opinions on the likely identity of the present nauplius and/or for supplying literature.

REFERENCES

Anderson, D.T. 1987. The larval musculature of the barnacle *Ibla quadrivalvis* Cuvier (Cirripedia, Lepadomorpha). *Proc. R. Soc. Biol. Sci.* B231: 313-338.

Bainbridge, V. & J. Roskell 1966. A re-description of the larvae of *Lepas fascicularis* Ellis and Solander with observations on the distribution of *Lepas* nauplii in the north-eastern Atlantic. In H. Barnes (ed.), *Some Contemporary Studies in Marine Science*. London: George Allen & Unwin.

Chun, C. 1896. Atlantis. Biologische Studien über pelagische Organismen. III. Die Nauplien der Lepaden nebst Bemerkungen über das Schwebvermögen der pelagisch lebenden Crustaceen. *Zoologica (Stuttgart)* 7: 77-106.

Colón-Urban, R., P.J. Cheung, G.D. Ruggieri & R.F. Nigrelli 1979. Observations on the development and maintenance of the deep sea barnacle *Octolasmis aymonini geryonophila* (Pilsbry). *Inter. J. Invert. Reprod.* 1: 245-252.

Dahms, H.-U. 1991. Usefulness of postembryonic characters for phylogenetic reconstruction in Harpacticoida (Crustacea, Copepoda). *Bull. Plankton Soc. Japan* Special Vol.: 87-104.

Dalley, R. 1984. The larval stages of the oceanic, pedunculate barnacle *Conchoderma auritum* (L.) (Cirripedia, Thoracica). *Crustaceana* 46: 39-54.

Grygier, M.J. 1987a. Nauplii, antennular ontogeny, and the position of the Ascothoracida within the Maxillopoda. *J. Crustacean Biol.* 7: 87-104.

Grygier, M.J. 1987b. New records, external and internal anatomy, and systematic position of Hansen's y-larvae (Crustacea: Maxillopoda: Facetotecta). *Sarsia* 72: 261-278.

Grygier, M.J. 1990. Early planktonic nauplii of *Baccalaureus* and *Zibrowia* (Crustacea: Ascothoracida) from Okinawa, Japan. *Galaxea* 8: 321-337.

Grygier, M.J. 1992. Laboratory rearing of ascothoracidan nauplii (Crustacea: Maxillopoda) from plankton at Okinawa, Japan. *Publ. Seto Mar. Biol. Lab.* 35: 235-251.

Grygier, M.J. 1993. Late planktonic naupliar development of an ascothoracidan crustacean (?Petrarcidae) in the Red Sea and a comparison to the Cirripedia. *Contr. Sci.* 437: 1-14.

Grygier, M.J. 1994. Developmental patterns and hypotheses of homology in the antennules of thecostracan nauplius larvae (Crustacea). *Acta Zool. (Stockh.)* 75: 219-234.

Huys, R. & G.A. Boxshall 1991. *Copepod evolution*. London: The Ray Society.

Itô, T. 1987. Proposal of new terminology for the morphology of nauplius y (Crustacea: Maxillopoda: Facetotecta), with provisional designation of four naupliar types from Japan. *Zool. Sci.* 4: 913-918.

Itô, T. 1990. Naupliar development of *Hansenocaris furcifera* Itô (Crustacea: Maxillopoda: Facetotecta) from Tanabe Bay, Japan. *Publ. Seto Mar. Biol. Lab.* 34: 201-224.

Itô, T. & M.J. Grygier 1990. Description and complete larval development of a new species of

Baccalaureus (Crustacea: Ascothoracida) parasitic in a zoanthid from Tanabe Bay, Honshu, Japan. *Zool. Sci.* 7: 485-515.

Izawa, K. 1987. Studies on the phylogenetic implications of ontogenetic features in the poecilostome nauplii (Copepoda: Cyclopoida). *Publ. Seto Mar. Biol. Lab.* 32: 151-217.

Kado, R. & R. Hirano 1994. Larval development of two Japanese megabalanine barnacles, *Megabalanus volcano* (Pilsbry) and *Megabalanus rosa* (Pilsbry) (Cirripedia, Balanidae), reared in the laboratory. *J. Exp. Mar. Biol. Ecol.* 175: 17-41.

Lang, W.H. 1976. The larval development and metamorphosis of the pedunculate barnacle *Octolasmis mülleri* (Coker, 1902) reared in the laboratory. *Biol. Bull.* 150: 255-267.

Lang, W.H. 1979. Larval development of shallow water barnacles of the Carolinas (Cirripedia: Thoracica) with keys to naupliar stages. *NOAA Tech. Rep. Natl. Mar. Fish. Serv. Circ.* 421: 1-39.

Mauchline J. 1977. The integumental sensilla and glands of pelagic Crustacea. *J. Mar. Biol. Ass. UK* 57: 973-994.

Moyse, J. 1987. Larvae of lepadomorph barnacles. *Crustacean Issues* 5: 329-362.

Newman, W.A. 1987. Evolution of cirripedes and their major groups. *Crustacean Issues* 5: 3-42.

Rose, M. 1929. *Nauplius pelagicus* nov. forme larvaire intéressante du plankton de la Baie d'Alger. *Trav. Publ. Stn. Aquicult. Pêche Castiglione* 1929: 109-130.

Sandison, E.E. 1950. *Nauplius longispinosus*, a new larval form of barnacle. *Trans. R. Soc. S. Afr.* 32: 301-313.

Walley, L.J. 1969. Studies on the larval development and metamorphosis of the cypris larva of *Balanus balanoides. Phil. Trans. R. Soc. London* (B)256: 237-280.

Walker, G. & V.E. Lee 1976. Surface structures and sense organs of the cypris larva of the barnacle, *Balanus balanoides. J. Zool., Lond.* 178: 161-172.

Walossek, D. 1993. The Upper Cambrian *Rehbachiella* and the phylogeny of Branchiopoda and Crustacea. *Fossils Strata* 32: 1-202.

Comparative morphology of the thoracican cyprid larvae: Studies of the carapace

Aleksey S. Elfimov
Department of Invertebrate Zoology, Moscow State University, Moscow, Russia

ABSTRACT

Cyprid morphology undoubtedly can provide valuable material for studies on the phylogeny of Cirripedia. In the Cirripedia Thoracica, among the features that may be useful for differentiation and systematics of cyprids are as follows: larval size, colour, carapace shape and ornamentation, and structure of appendages. This study reveals a multitude of morphological traits on the cyprid carapace within the Thoracica. However, the number of cypris species studied with the use of SEM seems insufficient for any comprehensive cladistic analysis. So, a variety of the carapace features and an *a priori* analysis of their polarity primarily are discussed here. Because of the intraspecific variability, cyprid size may be used mainly for a rough identification of larvae and very rarely for their taxonomy. Some features of carapace shape can be used for studies on cyprid systematics and for phylogenetic reconstructions. Carapace surface pattern and the armament of various cuticular structures (lattice and wheel organs, pores, pits, setae) undoubtedly can be used in studying cyprid evolution and in the phylogeny of the Cirripedia. Cyprids of many thoracican species have a distinct surface ornamentation on the carapace. Reticulated surface pattern seems to be the primitive state of the ornamentation in thoracican cyprids, and other patterns (such as honeycombed, ribbed, ridged, 'goose flesh', pitted, smooth) are advanced states of this character. Smooth carapace can be considered most advanced state of this character in thoracican cyprids.

1 INTRODUCTION

The comparative morphology of cyprids should form a significant part of the study of 'cirriped larval taxonomy' and phylogeny of Cirripedia. At present the cyprid larvae of approximately 80 species of Cirripedia Thoracica have been studied or illustrated. However, for many of them the descriptions are lacking in detail and only information on the size and general form of the cyprids has been presented. Most of these descriptions are not very helpful in differentiating species and in facilitating comparative study because of interspecific similarity and intraspecific variability in shape and size of cyprids. Standing (1980) studied seven species of barnacle cyprids

Table 1. Thoracican cyprids studied with SEM.

Pedunculata:	
Family Lepadidae	
Lepas sp.	Jensen et al. 1994
Lepas anserifera	Elfimov 1989
L. anatifera	Elfimov 1987a, 1989
L. australis	Elfimov 1989; Jensen et al. 1994
L.pectinata	Elfimov 1987a, 1989; Jensen et al. 1994
L. hillii	Elfimov 1987a, 1989
L. testudinata	Elfimov 1989
Dosima fascicularis	Elfimov 1987a, 1989; Al-Yahya 1991; Jensen et al. 1994
Conchoderma virgatum	Elfimov 1989
Family Poecilasmatidae	
Octolasmis lowei	Elfimov 1989
O. warwicki	Elfimov 1989
Family Heteralepadidae	
Heteralepas mystacophora	Elfimov 1984, 1986, 1989
H. microstoma	Elfimov 1989
Family Scalpellidae	
Lithotrya dorsalis	Dineen 1987
Capitulum mitella	Jensen et al. 1994
Pollicipes pollicipes	Al-Yahya 1991; Jensen et al. 1994
Scalpellum scalpellum	Svane 1986; Jensen et al. 1994
Ornatoscalpellum gibberum	Elfimov 1989
O. stroemii	Elfimov 1989; Jensen et al. 1994
Tarasovium cornutum	Elfimov 1989
Weltnerium scoresbyi	Elfimov 1989
Amigdoscalpellum vitreum	Elfimov 1989
Trianguloscalpellum gaussi	Elfimov 1989
T. liberum	Elfimov 1989
T. eugeniae	Elfimov 1989
Arcoscalpellum sergi	Elfimov 1989
Sessilia: Verrucomorpha:	
Family Verrucidae	
Verruca stroemia	Elfimov 1989; Al-Yahya 1991; Jensen et al. 1994
Sessilia: Balanomorpha:	
Family Chthamalidae	
Chthamalus montagui	Al-Yahya 1991; Jensen et al. 1994
C. stellatus	Al-Yahya 1991; Jensen et al. 1994
C. dalli	Elfimov 1989
Family Archaeobalanidae	
Elminius modestus	Al-Yahya 1991; Glenner & Hoeg 1993; Jensen et al. 1994
Solidobalanus hesperius	Elfimov 1989
Semibalanus balanoides	Walker & Lee 1976; Walker et al. 1987; Jensen et al. 1994
S. cariosus	Elfimov 1989
Acasta spongites	Al-Yahya 1991, Jensen et al. 1994
Megatrema (Boscia) anglica	Al-Yahya 1991, Jensen et al. 1994
Family Balanidae	
Balanus improvisus	Elfimov 1989, Glenner & Hoeg 1993, Jensen et al. 1994
B. crenatus	Elfimov 1989
B. rostratus	Elfimov 1989
B. amphitrite	Clare 1993

with light microscopy and concluded that the shape of the carapace is a more reliable indicator of species than size (particularly in regards some details of the lateral view – curved or broken dorsal margins, narrowed or pointed anterior and posterior ends, etc.). The structure of appendages – antennules, thoracic limbs, and caudal rami – also can be used in studies on comparative morphology of cyprids (Elfimov 1987a, 1989).

Standing (1980) also described the carapace surface of cyprids and further concluded that carapace sculpturing may be useful for cyprid differentiation and taxonomy. Some features of carapace surface (visible under the light microscope) were mentioned for other thoracican species by earlier authors – for the cyprids of *Lepas australis* and *Chirona hameri* (Darwin 1854), *Lepas pectinata* and *Dosima fascicularis* (Darwin 1854, Claus 1869), *Poecilasma kaempferi* (Hoffendahl 1904), *Balanus virgatum* (Kruger 1940), *Ibla idiotica* (Batham 1946a), *I.cumingi* (Karande 1974), *Scalpellum scalpellum* (Kaufman 1965), *Megabalanus psittacus* (Stefoni & Contreras 1979), *Balanus glandula* and *B.cariosus* (Strathman & Branscomb 1979), *Catomerus polymerus* and *Chamaesipho tasmanica* (Egan & Anderson 1989), and some others. Undoubtedly, the use of the scanning electron microscope (SEM) provides a valuable tool in extending our knowledge of the external morphological features of the cyprid larva (Lang 1979; Standing 1980).

Till recent times the use of SEM in studying thoracican cyprids was limited. Nott (1969) was among the first to use SEM for investigation of the cyprid antennule of *Semibalanus balanoides*. Then Walker & Lee (1976) studied with SEM the surface structures of the *S.balanoides* cypris. Svane (1986) showed a settling *Scalpellum scalpellum* cyprid at low magnification of SEM. Elfimov & Bogdanov (1984) used SEM for studying the carapace of some thoracican cyprids, and Elfimov (1986) described surface ornamentation for cypris larvae of *Heteralepas mystacophora* and briefly for lepadid cyprids (Elfimov 1984,1987a). Dineen (1987) made use of SEM in studying the carapace surface of *Lithotrya dorsalis* cyprids but considered that it was premature to speculate about the reliability of cypris carapace sculpturing in differentiating species until more SEM descriptions were available. During subsequent work, Elfimov (1989) using the SEM studied cyprid larvae of 28 species of Thoracica. The cyprids of 16 thoracican species were investigated and described by Al-Yahya (1991) and Jensen et al. (1994). Thus, at the present time the cyprids of 40 species of Thoracica have been studied with SEM (Table 1).

This study reveals a multitude of morphological traits of cyprid carapace within the Cirripedia Thoracica. Character state polarity and evolution of carapace features must be decided by a formal cladistic analysis. Obviously, however, an all out cladistic analysis must include characters from nauplii, cyprids and adults. Nevertheless, until this can be done I will attempt to suggest some hypotheses concerning the evolution of carapace features that could be subject to confirmation or falsification in such an analysis as outlined above.

2 MORPHOLOGICAL DIVERSITY OF THE CYPRID LARVAE

2.1 *Cyprid size*

Usually, the carapace length and height are used as size characteristics of the cyprid.

Among the Thoracica these vary considerably – from 440-480 and 210-260 μm, respectively, in *Chamaesipho columna* cyprids (Barker 1976) to 2600 and 1100 μm in *Lepas australis* (Jensen et al. 1994; Elfimov 1989). From the descriptive point of view it is convenient, probably, to distinguish three provisional size groups of larvae – small cyprids (with carapace length up to 600 μm), medium (length up to 1200 μm) and large (length above 1200 μm). The suborders Lepadomorpha and Balanomorpha are represented by larvae of all size groups, although in the former there are mostly large and medium size cyprids (with the greatest variability in the family Scalpellidae), and balanomorph larvae are of medium and small sizes.

Among the lepadomorph species the largest cyprids are in family Lepadidae – the length of *Lepas, Dosima* and *Conchoderma* larvae is more than 1100 μm and reaches 2110 μm in *L.testudinata* cyprids (Elfimov 1989) and 2600 μm in *L.australis* (Jensen et al. 1994). Most cyprid larvae in the families Poecilasmatidae and Heteralepadidae also have a large size (900-1400 μm), whereas all known Iblidae cyprids are small (450-600 μm). The length ranges are the widest in family Scalpellidae – from 420 μm in *Pollicipes polymerus* (Standing 1980) to 2260 μm in *Amigdoscalpellum vitreum* (Elfimov 1989).

The largest balanomorph cyprids are among the Archaeobalanidae – *Chirona hameri* (1454 μm length, Crisp 1962b), *Semibalanus balanoides* (up to 1385 μm, Crisp 1962a), *S.cariosus* (up to 1240 μm, Standing 1980). Cyprids of the Balanidae are smaller. Their length is usually in the range 550-900 μm. And the smallest larvae are in family Chthamalidae – 430-630 μm in length.

Only the larva of *Verruca stroemia* is known from suborder Verrucomorpha (its length is about 550 μm), so it is problematic to speculate about cyprids size in this suborder.

Of course, the size of a cypris carapace cannot be a reliable feature in differentiating species of cyprids and should be used cautiously in comparative cyprid study because it varies considerably depending on the environmental conditions (Pyefinch 1948; Crisp 1962a; Standing 1980).

The length/height ratio of larva is the more constant characteristic of species since this indicates the general form of the shell (elongate or rounded, 'short' or 'long' carapace, etc.). The length/height ratio of thoracican cyprids occurs within the limits of 1.66-2.73 (*Dosima fascicularis – Verruca stroemia*).

2.2 *Carapace shape*

The most common thing discussed in cyprid descriptions as a more reliable indicator of species is the lateral view of the carapace (carapace profile), even though the peculiarities of a dorsal view sometimes may also be useful for differentiating cyprids (Standing 1980). Larvae of different thoracican groups mostly differ in carapace shape and usually it is possible to discriminate the shape features common for larvae of a single taxonomic group (at least, at the suborder level).

The carapace profile (Fig. 1) of most lepadomorph larvae is characterized by the absence of 'angularity' and points (except for the posterior end). The frontal edge of the shell is straight or rounded, the dorsal margin evenly convex, and the ventral one is straight or slightly sinuosus. In dorsal view, the frontal end of the carapace as a rule is narrowly or broadly rounded; the posterior end is narrowly rounded or pointed.

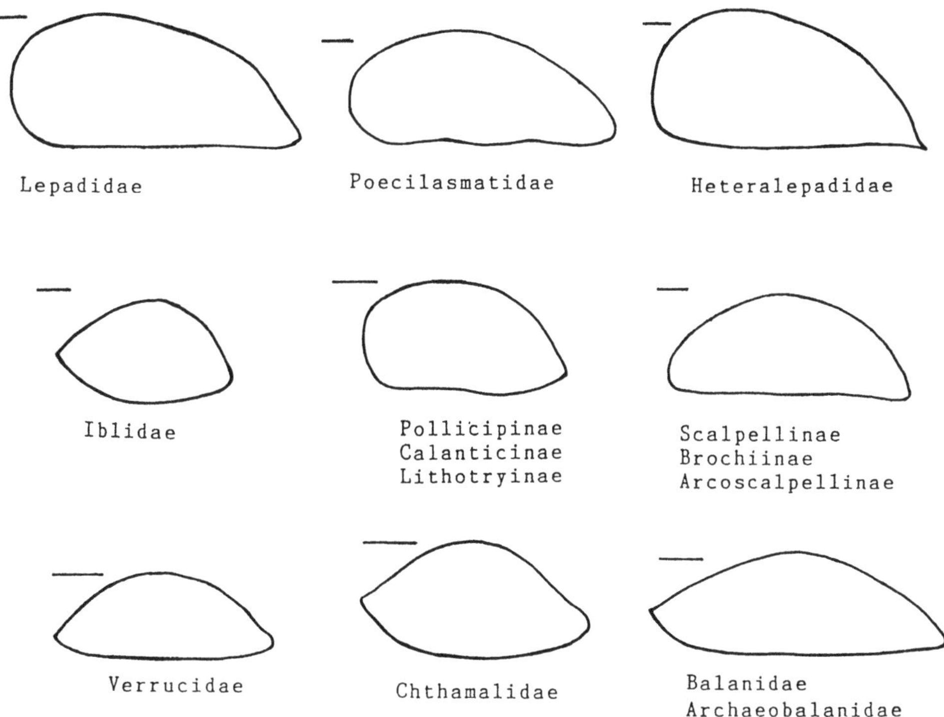

Figure 1. Generalized carapace profiles of thoracican cyprid larvae. Scale 100 μm.

In Lepadidae the cypris larvae of *Lepas* and *Conchoderma* differ in carapace form. The shell profile of *Conchoderma* cyprids has an almost straight frontal edge, a slightly convex dorsal margin, and a narrowly rounded posterior end the apical point of which is placed on the mid-longitudinal line of the carapace profile. In a lateral view of *Lepas* cyprids, the frontal edge is broadly rounded, the convex dorsal margin is lightly 'broken' in its posterior part, and the posterior end is sharpened.

In contrast to lepadid cyprids, carapace profiles of the poecilasmatid larvae have their maximum height in the middle part of shell rather than in the anterior (as in Lepadidae); the ventral margin is distinctly curved.

Carapace profiles of lepadid and heteralepadid cyprids are similar. However, the latter lack the 'angularity' of the posterio-dorsal edge, and the dorsal projection has its maximum width in the anterior part (not in the middle as in lepadid larvae) and tapers toward the posterior end.

In the family Scalpellidae, two groups of cyprids can be recognized from the carapace shape. First, in side view, the shell is relatively short and high; the anterior margin is broadly rounded, the posterior end is narrowly rounded; and the strongly convex dorsal edge reaches its highest point in the anterior half of the carapace. This group comprises cypris larvae of *Calantica spinosa* (Batham 1946b), *Capitulum mitella* (Yasugi 1937; Jensen et al. 1994), *Pollicipes polymerus* (Lang 1979; Standing 1980), *Pollicipes pollicipes* (Al-Yahya 1991), *Scalpellum scalpellum* (Kaufman 1965), *Scalpellum stearnsii* (Hoek 1907), *Barbaroscalpellum cochlearum* (Hiro 1933), *Lithotrya dorsalis* (Dineen 1987). Second, in side view, the shell is elongated, with the same rounding of the anterior and posterior ends; the 'apex' of evenly convex dorsal margin is located in the middle part of shell. This group includes cyprids of *Ornatosalpellum gibberum, O.stroemii* and most of the known larvae of the Arcoscalpellinae (Elfimov 1989).

The cypris of *Verruca stroemia* has an elongated carapace (in lateral view) with narrowly rounded or slightly pointed anterior and posterior ends.

Cyprid larvae of the suborder Balanomorpha also demonstrate certain common features in carapace shape – the shell profile is elongated (except in the family Chthamalidae), the ventral edge is straight or slightly convex, and the anterior and posterior ends are narrowly rounded or pointed (in lateral and dorsal views).

Chthamalid cyprids have a small short carapace with convex dorsal and ventral margins in lateral view; the highest point of the dorsal curve is located in the middle part of carapace. In *Chthamalus* cyprids, both the anterior and posterior ends of the carapace are narrowly rounded (or the anterior one is pointed), whereas in *Chamaesipho* larvae the anterior end is broadly rounded.

The carapace of coronuloid larvae is more elongated than in chthamalid cyprids, but its anterior end (in profile) is usually also pointed. The posterior one is narrowly rounded or pointed; the ventral margin is slightly convex or straight.

All known cyprids of the family Archaeobalanidae and most in the family Balanidae have a carapace with anterior and posterior ends narrowly rounded or pointed (in lateral view). However, there are two other groups of balanid cyprids with other shape features. The first group includes the cyprids with the anterior and posterior ends both broadly rounded (in lateral view) – *Balanus amphitrite variegatus* and *B. tintinnabulum* (Daniel 1958), *B. venustus* (Lang 1979). The second group consists of the species whose cyprids have a carapace with the anterior end broadly rounded and posterior one narrowly rounded (in profile) – *Balanus nubilis* (Standing 1980), *B. trigonus* (Barker 1976; Lang 1979), *B. improvisus* (Doochin 1951; Jones & Crisp 1954; Kuhl 1968; Standing 1980; Korn 1991).

So, the variety in carapace shape of the cyprid larvae enables some differentiation of species among larvae, especially during faunistic investigations of the relatively small areas inhabited by a few species of barnacles. However, in taxonomic and more detailed systematic research the shape characteristic should be used carefully (particularly, at the species and genus levels). The cuticular valves of the carapace are not so rigid as, for instance, in ostracodes, and some deformations of the shell that can alter its appearance can occur (but the specific features of shape, like angularities or points, remain). The shape of the carapace can vary in relation to the age of the larva (Kaufman 1965), the temperature at which the cyprid is reared (Lang 1979), and obviously also some other environmental factors and methods of specimen preservation and fixation. Such deformations don't affect the surface structures of the carapace.

2.3 *Surface ornamentation*

Some aspects of the morphological diversity, classification, and genesis of the thoracican cyprids sculpture and its significance in 'larval systematics' and phylogeny of the Cirripedia have previously been studied (Elfimov 1987a, 1989; Al-Yahya 1991; Jensen et al. 1994). Cyprids from eight thoracican families have been investi-

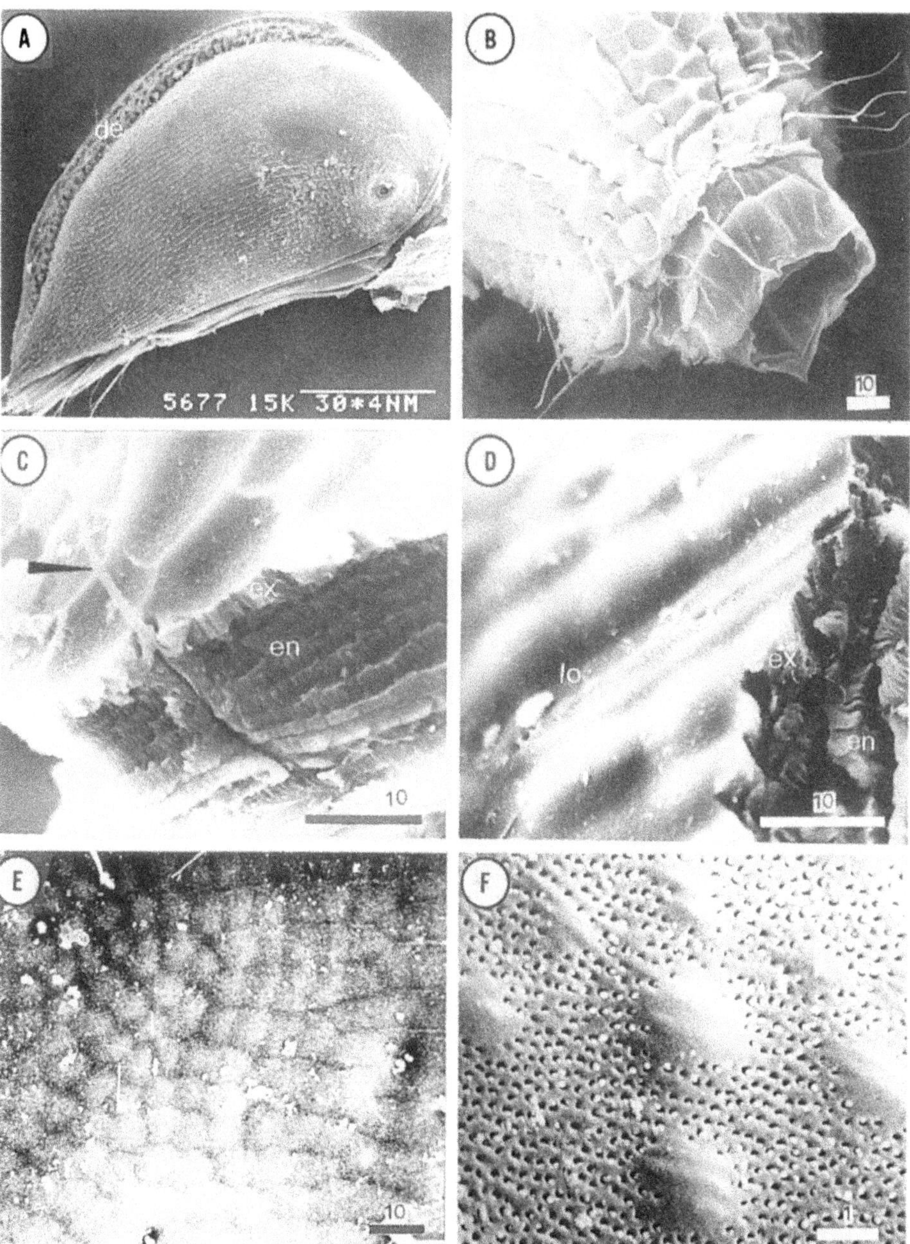

Plate 1. Cyprid carapace and elements of its ornamentation. (A-C) *Lepas pectinata.* (A) Whole larva, lateral view. Elements of macrosculpture – dorsal edge (*de*) and frontal horn (*fh*). Scale 300 µm; (B) Frontal horn; (C) Broken cuticle of the carapace. Elements of mesosculpture – polygonal meshes and seta (indicated); (D) *Dosima fascicularis.* Carapace cuticle broken across a lattice organ (*lo*); surface pattern formed by rounded knobs; (E) *Lepas anatifera.* Polygonal marking visible at high voltage of the SEM on smooth surface of cyprid cuticle; each mesh of the marking presumably corresponds to one cell of hypodermis producing a cuticle; (F) *Heteralepas microstoma.* Carapace sculptured by small nodules (mesosculpture) with slight ribs on their surface (microsculpture). *de* – dorsal ridge; *en* – endocuticle; *ex* – exocuticle; *fh* – frontal horn. Scale µm.

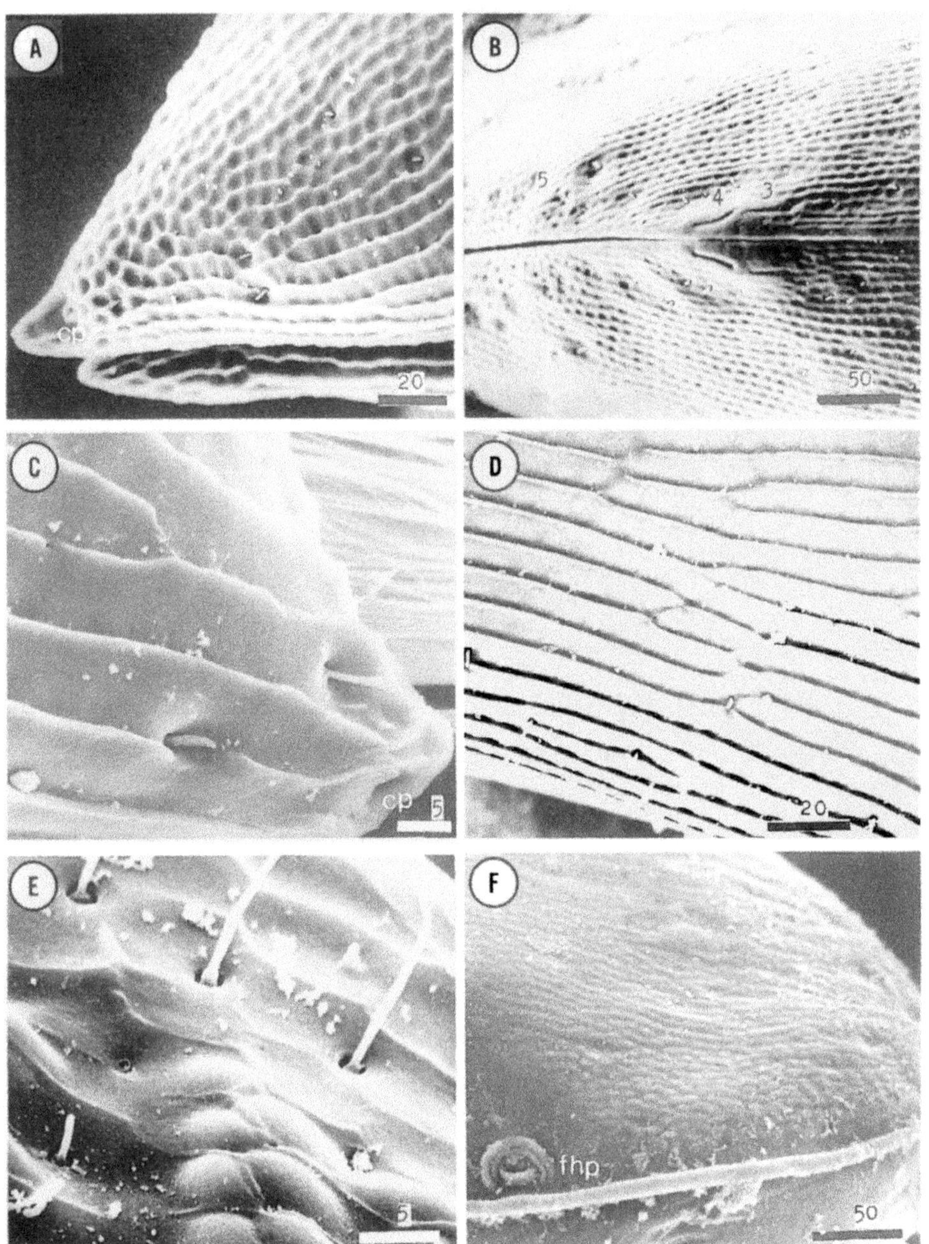

Plate 2. Surface patterns in *Lepas* and *Octolasmis* cyprids. (A) *L.anserifera.* Posterior end of carapace; (B) *L.anatifera.* Postero-dorsal area. Lattice organs (3-5th pairs) indicated; (C) *L.anatifera.* Posterior end of left valve with caudal pore (*cp*) and two setae; (D) *L.hillii.* Ventral part of valve; (E) *L.pectinata.* Surface sculpture on the dorsal edge (reticulation, knobs, setae, pores); (F) *O.lowei.* Anterior part of right valve. *cp* – caudal pore; *fhp* – frontal horn pore. Scale μm.

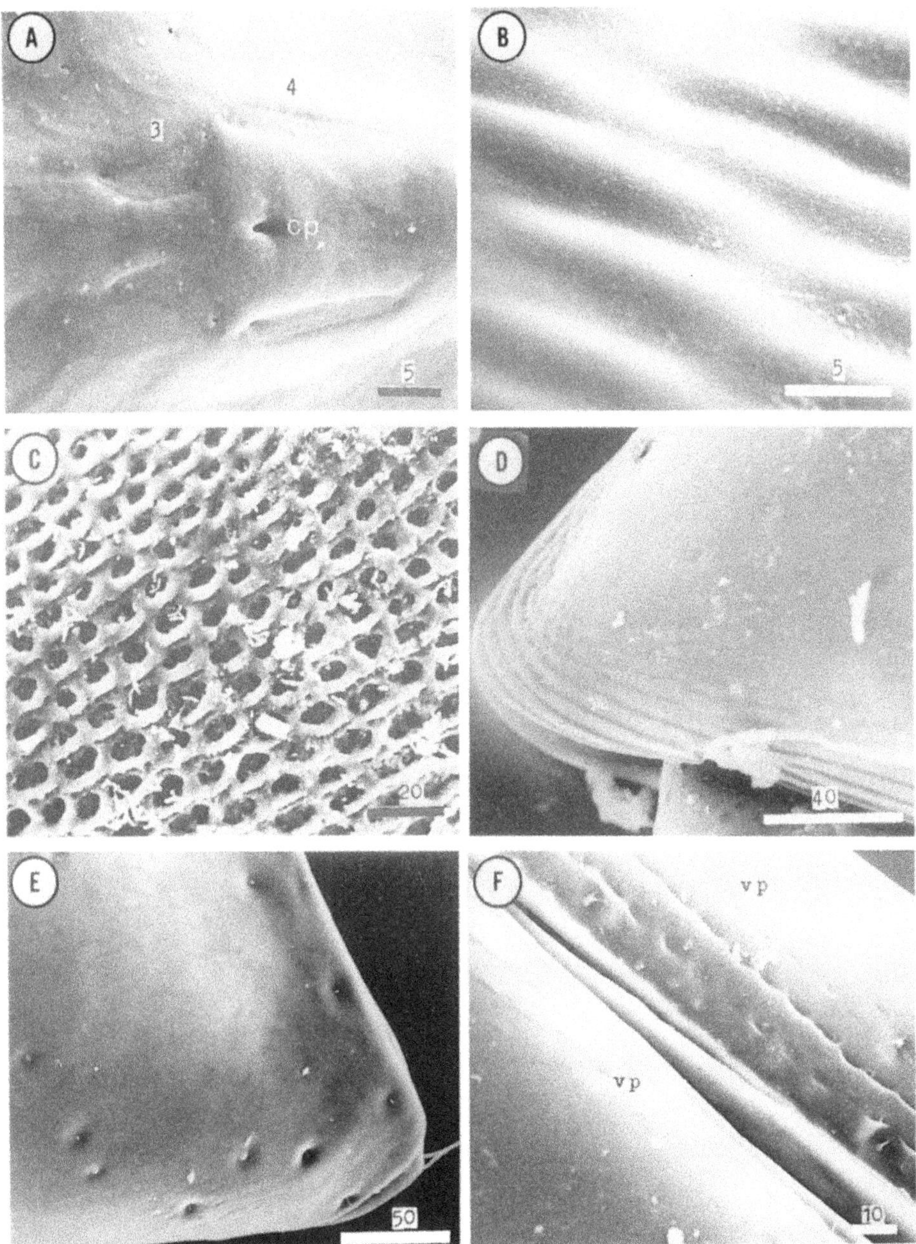

Plate 3. Carapace sculpturing in scalpellid and balanomorph cyprids. (A) *Trianguloscalpellum liberum.* Dorsal area with lattice organs of 3d and 4th pairs. Note a fine reticulation on the carapace. Posterior end right; (B) *Weltnerium scoresbyi.* Lateral area; (C) *Trianguloscalpellum gaussi.* Lateral area; (D) *Chthamalus dalli.* Posterior end of right valve; (E) *Semibalanus cariosus.* Posterior end; (F) *Balanus improvisus.* Ventral surface of valves. *cp* – central pore; *vp* – ventral pore. Scale μm.

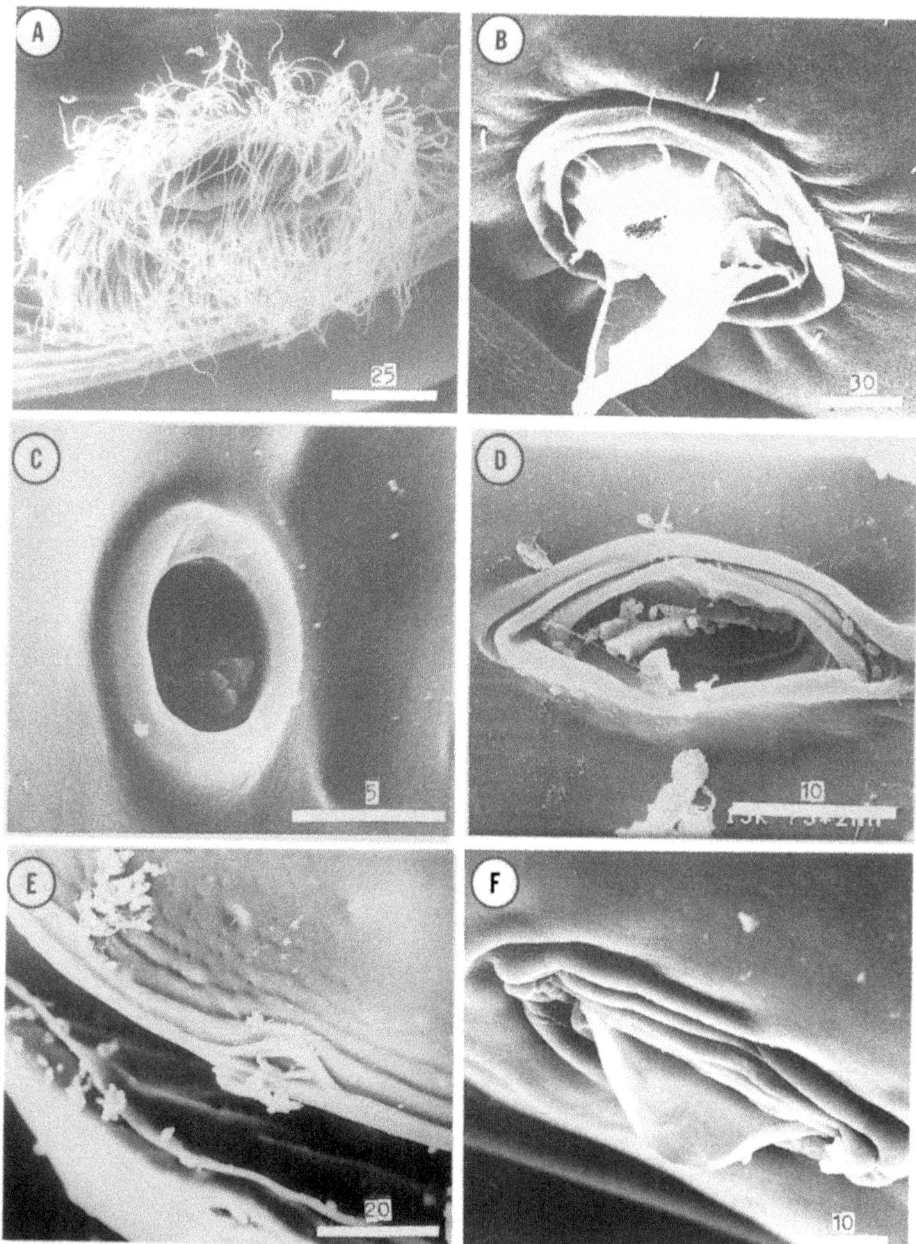

Plate 4. Frontal horn pores of cyprid larvae. (A) *Lepas sp.*(presumably *L.anatifera,* from plankton sample). Right pore; (B) *Conchoderma virgatum.* Left pore; (C) *Trianguloscalpellum liberum.* Right pore; (D) *Verruca stroemia.* Left pore; (E) *Chthamalus dalli.* Anterior end of the carapace right; (F) *Semibalanus cariosus*. Left pore. Scale μm.

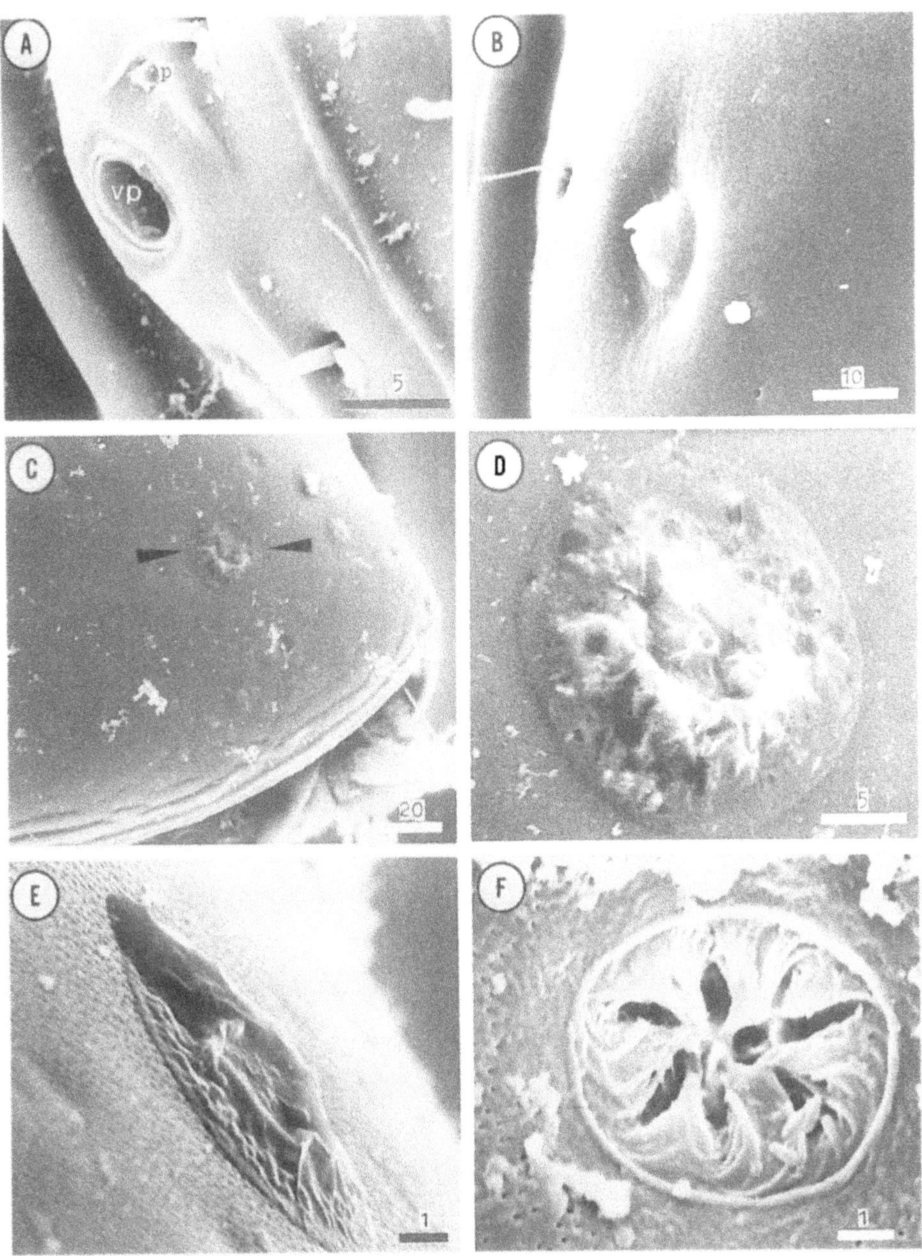

Plate 5. Ventral pores, papillae and 'weel organs' on the cyprid carapace. (A) *Lepas anatifera.* First ventral pore (vp) and papilla (p) on the anterio-ventral part of left valve; (B) *Conchoderma virgatum.* Fronto-ventral cuticular papilla; (C-E) 'Weel organs' in the cyprids of *Chthamalus dalli,* (C) Anterior part of right valve; (D) Left anterior 'wheel organ'; (E) Left anterior 'wheel organ' in oblique view; (F) *Heteralepas mystacophora.* 'Wheel organ' on the posterio-ventral part of the valve. Scale μm.

gated (Table 1). This has allowed me to delineate the variety of the carapace ornamentation within the suborder Thoracica (Elfimov 1989). Among the species that have been investigated were cypris larvae from three genera of the family Lepadidae and six species of the genus *Lepas*. This has consequently allowed us to speculate about probable evolutionary trends in transformations of sculpture within some small taxonomic groups. Also, many specimens of *Dosima fascicularis* and *Heteralepas mystacophora* cyprids were studied in detail. So it was possible to estimate the degree of the variability in ornamentation among the cyprids of a single species.

Obviously, it is premature to discuss the variability, origin and evolution of cyprid carapace ornamentation in detail, since the number of thoracican species that have been studied from this point is not as many as compared, for instance, with the ostracodes. For the latter, the various aspects of shell morphology and sculpturing have been analysed much more thoroughly (e.g. see Kessling 1951; Hartmann 1966, 1982; Benson 1974, 1981; Liebau 1975, 1977, 1982; Shornikov 1981, 1988, 1989; Shornikov & Michailova 1990).

Different criteria are used in classification of surface structures in ostracodes, e.g., location on the shell, size and shape of the elements, their origin and function, and relation with various organs. Since not many species of barnacle cyprids have been investigated with SEM, Elfimov (1989) has sorted the cyprid surface structures mainly from a descriptive point of view. In these descriptive terms, three levels of the sculpturing were distinguished. First is the *macrosculpture* – the structures that alter the carapace shape; e.g. the frontal horns and dorsal ridge in *Lepas pectinata* cypris (Pl. 1a,b);

Second is the *mesosculpture* – mainly, the structures that form a surface pattern to the carapace (polygonal, pitted, wavy, etc.) and do not alter the general shape of the larva (lateral mesosculpture in terms of ostracodology). Size and distribution of these structures are presumably due to the epithelial cells of hypodermis that produce a cuticle (Neville 1975). Therefore, it is possible to distinguish the 'units' in the surface pattern (e.g., meshes – Pl. 1b,c; tubercles – Pl. 1d), the transverse sizes or the distances between them (apprx. 6-8 μm) correspond to the diameter of the the polygonal markings (Pl. 1e) visible on the cuticle in many crustaceans and other arthropods (e.g. Hinton 1970; Halcrow 1993).

Another kind of mesosculpture includes the cuticular elements related to internal larval organs (various pores, lattice and wheel organs, setae, papillae – e.g. Pl. 1c,d). Most mesosculptural structures are visible under low SEM magnification (<1,000x) and some of them even under the high magnification of light microscope (especially the surface patterns. e.g. see Strathman & Branscomb 1979).

Third is the *microsculpture* – the fine structures that do not alter the surface design and can be distinguished only under the high SEM magnification (>2,000x). Ordinarily these sculptural elements are placed on the surface of the mesosculptural elements, for instance the slight ribs on the surface of nodules of *Heteralepas microstoma* cyprids (Pl. 1f).

Unlike the Ostracoda, in the cyprid larvae, macrosculpture and microsculpture are presented by a few elements. The greatest diversity of cuticular structures occurs at the mesosculptural level.

Carapace ornamentation varies in the different taxonomic groups of the Thoracica and also on the different parts of the larva carapace. The marginal sculpture is more distinctive than the ornamentation on the lateral areas of valves. For example, even the generally smooth carapace of the cyprids of *Chthamalus dalli* and *Balanus improvisus* ordinarily has some ornamentation along its edges (Pl. 3d,f). Also, many of the surface structures related to internal organs (lattice and wheel organs, pores, papillae, setae) tend to be situated on the marginal areas of valves (Walker & Lee 1976; Walker et al. 1987; Elfimov 1986, 1989; Al-Yahya 1991; Jensen et al. 1994).

Among the Thoracica, the cypris larvae of the Lepadomorpha are the most ornamented, particularly in the Lepadidae and Scalpellidae. All lepadid cyprids have clearly distinct surface patterns – honeycombed (Pl. 1b,c and 2a), cellular (Pl. 2b), ribbed and costulated (Pl. 2c,d). Both reticulation and knobs (within reticulae) are present on the dorsal ridge of *L. pectinata* larva (Pl. 2e). Carapace of the *Dosima fascicularis* cypris is covered by similar knobs (Pl. 1d) that form a 'goose-flesh' texture (Jensen et al. 1994), as in the cyprids of *Octolasmis* (Pl. 2f). The carapace of the cyprids of *Conchoderma* is smooth.

The carapace of heteralepadid larvae is smooth (*H. mystacophora* – Elfimov 1986) or covered by small nodules of 1.5 µm in diameter (*H.microstoma* – Pl. 1f).

The cyprid larvae of *Lithotrya, Calantica,* and *Pollicipes* are the most 'interesting' from the aspect of ornamentation since they demonstrate various types of mesosculpture (Dineen 1987, Al-Yahya 1991, Jensen et al. 1994). It might be worthwhile investigating the cyprid surface structures in other species of these scalpellids. Most cyprids of the subfamily Arcoscalpellinae have a generally smooth (Pl. 3a – note a fine reticulation) or slightly wavy (Pl. 3b) carapace (Elfimov 1989; Jensen et al. 1994). However, an 'enigmatic' surface pattern was found in the cyprid of *Trianguloscalpellum gaussi* (Pl. 3c).

The larvae of the family Chthamalidae are usually sculptured by small funnel-shaped pits (Al-Yahya 1991; Jensen et al. 1994) and grooves (Elfimov 1989), particularly on the ventral surface of the valves (Pl. 3d and 5a).

Most cyprids of the Archaeobalanidae and Balanidae have a generally smooth carapace (Pl. 3e), although the shallow depressions and pits were found in the cyprids of *Semibalanus balanoides* (Walker & Lee 1976), *Balanus improvisus* (Pl. 3f), *Elminius modestus* and *Megatrema anglica* (Al-Yahya 1991; Jensen et al. 1994).

Various cuticular structures related to specific internal organs of larvae are also elements of surface ornamentation, mainly of the mesosculptural level. The most interesting of these are the lattice organs (Elfimov 1984, 1986, 1987a,b, 1989; Al-Yahya 1991; Jensen et al. 1994). These structures are located on the dorsal surface of the valves symmetrically in relation to the hinge line (Pl. 2b and 3a). Most larvae have 5 pairs of lattice organs, 2 anterior and 3 posterior, the last pair is situated at some distance from the 3rd and 4th ones. The superficial part of each organ (Pl. 1d) is usually represented by an area of large-pored cuticle of 15-35 µm length and 1-3 µm width (the central part or lattice field) surrounded by non-porose thickened cuticle (peripheral part). A large pore is located at one end of the lattice field, while at the other end the cuticle of the field 'dives' underneath the edge of the periferal part. Most of these structures are arranged around the place where the valves diverge. The organs of the first and second pairs have the large pore at their anterior end, and the organs of the third and fifth pairs at their posterior end. They are presumably chemoreceptive structures.

The frontal horn pores vary in size, shape, structure and arrangement on the valves of the larvae from the different groups of the Thoracica. Three basic types of the frontal horn pores can be distingushed – lepadoid, scalpelloid, and balanoid (Elfimov 1989):

Lepadoid pores (Pl. 4a) are relatively large openings with the thin and soft cuticle on their edges. The pore is surrounded by setae (of different numbers) and located in the lower part of the frontal margin of the valve. In the *L. pectinata* cypris, the pores are situated at the tip of the frontal horns (Pl. 1b), and in *L. australis* they are on the small cone-shaped protuberance of the valve. Such pores are characteristic for lepadid cyprids.

Scalpelloid pores (Pl. 4c) are small round openings with a thickened rim. They are located on the lower lateral surface of the anterior part of the carapace. These pores are found in most of the Arcoscalpellinae larvae (Elfimov 1989).

Balanoid pores (Pl. 4f) are longitudinally extended openings. The edge of the pore is formed by folded cuticle with a flap. Pores are situated on the ventral surface of the anterior part of the carapace close to the ventral margin of valve. They are present in the cyprids of the Balanidae and Archaeobalanidae.

The frontal horn pores of the cyprids of *Conchoderma virgatum* resemble those of the lepadids (Pl. 4b). Both of the known of heteralepadid cyprids have small frontal horns (Elfimov 1986, 1989). The pores of the larvae of *Verruca stroemia* and *Chthamalus dalli* resemble the balanoid type (Pl. 4d,e, respectively).

In some cyprids, the caudal and ventral pores of 5-6 μm in diameter and conical papillae of 2.0-2.4 μm height were also found (Elfimov 1987a,b). In the *Lepas* cyprids, five ventral pores and five papillae are located along the ventral margin of the valve (Pl. 5a,b). The caudal pore is situated at the posterior end of the valve (Pl. 2c,d).

For the cyprid larvae of *Heteralepas mystacophora* and *Chthamalus dalli* the wheel-shaped structures ('wheel organs') were also described (Elfimov 1986, 1989). These are round openings 6-8 μm in diameter with a thickened rim. The central part of these structures is formed by a thin (and probably soft) cuticle with radial folds (Pl. 5c-f). A mechanoreceptive function was suggested (the thin cuticle, like a membrane, might be sensitive to water pressure).

Larvae of different species also differ in the number, size, and arrangement of setae on the carapace (Walker & Lee 1976). The setae are 3-20 μm long and are concentrated, as with the other sensory organs, along the margins of valves and on the posterior end of carapace. In the lepadid cyprids, they form rows along the frontal and anterio-dorsal margins of valves. In the *Heteralepas* larvae, the frontal setae are located within pits that form a row along the frontal edge of valve (about 15 pits in all). Each pit has one seta on its posterior wall (Elfimov 1986).

3 PHYLOGENETIC IMPLICATIONS OF CYPRIDS FEATURES

The examination of cyprid structures undoubtedly can provide valuable material for studies on the phylogeny of the Cirripedia. In the Thoracica, among the features that may be useful for the differentiation and systematics of cyprids are as follows: size and relative size of larva, colour, carapace shape and ornamentation, and structure of appendages (Lang 1979; Standing 1980; Elfimov 1987a, 1989; Al-Yahya 1991;

Jensen et al. 1994). Because of the variability of some of these features (caused by the effect of environment, for instance), not all of them have an equal significance for studying evolution of larvae and phylogeny of the Cirripedia.

Also, the number of species studied in detail in regard to the cypris seems insufficient for comprehensive cladistic analysis including the building of cladograms and the determination of true phylogenetic relationships among thoracican cyprids. So, *a priori* analysis concerning the polarity of the carapace features and the hypothetical trends in their evolution are primarily discussed here.

Different criteria for character polarization were used (mainly, outgroup comparison and ontogeny). The Acrothoracica might be chosen as the outgroup. Their origin from a pre-pedunculate stock (not from pedunculate barnacles) was assumed (Newman 1987). Also, a clear trend in their evolution is postulated (Tomlinson 1987) accordingly to which some directions in the evolution of their cyprids were also suggested (Elfimov 1989). Unfortunately, only two species of acrothoracican cyprids have been studied with the SEM (Jensen et al. 1994). Because of that, the information on other groups of the Thecostraca was employed in determining the polarization of character states. Size frequency distribution in a population of cyprids results from both genotypic and phenotypic variability. Due to intraspecific variability, cyprids size may be used only for a rough identification of larvae and very rarely for their specific taxonomy. Unlike rhizocephalan cyprids, which demonstrate a sexual dimorphism in carapace length within a species (e.g., Walker 1985), analysis of size distribution in thoracican cyprids may be interesting basically from the viewpoint of the larval ecology.

Sometimes larval size can be considered a distinguishing feature for all cyprids of a taxon, e.g., in the Lepadidae (large cyprids) and Chthamalidae (small cyprids). Undoubtedly having ecological significance (Moyse & Knight Jones 1967; Lucas et al. 1979; Moyse 1987), the large size of the *Lepas* cyprids along with other features of larval development (Moyse 1987) reflects the peculiar pathway of lepadids specialization. The small size of chthamalid larvae is presumably due to their intertidal habitat (Korn 1988, this volume) and can also be discussed as one of the aspects of their evolution.

Remarkably, most known cypris larvae of the Acrothoracica have a carapace length from 250-600 μm and only two species are in the range of 820-870 μm – *Weltneria spinosa* (Tomlinson 1969; Jensen et al. 1994) and *W.zibrowii* (Turquier 1985). The genus *Weltnerium* is thought to be the most 'primitive' among the Acrothoracica (Newman 1987; Tomlinson 1987) and their cyprids might also be believed to have some primitive features. Cypris larvae of the Rhizocephala are small too – from 50-400 μm (Glenner et al. 1989). Probably the small size of cyprids is one of the results of parasitic evolution of the Acrothoracica and Rhizocephala.

The shape of the carapace as well as its size is often discussed from the point of view of larval habitat and can also have an ecological significance. For example, a rounded (even almost ball shaped) shell of *Dosima fascicularis* cypris as well as some juvenile features (Memmi & Elfimov 1987) can be explained by the conditions of larval settlement. Cyprids form their own float rather than settle on a stable substratum (Zevina 1982). It follows that there is no orientation in cypris settlement. Consequently, 'orientation' in the shape of the larva is absent too – height and width of the carapace are approximately equal and differ little from the length (Elfimov

1988). On the other hand, the narrow and elongated shell of balanid cyprids is more 'suitable' in terms of orienting cyprids during settlement on a hard and rough substratum in relation to bottom contour and water movements (Crisp 1984).

Nevertheless, despite the intraspecific variability, some features of carapace shape (mainly, at the anterior and posterior ends in lateral view) can be used for studies on cyprid systematics and phylogenetic reconstructions. Evenly rounded ends of the carapace could be ancestral features, since they occur in both the putative primitive acrothoracicans (Tomlinson 1969) and the ascothoracidan larvae (e.g. Grygier 1983).

Carapace sculpturing and its armament of pores and sensilla can be undoubtedly used in the studying the evolution of the cyprid and the phylogeny of the Cirripedia (Elfimov 1989; Al-Yahya 1991; Jensen et al. 1994). The morphology, variability, and phylogenetic implications of cuticular structures were investigated with the SEM in various crustaceans (e.g., Liebau 1977; Klepal & Kastner 1980; Klepal 1983; Bresciani 1986; Halcrow & Bousfield 1987). The intraspecific variations in surface ornament (Hartmann 1982) and the interspecific convergent similarities (Halcrow & Bousfield 1987) have been shown. Nevertheless, Halcrow & Bousfield (1987) concluded that the trends in the surface ultrastructure within the Amphipoda tend to conform generally but not rigidly with the phyletic groupings. A similar conclusion was obtained by Klepal & Kastner (1980) on peracaridans. In ostracodes, Hartmann (1966) concluded that the surface ornamentation of shell valves is mainly a species criterion and may not be used for higher level taxonomy. However, further investigations in ostracode ornamentation (e.g., Liebau 1977; Shornikov & Michailova 1990) have demonstrated the potential of using the carapace sculpture in reconstructing phylogeny of ostracodes.

Cyprids of many thoracican species have a distinct surface ornamentation on the carapace while most known larvae of the Acrothoracica and Rhizocephala are generally smooth. This prompts the question, what kind of carapace was on the ancestral type – smooth or sculptured. Presumably, the latter condition can be argued as plesiomorphic in the thoracican cyprids because of the presence of sculpturing in the larvae of Facetotecta (Schram 1970a,b; Grygier 1987; Ito 1987) and some Ascothoracida (Grygier 1988, 1990, 1993; Ito & Grygier 1990). 'Brick-work' sculpturing of the naupliar carapace in *Ibla cumingi* was also mentioned (Karande 1974; Gaonkar & Karande 1980).

The next problem is to decide what type of lateral mesosculpture can be considered plesiomorphic for thoracican cyprids. The honeycombed surface pattern of facetotectan cyprids (e.g., Schram 1970a), 'concentric ridge pattern with many oblique connections' in petrarcid nauplii (Grygier 1993:3), and honeycombed sculpturing in some ascothoracidan larvae (Grygier 1988; Itô & Grygier 1990) allow us to assume a reticulated pattern to be most primitive in the Thoracica. Reticulated or honeycombed ornament often occur in other groups of the Crustacea such as the Ostracoda (e.g. Liebau 1977), Cladocera (e.g. Alonso 1990), Amphipoda (e.g., Halcrow & Bousfield 1987), and this might confirm plesiomorphy of a reticulated pattern in cirriped cyprids. Moreover, reticulated ornament is considered a primitive state in ostracodes (Liebau 1977; Shornikov 1981) and was found in some of the Cambrian ostracode species (Shornikov & Michailova 1990).

The results of other studies on transformation of sculpture elements in crustacean (Liebau 1977, 1982; Klepal & Kastner 1980; Klepal 1983; Shornikov 1988) reveal that a general trend in the evolution of multi-elemental patterns or sets of serial structures might be as follows: in the primitive stage there might be numerous elements that vary in regard to number and arrangement, then a fixation of their number and distribution is observed, and then finally a reduction is possible. Smoothing of the carapace sculpture may be one of the consequences of this trend. Surface ornament in thoracican cyprids seems to smooth out too. It can be demonstrated by the 'set' of lepadid cyprids that can be arranged in transformation series from *L. pectinata* larva with reticulated and honeycombed ornament to the smooth carapace of the *Conchoderma virgatum* cypris (Elfimov 1987a, 1989). So, reticulated surface patterns can be considered as a primitive state of the ornamentation in thoracican cyprids, and other patterns (honeycombed, ribbed, ridged, 'goose flesh', pitted, smooth) are advanced states of this character.

Other surface structures on the cyprid carapace may also provide valuable material of phylogenetic and evolutionary significance. Issues of plesiomorphy and apomorphy, and phylogenetic implications of lattice organs were discussed by Jensen et al. (1994). A variety of issues for frontal horn pores concerns their structure and placement. In the absence of sufficient information on the structure of frontal pores in other groups of Thecostraca, it seems more likely to employ the ontogenetic criterion to make a decision about polarity of this character. By this criterion, frontal horns or their rudiments in thoracican cyprids (*Lepas pectinata, L. australis, Heteralepas mystacophora, H. microstoma*) might be considered plesiomorphic for the Thoracica because of their presence in nauplii. The frontal pores of 'balanoid' and 'scalpelloid' types are more simple in their structure (at least, externally) and probably represent the next stage of pore differentiation. In rhizocephalan cyprids, frontal pores are situated near the antero-ventral margin of the carapace, as in some larvae of the balanids and *Verruca stroemia*. It is noteworthy that most of the rhizocephalans that hatch as cyprids lack the frontal horn pores (Glenner et al. 1989; Andersen et al. 1990). Some scalpellids also hatch as cyprids (Barnard 1924; Zevina 1982) and all cyprids dissected from adults have very simple frontal pores, i.e., just the opening with a thickened margin without any surrounding setae and flaps (Elfimov 1989).

The number and arrangement of ventral and caudal pores in lepadid cyprids suggest their homology with marginal gland spines of nauplii (Lang 1979; Moyse 1987), which in turn are supposed to be homologous to the marginal processes of petrarcid nauplii (Grygier 1990, 1993).

In *Weltnerium scoresbyi* frontal horn pore and ventral pores are very similar (the frontal pore is a bit larger). This similarity might be evidence for the homology of frontal horn pores to other ventral pores and their common origin from naupliar marginal pores.

Obviously, further investigations of the structure and number of various cyprid pores must be done. These structures can be expected to offer valuable material for phylogenetic reconstructions in the Cirripedia. Walker & Lee (1976) described small pores (0.5 m in diameter) on the carapace of the *Semibalanus balanoides* cypris. These pores are regularly arranged and can be traced back to those evident on the nauplius. The number of ventral pores (excluding frontal horn pores) varies in different cyprids species – five pairs in lepadids, six pairs in *Verruca stroemia,* and seven pairs in *Weltnerium scoresbyi* (Elfimov 1989). In ascothoracidan nauplii, pores are gradually added around the margin of the dorsal shield as the larvae molt (Grygier

1993). So, the small number of ventral pores on the cypris could be considered as a primitive state.

Determining the polarity of some other surface structures (such as the frontal pits in the heteralepadid larvae, or wheel organs) seems at the moment impractical since these structures are found in very few species of cyprids.

ACKNOWLEDGMENTS

I express my sincerest appreciation to Prof. Frederick R. Schram, Dr. John Moyse, Dr. Jens T. Hoeg, and some annonymous reviewers for a valuable help in improving this manuscript. I thank Prof. G.B. Zevina and Dr. O. Korn for their advice and assistance in the course of my work. Special thanks are given to Dr. M.J. Grygier who translated into English the abstract of my PhD thesis and made helpful comments on that thesis. Portions of this study were done as a graduate and postgraduate student at the Moscow University (Department of Invertabrate Zoology), with an postdoctoral fellowship from the Central European University (Budapest, Department of Environmental Sciences and Policy), and at the University College of Swansea (Marine, Environmental and Evolutionary Research Group). I thank the Laboratory of Electron Microscopy (Moscow University) for technical assistance.

REFERENCES

Alonso, M. 1990. *Estatheroporus gauthieri,* new genus, new species (Cladocera: Chydoridae), from Mediteranean countries. *J. Crust. Biol.* 10: 148-161.

Al-Yahya, H.A.H. 1991. Studies on cirripede larvae with special reference to external features of cyprids from five families. Ph. D. Thesis. University of Wales.

Andersen, M.L., M. Bohn, J.T. Hoeg & P.G. Jensen 1990. Cypris ultrastructure and adult morphology in *Ptychascus barnwelli n.sp.* and *P. glaber* (Crustacea: Cirripedia: Rhizocephala), parasites on semiterrestrial crabs. *J. Crust. Biol.* 10: 10-28.

Barker, M.F. 1976. Culture and morphology of some New Zealand barnacles (Crustacea: Cirripedia). *N.Z.J. Mar. Freshwater Res.* 10: 139-158.

Barnard, K.H. 1924. Contributions to the Crustacean fauna of South Africa,N7, Cirripedia. *Ann. South Afr. Mus. Edinburgh* 20: 1-103.

Batham, E.J. 1946a. Description of female, male and larval forms of a tiny stalked barnacle *Ibla idiotica n.sp. Trans. Roy. Soc. N.Z.* 75: 347-356.

Batham, E.J. 1946b. *Pollicipes spinosus* Quoy and Gaimard. 2: Embryonic and larval development. *Trans. Roy. Soc. N.Z.* 75: 405-418.

Benson, R.H. 1974. The role of ornamentation in the design and function of the ostracode carapace. *Geosci. Man Baton Rouge.* 6: 47-57.

Benson, R.H. 1981. Form, function, and architecture of ostracode shells. *Annu. Rev. Earth and Planet . Sci.* 9: 59-80.

Bresciani, J. 1986. The fine structure of the integument of free-living and parasitic copepods. A review. *Acta Zool.*(Stockholm) 67: 125-145.

Clare, A.S. 1993. Invertebrate larval settlement. *Mar. Biol. Ass. UK Ann. Rep.* 1993: 32-34.

Claus, C. 1869. Die Cypris-ahnliche Larve (Puppe) der Cirripedien und ihre Verwandlung in das festsitzende Thier. Ein Beitrag zur Morphologie der Rankenfussler. *Schr. Ges. Naturw. Marburg, Suppl-Heft* 5: 1-17.

Crisp, D.J. 1962a. The planctonic stages of the *Balanus balanoides* (L.) and *Balanus balanus* (L.) from the temperate waters. *Crustaceana* 3: 207-221.

Crisp, D.J. 1962b The larval stages of *Balanus hameri* (Ascanius, 1767). *Crustaceana* 4: 123-130.

Crisp, D.J. 1984. Overview of research on marine invertebrate larvae, 1940-1980. In J.D. Costlow & R.C.Tipper (eds.), *Marine Biodeterioration: An Interdisciplinary Study*: pp. 103-126. Annapolis, Maryland: Naval Institute Press.

Daniel, A. 1958. The development and metamorphosis of three species of sessile barnacle. *J. Madras Univ.* 28: 23-47.

Darwin, C. 1854. *A monograph of the subclass Cirripedia with figures of all the species. The Balanidae, the Verrucidae etc.* London: Ray Soc.

Dineen, J.F. 1987. The larval stages of *Lithotrya dorsalis* (Ellis et Sollander, 1786): a burrowing thoracican barnacle. *Biol. Bull.* 172: 284-298.

Doochin, H.D. 1951. The morphology of *Balanus improvisus* Darwin and *B.amphitrite niveus* Darwin during initial attachment and metamorphosis. *Bull. Mar. Sci. Gulf Caribb.* 1: 15-39

Egan, E.A. & D.T. Anderson 1989. Larval development of the chthamalid barnacles *Catomerus polymerus* Darwin, *Chamaesipho tasmanica* Foster and Anderson and *Chthamalus antennatus* Darwin (Crustacea, Cirripedia). *Zool. J Linn.Soc.* 95: 1-28.

Elfimov, A.S. 1984. Morphology of the shell of the cyprid larvae (Cirripedia, Thoracica). In *Proc. 15th Conf. Young Sci. Biol. Fac. Moscow Univ.* 2: 128-132. Moscow: MSU (in Russian).

Elfimov, A.S. 1986. Morphology of the carapace of the cyprid larva of the barnacle *Heteralepas mystacophora. Biol. Morya* 3: 30-34 (*The Soviet Journal of Marine Biology* 12: 152-156).

Elfimov, A.S. 1987a. Comparative morphological study of the cyprid larvae of the genus *Lepas* (Cirripedia, Lepadomorpha). In *Proc. 18th Conf. Young Sci. Biol. Fac. Moscow Univ.* 3: 2-6. Moscow: MSU (in Russian).

Elfimov, A.S. 1987b. Diversity of sensory organs of the cyprid larvae and their adaptation to settlement conditions. In *Proc. 3rd All-Union Conf. on Biodeterioration, Donetzk* 2: 235-236. Moscow: Ac.Sci. (in Russian).

Elfimov, A.S. 1988. Adaptations of the thoracican cyprid larvae to habitat conditions. In *Proc. 3rd All-Union Conf. on Marine Biology, Sevastopol* 1: 6-7 (in Russian).

Elfimov, A.S. 1989. The cyprid larvae of cirripedes and their significance in the formation of fouling. Ph.D.Thesis, Moscow University (in Russian).

Elfimov, A.S. & A.G. Bogdanov 1984. The use of the scanning electron microscopy methods in the study of the cirriped larvae. In Proc. *4th All-Union Symposium on Scanning Electron Micro scopy.* 1: 156. Moscow: Ac. Sci. (in Russian).

Gaonkar S.N. & A.A. Karande 1980. Observations on the life history of the pedunculate barnacle *Ibla cumingi J. Bombay Nat. Hist. Soc.* 76: 305-310

Glenner, H., J.T. Hoeg, A. Klysner & B. Brobin Larsen 1989. Cypris ultrastructure, metamorphosis and sex in seven families of rhizocephalan barnacles (Crustacea: Cirripedia: Rhizocephala). *Acta Zool.*(Stockholm) 70: 229-242.

Grygier, M.J. 1983. A novel planktonic ascothoracid larva from St.Croix (Crustacea). *J. Plank. Res.* 5: 197-202.

Grygier, M.J. 1987. New records,external and internal anatomy, and systematic position of Hansen's y-larvae (Crustacea: Maxillopoda: Facetotecta). *Sarsia* 72: 261-278.

Grygier, M.J. 1988. Larval and juvenile Ascothoracida (Crustacea) from the plankton. *Publ. Seto Mar. Biol. Lab.* 33: 163-172.

Grygier, M.J. 1990. Early planktonic nauplii of *Baccalaureus* and *Zibrowia* (Crustacea: Ascothoracida) from Okinawa, Japan. *Galaxea* 8: 321-337.

Grygier, M.J. 1993. Late planktonic naupliar development of an ascothoracidan crustacean (? Petrarcidae) in the Red Sea and a comparison to the Cirripedia. *Nat. Hist. Mus. Los Angeles County. Contrib. in Science.* 437: 1-14.

Halcrow, K. 1993. Pore canal systems and assocoated cuticular microstructures in amphipod crustaceans. In M.N. Horst & J.A. Freeman (eds.), *The Crustacean Integument. Morphology and Biochemistry*: pp. 39-77. London: CRC Press.

Halcrow, K. & E.L. Bousfield 1987. Scanning electron microscopy of surface microstructures of some gammaridean amphipod crustaceans. *J. Crust. Biol.* 7: 274-287.

Hartmann, G. 1966. Ostracoda. In *Bronns Klassen und Ordnungen des Tierreichs*. Leipzig: Acad. Verl.

Hartmann, G. 1982. Variation in surface ornament of the valves of three ostracod species from Australia. In R.H. Bate, E. Robinson & L.M. Shepperd (eds.), *Fossil and recent ostracods*, pp. 365-380. Chichester: Horwood.

Hinton, H.R. 1970. Some little known surface structures. In H.R. Hepburn (ed.), *Insect ultrastructure*, pp. 41-58. Oxford: Blackwell Scientific Publication.

Hiro, F. 1933. Report on the Cirripedia collected by the surveying ships of the imperial fisheries experimental shelf bordering Japan. *Rec. Oceanogr. Works Japan, Tokyo* 5: 11-84.

Hoek, P.P.C. 1907. The Cirripedia of the Siboga-Expedition. A. Cirripedia pedunculata. *Siboga Exp. Reports* 31: 1-127.

Hoffendahl, K. 1904. Beitrag zur Entwicklungsgeschichte und Anatomie von *Poecilasma aurantium* Darwin. *Z. Jahrb. Abth. Morph.* 20: 363-398.

Itô, T. 1987. Three forms of nauplius y type VIII larvae (Crustacea: Facetotecta) from the North Pacific. *Publ. Seto Mar.Biol.Lab.* 32: 141-150.

Itô, T. & M.J. Grygier 1990. Descriptions and complete larval development of a new species of *Baccalaureus* (Crustacea: Ascothoracida) parasitic in a zoanthid from Tanabe Bay, Honshu, Japan. *Zool. Sci.* 7: 485-515.

Jensen, P.G., J. Moyse, J.T. Hoeg & H. Al-Yahya 1994. Comparative SEM studies of lattice organs: putative sensory structures on the carapace of larvae from Ascothoracida and Cirripedia (Crustacea Maxillopoda Thecostraca). *Acta Zool.* (Stockholm) 75: 125-142.

Jones, L.W.G. & D.J. Crisp 1954. The larval stages of the barnacle *Balanus improvisus* Darwin. *Proc. Zool. Soc. London* 123: 756-780.

Karande, A. 1974. Development of the pedunculate barnacle *Ibla cumingi* Darwin. *Indian J. Mar. Sci.* 3: 173-177.

Kaufmann, R. 1965. Zur Embryonal-und Larvalentwicklung von *Scalpellum scalpellum* L. (Crustacea Cirripedia). *Z. Morphol. Okol. Tiere* 55: 161-232.

Kesling, R.V. 1951. Terminology of ostracod carapaces. *Contrib. Mus. Paleontol. Mich. Univ.* 9: 93-171.

Klepal, W. 1983. Morphogenesis and variability of cuticular structures in the genus *Ibla* (Crustacea, Cirripedia). *Zool. Scripta* 12: 115-125.

Klepal, W. & R.T. Kastner 1980. Morphology and differentiation of nonsensory cuticular structures in Mysidacea, Cumacea and Tanaidacea (Crustacea: Peracarida). *Zool .Scripta* 9: 271-281.

Korn, O.M. 1988. Larval morphology and phylogenetic system of barnacles of the order Thoracica. *Zool. Zhur.* 67: 1644-1659 (in Russian).

Korn, O.M. 1991. Larvae of the barnacle *Balanus improvisus* in theSea of Japan. *Biol. Morya* 1: 2-62 (in Russian).

Kruger, P. 1940. Cirripedia. In *Bronn's Klassen und Ordnungen des Tierreiches.* Leipzig: Academische Verlagsgsellschaft.

Kuhl, H. 1968. Die Beeinflussung der Metamorphose von *Balanus improvisus* Darwin durch Giftstoffe. *Proc. 2nd Intern. Congr. for Marine Corrosion and Fouling , Athens.* 383-390.

Lang, W.H. 1979. Larval development of shallow water barnacles of the Carolinas (Cirripedia: Thoracica) with keys to naupliar stages. *NOAA Techn. Rep. NMFS* Circular 421: 1-39.

Liebau, A. 1969. Homologisierende Korrelationen von Trachyleberididen-Ornamenten (Ostracoda, Cytheracea). *N. Jb. Geol. und Palaontol. Monatsh.* 7: 390-402.

Liebau, A. 1975. The left-right variation of the ostracode ornament. *Bull. Amer. Paleontol.* 65(282): 77-86.

Liebau, A. 1977. Carapace ornamentation of the Ostracode Cytheracea: Principles of evolution and functional significance. In H.Loffler & D.Danielopol (eds.), *Aspects of ecology and zoogeography of recent and fossil Ostracoda*: pp. 107-120. The Hague: Junk.

Liebau, A. 1982. Sculpture evolution in the Trachyleberididae and the discrepancies between 'Rib pattern taxonomy' and 'Mesh pattern taxonomy'. In *Eight Intern. Symp. on Ostracoda: Progr. and Abstr.*, p.66. Houston.

Lucas, M.I., G. Walker, D.L. Holland & D.J. Crisp 1979. An energy budget for the free-swimming and metamorphosing larva of *Balanus balanoides* (Crustacea, Cirripedia). *Mar. Biol.* 55: 221-229.

Memmi, M.P. & A.S. Elfimov 1987. Functional morphology of the capitulum cuticular cover of goose barnacles of the genus *Dosima* (Cirripedia, Lepadomorpha). *Dokl. Akad. Nauk SSSR* 297: 506-508 (in Russian).

Moyse, J. 1987. Larvae of lepadomorph barnacles. *Crustacen Issues* 5: 329-362.

Moyse, J. & E.W. Knight-Jones 1967. Biology of cirripede larvae. In *Proc. Symp. Crustacea, Ernaculam, 1965. Mar. Biol. Soc. India* 2: 595-611.

Neville, A.C. 1975. *Biology of the Arthropod Cuticle.* Berlin: Springer.

Newman, W.A. 1987. Evolution of cirripedes and their major groups. *Crustacean Issues* 5: 3-42.

Nott, J.A. 1969. Settlement of the barnacle larvae: surface structure of the antennular attachment disc by scanning electron microscope. *Mar. Biol.* 2: 248-251.

Pyefinch, K.A. 1948. Methods of identification of the larvae of *Balanus balanoides* (L.), *B.crenatus* Brug. and *Verruca stroemia* (O.F.Muller). *J. Mar. Biol. Assoc. UK* 27: 451-463.

Schram, T.A. 1970a. Marine biological investigations in the Bahamas.14. Cypris Y, a later developmental stage of Nauplius Y Hansen. *Sarsia* 44: 9-24.

Schram, T.A. 1970b. On the enigmatical larva Nauplius Y I Hansen. *Sarsia* 45: 53-68.

Shornikov, E.I. 1981. *Ostracods Bythocytheridae of far-eastern seas.* Moscow: Nauka (in Russian).

Shornikov, E.I. 1988. The pathways of morphological evolution of Bythocytheridae. In T. Hanai, N. Ikeya & K. Ishizaki (eds.), *Evolutionary Biology of Ostracoda,* pp. 951-965. Tokyo: Kodansha Ltd.

Shornikov, E.I. 1989. Ostracods of the family Bythocytheridae: Comparative morphology, pathways of evolution, systematics. D. Sc. Thesis. Leningrad (in Russian).

Shornikov, E.I. & E.D. Michailova 1990. *Ostracods Bythocytheridae at the early stage of development. Comparative morphology, palaeoecology and evolutionary pathways.* Moscow: Nauka (in Russian).

Standing, J.D. 1980. Common inshore barnacle cyprids of the Oregonian faunal province (Crustacea: Cirripedia). *Proc. Biol. Soc. Wash.* 93: 1184-1203.

Stefoni, D.L. & G.T. Contreras 1979. Estudio descriptivo comparado de los estados larvarios tempranos y cypris de Balanomorfos Chilenos. *Acta Zoologica Lilloana* 35: 547-561.

Strathmann, R.R. & E.S. Branscomb 1979. Adequacy of cues to favorable sites used by settling larvae of two intertidal barnacles. In S.E. Stancyk (ed.), *Reproductive Ecology of Marine Invertebrates,* pp. 77-89. Columbia: Univ. South Carolina Press.

Svane, I. 1986. Sex determination in *Scalpellum scalpellum* (Cirripedia: Thoracica: Lepadomorpha), a hermaphroditic goose barnacle with dwarf males. *Mar. Biol.* 90: 249-253.

Tomlinson, J.T. 1969. The burrowing barnacles (Cirripedia: Order Acrothoracica). *Bull. U.S. Nat. Mus.* 296: 1-162.

Tomlinson, J.T. 1987. The burrowing barnacles (Acrothoracica). *Crustacean Issues* 5: 63-71.]

Turquier, Y. 1985. Cirripedes Acrothoraciques des cotes occidentales de la Mediterranee et de l'Afrique du Nord: II. *Weltneria zibrowii* n.sp. *Bull. Soc. Zool. Fr.* 110: 170-189.

Walker, G. 1985. The cypris larvae of *Sacculina carcini* Thompson (Crustacea: Rhizocephala). *J. Exp. Mar. Biol. Ecol.* 93: 131-145.

Walker, G. & V.E. Lee 1976. Surface structures and sense organs of the cypris larva of *Balanus balanoides* as seen by scanning and transmission electron microscopy. *J. Zool.* 178: 161-172.

Walker, G., A.B. Yule & J.A. Nott 1987. Structure and function of balanomorph larvae. *Crustacean Issues* 5: 307-328.

Yasugi, R. 1937. On the swimming larvae of *Mitella mitella* (L.). *Bot. Zool. Tokyo* 5: 792-796.

Zevina, G.B. 1981. Barnacles of the suborder Lepadomorpha (Cirripedia: Thoracica) of the World Ocean. 1 Family Scalpellidae. *Opredeliteli Fauni SSSR* 127: 1-398 (in Russian).

Zevina, G.B. 1982. Barnacles of the suborder Lepadomorpha (Cirripedia Thoracica) of the World Ocean. 2. *Opredeliteli Fauni SSSR* 127: 1-223 (in Russian).

Attachment organs in cypris larvae: Using scanning electron microscopy

John Moyse
Marine and Environmental Research Group, School of Biological Sciences, University of Wales, Swansea, UK
Jens T. Høeg & Peter Gram Jensen
Department of Cell Biology and Anatomy, Institute of Zoology, University of Copenhagen, Copenhagen, Denmark
Hamad A.H. Al-Yahya
Biology Department, King Saud University, Riyadh, Saudi Arabia

ABSTRACT

Studies on settlement in barnacles, including the morphology of the antennular attachment organ in the cypris larva have until now been based almost exclusively on *Semibalanus balanoides* and a few other 'model' barnacles. In an attempt to exemplify a comparative approach, we have used mainly scanning electron microscopy to study the antennular attachment organ of cypris larvae from a wide selection of species, representing all three orders of the Cirripedia. Our study demonstrates that the attachment organs vary in several respects and can differ appreciably from those of *S. balanoides*. No high level taxonomic groups seem to exhibit a unique (apomorphic) attachment organ, since the structure differs both within the Rhizocephala and between the various taxa of the Thoracica. Some widespread similarities may represent either true homologies or convergent evolution due to precise functional needs. For instance, the familiar radially symmetrical form characterizes species inhabiting the rocky intertidal, but with the present assemblage of species we cannot entirely refute the possibility that it also represents a true synapomorphy. Some of the variation clearly represents unique apomorphies at low taxonomic levels linked to very special functional requirements. The neustonic species *Dosima fascicularis* has a large flexible attachment disc, perhaps adapted for production of its float. The balanoid coral commensal *Megatrema anglica* has a unique, long and pointed attachment organ, presumably adapted for penetrating the coral coenosarc before attaching to its host's skeleton. Surprisingly, however, the rhizocephalan *Clistosaccus paguri,* which employs its antennule for penetrating through host integument, exhibits no comparable specialization of its attachment organ. In general, we did not identify any unique traits that characterize the parasitic Rhizocephala apart from the absence of the velum. Despite the considerable variation found in attachment organ morphology, the basic homology of its structure remains unchallenged. To substantiate this conclusion, we also studied the cyprid-like larvae of the Ascothoracida to demonstrate that their antennular attachment mechanism has little similarity to the one in the Cirripedia. This once again emphasizes the apomorphic nature of the cirripede cyprid and the basic monophyly of the whole taxon. We emphasize that larval characters studied by ultrastructural methods, including attachment organ morphology, will play an integral part in any future analysis of cirripede phylogeny.

1 INTRODUCTION

The barnacle cypris larva exhibits a unique series of apomorphic specializations associated with its role as the settlement stage in ontogeny. These include adaptations of the four segmented antennules used in exploratory walking and final attachment.

Twenty-five years ago a series of studies under the direction of the late Professor D.J. Crisp at the Menai Bridge Marine Science Laboratories revealed that cypris larvae attach using the third antennular segment as an adhesive pad rather than a suction cup as previously supposed. The research was based almost exclusively on *Semibalanus balanoides* (Thoracica: Balanomorpha). The classic studies by Nott (1969), Nott & Foster (1969) and Walker (1973) of *S. balanoides* established our present understanding of form and function of the attachment organ on the third antennular segment. Prior to these TEM and SEM studies it was widely assumed that the attachment organs functioned as suction cups (Saroyan et al. 1968). Nott and Foster, however, showed that the ultrastructure of the third segment is largely incompatible with a suction mechanism, being almost devoid of muscles other than those operating the sensory fourth segment. Their view that the attachment organs initially act as adhesive pads is now generally accepted. Feuerborn (1933) had already previously shown that the cement gland empties on the attachment disc for permanent attachment.

Although a few recent studies attempted to revive the suction cup hypothesis (e.g., Lindner 1984), ingenious experiments have now demonstrated that suction cannot account for the force required to remove temporarily attached cyprids (Yule & Crisp 1983; Yule & Walker 1984a; and review by Yule & Walker 1987). Two kinds of gland operate during settlement. Throughout the period of exploratory walking, secretions from numerous, unicellular antennulary glands account for the temporary attachment; final and irreversible attachment takes place by secretion from the pair of multicellular cement glands (Nott & Foster 1969; Walker 1973; Yule & Walker 1984b; Høeg 1985a). Parallel to these findings, extensive fundamental studies partly driven by the needs of the antifouling industry, have pursued behavioural and biochemical aspects of barnacle settlement mostly on *Semibalanus balanoides* and a few species of *Balanus* (Clare this volume, Walker this volume).

The studies on these few 'model species' have contributed immensely to our knowledge of barnacle settlement biology. However, the substratum and ecological conditions associated with settlement vary extensively within the Cirripedia, from rock surfaces in intertidal species, over sharks, whales, hydroids, corals and crustaceans in the many epibiotic and parasitic species, to boring into limestone as occurs in the Acrothoracica. One would accordingly expect a phylogenetically and functionally interesting variation in attachment organ morphology of the cypris larvae. Darwin (1852) illustrated a *Lepas* cyprid in considerable detail and noted that cypris antennules vary from species to species, but surprisingly few studies since then have dealt with cypris morphology in other species. In thoracicans other than *S. balanoides,* the few existing accounts of antennular morphology are based on light microscopy and have largely been neglected, e.g., Nilsson-Cantell (1921). Even today, descriptions of cyprids normally stand as adjuncts to descriptions of nauplii without any useful morphological information whatsoever. Some authors, e.g., Standing (1981) and Miller & Roughgarden (1994), have attempted to make simple comparisons but generally aimed only at mere identification of larvae from plankton samples rather than any deeper biological understanding. Among cyprids of the parasitic barnacles (Rhizocephala) in contrast, we now have a very solid understanding of morphological variation as a result of recent studies by Walker (1985) and by a group of workers under the direction of Jens Høeg in Copenhagen (e.g., Høeg 1985a,b, 1987a,b, Glenner et al. 1989; Jensen et al. 1994a). These studies revealed that among rhizocephalan barnacles antennular morphology varies in many interesting details, although the cyprids can still readily be compared with those in the Thoracica (Høeg 1992a,b).

In a previous paper we compared carapace sense organs ('lattice organs') of cyprids from a large selection of cirripede species to demonstrate the use of cyprid morphology in understanding cirripede evolution (Jensen et. al. 1994b). Otherwise, the unpublished doctoral thesis of Elfimov (1989) provides the only other thorough comparative study of thoracican cyprids (see Elfimov, this volume).

The present study investigates the morphology of the antennular attachment organ from a wide range of species representing all three orders of the Cirripedia. For comparison, we also include the cyprid-like larvae of the Ascothoracida, a putative sister group to the Cirripedia (Grygier 1987; Høeg 1992a; Spears et al. 1994).

2 MATERIALS AND METHODS

Rearing and SEM study took place independently by two teams, JM and HAY in

Table 1. List of species studied by SEM and source of ovigerous adults from which the cyprids were reared or dissected. *Lepas australis* were dredged.

Species	Family	Source
Acrothoracica		
1. *Trypetesa lampas* (Hancock 1849)*	Trypetesidae	West coast of Sweden
Thoracica		
2. *Lepas australis* (Darwin 1852)	Lepadidae	Unknown
3. *Dosima fascicularis* (Ellis & Solander 1786)#	Lepadidae	South Wales
4. *Ornatoscalpellum stroemii* (Sars 1859)	Scalpellidae	Iceland
5. *Pollicipes pollicipes* (Gmelin 1789)#	Scalpellidae	Brittany
6. *Capitulum mitella* (L.), 1767#	Scalpellidae	Hong Kong
7. *Elminius modestus* (Darwin 1854)#	Archaeobalanidae	South Wales
8. *Balanus improvisus* (Darwin 1854)	Balanidae	Oslo Fjord
9. *Chthamalus montagui* (Southward 1976)#	Chthamalidae	South Wales
10. *Acasta spongites* (Poli 1791)#	Archaeobalanidae	West Wales
11. *Megatrema anglica* (Sowerby 1823)#	Pyrgomatidae	West Wales
12. *Verruca stroemia* (Muller 1776)#	Verrucidae	South Wales
Rhizocephala		
13. *Briarosaccus tenellus* (Boschma 1970)	Peltogastridae	West coast of Canada
14. *Clistosaccus paguri* (Lilljeborg 1860)*	Clistosaccidae	West coast of Sweden
15. *Mycetomorpha vancouverensis* (Potts 1911)	Mycetomorphidae	North Pacific
Ascothoracida		
16. *Ulophysema oeresundensis* (Brattström 1936)	Dendrogastridae	The Sound, Denmark

Swansea, and JTH and PGJ in Copenhagen. The cyprids marked with '#' in Table 1 were laboratory reared at the School of Biological Sciences, University of Wales, Swansea, by JM and HAY using the methods of Stone (1989). The resulting advanced stage VI nauplii were isolated in filtered seawater containing antibiotics. To avoid contamination with detritus and bacteria the cyprids were removed as soon as possible to further clean seawater and anaesthetized with menthol. Fixation (in Bouins) was performed when the antennules were suitably relaxed. The cyprids were dehydrated through acetone, diethyl ether and air dried in vacuum. They were coated with gold in a Polaron MK2 sputter coater and viewed in a Jeol SCM 35C SEM operated at 30kV.

The species marked '*' in Table 1 were laboratory reared by JTH and PGJ at the Kristineberg Marine Biological Station on the west coast of Sweden. Rearing followed the method of Høeg (1984) and Svane (1986). Cyprids were fixed in glutaraldehyde in seawater and postfixed in 1% OsO_4 followed by transfer through an ethanol series and storage at 70%. Specimens were then gradually transferred to 70% acetone and fully dehydrated in a series of that solvent. This was followed by critical point drying using CO_2, mounting on aluminium stubs, sputter coating with gold, and examination in a Jeol JSM 840 SEM operated at 5 or 15kV. We received cyprids of *Briarosaccus tenellus* (reared), *B. improvisus* (reared) and *L. australis* (dredged from plankton) by courtesy of Prof. T. Shirley, Univ. of Alaska, Dr. H. Hovde, Balanus Tox Test, Oslo, and Prof. W.A. Newman, Scripps Institution of Oceanography respectively. Cyprids of *Ornatoscalpellum stroemii* were dissected from fixed, ovigerous adults kindly provided by the ZMUC. PGJ collected larvae of *Ulophysema oeresundense* from live adult parasites dredged in the Sound, Denmark.

3 RESULTS

3.1 *General findings and nomenclature*

Figure 1, redrawn from Nott & Foster (1969), illustrates key features of the third and fourth antennular segments in cirripede cyprids. Plates 1-13 SEM micrographs of the attachment organs in the species studied here. Line drawings (Fig. 2) diagrammatically represent some of the distinctive features in selected species. Numerical information is set out in Table 2.

Our account follows the anatomical terminology adopted by Nott & Foster (1969) for the various structures observed for *Semibalanus balanoides* antennules (see Fig. 1).

Nott & Foster (1969) and Walossek et al. (submitted) explain in detail how the attachment organ originates as a bulge distally on the elongate second antennular segment. The fourth segment, terminal in the nauplii, becomes displaced laterally in the cyprid so the attachment organ ends up as the functionally terminal part of the antennule.

In the following account on single species we emphasize the morphology of the third antennular segment, the shape of its attachment disc and the intricacies of velum, skirt and cuticular villi. We refer to the first, second and fourth segments only where necessary for understanding the structure or functioning of the third.

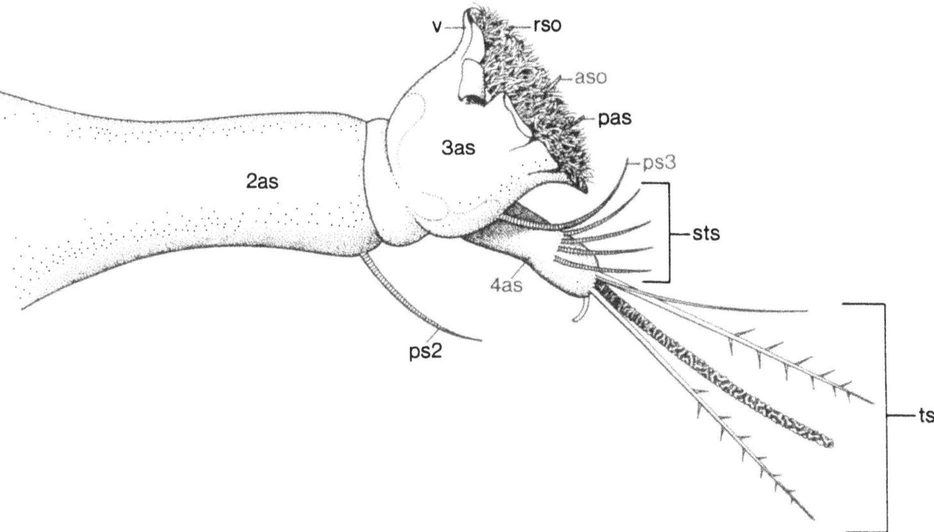

Figure 1. *Semibalanus balanoides*. Schematic drawing of antennule; showing the principal features of the 3rd and 4th segments. Redrawn after Nott & Foster (1969).

We give details on sense organs (sensilla) where observed, but emphasize that SEM alone cannot describe them adequately nor even fix their number, since many may remain obscured by the carpet of cuticular villi on the attachment disc. The same holds true for gland openings. TEM remains the preferred technique for counting and studying these structures (see Pl. 1C). All sense organs on the third segment represent setae (sensilla) and should ideally be called accordingly. However, such a change of Nott & Foster's (1969) nomenclature would result in identical names for the postaxial sense organ on the attachment disc and the postaxial seta 3 on the side of the third segment. To avoid such confusion we chose to keep their original nomenclature (Fig. 1).

3.2 *Results from single species*

3.2.1 *Trypetesa lampas*

Trypetesa lampa (Hancock 1849) (Pl. 1) is a boring commensal in hermit crab inhabited gastropod shells. Cypris larva (c. 590 μm long).

Antennules (Pl. 1A): Second segment relatively long and slender, third segment elongate with the attachment disc on the functionally ventral side. Fourth segment positioned laterally and about one third up from the proximal end of the third segment.

Attachment disc: (Pl. 1B): Oblong, c. 20 μm long and with a maximum width of c. 9 μm. A few small circular pores (antennulary gland openings) seen between the villi. A system of narrow grooves possibly containing the openings of the cement canal seen near the axial organ.

Skirt: Consists of a single unbroken flap. The outer surface of the skirt completely smooth, inner surface with a few scattered villi.

Table 2. Measurements of antennular features in cyprids.

Species	Carapace length µm[(#)]	Range µm	Number in sample	Antennule 3rd article length	Antennule 3rd article diameter	Attachment disc diameter	Attachment disc area	Number of microvilli per µm²
T. lampas	590	520-630	195	22	13	11*	190	5.4
L. australis	2580	2430-2820	10	160	108	71*	8760	7.0
D. fascicularis	940	890-980	8	14	67	68	3630	2.2
O. stroemii	910	850-950	50	70	33	33*	1560	5.3
P. pollicipes	440	390-470	25	24	28	25	490	8.1
C. mitella	460	-	2	37	24	23	420	10.2
E. modestus	510	470-550	50	20	19	23	420	9.3
S. balanoides	830	810-880	50	27	33	30	710	7.4
B. improvisus	600	570-640	10	16	21	20	310	8.3
C. montagui	430	390-470	50	19	27	24	450	7.5
A. spongites	500	470-540	25	15	14	21	350	5.3
M. anglica	570	540-620	15	41	12	3.4	85	4.8
V. stroemia	540	530-570	15	18	19	29	660	3.7
B. tenellus	-	410-500	-	25	23	6*	140	3.4
C. paguri	170	160-190	122	4.5	3	2.7*	16	5.2
M. vancouverensis	410	390-430	6	7	10	8*	160	3.3

[(#)]Mean or from single measurement. * These measurements represent the width of the attachment disc since they are elongate rather than circular in those species.

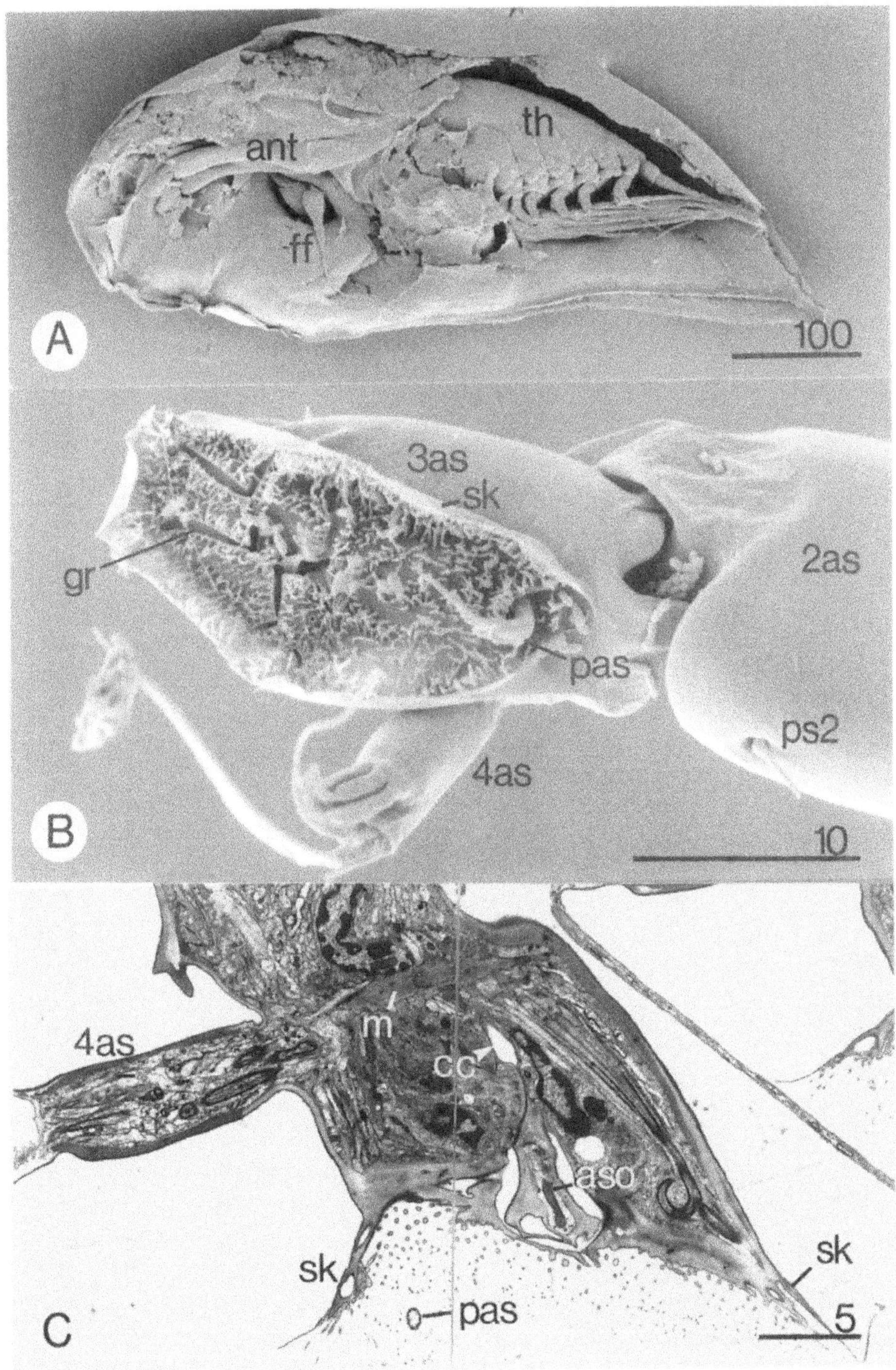

Plate 1. *Trypetesa lampas*, cyprid. (A) Entire cyprid opened showing the swimming position of the antennules; (B) Third antennular segment with the attachment disc exposed. The distinct grooves probably serve to distribute cement over the disc at irreversible settlement; (C) TEM of third and 4th segment. Note the cement canal.

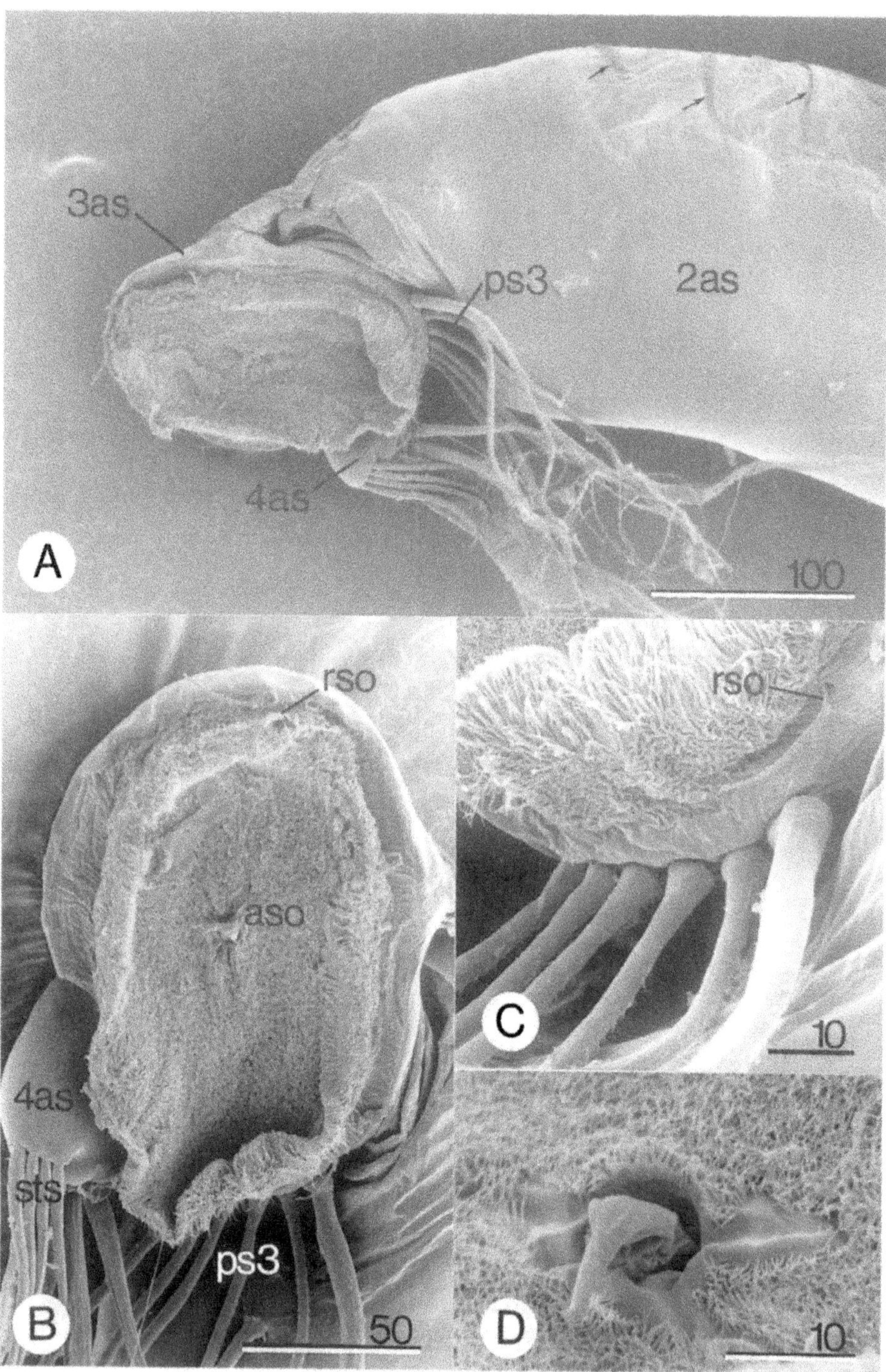

Plate 2. *Lepas australis*, cyprid. (A) Right antennule in mesial view; (B) Attachment disc. Note the entirely exposed radial sense organs (rso); (C) Postaxial side of proximal part of the third segment showing the six postaxial setae; (D) Axial sense organ.

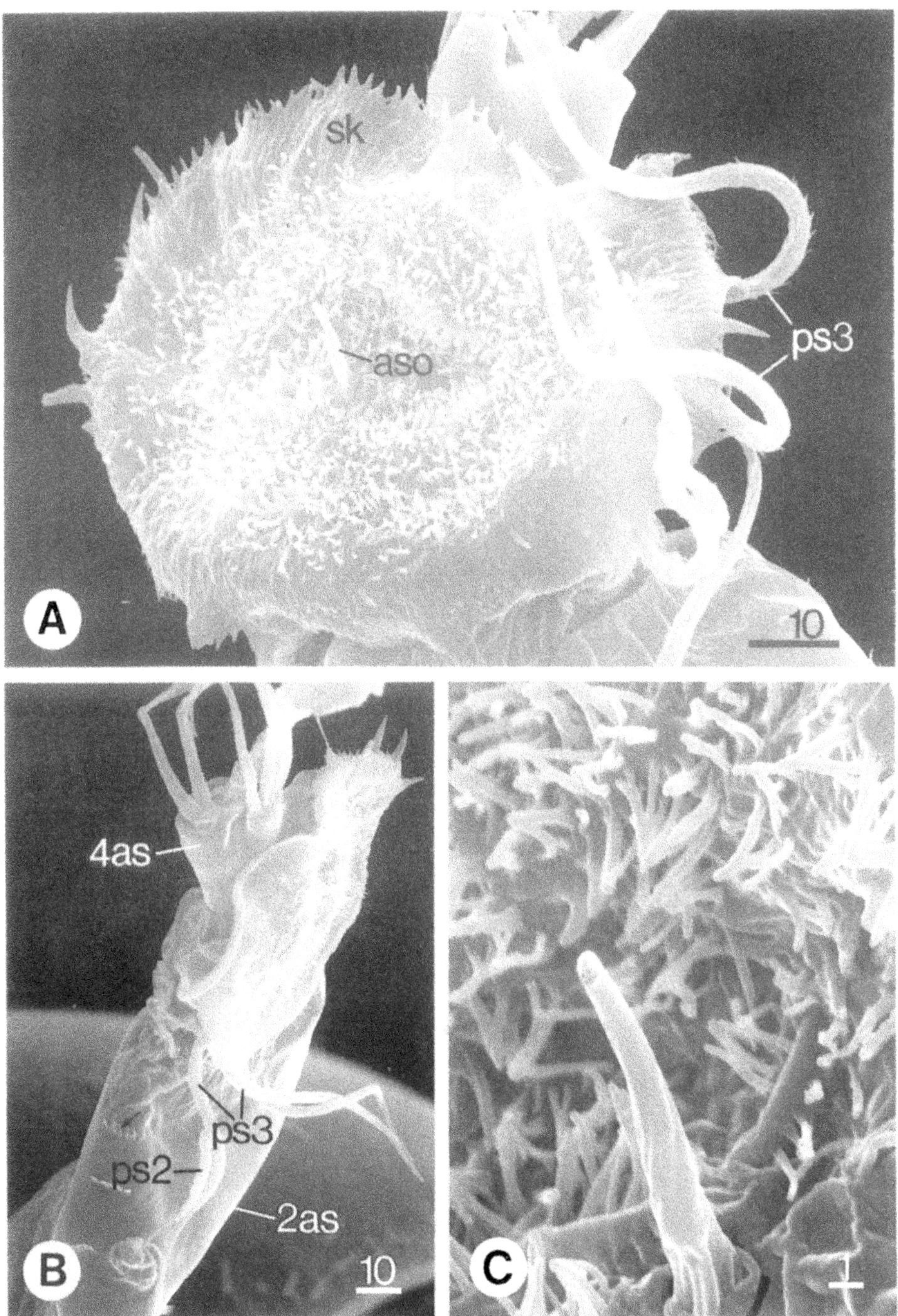

Plate 3. *Dosima fascicularis*, cyprid. (A) Attachment disc showing villi mainly in centre of disc, thin outer skirt devoid of normal villi furnished with fine, radially oriented, cuticular processes. Four postaxial setae; (B) Lateral view of antennule showing shallow shape of third segment; (C) Centre of disc with cuticular villi and axial sense organ.

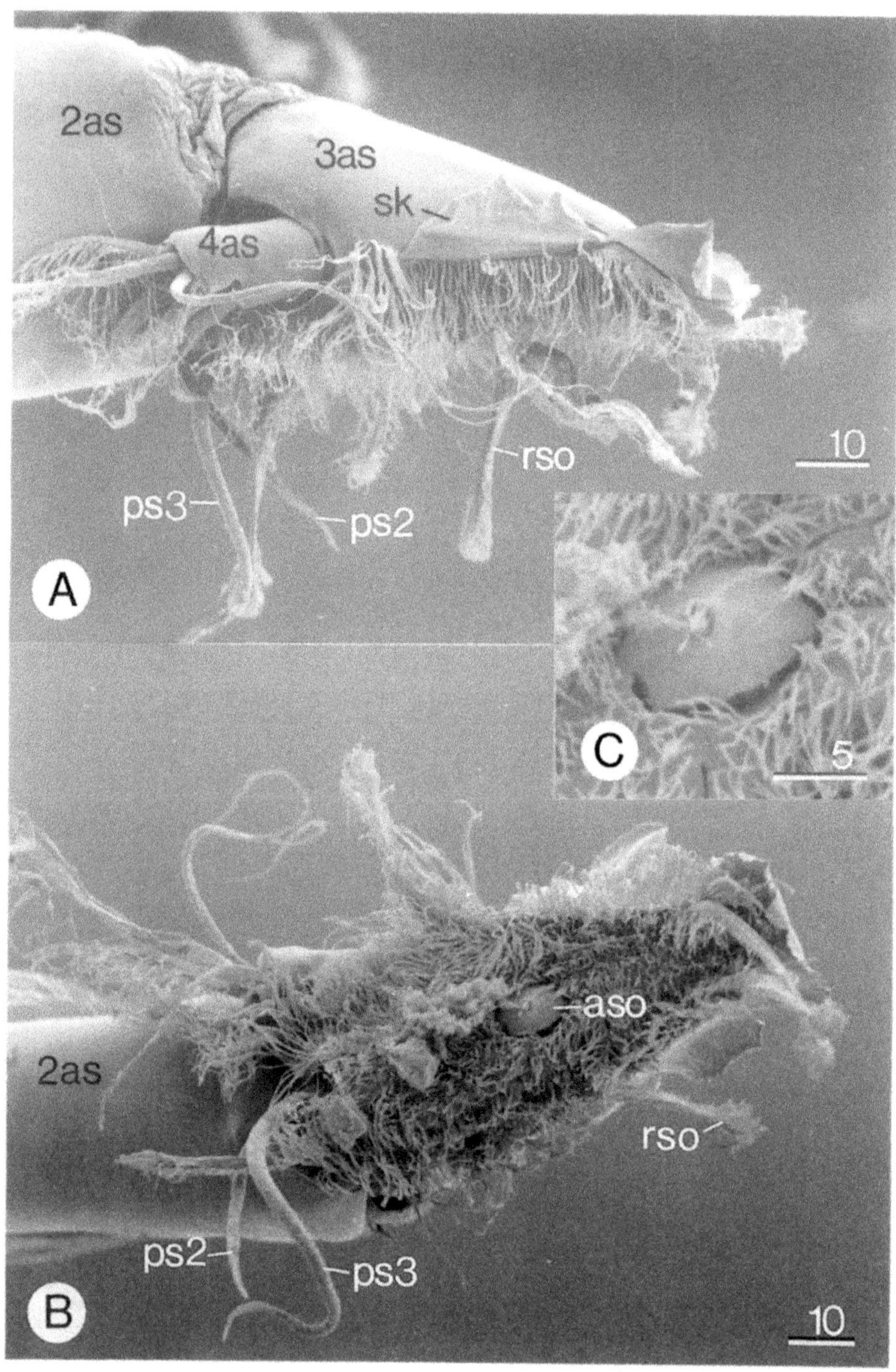

Plate 4. *Ornatoscalpellum stroemii*, cyprid. (A) Third antennular segment of right antennule, lateral view. Note the special morphology of the radial sense organs (rso); (B) Attachment disc. The postaxial sense organ was not identified, probably because of the dense mat of long cuticular villi; (C) Close up of the axial sense organ.

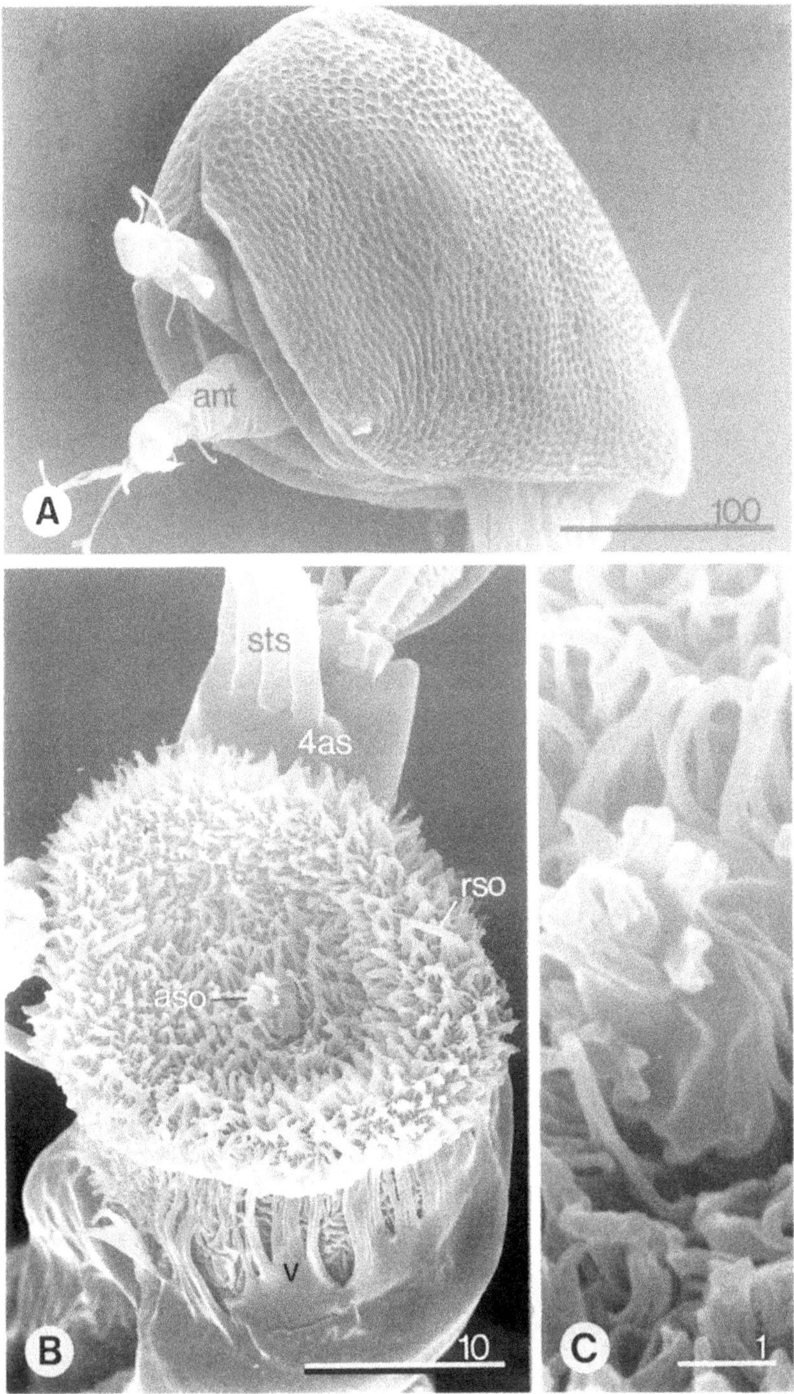

Plate 5. *Pollicipes pollicipes*, cyprid. (A) Entire cyprid showing anterior gape of valves and emergence of antennules; (B) Third segment of antennule showing filamentous velum below edge of disc; wrinkled wall of segment visible between velum elements; (C) Centre of disc with axial sense organ.

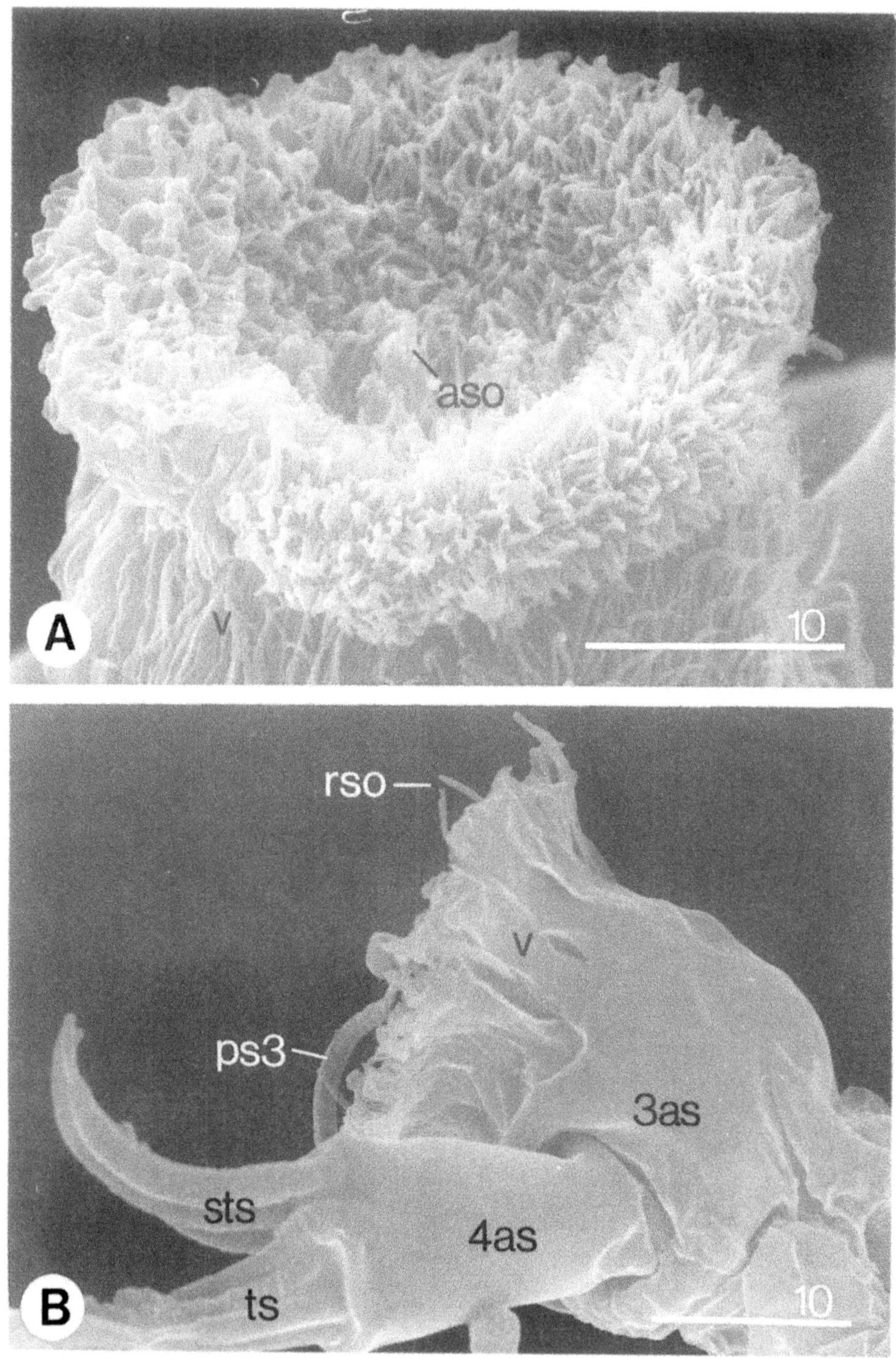

Plate 6. (A) *Capitulum mitella*, cyprid. Attachment disc; (B) *Elminius modestus*, cyprid. Third and fourth segment.

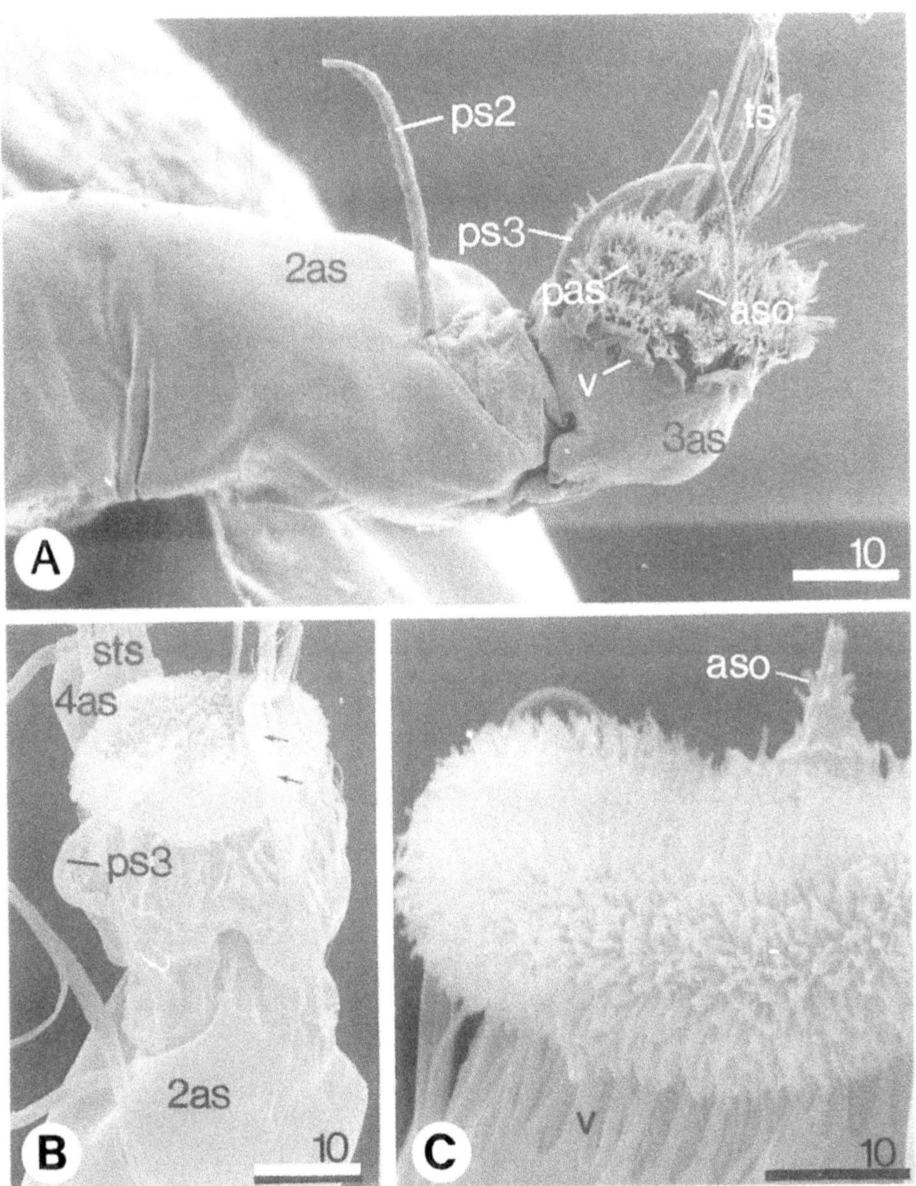

Plate 7. (A) *Balanus improvisus*, cyprid. Antennule and attachment disc; (B) *Chthamalus montagui*, cyprid. Antennule showing velum; (C) *Chthamalus montagui*, cyprid. Detail of attachment organ showing cuticular villi on reflexed disc, filamentous velum and prominent axial sense organ.

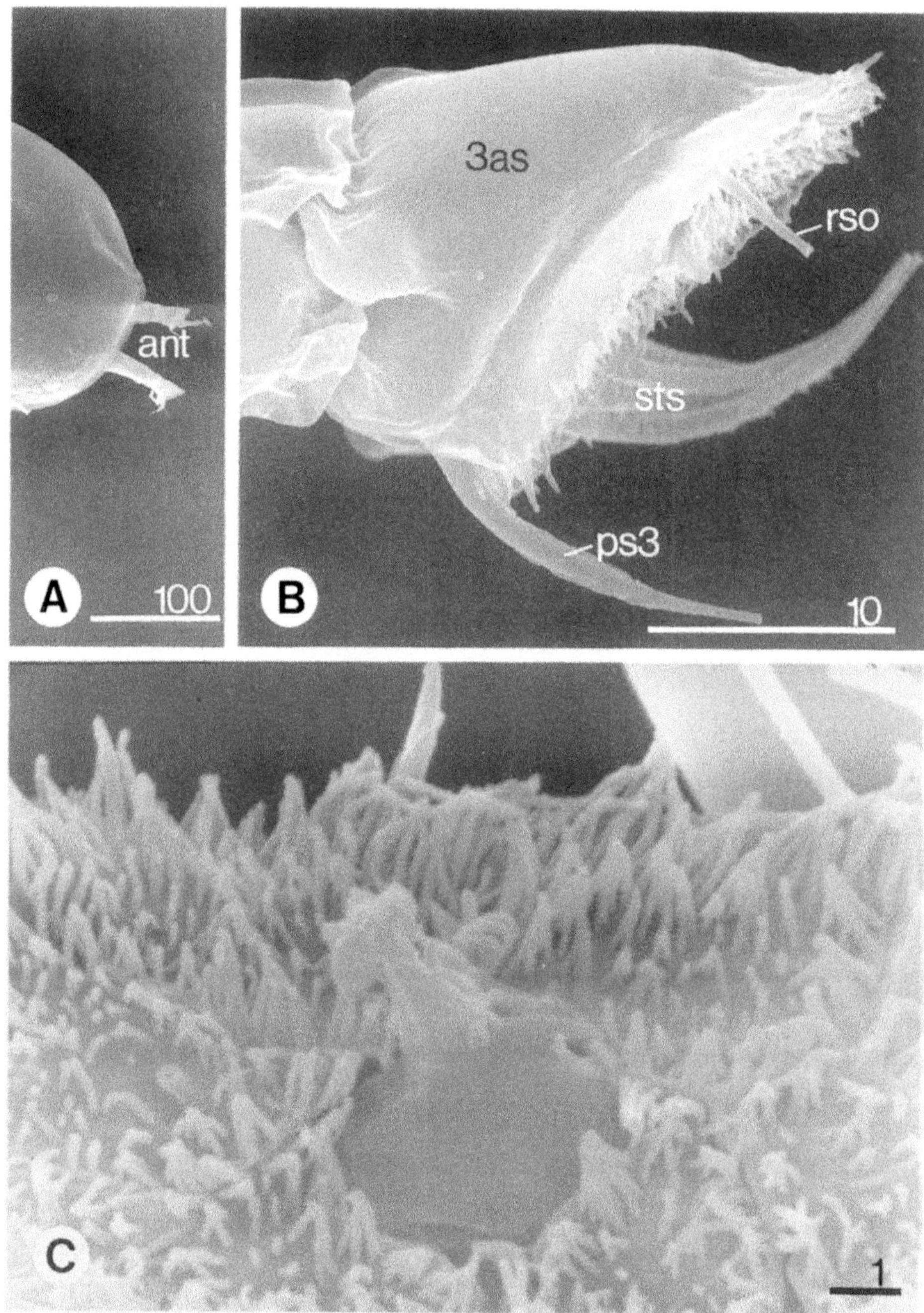

Plate 8. *Acasta spongites*, cyprid. (A) Indicating very small antennules; (B) Third segment detail; (C) Centre of disc with axial organ, postaxial sense organ, radial sense organs and cuticular villi.

Villi: Relatively few and not dense, thus revealing gland pores. Individual villus less than 2 µm long.

Axial sense organ: Indistinct in SEM, but seen clearly in TEM (Pl. 1C). Situated centrally on the attachment disc.

Postaxial sense organ: A large and strongly tapering seta situated just inside the velum near the postaxial margin of the attachment disc. Considered to be homologous with the postaxial sense organ of *Semibalanus balanoides* as described by Nott & Foster (1969).

Radial sense organs: Only two observed, situated near the distal margin of the attachment disc, in a position matching the position of the two large radial sense organs of *Lepas australis* (see below). Both less than 5 µm long, isodiametric, smooth, and with a distinct terminal pore.

Postaxial setae: Not present, and no setae on the third segment outside the attachment disc.

3.2.2 *Lepas australis*

Lepas australis (Darwin 1852) (Pl. 2) is a warm water cosmopolitan form attached to floating objects in the neuston. Cypris larva (c. 2600 µm long).

Antennules (Pl. 2A): By far the largest antennule of the species studied herein. The second segment relatively short and voluminous with an area of thin cuticle with four transverse rows of spinules on the dorsal side of the article. The third segment is elongate and relatively flat. The attachment disc is positioned on the functionally ventral side of the segment; fourth segment positioned laterally close to the second segment.

Attachment disc (Pl. 2B): By far the largest attachment disc, oval in outline, c. 70 µm by c. 140 µm. The whole disc surface slightly depressed with markedly raised margins, which seem to possess some flexibility especially at the proximal end of the disc.

Velum: No velum, so the radial sense organs are, contrary to what is seen in *S. balanoides*, exposed in their full length. The cuticle has elaborate folds in a zone between the margin of the attachment disc and the bases of the radial sense organs (Pl. 2C).

Villi: Densely packed, only a small area around the axial sense organ without villi. Individual villus narrow, c. 1 µm long near the axial sense organ but reaching c. 10 µm near the margin of the attachment disc.

Axial sense organ (Pl. 2D): Very distinct, situated in a circular depression in the very centre of the attachment disc. With a basal diameter of c. 5 µm tapering to c. 1 µm at the blunt and open-ended tip.

Postaxial sense organ: Not observed, if present, very indistinct and hidden between the villi.

Radial sense organs: Eight found along the margin of the attachment disc below the above mentioned folds of the cuticle. All situated in narrow depressions and c. 12-15 µm long except for the distal two which reach a length of c. 20 µm. These two also situated nearer to the margin of the attachment disc than the remaining radial sense organs.

Postaxial setae (Pl. 2C): Six present in a straight row, each more than 130 µm long, with distinctly swollen bases and covered with small spinules except for the proximal 10-20 µm.

3.2.3 *Dosima fascicularis*

Dosima fascicularis (Ellis & Solander 1786) (Pl. 3) is a warm water cosmopolitan species from the neuston. Cypris larva (c. 940 µm long).

Antennules (Pl. 3B): Relatively long and slender emerge towards the anterior end of the ventral border of the carapace valves, posterior to the frontal horn pores. The closed valves meet in mid-ventral position but gape anteriorly and posteriorly where the antennules and thoracic appendages respectively emerge. The first segment normally remains inside the carapace. The second segment is long, slender and parallel-sided.

Third segment (Pl. 3A): The attachment surface is exceptionally large (diameter c. 75 µm and area of c. 4400 $µm^2$), dwarfing and obscuring the third segment proper. Unlike some other species it is neither dished nor inflated but relatively flat. It has a regular disc shape with a very thin, flexible and slightly wavy outer region (Pl. 3D) here called the skirt, comprising outer 45% radially. It has a delicate frilly edge.

Attachment disc: The disc is subtended by a relatively small, low, and truncated cone-shaped structure, forming the body of the third segment. The fourth segment is large and protrudes distally, well beyond the third segment. In preserved material the attachment organ is uniquely angled such that its surface is in the same plane as the longitudinal axis of the appendage (Pl. 3A).

Velum: None observed in the material examined. If present must be rudimentary.

Villi (Pl. 3A and C): Typical villi are concentrated in the central area around the axial sense organ but are not as densely packed (c. 2.2 per $µm^2$) as in *Chthamalus* (7.5 per $µm^2$). The skirt is almost devoid of villi, but is marked by five radial striations which tend to rise peripherally from the surface as delicate tapering threads.

Axial sense organ (Pl. 3C): Very long (11.8 µm) standing well proud of the surrounding villi; unbranched and smooth sided, with an open-ended tube-like distal section.

Radial sense organs (Pl. 3A and B): Eight of these were observed, some open-ended, all unusually long (about 8 µm).

Postaxial sense organ: Indistinct.

Postaxial setae 3 (Pl. 3A and B): Four of these were observed; very long (c. 65 µm), covered with delicate spinules. All arch over the attachment surface.

3.2.4 *Ornatoscalpellum stroemii*

Ornatoscalpellum stroemii (Sars 1859) (Pl. 4) is a species that is found in the Atlantic, sublittoral to deep sea, attached to cnidarians or marine sponges. Cypris larva (c. 914 µm long).

Antennules: Second segment relatively short and tapering. Third segment elongate with the attachment disc on the functionally ventral side. The fourth segment positioned laterally near the second segment.

Attachment disc (Pl. 4B): Elongate oval, narrowing towards the distal end. Grooves as in *T. lampas* seen near the axial sense organ (Pl. 4C).

Skirt (Pl. 4A): Conspicuous, consisting of a number of flaps which may reach a length of almost 10 µm.

Villi: Numerous, relatively short near the centre of the disc exposing the disc itself; but more than 10 µm in length near the periphery.

Axial sense organ (Pl. 4C): An arched area in the centre of the disc with a few very short villi and a papilla at the apex.

Postaxial sense organ: Not seen, but the very long villi at the periphery of the attachment disc may easily have obscured this.

Radial sense organs: Very conspicuous, relatively the largest radial sense organs seen in the species studied here, reaching a length of almost 30 μm and with short cuticular villi distally.

Postaxial setae: Large, more than 50 μm tapering towards the distal end, surface smooth.

3.2.5 *Pollicipes pollicipes*

Pollicipes pollicipes (Gmelin 1789) (Pl. 5) was collected from the midtidal zone of surf beaten rocky shores in Brittany, France. Cypris larva (c. 450 μm long).

Antennules (Pl. 5A): Project anteriorly through wide gap between valves.

Third segment: Cup-shaped, with attachment disc at right angles to axis of the appendage. The maximum diameter of the third segment is c. 28 μm with the length of c. 24 μm.

Attachment disc (Pl. 5B): Circular and at right angles to the axis of the segment; diameter c. 25 μm and area about 490 μm. Slightly dished centrally; edges slightly reflexed.

Velum (Pl. 5B): Filamentous, with groups of long filaments united proximally into flat tangential plates. Walls of third segment (where visible between velar filaments) thrown into elaborate folds.

Villi (Pl. 5B and C): Very numerous; average length of a villus 4.2 μm. Villi extend to and protrude laterally at the margin of the disc.

Axial sense organ (Pl. 5C): Very short, (about 2.8 μm in length) with numerous filaments arising near top of a buttressing wall.

Radial sense organs (Pl. 5B): Three were observed; open-ended; quite long (length about 7 μm).

Postaxial sense organ: Similar to radial sense organs but shorter.

Postaxial seta 3 (Pl. 5A): Originates proximally to disc and curves away from the attachment disc; length about 24 μm.

3.2.6 *Capitulum mitella*

Capitulum mitella (L.), 1758 is tropical in intertidal splash zones (Hui 1983). Cypris larva (c. 470 μm long) (Pl. 6A).

Antennules: Emerge through pronounced anterior (as opposed to ventral) gap between carapace valves.

Third segment: Small, radially symmetrical, cylindrical; diameter about 25 μm, length about 35 μm.

Attachment disc (Pl. 6A): Circular and at right angles to axis of segment. Edges strongly inflated. Slightly dished centre.

Velum: Elaborately filamentous – perhaps several layers but not uniformly around whole segment.

Villi: Very numerous, average length apparently at least 6 μm; protrude laterally and even proximally at inflated rim.

Sense organs: Axial sense organ moderately developed with tubular terminal region.

Radial sense organs: Only two or possibly three observed.

Postaxial sense organ: Not observed.

Postaxial seta 3: Short and slender; not arching over disc.

3.2.7 *Elminius modestus*

Elminius modestus (Darwin 1854) (Pl. 6B) was collected from the rocky intertidal. Cypris larvae (c. 513 µm).

Antennules: Emerge from the carapace anteriorly but ventrally to the sharp angle between dorsal and ventral borders of valves.

Third segment (Pl. 6B): Bell-shaped, about 20 µm in length and 19 µm in diameter; plane of disc at right angles to segment axes i.e. regularly radially symmetrical, apart from fourth segment.

Attachment disc: Disc-shaped flat, maximum diameter about 23 µm and area about 415 μm^2, wider than rest of third segment

Velum (Pl. 6B): Consisting of well marked overlapping cuticular flaps, minutely lobed at distal edge – but no extension into filaments.

Villi: Very numerous.

Axial sense organ: Smooth conical, flask-shaped, very long (c. 8.5 µm); open ended.

Radial sense organs: About 6 counted; long (c. 4 µm) open-ended, some emerge within the rim of velar flaps.

Postaxial sense organ: Not observed.

Postaxial seta 3: Long (c. 27 µm) smooth, open ended; curved towards axial sense organ.

3.2.8 *Balanus improvisus*

Balanus improvisus (Darwin 1854) (Pl. 7A) occurs in the rocky intertidal. Cypris larva (c. 600 µm long).

Antennules: Second segment shaped as a relatively short tube; very extensive arthrodial membrane between the 2nd and 3rd segments, presumably giving the latter a large freedom of movement. Third segment bell-shaped and with a diameter comparable to the diameter of the second segment.

Attachment disc: Circular and positioned at the functionally distal end of the third segment.

Velum: As in *S. balanoides* it consists of a number of overlapping cuticular flaps inserting on the sides of the third segment well below the level of the attachment disc itself.

Villi: The whole surface of the attachment disc covered with a dense mat of villi concealing the cuticular surface of the disc itself.

Axial sense organ: Smooth conical, flask-shaped, very long, extending well beyond the cuticular villi. Positioned in the very centre of the attachment disc.

Postaxial sense organ: Simple, protruding c. 4 µm beyond the villi and positioned halfway between the axial sense organ and the rim.

Radial sense organs: None seen, probably obscured by the villi.

Postaxial seta: Relatively stout and blunt, c. 20 µm long with a few small spinules on the side facing the attachment disc.

3.2.9 *Chthamalus montagui*

Chthamalus montagui (Southward 1976) (Pl. 7B andC) was collected from the high rocky intertidal. Cypris larva (c. 430 µm long).

Antennules (Pl. 7B): Emerge anteriorly.

Third segment: Cylindrical and aligned with second;

Attachment disc (Pl. 7B and C): Circular, inflated and at right angles to axis of segment, edges strongly reflexed.

Velum (Pl. 7B and C): Filamentous, the numerous seta-like elements arranged in a single row and pressed loosely against the sides of the segment. Filaments tend to be united at their bases into flaps; maximum length of filaments about 6 µm. Overall total number of filaments in one specimen estimated at 90.

Villi (Pl. 7C): Very numerous; density of villi is 7.5 per μm^2; villi protrude laterally at swollen edges of disc.

Axial sense organ (Pl. 7C): Protruding, short with numerous truncated side-filaments. Tip castellated and open-ended.

Postaxial sense organ (Pl. 7C): Long (perhaps two)

Radial sense organs: In one particular specimen 3 were counted, open-ended, long (3.5 µm) each three times length of a villus.

Postaxial seta 3 (Pl. 7C): Rough, covered with some short setules, open-ended tapering, about 24 µm in length – sited close to axial sense organ.

3.2.10 *Acasta spongites*

Acasta spongites (Poli 1791) (Pl. 8) is a commensal in the sponge, *Dysidea fragilis* (Montagu). Cypris larva (c. 500 µm long).

Antennules (Pl. 8A): Emerge anteriorly.

Third segment and attachment disc (Pl. 8B): Attachment surface at a sharp angle to axis of appendage, making the third segment 'hoof-shaped', length between about 9 µm along ventral side and about 21 µm at dorsal aspect; maximum diameter of third segment about 14 µm; angled disc with maximum diameter of about 23 µm.

Attachment disc (Pl. 8B): Rounded, small (c. 21 µm mean diameter; with area of 350 μm^2; angle between attachment disc and third segment axis is 47°.

Velum (Pl. 8B): Cuticular flaps overlap each other, well marked, thickened rim; no extension into filaments.

Villi (Pl. 8C): Numerous; vary in length; maximum length about 2 µm; short and less concentrated villi around axial sense organ; mean density of villi 5.3 per μm^2.

Axial sense organ (Pl. 8C): Long (3.7 µm), large smooth base, slender tip with few spines; wide at base, tapering at distal end; open-ended.

Radial sense organs (Pl. 8C): Very long (c. 5 µm), open ended; about 7 of these counted.

Postaxial sense organ (Pl. 8C): Short and stout.

Postaxial seta 3 (Pl. 8B): Short (c.25 µm) furnished with few spinules; wide at base, tapering at distal end; open-ended.

3.2.11 *Megatrema anglica*

Megatrema anglica (Sowerby 1823) (Pl. 9) is a commensal of the ahermatypic coral, *Caryophyllia smithii* Stokes & Broderip. Cypris larva (c. 570 µm long).

Antennules: Highly distinctive; emerge through anterior gape between cypris valves. First segment deep and heavy. Second segment wide at base, tapering to joint with third.

Third segment (Pl. 9B): This segment has a peculiar shape; sharply conical; sides angled at about 19°; mean diameter 12 µm; mean length 41 µm.

Attachment disc (Pl. 9C): Not disc-shaped, but greatly elongated; angle between attachment surface and third segment axis much less than 90° (26°); attachment surface has only very small area (c. 85 µm^2); surrounded on either side by symmetrically arranged velar flaps (Pl. 9B).

Velum (Pl. 9C): Consists of twelve flaps which vary in width from about 7 µm near the tip to over 10 µm near the base, with three smaller ones transversely at proximal end; all flaps have a well-marked, slightly thickened rim with no extensions into filaments; each lateral flap overlaps the next more proximal flap.

Villi (Pl. 9C): Numerous; vary in length being progressively longer proximally, but their diameter is relatively constant at about 0.15 µm; most arise laterally from within bounding velum; the central area of the attachment surface is devoid of villi; orientation of larger villi towards distal end, others directed towards midline.

Axial sense organ: Not observed.

Radial sense organ (Pl. 9C): Three pairs observed. The two distal pairs very short (1 µm long), stout, paired structures arising from within velum, open-ended, curved, strong, and fang-like. Third pair, sited proximally to others much longer and flexible.

Postaxial sense organ: Not observed.

Postaxial seta 3 (Pl. 9C): Short (c. 20 µm), open-ended, setulate, arises immediately proximal to villous area.

3.2.12 *Verruca stroemia*

Verruca stroemia (Müller 1776) (Pl. 10) is a form that is sublittoral, attached to a variety of biotic and abiotic substrata. Cypris larva (c. 540 µm long).

Antennules (Pl. 10A): Emerge from carapace valves in a ventral position posterior to the frontal horn pores.

Third segment (Pl. 10A and C): Directed ventrally by virtue of the angle between the attachment disc and the axis.

Attachment disc (Pl. 10A): Dish-shaped, with a diameter of about 29 µm and area of 660 µm. Edges of attachment disc produced as a very thin and flexible cuticular extension or 'skirt' (Pl. 10A and C). The skirt has a frilly edge.

Velum: No obvious velum observed.

Villi (Pl. 10B and C): Low density; about 3.7 per µm^2.

Axial sense organ (Pl. 10B and C): Large and prominent, wide at base, smooth and long (c. 5.3 µm) with single open-ended tube.

Radial sense organs (Pl. 10C): Five identified, open-ended, quite long (5.5 µm in length) orientated peripherally.

Postaxial sense organ: Not observed.

Postaxial seta 3 (Pl. 10A): Well-developed.

3.2.13 *Briarosaccus tenellus*

Briarosaccus tenellus (Boschma 1970) (Pl. 11) is parasitic on abdomen of the lithodid *Hapalogaster mertensii*, intertidal to 55 m. Cypris larva (c. 400 µm long).

Antennules (Pl. 11A): Second segment long and slender compared to other kentrogonids (Walker 1985; Glenner et al. 1989). Third segment very elongate with the attachment disc on the functionally ventral side of the segment. Fourth segment positioned laterally almost halfway between the second segment and the distal end of the attachment disc.

Attachment disc (Pl. 11C): Very elongate, c. 25 µm long and only c. 6 µm wide, appearing slightly wider at the distal end. The spinous process found in many other rhizocephalan cyprids is absent in males of this species.

Skirt (Pl. 11B): The whole attachment disc is enclosed by a narrow but distinct and continuous skirt which seems to be a little larger in the distal part of the attachment disc.

Villi: The proximal half of the attachment disc densely packed with villi. The distal half with less densely packed villi and a central area almost without villi. Individual villus c. 1-2 µm long.

Axial sense organ: Rather inconspicuous, situated in the area without villi.

Postaxial sense organ (Pl. 11B and C): Positioned centrally in the proximal half of the attachment disc. About 5 µm long, strongly tapering and without any ornamentation.

Radial sense organs: Not seen; so if present, they are small and difficult to distinguish from the cuticular villi of the attachment organ.

Postaxial setae: No postaxial seta identified proximally to the attachment disc. Instead the male cyprid possess a large aesthetasc just outside the skirt at the proximal edge of the attachment organ.

3.2.14 *Clistosaccus paguri*

Clistosaccus paguri (Lilljeborg 1860) (Pl. 12) is parasitic on the abdomen of various hermit crabs (Høeg 1982). Cypris larva (c. 175 µm long).

Antennules (Pl. 12A): The antennules of *C. paguri*, including the attachment organ, were described and documented by Glenner et al. (1989). In terms of setation this species possesses the most reduced antennules and attachment apparatus among the species studied here. The third segment is hoof-shaped with the attachment disc at the functionally distal end of the segment. Fourth segment positioned proximally to the attachment disc.

Attachment disc (Pl. 12B): Minute, 7 µm long and with max. width of 3 µm. Large central area almost completely without villi, not obscured by secreted cement as suggested by Glenner et al. (1989). Large spinous process present at the distal end of the attachment disc.

Villi: The rather few, short villi restricted to the margin of the attachment disc.

Axial sense organ: No axial sense organ identified on the attachment disc.

Postaxial sense organ (Pl. 12B): A single, long (c. 3 µm) and distally tapering seta, termed axial seta by Glenner et al. (1989) could be homologous to the postaxial sense organ of the thoracican cyprids.

Radial sense organs: None observed. No single ‘villus’ in the periphery of the attachment disc fits this designation, so probably absent altogether.

Postaxial seta: Absent, since no seta inserts proximally outside the attachment disc, neither a simple seta nor an aesthetasc.

3.2.15 *Mycetomorpha vancouverensis*

Mycetomorpha vancouverensis (Potts 1911) (Pl. 13A) is parasitic on abdomen of Caridea. Cypris larva (c. 430 μm long).

Antennules: The antennules of this species including the attachment organ were described and documented by Glenner et al. (1989). The third segment is rather elongate but seems to have the attachment disc in a functionally distal position as in *C. paguri*. Fourth segment positioned laterally close to the second segment.

Attachment disc: Oval in shape, the distal end narrower than the proximal, with a distal spinous process and with grooves of the cement canal readily visible.

Skirt: Continuous around the disc and more distinct than the one found in *C. paguri*.

Villi: Relatively few, also cuticular disc itself always exposed. The villi of equal length; evenly distributed but absent around the axial organ.

Axial sense organ: Simple, tube form and slightly tapering, positioned in the anterior end of the central area, c. 5 μm behind the spinous process.

Postaxial sense organ: Conspicuous, c. 10 μm long, extending far beyond the villi.

Radial sense organs: None seen.

Postaxial seta 3: As in *B. tenellus*, no simple postaxial seta, but a large aesthetasc found approximately in the same position.

4 COMPARISON OF SPECIES

4.1 *Shape of the third segment*

As seen in the SEM micrographs the shape of the third segment and the outline and orientation of the attachment disc vary considerably (Fig. 2) both within and between the three orders here studied. We can therefore conclude that the morphology epitomized by *Semibalanus balanoides* by no means represents a generalized condition in the Cirripedia, nor is it likely to represent the plesiomorphic ground pattern.

Some species possess a bell- or cup-shaped third segment with an almost circular attachment disc, orientated at right angles to the longitudinal axis of the third segment. The attachment disc is therefore located at the functionally distal end of the antennule. Because the length of the third article is approximately the same as its diameter it can be described as isodiametric. This morphology exists in the balanomorphs *Semibalanus balanoides* (see Nott & Foster 1969), *Elminius modestus* (Pl. 6B), *Balanus improvisus*, and *Chthamalus montagui*, and in the 'scalpellids' *Pollicipes pollicipes* and *Capitulum mitella*. However, another balanomorph, the sponge inhabiting *Acasta spongites*, has an attachment disc with a somewhat oval attachment outline and much more strongly angled disc surface. This hoof-shaped third segment represents a morphological transition to the elongate third segments described below. The lepadids *Lepas australis* and *Dosima fascicularis* both possess a very shallow third segment in the form of a low inverted cone with its attachment surface orientated parallel to the longitudinal axis of the antennule.

At another extreme, the third segment is not isodiametric but has an elongate or cylindrical form with a distinctly oval or at times even rectangular attachment disc

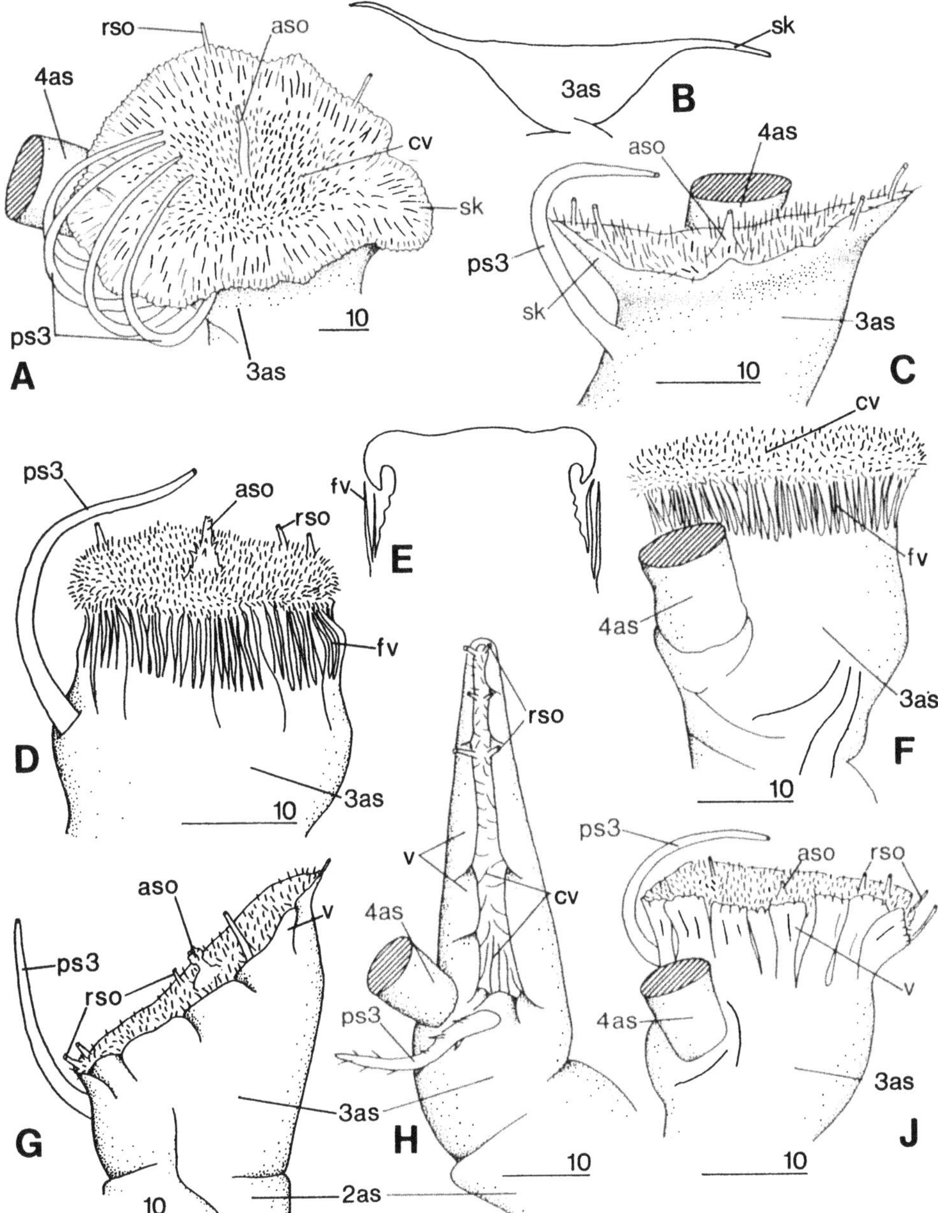

Figure 2. Diagrammatic representation of principal features of third segments. (A) *Dosima fascicularis;* (B) Hypothetical section of third segment of A; (C) *Verruca stroemia*; (D) *Chthamalus montagui*; (E) Hypothetical section of third segment of D; (F) *Pollicipes pollicipes*; (G) *Acasta spongites*; (H) *Megatrema anglica*; (J) *Elminius modestus*. Scalebars in μm.

oriented more or less ventrally in the extended antennule. We found this morphology in the acrothoracican *Trypetesa lampas* (Pl. 1B), the rhizocephalans *Briarosaccus tenellus* and *Mycetomorpha vancouverensis* (Pl. 11 and 13A), and in the scalpellid *Ornatoscalpellum stroemii* (Pl. 5). Moreover, Glenner et al. (1989) found this shape in all Rhizocephala Kentrogonida and some Akentrogonida. Surprisingly, however, some akentrogonids such as *Clistosaccus paguri* (Pl. 12) and *Sylon hippolytes* (see Glenner et al. 1989: Fig. 33) display a hoof-shaped third segment reminiscent of the condition in *Acasta*. The balanomorph coral barnacle *Megatrema anglica* possesses an extremely elongated, parallel-sided attachment organ, with a length/breadth ratio of approximately twelve and with the rim obscured by the velum (Pl. 9B).

4.2 *Velum and skirt*

Configuration of the attachment disc rim varies extensively between species. In most species a raised rim (velum or skirt) of very thin cuticle surrounds the attachment disc and when present, is always peripheral to the radial sense organs.

We call this structure a velum *sensu* Nott & Foster (1969), when it inserts about halfway down the sides of the third segment and thus removed from the actual rim of the attachment disc itself (e.g. Fig. 1 and Pl. 13). A skirt is less extensive in size and rooted almost immediately peripheral to the villus-covered area so it appears as part of the margin of the disc itself.

Semibalanus balanoides has a velum consisting of a series of tangential and slightly overlapping flaps, whose free distal edges have a minutely frilly margin (Nott 1969, Fig. 1). We found a similar, but not necessarily frilly velum in *Elminius modestus, Balanus improvisus, B. amphitrite, Acasta spongites* and, perhaps surprisingly, *Megatrema anglica*, whose attachment organ deviates considerably in several other respects (Pl. 6B, 8B and 9B). Such a velum of flaps seems to characterize the Balanidae and may represent an apomorphy.

In *Capitulum mitella, Pollicipes pollicipes*, and *Chthamalus montagui* a row of numerous tapering filaments replaces the cuticular flaps. Since these filaments also insert somewhat down the side of the third segment we consider them as a type of velum. When looking at the attachment disc face on, the reflexed edge of the disc often obscures this type of velum (Pl. 5B, 6A, 7C and 13).

A typical skirt occurs in *Trypetesa lampas*, most Rhizocephala such as *Briarosaccus tenellus* and *Mycetomorpha vancouverensis*, and in the scalpellid *Ornatoscalpellum stroemii* and in *Verruca stroemia*. In these species it consists of a rim of thin cuticle almost immediately peripheral to the villus-covered attachment disc so it appears to stand on the margin of the disc. In *O. stroemii* the skirt consists of several, slightly overlapping flaps just as in a velum, but in *T. lampas* and the Rhizocephala the skirt is perfectly continuous except at the extreme proximal end of the disc where it is absent altogether.

A remarkable skirt-like structure exists in the lepadid *Dosima fascicularis*, but probably arose convergently with other skirts (Pl. 2A). Another lepadid, *Lepas australis*, has neither a velum nor a skirt, leaving the radial sense organs fully exposed to view. Again, in *L. australis* a naked peripheral zone circumscribes the attachment disc, separated from the central villus-covered area by a distinctly raised ridge with villi on both inner and outer sides (Pl. 2B). Proximal to this is a ridge which develops into a veritable girdle. These unique features may well serve the same function as the velum seen in other species.

4.3 *The rim of the attachment disc*

In species with a velum, the attachment disc itself habitually has a strongly reflexed rim, as illustrated for *Semibalanus balanoides* by Nott (1969) and Nott & Foster (1969: Figs 2 and 3). This reflexed posture results in a convex rim, of inflated appearance as also seen in *Capitulum mitella, Pollicipes pollicipes* and *Chthamalus montagui* (Fig. 1, Pl. 1and 4). Cuticular villi populate the disc surface but never extend around the rim on to the lateral walls of the segment.

In species with a skirt, the villus-covered area usually extends to its very base. Some villi may even insert on the inner side of the skirt but none are found beyond its rim. *Dosima fascicularis* displays a very unusual distribution of cuticular structures on the attachment disc. The skirt has almost no villi (which are confined to the centre of the disc) but is marked by fine radial striations which tend to rise peripherally from the surface in delicate tapering threads (Pl. 3A).

4.4 *Sense organs*

As described by Nott & Foster (1969) in *Semibalanus balanoides*, the attachment disc typically carries three types of sensory setae (called organs), viz., an axial sense organ, a postaxial sense organ, and eight (or up to eight) radial sense organs. In addition, most species carry a sensory seta, the postaxial seta 3, on the side of the third segment beyond the perimeter of the attachment disc. The straight line passing through the postaxial seta 3, the postaxial sense organ and the axial sense organ corresponds to the plane in which the pre- and post-axial series of setae in nauplii are deployed (Fig. 1). The setation pattern of this third segment seems, with few modifications, to hold true throughout the Cirripedia (Glenner et al. 1989), but we emphasize again that SEM alone remains inadequate to count structures on the attachment disc accurately.

4.4.1 *The axial sense organ*

This organ occurs at the approximate center of the attachment disc and frequently has a species-specific shape. Nott & Foster (1969) regarded axial sense organs as chemosensory receptors for testing the substratum prior to settlement.

They often consist of a cone shaped base which abruptly tapers to an open-ended, tubular seta, which usually reaches out above the general level of the cuticular villi. Both the basal cone and the tubular seta may be adorned with cuticular villi (Pl. 7C). In most Rhizocephala and in *Trypetesa lampas*, the axial sense organ remained obscured by villi on the disc although we detected it on TEM sections (Pl. 1C). In the Rhizocephala, we can now see that the seta called axial sense organ by Glenner et al. (1989) must represent the postaxial sense organ. In fact, Plate 13A of *Mycetomorpha vancouverensis* shows both the true axial organ and the postaxial organ, called axial by these authors.

Similarly, we believe that the open-ended structure called spinous process by Glenner et al. (1989: Figs 21 and 26) in cyprids of *Peltogaster* and *Peltogasterella* truly represents the axial organ (but not the spinous process in other species described by these authors).

4.4.2 *Radial sense organs*

These sensilla occur at the periphery of the disc proper but always inside the skirt or velum. Nott & Foster (1969) found eight in *Semibalanus balanoides* using TEM and using this technique we have seen a comparable number in rhizocephalan cyprids. Two of the organs habitually stand proud above the carpet of cuticular villi situated at the 'distal' end of the disc and on each side of the segment axis, but most of the radial organs habitually remain obscured by villi. Occasionally one radial sense organ (or seta) appears inset from the edge, giving a false impression of a postaxial sense organ.

4.4.3 *Postaxial sense organ*

The true postaxial sense organ is a seta inserting on the disc between the axial organ and the proximal edge. Although at times difficult to detect, we found it with ease in most Rhizocephala, where it is unusually long.

4.4.4 *Postaxial seta 3*

This sensory seta inserts clear of the attachment disc and we could identify it in all species of Thoracica. Notably however, it is lacking in cyprids of the Acrothoracica, unless what we call postaxial organ in Plate 1B truly represents the postaxial seta 3. Rhizocephalan cypris larvae always lack a simple postaxial seta 3.

However, in approximately the same position male cyprids of the Rhizocephala Kentrogonida and *Mycetomorpha vancouverensis* carry a large aesthetasc, whereas a seta (simple or aesthetasc) is altogether lacking from this position in cyprids of both female Kentrogonida and all Akentrogonida except *M. vancouverensis* (Walker 1985; Glenner et al. 1989; Andersen et al. 1990). We therefore suggest that this male specific aesthetasc represents the homologue of the postaxial seta 3 in the Thoracica. It inserts in approximately the right position. Moreover, Plate 13A shows that in *M. vancouverensis* the aesthetasc occurs together with both a postaxial organ and an axial organ on the attachment disc, and the same holds true for *C. paguri* (Glenner et al. 1989: Figs 17 and18; with their axial organ reinterpreted as postaxial). This obviously refutes a homology between the rhizocephalan aesthetasc and either of these structures. Nott & Foster (1969) showed that the postaxial seta 3 originates from seta 11 or 12 in the naupliar antennule. Unfortunately, however, rhizocephalan nauplii have reduced antennular setation and lack both seta 11 and 12 (Walossek et al. submitted). Hence we cannot support our homologization scheme of the third segmental aesthetasc with ontogenetic arguments.

The lepadids *Lepas australis* and *Dosima fascicularis* provide a special problem. In place of a single postaxial seta 3, they carry 6 and 4 setae respectively in an identical position. Multiplication of setae has probably occurred many times in crustacean evolution and it is easy to imagine a genetic basis for such a change. Thus we do not hesitate to homologize the 6 and 4 setae in *Lepas* and *Dosima* with the single postaxial seta in other Cirripedia. However, it remains entirely open to speculation what adaptive value, probably linked to a neustonic life style, caused the evolution of this presumably apomorphic trait.

4.5 *Glands*

Trypetesa lampas has a low density of villi, and our micrographs therefore reveal a system of grooves on the attachment disc at the bottom of which the cement gland has its exit pore(s). A few of our micrographs similarly show the simple exit pores of the unicellular, antennulary glands.

In some Rhizocephala, Glenner et al. (1989) found a so-called spinous process at the distal end of the attachment surface. Using TEM, Høeg (1987a, 1990) showed that this process is located within the exit pore of a specialized, large unicellular gland. We found no similar structure in the non-rhizocephalan cyprids, and the detailed TEM account of Nott & Foster (1969) shows that the gland cell in question is also lacking, at least from *Semibalanus balanoides*. We therefore suggest that this rhizocephalan process and associated gland represents an apomorphy.

5 FUNCTION AND PHYLOGENY

This survey, based on 15 species from 11 families of the Cirripedia, reveals significant variation in the morphology of cyprid attachment organs, and provides pointers to useful areas of future study. These include functional adaptations to the ecology of the adults to which they give rise and character states useful for hennigian analysis of phylogeny.

5.1 *The velum and its significance*

A filamentous velum is seen here in species of Scalpellidae and Chthamalidae, i.e. spanning the supposedly fundamental division of barnacles into pedunculate and sessile. On the other hand, a velum of tangential flaps is seen in all the species of Balanidae studied. The two types of velum are undoubtedly homologous. Indeed *Chthamalus montagui* has a partly intermediate condition, with the bases of filaments sometimes combined into flap-like groups (Pl. 4D).

Projecting the velum character into accepted phylogeny (Newman 1987) requires that the velum of overlapping flaps as seen in the Balanidae is apomorphic. This distinctiveness of the balanomorph velum runs parallel to another advanced feature of balanomorph larvae. This is the loss of the distinctive triplet of finely setulate plumose setae on the antennal endopod seen in many planktotrophic nauplii. These special setae are present in chthamalid nauplii, planktotrophic scalpellid nauplii and also in verrucids, lepadids and ascothoracids (Grygier 1993) but absent in the Balanidae (Moyse 1987; Korn 1988) and Coronulidae (Grygier 1993). It is postulated that the plumose setae triplet and filamentous velum together represent ground plan characters for all thoracican larvae whereas the balanid condition represents a modification. It would be instructive to determine which type of velum, if any, is present in the coronulids.

Functionally both types of velum support the rim of the disc while providing flexibility for changes in diameter during temporary attachment and probably adjust to local distortion on uneven substratum.

No velum has been observed in Acrothoracica, Rhizocephala, Verrucomorpha or even in the Lepadidae so far studied.

A striking feature associated with the velum of *Pollicipes pollicipes* concerns the wrinkled appearance of the wall of the third segment, evident in places between the filaments (Pl. 5B). These specimens had been treated with anaesthetic so the wrinkling is assumed to represent the relaxed state. We may postulate that during settlement the third segment inflates due to haemocoelic fluid under hydraulic pressure. This hypothesis explains other anomalies. Inflation might elevate the rim of the attachment disc bringing the reflexed edges of the disc into a more appropriate position during settlement as well as taking up the slack in the wrinkled sides of the third segment. However, we have no behavioural evidence for this hypothesis.

5.2 *The skirt*

We and Glenner et al. (1989) found this structure in the Thoracica, Acrothoracica, and Rhizocephala. Velum and skirt never occur together which could indicate that they represent different elaborations of the same basic structure. However, we find it more plausible that velum and skirt represent independent characters and that the latter arose as an elaboration of the attachment disc rim. Comparison with established phylogeny indicates that velum shape in various species can be set out in a convincing transformation series. What we call a skirt of overlapping flaps in *Scalpellum scalpellum* (unpublished) and *Ornatoscalpellum stroemii* (Pl. 4) do not resemble skirts in other taxa and could in fact represent a velum moved very close to the attachment disc.

The functional significance of the specialized skirt in *Dosima fascicularis* remains disputable and is not a feature common to all lepadids. *Lepas australis* cyprids lack a skirt and like *Conchoderma* cyprids (Elfimov, pers. comm.) they have dense cuticular villi to the reflexed edge of the disc. Perhaps the *Dosima* skirt relates to the unique habit of secreting its own oceanic floating device. Precise settlement behaviour and metamorphosis of *Dosima* remain unknown. Specimens sometimes attach to floating Phaeophyceae or even small logs, but we have no evidence that it can actually metamorphose in the absence of external substratum. Many small, apparently isolated specimens examined by one of us (JM) had all attached to very small pieces of flotsam such as chaff or feather barbules. We tentatively suggest that the skirted *Dosima* disc evolved as an adaptation for wrap-around settlement on small convex material (including floats of conspecifics) or as part of the special mechanism for the process of float formation using its own cement secretion. It is even possible that the centre and skirt of the disc of *Dosima* serves in different, apomorphic, adhesion mechanisms.

In most other species the skirt is less extensive and its function remain unclear. It obviously provides a flexible boundary which defines an almost closed chamber beneath the attachment disc when it contacts the substratum. This could either facilitate chemoreception during walking or assist in confining cement to beneath the disc during final attachment.

We mentioned earlier that the radially symmetrical and isodiametric shape of the third article as typified by *Semibalanus balanoides* is probably not the primordial plan. We now suggest it is an adaptation to intertidal life. We pursue this suggestion by first looking again at the contours of the attachment surface and the density of villi.

5.3 *Contours of the attachment disc*

A feature of fully intertidal species including especially *Pollicipes pollicipes, Capitulum mitella,* and *Chthamalus montagui* concerns the inflated shape of the rim of the disc (Pl. 5B, 6A and 7C). This is not an artifact, since we observe it in live and in relaxed specimens. Even *Semibalanus balanoides* has a slightly reflexed rim (curved downward) so that in the TEM section (Nott & Foster 1969: Fig. 2) the outermost villi project radially. Such a situation has apparently progressed further in the three species mentioned above, so that reflexion of the rim rather than inflation is the probable explanation of the rounded contour (Fig. 2J). The curvature seen in these intertidal species appears to increase the area available for villi, although the relationship is probably more complicated.

5.4 *Density of villi*

Nott & Foster (1969) suggested that the cuticular villi covering the attachment disc provide the large surface area necessary for optimum action of the adhesive secretions used in temporary attachment. From this we can infer a positive but not necessarily direct relationship between the density of villi and adhesive force. Table 2 gives some counts of villi per unit area of attachment disc in different species. The accuracy of such counts (taken from measured areas on SEM photographs) remains somewhat unreliable, especially at higher densities when clumping may occur and result in lower counts than reality. At low densities where the origins of villi emerging from the disc surface are visible, e.g. the rhizocephalan *Clistosaccus paguri* and the acrothoracican *Trypetesa lampas*, counting is demonstrably repeatable.

However, many genera e.g., *Chthamalus, Pollicipes, Capitulum,* and *Lepas,* had such a high density of villi, most surface structures such as gland openings and sensory setae were obscured from view. Many of the higher counts of villi may therefore be too low.

Taking the figures in Table 2 at face value clearly indicates a higher density in the intertidal species of this study. Presuming a relation between villus density and adhesive force, this correlates with the higher wave action forces likely to be met here than sublittorally.

The density of villi seems to exhibit but little correlation to large scale taxonomy. For example, peltogastrid rhizocephalans including *Briarosaccus tenellus* have a relatively high density of villi, whereas the density is low in *Clistosaccus paguri.*

5.5 *Symmetry of the attachment organ and ecology*

The fully radially symmetrical condition in which the attachment surface is accurately circular and at right angles to the longitudinal axis of the segment is widely dispersed. It is found in the archaeobalanids *Semibalanus balanoides* and *Elminius modestus*, and in the balanids *Balanus improvisus* and but also in *Chthamalus montagui*, in *Capitulum mitella* and *Pollicipes pollicipes*. These species all have a very high density of cuticular villi on the reflexed edges to the disc. They also share the ecological feature of settlement in conditions of high wave action in the mid- or upper-intertidal zone. Speculation suggests that during the exploratory phase of cypris settlement, entailing temporary attachment, wave impact forces may impinge from any direction parallel to the substrate. Such forces have been well studied and incorporate friction and drag elements (Jones & Demoutropoulos 1968). Dislodgement is probably best resisted by such radial symmetry. The very flexible joint between second and third segments of the antennule permit passive flexure at this point when the cyprid body is subject to wave drag in alternating directions; the third article meanwhile staying attached. Thus the radially symmetrical third segment might be interpreted as a as a convergence in different families to settling in high energy situations such as intertidal habitats to which all six of the above species penetrate. However, although both *Capitulum* and *Pollicipes* belong to the Scalpellidae or at least the Scalpelloidea, neither of these grouping are with any certainty monophyletic. In fact, we have evidence from adult morphology that either or both of these two genera stand phylogenetically closer to the Balanomorpha than to other 'Pedunculata' (Glenner et al. 1995). It follows that an alternative explanation would see the bell-shaped, radially symmetrical third segment as a true synapomorphy coupled to a high intertidal habitat. Traits in nauplius morphology, on the other hand, indicate that *Capitulum mitella* actually branched off very early in thoracican phylogeny (Moyse 1987). Tests of these alternative hypotheses should focus on attachment organ morphology in taxa that branch off early on the balanomorphan tree, viz., *Eochionelasmus, Pachylasma*, and *Catophragmus*.

In contrast to radial symmetry, the third segment of many cyprids has the attachment surface sloped at less than a right angle to the segment axis, leading to bilateral symmetry. Darwin (1854) referred to these as hoof-shaped. In the present study, *Acasta spongites* and *Verruca stroemia* from low energy, sub-littoral habitats, and the parasitic *Clistosaccus paguri* provide good examples of such morphology.

5.6 *The attachment organ used in penetration*

Megatrema anglica definitely exemplifies one of the most specialized attachment organs in the present study. This species settles by penetrating the coenosarc of its host *Caryophyllia smithi* to attach to the coral skeleton (Moyse 1971). The exceptionally pointed shape of the third segment (Pl. 9B) and consequent narrow attachment disc must represent an adaptation for penetration, in this species a process likely more critical than attachment itself. Future studies should investigate whether all coral barnacles that attach to their host skeleton by penetrating the soft tissue share such antennular morphology. The method of penetration remains unknown, but we note that structures, presumed to be homologous of radial sense organs, have a robust and fang-like shape (Pl. 9C) perhaps aiding traction of the penetrating antennule.

Cyprids of the rhizocephalan *Clistosaccus paguri* similarly use one of the antennules to penetrate through the integument of the hermit crab or virgin externa on which they have settled (Høeg 1985b, 1990). Surprisingly, however, the attachment organ of *C. paguri* cyprids exhibits no obvious adaptations for such penetration, which lends credence to Høeg's (1985b) postulate that penetration of the *C. paguri* is aided by secretions serving to dissolve the underlying integument. For comparison, most other rhizocephalans have distinctly bilaterally symmetric attachment organs (Glenner et al. 1989).

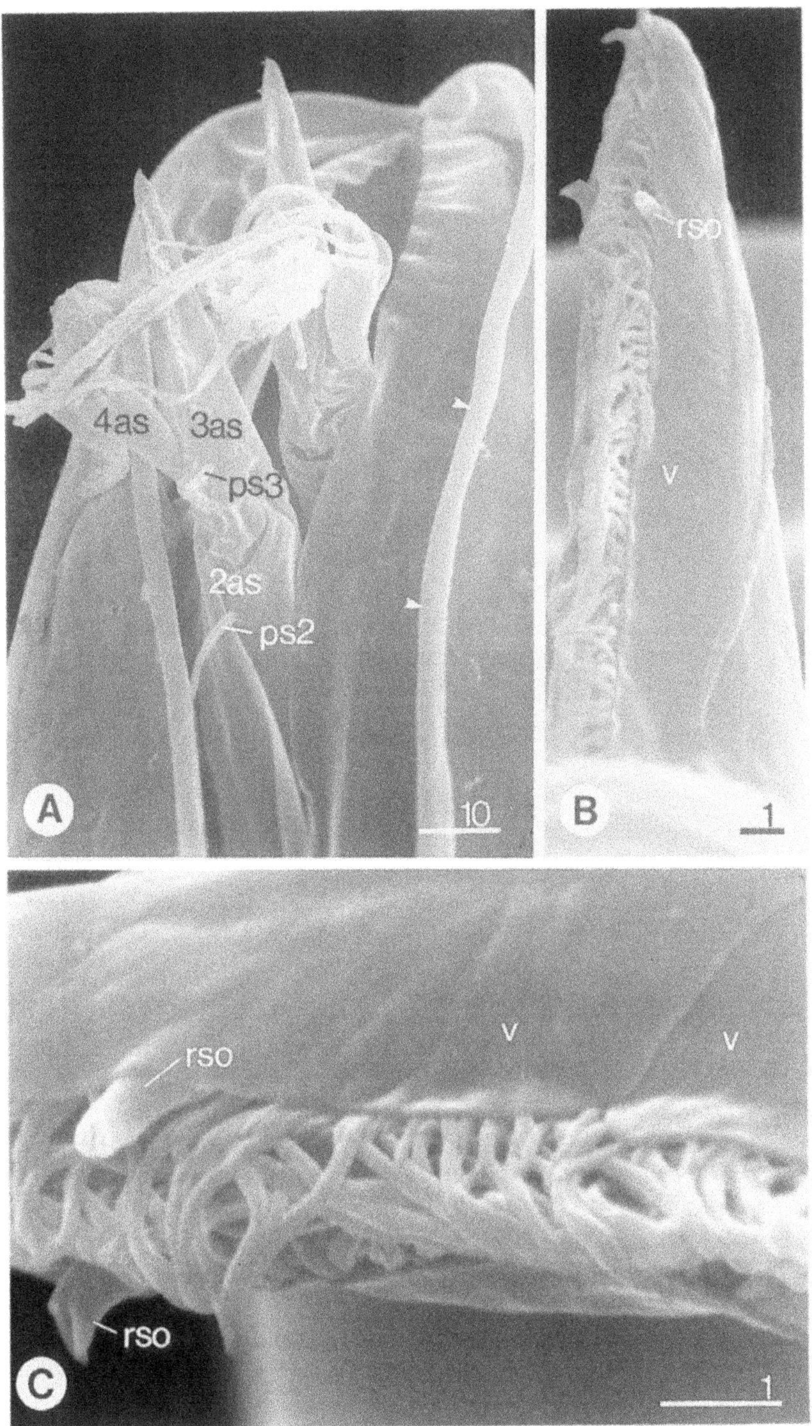

Plate 9. *Megatrema anglica*, cyprid. (A) Antennules emerging between carapace valves; (B) Tip of pointed third segment; (C) Detail of 'attachment surface' showing villi, velar flaps, and fang-like radial sense organs.

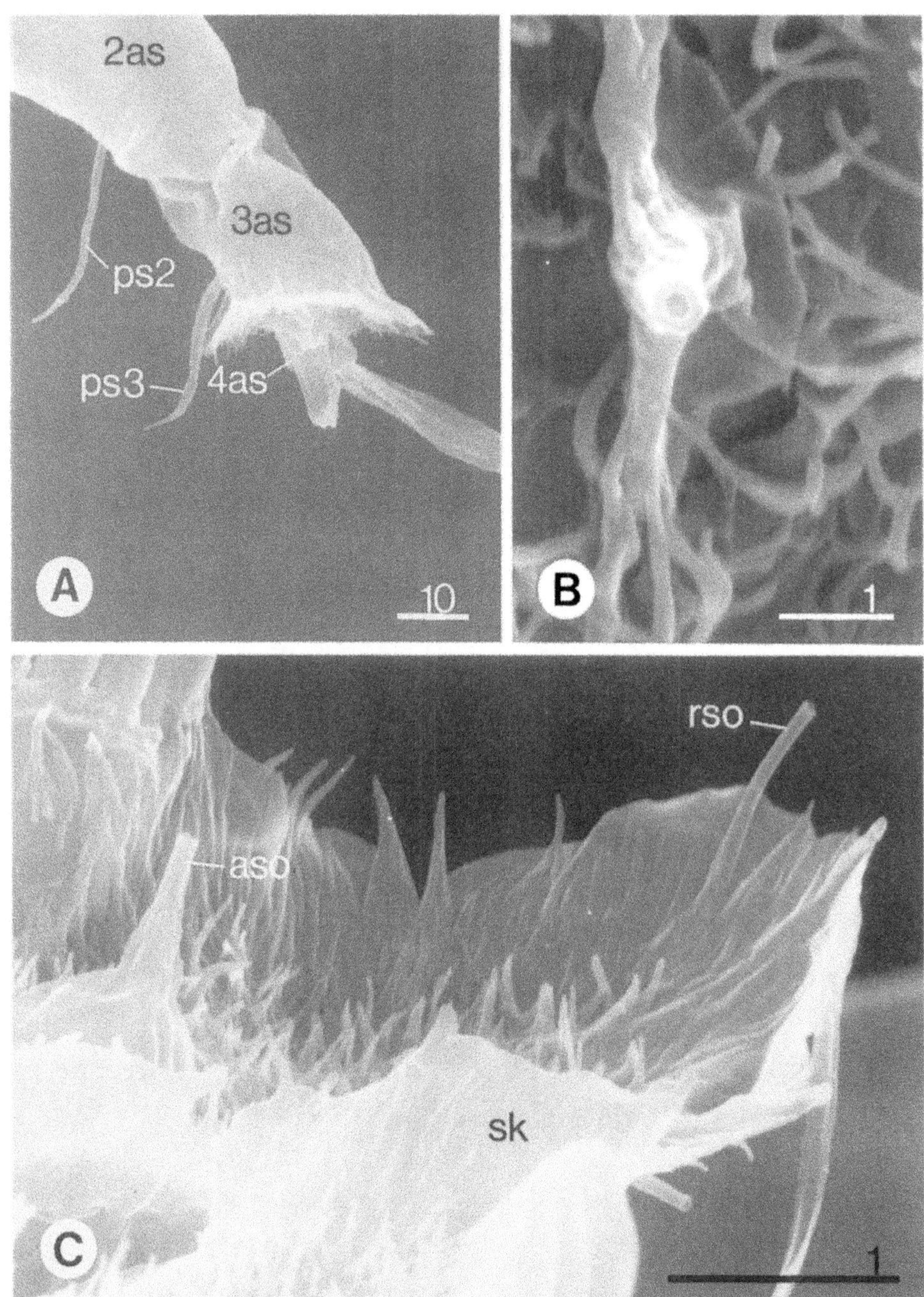

Plate 10. *Verruca stroemia*, cyprid. (A) Antennule showing angled attachment surface and disc extended as skirt; (B) End on view of axial sense organ; (C) Detail of part of disc showing skirt devoid of cuticular villi; axial and radial sense organs.

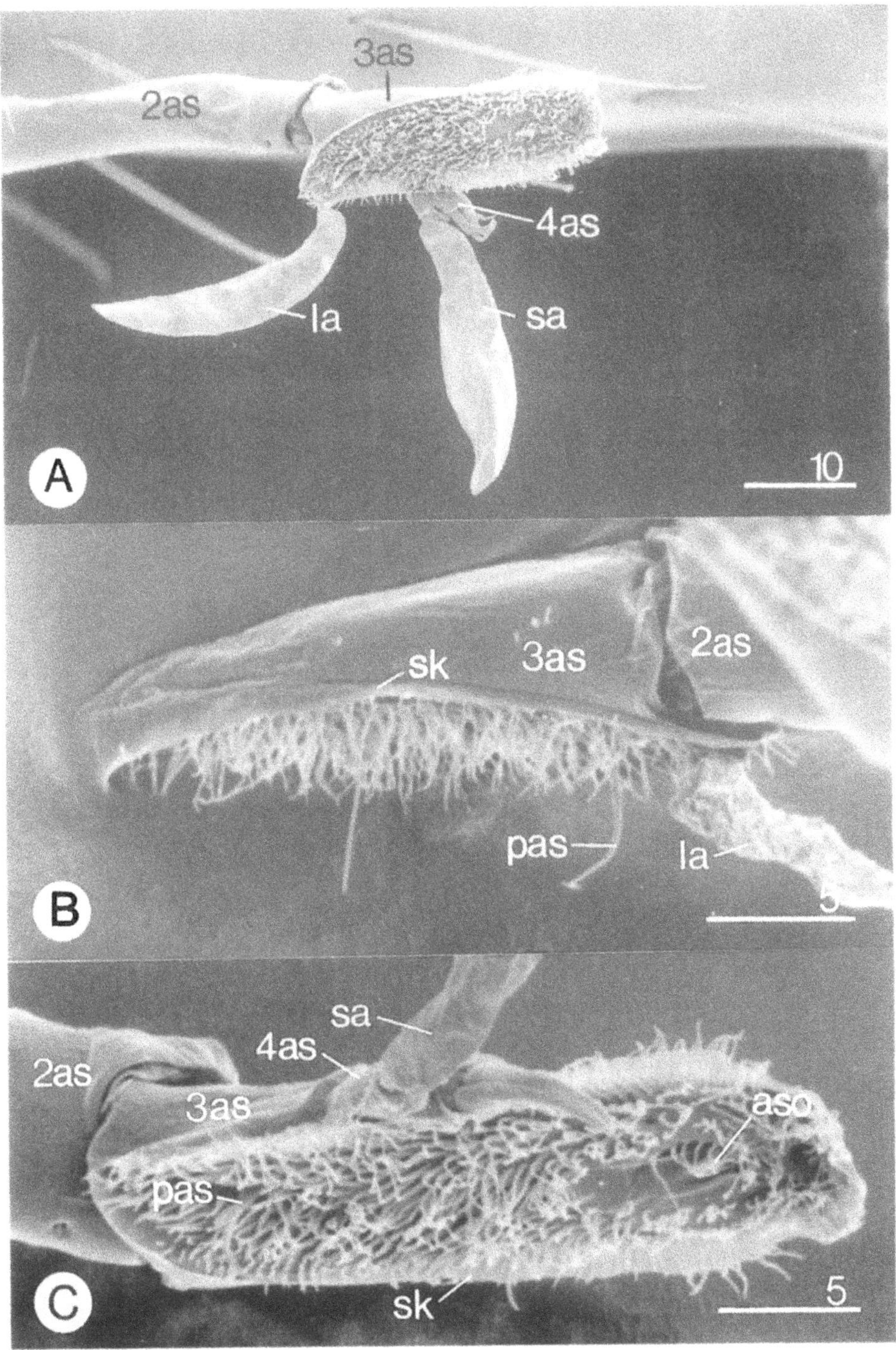

Plate 11. *Briarosaccus tenellus*. (A) Male cyprid, left antennule showing 3rd, 4th, and the distal part of the 2nd segment. Note that the large aesthetasc on the third segment is not bifurcated as it is in the other peltogastrids; (B) Female cyprid, right third antennular segment in median view; (C) Female cyprid, attachment disc.

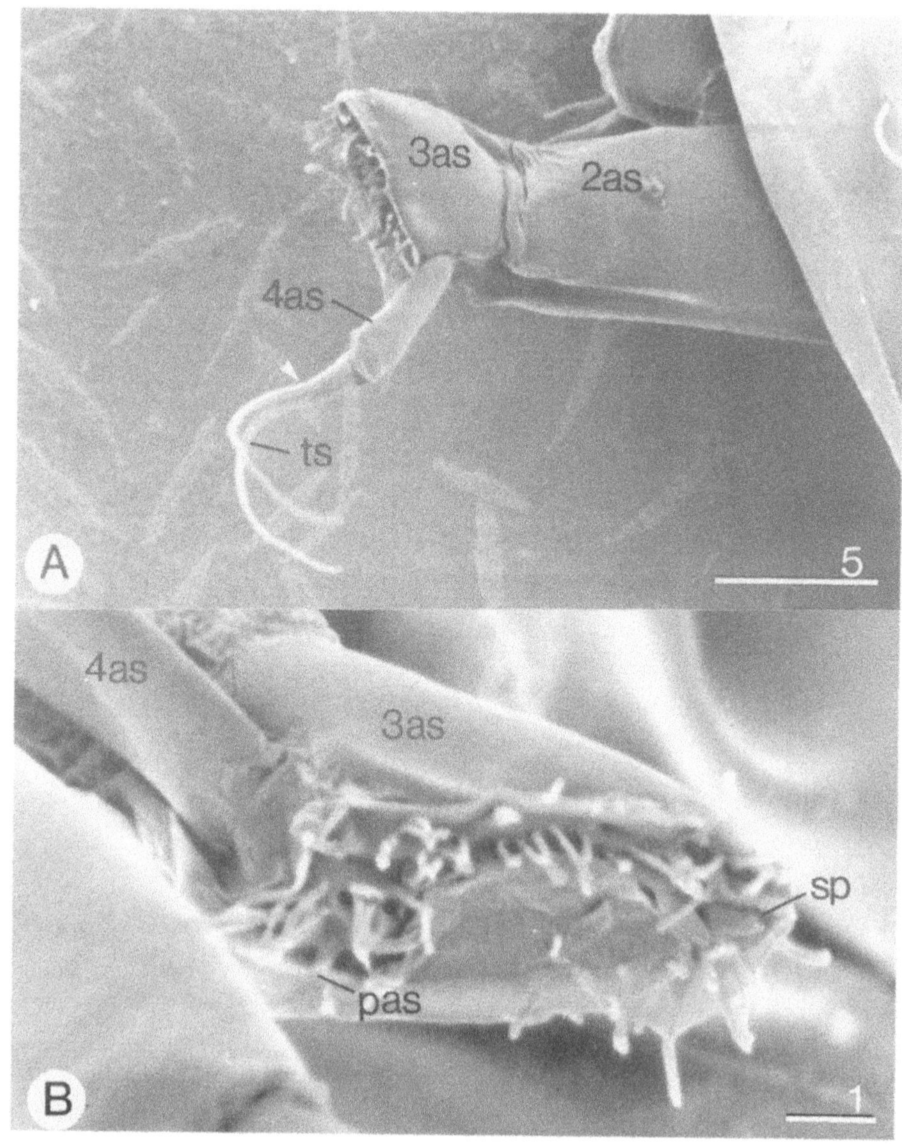

Plate 12. *Clistosaccus paguri*, cyprid. (A) Left antennule, lateral view; (B) Third antennular segment. Note the few peripheral villi.

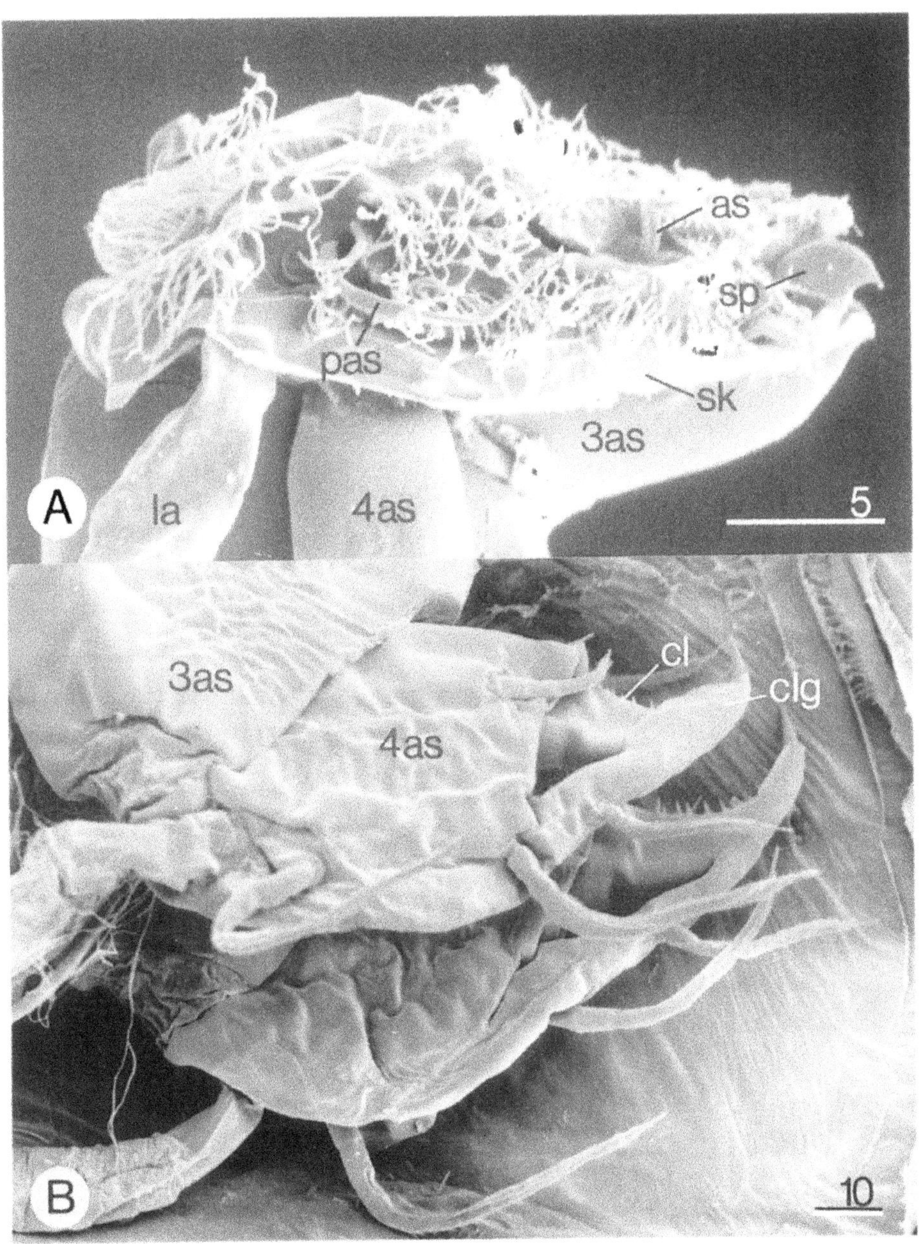

Plate 13. (A) *Mycetomorpha vancouverensis*, cyprid. Third antennular segment showing the attachment disc; (B) *Ulophysema oeresundense*, 2nd ascothoracid larva, female. Fourth antennular segment with claw (cl) and claw guard (clg).

6 CONCLUSIONS

Our study has revealed that the structure of the attachment organ varies through the Cirripedia and highlights the obvious fact that we cannot reliably generalize from the morphology of a single species to a larger taxon. Some of the variation here described obviously represent apomorphic specializations linked to functional requirements peculiar to certain taxa. Other traits, such as the isodiametric third segment may have evolved convergently due to similar ecological situations or represent true homologies.

Obviously, no single character or character set will suffice to unveil the phylogeny of a taxon. It is possible to map a single character onto an established phylogeny. This will lead to polarization of the character states and open up for suggesting a scenario for the evolution of the trait in question. Alternatively, we may, in the absence of a well established phylogeny, combine the trait under study with other characters and use them as input in a full fledged phylogenetic analysis.

Unfortunately, we face the latter situation in Cirripedia. Many of the taxa traditionally recognized, such as Pedunculata, Scalpellidae, and Sessilia may well be para- or even polyphyletic groups and of no value in a phylogenetic discussion (Glenner et al. 1995). Only by using a large number of well-studied characters from all ontogenetic stages can we ever hope to unveil evolution in a group so complex as the Cirripedia, spanning from intertidal filter feeders over a multitude of commensals to the most highly specialized parasites amongst higher metazoans. We emphasize that larval traits studied by ultrastructural methods in both nauplii and cyprids must form an integral part of such an across-the-board analysis of cirripede phylogeny. Again, such a multidisciplinary approach stands squarely in the tradition of Prof. W.A. Newman to whom this volume is dedicated.

ACKNOWLEDGMENTS

JM and HAY are grateful to Professor John Beardmore of the School of Biological Sciences for the use of facilities. They also wish to extend thanks to Dr. M. Fordy for help with SEM studies. JH and PGJ thank the Zoology Museum of the University of Copenhagen for providing SEM facilities and specimens of *Ornatoscalpellum stroemii*. They also take this opportunity to acknowledge the late Mr. B.W. Rasmussen for cheerful assistance and for having maintained these facilities in perfect working order. We wish to thank Dr. Aleksey Elfimov and H. Glenner for useful discussion. Dr. Edmond Hui supplied live specimens of *Capitulum mitella*, Prof. W. A. Newman supplied cyprids of *Lepas australis*, and Dr. H. Hovde supplied the cyprids of *Balanus improvisus*. We are most grateful to these colleagues for their assistance. JTH acknowledge Grant no. 11-9652 from the Danish National Science Research Council defraying travel expenses for two visits to Swansea.

REFERENCES

Andersen, M.L., M. Bohn, J.T. Høeg & P.G. Jensen 1990. Cypris ultrastructure and adult morphology in *Ptychascus barnwelli* n. sp. and *P. glaber* (Crustacea: Cirripedia: Rhizocephala), parasites on semiterrestrial crabs. *J. Crust. Biol.* 10: 20-28.

Darwin, C. 1852. *A monograph on the sub-class Cirripedia, with figures of all the species. The Lepadidae; or, pedunculated cirripedes*. London: Ray Society.

Darwin, C. 1854. *A monograph on the sub-class Cirripedia, with figures of all the species. The*

Balanidae, (or sessile cirripedes); the Verrucidae, etc., etc., etc. London: Ray Society.

Elfimov, A.S. 1989. *The cypris larvae of cirripedes and their significance in the formation of fouling.* Candidate of Science thesis. Moscow State University.

Feuerborn, H. 1933. Das cyprisstadium des Süsswasserrhizocephalen *Sesarmoxenos. Verh. Dtsch. Zool. Ges.* 35: 127-138.

Glenner, H., J.T. Høeg, A. Klysner & B. Brodin Larsen 1989. Cypris ultrastructure, metamorphosis and sex in seven families of parasitic barnacles (Crustacea: Cirripedia: Rhizocephala). *Acta Zool. (Stockholm)* 70: 229-242.

Glenner, H. M.J. Grygier, J.T. Høeg, P.G. Jensen, & F.R. Schram 1995. Cladistic analysis of the Cirripedia Thoracica. *Zool. J. Linn. Soc.* 114: 365-404.

Grygier, M.J. 1987. New records, external and internal anatomy, and systematic position of Hansen's Y-larvae (Crustacea: Maxillopoda: Facetotecta). *Sarsia* 72: 261-278.

Grygier, M.J. 1993. Late planktonic naupliar development of an ascothoracidan crustacean (?Petracidae) in the Red Sea and a comparison to the Cirripedia. *Contrib. Sci. Nat. Hist. Mus. Los Angeles County* 437: 1-14.

Hawkes, C.R., T. R. Meyers & T.C. Shirley 1985. Larval biology of *Briarosaccus callosus* Boschma (Cirripedia: Rhizocephala). *Proc. Biol. Soc. Wash.* 98: 935-944.

Høeg, J.T. 1982. The anatomy and development of the rhizocephalan barnacle *Clistosaccus paguri* Lilljeborg and relation to its host *Pagurus bernhardus* (L.). *J. Exp. Mar. Biol. Ecol.* 58: 87-125.

Høeg, J.T. 1984. A culture system for rearing marine invertebrate larvae and its application to larvae of rhizocephalan barnacles. *J. Exp. Mar. Biol. Ecol.* 84: 167-172.

Høeg, J.T. 1985a. Cypris settlement, kentrogon formation and host invasion in the parasitic barnacle *Lernaeodiscus porcellanae* (Müller) (Crustacea: Cirripedia: Rhizocephala). *Acta Zool. (Stockholm)* 66: 1-45.

Høeg, J.T. 1985b. Male cypris settlement in *Clistosaccus paguri* Lilljeborg (Crustacea: Cirripedia: Rhizocephala). *J. Exp. Mar. Biol. Ecol.* 89: 221-235.

Høeg, J.T. 1987a. Male cyprid metamorphosis and a new male larval form, the trichogon, in the parasitic barnacle *Sacculina carcini* (Crustacea: Cirripedia: Rhizocephala). *Phil. Trans. R. Soc. Lond.* (B)317: 47-63.

Høeg, J.T. 1987b. The relation between cypris ultrastructure and metamorphosis in male and female *Sacculina carcini* (Crustacea, Cirripedia). *Zoomorphology* 107: 299-311.

Høeg, J.T. 1990. 'Akentrogonid' host invasion and an entirely new type of life cycle in the rhizocephalan parasite *Clistosaccus paguri* (Thecostraca: Cirripedia). *J. Crust. Biol.* 10: 37-52.

Høeg, J.T. 1992a. The phylogenetic position of the Rhizocephala: Are they truly barnacles?. *Acta Zool. (Stockholm)* 73: 323-326.

Høeg, J.T. 1992b. Rhizocephala. In F.W. Harrison & A. G. Humes (eds.), *Microscopic Anatomy of Invertebrates*, Vol. 9. Crustacea, pp. 313-345. New York: Wiley-Liss Inc.

Hui, E. 1983. Observations on cirral activity in juvenile Pollicipoid barnacles (Cirripedia, Lepadomorpha). *Crustaceana* 45: 317-318.

Jensen, P.G., J.T. Høeg, S. Bower & A.V. Rybakov 1994. Scanning electron microscopy of lattice organs in cyprids of the Rhizocephala Akentrogonida (Crustacea Cirripedia). *Can. J. Zool.* 72: 1018-1026.

Jensen, P.G, J. Moyse, J.T. Høeg & H. Al-Yahya 1994. Comparative SEM studies of lattice organs: putative sensory structures on the carapace of larvae from Ascothoracida and Cirripedia (Crustacea Maxillopoda Thecostraca). *Acta Zool. (Stockholm)* 75: 125-142.

Jones, W.E. & A. Demoutropoulos 1968. Exposure to wave action: measurement of an important parameter on rocky shores on Anglesey. *J. Exp. Mar. Biol. & Ecol.* 2: 46-63.

Korn, O.M. 1988. Larvae morphology and phylogenetic system of barnacles of the order Thoracica [In Russian w. English abstract]. *Zool. Zh.* 67: 644-651.

Lindner, E. 1984. The attachment of macrofouling invertebrates. *Proc. Symp. Mar. Biodeterior.* 1984: 183-201.

Miller, K.M. & J. Roughgarden 1994. Descriptions of the larvae of *Tetraclita rubescens* and *Megabalanus californicus* with a comparison of the common barnacle larvae of the central California coast. *J. Crust. Biol.* 14: 579-600.

Moyse, J. 1971. Settlement and growth pattern of the parasitic barnacle *Pyrgoma anglica*. In D.J. Crisp (ed.), *Fourth European Marine Biology Symposium*, pp. 125-142. Cambrigde: Cambridge University Press.

Moyse, J. 1987. Larvae of lepadomorph barnacles. *Crustacean Issues* 5: 329-362.

Newman, W.A. 1987. Evolution of cirripedes and their major groups. *Crustacean Issues* 5: 3-42.

Nilsson-Cantell, C-A. 1921. Cirripedien-Studien. Zur Kenntnis der Biologie, Anatomie, und Systematik dieser Gruppe. *Zool. Bidr. Upps.* 7: 6-395 + Pl.

Nott, J. 1969. Settlement of barnacle larvae: surface structure of the antennular attachment disc by scanning electron microscopy. *Mar. Biol.* 2: 248-251.

Nott, J. & B. Foster 1969. On the structure of the antennular attachment organ of the cypris larva of *Balanus balanoides* (L.). *Phil. Trans. R. Soc. Lond.* (B)256: 115-134.

Saroyan, J.R., E. Lindner & C.A. Dooley 1968. Attachment mechanism of barnacles. In Anonymous (ed.), *Proceeding of the 2nd International Congress of Marine Corrosion and Fouling*, pp. 495-512 Athens: Technical Chamber of Greece.

Spears, T., L.G. Abele & M. A. Applegate 1994. Phylogenetic study of cirripedes and selected relatives (Thecostraca) based on 18S rDNA sequence analysis. *J. Crust. Biol.* 14: 641-656.

Standing, J. 1981. Common inshore barnacle cyprids of the Oregonian fauna province (Crustacea: Cirripedia). *Proc. Biol. Soc. Wash.* 93: 1184-1203.

Stone, C.J. 1989. A comparison of algal diets for cirripede nauplii. *J. Exp. Mar. Biol. Ecol.* 132: 17-40.

Svane, I. 1986. Sex determination in *Scalpellum scalpellum* (Cirripedia Thoracica Lepadomorpha), a hermaphroditic goose barnacle with dwarf males. *Mar. Biol.* 90: 249-253.

Walker, G. 1973. The early development of the cement apparatus in the barnacle species, *Balanus balanoides* (L.) (Crustacea: Cirripedia). *J. Exp. Mar. Biol. Ecol.* 12: 305-314.

Walker, G. 1985. The cypris larvae of *Sacculina carcini* Thompson (Crustacea: Cirripedia: Rhizocephala). *J. Exp. Mar. Biol. Ecol.* 93: 131-145.

Walossek, D., J.T. Høeg & T.C. Shirley. in press. Larval development of the rhizocephalan cirripede *Briarosaccus tenellus* (Maxillopoda: Thecostraca) reared in the laboratory: A scanning electron microscopy study. *Hydrobiologia.*

Yule, A.B. & D.J. Crisp 1983. Adhesion of cypris larvae of the barnacle, *Balanus balanoides,* to clean and arthropodin treated surfaces. *J. Mar. Biol. Ass. UK* 63: 261-271

Yule, A.B. & G. Walker 1984a. The temporary adhesion of barnacle cyprids: Effects of some differing surface characteristics. *J. Mar. Biol. Ass. UK* 64: 429-439.

Yule, A.B. & G. Walker 1984b. The adhesion of the barnacle, *Balanus balanoides*, to slate surfaces. *J. Mar. Biol. Ass. UK* 64: 147-156.

Yule, A.B. & G. Walker 1987. Adhesion in barnacles. *Crustacean Issues* 5: 389-404.

APPENDIX 1: LIST OF ABBREVIATIONS

2as:	second antennular segment,	la:	aesthetasc,
3as:	third antennular segment,	pas:	postaxial sense organ,
4as:	fourth antennular segment,	ps2:	postaxial seta II,
ant:	antennule,	ps3:	postaxial seta III,
aso:	axial sense organ,	rso:	radial sense organ,
cc:	cement canal,	sa:	subterminal aesthetasc,
cl:	claw,	sk:	skirt,
clg:	claw guard,	sp:	spinous process,
cv:	cuticular villi,	sts:	subterminal setae of fourth segment,
ff:	frontal filaments,	th:	thorax,
fv:	filamentous velum,	ts:	terminal setae of fourth segment,
gr:	groove in attachment disc,	v:	velum.

Cuticular structures in the males of Scalpellidae (Cirripedia Thoracica): A character analysis

W. Klepal & H.L. Nemeschkal
Institut für Zoologie, Universität Wien, Wien, Austria

ABSTRACT

In order to consider males in the framework of the systematics of cirripedes, characters have to be detected that are present in all males irrespective of their state of development. This criterion implicitly leads to the cuticular structures, which were tested for their applicability as systematic features. The analyses revealed that they are valuable for this purpose at lower taxonomic levels. However, they are inapplicable when defining higher systematic categories. Because of their obviously high functional burden, several cases of parallel evolution must be assumed.

1 INTRODUCTION

In early times, Cirripedia Thoracica were considered to be exclusively hermaphroditic. Therefore, males have been fascinating ever since they were first discovered. Darwin (1851) described the males (associated with females) and complemental males (associated with hermaphrodites) of *Ibla* and *Scalpellum* (then both Lepadomorpha), and he compared the males of the various species with each other and also with the hermaphrodites and females to which they are attached. He was the first author who tried to classify the males by themselves. He distinguished two groups by their body shape: one with the capitulum and peduncle clearly discernible, and the other sac-like animals. For classifying the latter group Darwin used the number of plates, noticing that the males with a distinct capitulum and peduncle differ much more from the sac-like form than they do from their respective hermaphrodites (females). On the other hand, Darwin pointed out that the pedunculate males of *Scalpellum peronii* and *S. villosum* resemble each other more than do the two hermaphrodite forms. He realized that it is very difficult to weigh the value of the differences in the different parts of species. Pilsbry (1908) suggested including the dwarf males in the systematics of the Cirripedia. This idea was followed up by Nilsson-Cantell (1921) who soon realized that the cirri and plates are variable characters in the males even of closely related genera. Nevertheless, Nilsson-Cantell (1932) distinguished four groups of sac-like males by the design of the plates and cirri. Since then dwarf (complemental) males were mentioned whenever they occurred in any newly described species. Their shape, the number of plates, and/or cirri was recorded, but nobody ever tried seriously to study the taxonomy of the cirripedes based on males. This was partly due to the lack of suitable methods. Now scanning electron microscopy (SEM) provides a useful tool for such an investigation. Although males are now also known in Balanomorpha (Henry & McLaughlin 1965, 1967; Dayton et al. 1982; Foster 1983; Crisp 1983; Hui & Moyse 1984) this study deals with those of the Pedunculata Scalpellidae only. Characters that are present in all males, irrespective of their degree of morphological simplification, will be investigated.

2 CHARACTERS USED IN THE SYSTEMATICS OF CIRRIPEDIA THORACICA

In the hermaphrodites and females of the Pedunculata, the plates mainly on the capitulum and, in the fully grown hermaphrodites of the Sessilia, the wall plates and the opercular valves are of taxonomic importance. Any key for their systematics is based on these fairly stable structures. In addition, body appendages like cirri, mouthparts, caudal appendages, and penes in the hermaphrodites are used for the identification of species. An attempt is made to use the same characters for the systematics of the males. From the males described in literature (Klepal 1987) it is obvious that most authors had the body shape in view, as well as the presence, number, size, and sometimes the shape of the plates (or valves) (Pl. 1). All other information on the shape of the body, the presence of cirri and their number, the presence and condition of the alimentary canal, and the existence of a penis or a penis-thorax is scarce or does not exist at all. It is interesting that all authors paid attention to the cuticular structures of the males, although they were not described for every single species investigated. The terms used for the various cuticular structures are different depending on the author. Rarely was a drawing made, so usually one cannot compare this with the description and with our own SEM-based investigations (mostly on other species).

3 CUTICULAR STRUCTURES

3.1 *Definitions and transitional stages*

In this paper, all cuticular structures (except the plates) on the outer surface of the capitulum and peduncle, or the whole outer surface of sac-like males, are considered (Pl. 2 and 3). These include the non-sensory cuticular structures as well as structures in which a sensory function is possible but has not been demonstrated yet. The structures will be listed according to their level of complexity.

3.1.1 *Definitions*

Fringe: Solid, filiform cuticular projection of the same diameter throughout its length, with a rounded tip. It is often bent and thus appears flexible (Pl. 2A).

Seta: Hollow, flexible cuticular structure, relatively long and conical. It is possible, but still unproven, that this structure is sensory (Pl. 2B). The seta is inserted well below the external surface of the exoskeleton and has thus an infracuticular articulation (Jacques 1989).

Scale: Sheet-like structure with a relatively wide, often trapezoid-shaped basis and a sometimes rounded, free apical pole with fringes. The fringes are often shorter than the basis. Mainly because of the fringes the scales appear flexible (Pl. 2C).

Tooth or spine: Solid conical cuticular projection with a pointed tip. It is usually straight and thus appears stiff (Pl. 2D). A tooth is relatively short (smaller than 6μm) and a spine is longer (more than 6 μm).

Comb: A series of tooth-like structures united by a common basis; it appears stiff (Pl. 3A).

3.1.2 *Transitional stages*

Transitional stages are found between:

Fringe-spine (tooth): In this case, individual fringes are substructures of spines (teeth). The cuticular structure has a single or a bifid tip and a conical base, and, in the case of spines, sometimes with a definite shoulder (Pl. 3B).

Fringe-scale: A number of fringes are united at their bases. In comparison with a typical scale, the fringes are further apart at their free ends. The width of the basis is variable along the structure (Pl. 3C).

Scale-comb: A cuticular structure with both filiform and cone-shaped projections on a common basis (Pl. 3D).

Tooth-like structures of the comb: These may be true teeth or they may consist of fused fringes.

3.2 *Genesis of the cuticular structures*

We hypothesise the following evolution of the cuticular structures (Fig. 1). The

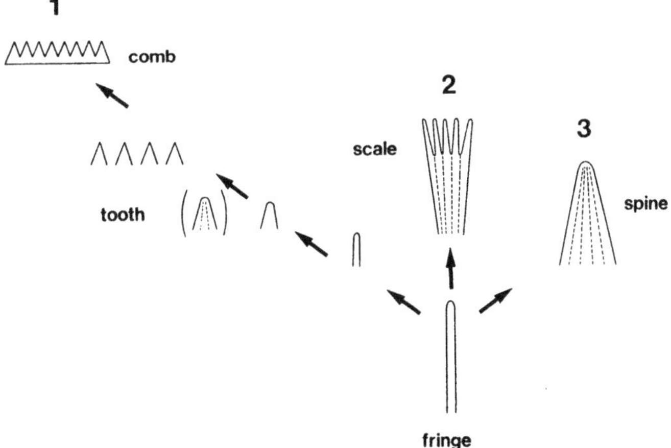

Figure 1. The proposed anagenetic change of cuticular structures as a result of detailed morphological studies. (1) A tooth arises by shortening of a fringe and widening of its basis. Serially arranged teeth fuse at their bases and form a comb. (2) Partial fusion of several fringes gives rise to a scale. (3) Complete fusion of several fringes forms a spine.

original structure is the fringe. One path of differentiation consists in the shortening and further in the fusion of this structure. Thus the typical, long, and narrow fringe is reduced to a short one. By broadening of the basis, the short fringe becomes a tooth. Several teeth, serially arranged, form a row of teeth. Fusion of several teeth along a common basis forms a comb. An alternative way to form teeth is by fusion of 2-3 short fringes. Teeth formed that way are arranged in a row, and eventually a comb is formed as described above. The second path of differentiation consists in the partial fusion of the fringes along their basis to at least their midlength. This results in a scale. A third path of differentiation is the complete fusion of several fringes and the formation of a spine. A tooth may also be formed by the reduction of a spine.

In the genesis of the cuticular structures, two categories of characters are involved: 1) Degree of fusion and 2) Length. Four degrees of fusion may be distinguished: a) Free structure, b) Fusion along the basis, c) Fusion along the basis and for much of the length of the structures, and d) Complete fusion along their length. Two character states of length are discernible: a) Long and b) Short. The assumption that the fringe is the original cuticular structure of the Pedunculata is supported first by their being the main and only definite non-sensory cuticular structures of the Calanticinae, an ancestral subfamily within the Scalpellidae, and second by them being substructures of teeth, and as a consequence combs, scales, and spines, in the other species investigated.

3.3 *Description of scalpellid males*

3.3.1 *Kinds and characters of cuticular structures*

The kinds of cuticular structures and their transitional stages in the males of various species investigated are listed in Table 1. In the Calanticinae, there are only two kinds of cuticular structures. In the Scalpellinae there are three. In the Meroscalpellinae there are also three, and in one species, *Gymnoscalpellum tarasovi*, there are five kinds of cuticular structures and two transitional stages. In the Arcoscalpellinae, there are three to five kinds of cuticular structures in the species investigated. Most transitional stages in this subfamily are between scales and combs and between fringes and scales, but there are also some between fringes and teeth.

Not only the kinds of structures present are important for the identification of the species, but also their detailed form. In the case of a simple structure, its dimensions (length/width) have to be recorded and in the case of a more complex structure apart from the dimensions of the structure as a whole (length/width/width of basis), the type, number, dimensions of, and distance between substructures have to be considered. A comparison of single teeth and combs shows that the single teeth decrease in their dimensions from the Scalpellinae, through the Meroscalpellinae, to the Arcoscalpellinae. Likewise the basis of the combs is widest in the Scalpellinae, less wide in the Meroscalpellinae, and narrowest in the Arcoscalpellinae (Pl. 4). The length of the teeth and their number in the combs are about the same in the Scalpellinae and Arcoscalpellinae. However, in the Arcoscalpellinae the teeth of the combs are considerably narrower than in the other subfamilies. In the Meroscalpellinae, there are fewer but distinctly longer and wider teeth in the combs than in the other two subfamilies investigated. A comparison of the fringes shows that they are longest and widest in the Calanticinae and Meroscalpellinae, while in the Arcoscalpelli-

Table 1. Kinds of structures in the species investigated.

	Cuticular structures					
	Fringes	Setae	Scales	Spines	Teeth	Combs
Calanticinae						
Calantica mortenseni	+	+				
Smilium peronii	+	+				
Scalpellinae						
Ornatoscalpellum stroemii		+			+	+
Ornatoscalpellum gibberum		+			+	+
Meroscalpellinae						
Gymnoscalpellum tarasovi	+	+	+		+	+
Litoscalpellum discoveryi		+			+	+
Litoscalpellum aurorae		+			+	+
Arcoscalpellinae						
Arcoscalpellum michelottianum			+	+	(+)	
Arcoscalpellum tritonis	+	+			+	+
Teloscalpellum ventricosum	(+)	+	+		+	+
Trianguloscalpellum darwinii	(+)	+	(+)		+	+
Tarasovium cornutum		+			+	+
Weltnerium bouvieri		+			+	+
Weltnerium convexum		+			+	+
Weltnerium nymphocola		+			+	+

	Transitional stages Fringes-Scales	Fringes-Spines	Fringes-Teeth	Scales-Combs
Calanticinae				
Calantica mortenseni				
Smilium peronii				
Scalpellinae				
Ornatoscalpellum stroemii				
Ornatoscalpellum gibberum				
Meroscalpellinae				
Gymnoscalpellum tarasovi		[+]		+
Litoscalpellum discoveryi				
Litoscalpellum aurorae				
Arcoscalpellinae				
Arcoscalpellum michelottianum	+			
Arcoscalpellum tritonis			(+)	+
Teloscalpellum ventricosum	(+)			+
Trianguloscalpellum darwinii				+
Tarasovium cornutum				
Weltnerium bouvieri				
Weltnerium convexum				
Weltnerium nymphocola				+

(+) = structure not in typical expression; [+] = transition via scale.

nae they are only about a quarter of the length, and in the Scalpellinae they are very short (about 1/10 to 1/14 of those in the Calanticinae and Meroscalpellinae).

3.3.2 *Distribution and arrangement of cuticular structures*

The cuticular structures are distributed both on the capitulum and the peduncle in the Calanticinae; and on the capitular portion and far down towards the peduncular pole of the sac-like males of the Scalpellinae, Meroscalpellinae, and of some species of Arcoscalpellinae (Pl. 1). In the Calanticinae, fringes are arranged in rows, which are closer together on the peduncle than on the capitulum. In the sac-like males, fringes are arranged irregularly on the capitular pole of the body. Setae are on the capitulum of the Calanticinae either on plates, or separately. On the plates in this subfamily and on the sac-like males in the other subfamilies the setae are usually in pits. Whether these setae are sensory remains to be investigated. In some species (e.g. *Teloscalpellum ventricosum, Ornatoscalpellum gibberum*), the plates recede from these fringe-like cuticular structures. In others (e.g. *Trianguloscalpellum darwinii, Weltnerium bouvieri* and *Weltnerium convexum*), the plates have disappeared completely and only the setae arranged in round or oblong groups are left. In *Tarasovium cornutum,* there are only two groups of short and relatively thin setae. In all species, the antennulae of the males carry fringes and setae. The gradual recession of the plates and the decrease in number, length and diameter of the setae (Pl. 5) are interpreted as a sign of reduction, which thus affects cuticular structures as well as organs and appendages of the males.

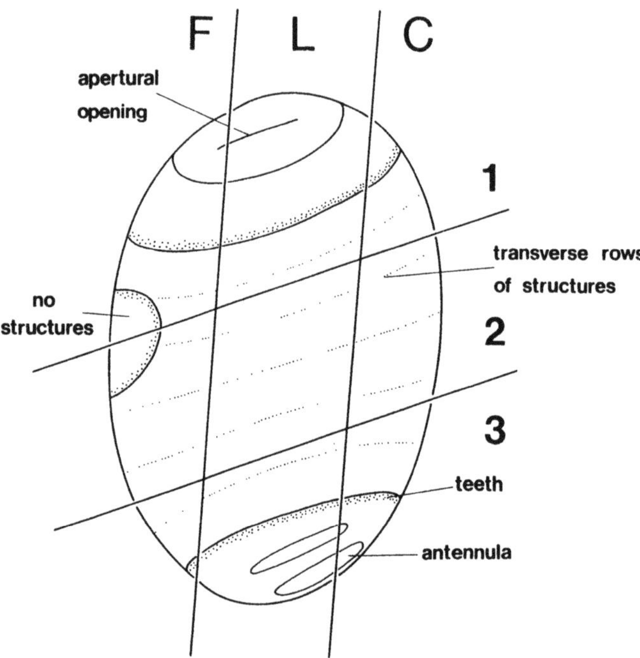

Figure 2. Schematic drawing of sac-like male. 1: top, 2: middle, 3: bottom region, F: frontal, L: lateral, C: caudal side; note bare patches on the capitular pole, especially around the apertural opening, and on the peduncular pole around the antennulae.

In the sac-like males, the area immediately around the apertural opening on the capitular pole is bare, with the exception of *Gymnoscalpellum tarasovi* and *Arcoscalpellum michelottianum,* where there are scales next to the apertural opening. In some species, there is a separate bare patch on the capitular pole. On the peduncular pole next to the antennulae, there is also a bare area of different extent in most species investigated (Fig. 2). Scales and combs are distributed on the capitular and the peduncular pole and they are always arranged in rows, mostly in a transverse direction, but in some species also in a vertical direction around the apertural opening (e.g. in *Arcoscalpellum tritonis*). Spines and teeth are also on the capitular and peduncular poles, arranged in rows. In the transition zone, between the rows of combs and teeth and the bare areas, the teeth are irregularly arranged. The few transitional stages of cuticular structures on any body surface have the same kind of arrangement as the structures between which they are morphologically intermediate. Of the species investigated so far the greatest variability in the distribution of cuticular structures is noted in the Meroscalpellinae. In the Arcoscalpellinae, all species (except *Arcoscalpellum michelottianum*) have a similar type of distribution and arrangement of the structures, but their orientation may be different.

3.3.3 *Species interrelationship*

The intention of the present study is a pure character analysis open to future interpretations, putting emphasis either on ecological perspectives, or phylogenetic reconstruction. The analysis must be a preliminary one because a single character system, e.g. cuticular structures, cannot be the sole basis for ecological investigations, nor for phylogenetic reconstructions. As a consequence, we present some alternative analyses between the two theoretical extremes, phenetics and phylogenetic systematics, to provide different options for interpretation. This study can only indicate trends. Because of the small number of species investigated by us and the limited information still available it is not possible yet to attempt a revision of the existing system of the Scalpellidae on the basis of the males. This can only be accomplished once a larger number of males is studied.

Before the analysis, the characters of the cuticular structures were entered and coded in a character state matrix for the investigated species (Tables 1 and 2). These species are the operational taxonomic units (= recent species, OTUs).

From the phenetic point of view OTUs are ordered by overall similarity (Fig. 3). The OTU-clusters of the UPGMA-dendrogram are very similar to the assemblages in the Wagner network (Fig. 4).

A Wagner network is suited to form a transition from a purely numerical taxonomic treatment of data to a phylogenetic analysis. On the one hand, Q-cluster analyses are optionally based on Manhattan-distances (Sneath & Sokal 1973), whereas the Wagner network is obligatorily based on this type of distance measure (Farris 1970). On the other hand, a Wagner network is conceptually much closer to the constructions of cladograms than of phenograms. Our detailed character analysis was made by constructing such a Wagner network (Sneath & Sokal 1973: 325) using a program package written in AMIGA-Basic by the author HLN). This network is an unrooted tree. The nodes in the tree represent OTUs and/or HTUs (hypothetical taxonomic units = hypothetical ancestral species). The distances between immediately linked nodes represent phenetic distances. These delineate the changes in character states

Table 2. Matrix of character states.

OTU 1	*Weltnerium convexum*	OTU 9	*Litoscalpellum aurorae*
OTU 2	*Weltnerium nymphocola*	OTU 10	*Litoscalpellum discoveryi*
OTU 3	*Weltnerium bouvieri*	OTU 11	*Gymnoscalpellum tarasovi*
OTU 4	*Teloscalpellum ventricosum*	OTU 12	*Ornatoscalpellum stroemii*
OTU 5	*Tarasovium cornutum*	OTU 13	*Ornatoscalpellum gibberum*
OTU 6	*Arcoscalpellum michelottianum*	OTU 14	*Calantica mortenseni*
OTU 7	*Arcoscalpellum tritonis*	OTU 15	*Smilium peronii*
OTU 8	*Trianguloscalpellum darwinii*		

Per region two characters were recognized, i.e. fusion (f) and length (l). A casual difference between the states of a character in a specific region is expressed in two categories for this character, e.g. f1 and f2. The first category (f1) represents the most abundant state whereas the second category (f2) expresses the rare state. The case of exclusively one state per region is expressed by an equivalent value of state in both categories f1 and f2.

OTU: operational taxonomic unit = recent species; HTU: hypothetical taxonomic unit = hypothetical ancestral species; f: degree of fusion (1: free, 2: basally fused, 3: totally fused); l: length (1: short, 2: long); 0: no structures; F 1,2,3: frontal side in region 1,2,3 of the body; L 1,2,3: lateral side in region 1,2,3; C 1,2,3: caudal side in region 1,2,3.

	Region F 1				L 1				C 1				F 2				L 2	
	f1	f2	l1	l2	f1	f2	l1	l2	f1	f2	l1	l2	f1	f2	l1	l2	f1	f2
OTU																		
1	2	2	1	1	2	2	1	1	2	2	1	1	2	2	1	1	2	2
2	2	1	1	1	2	2	1	1	2	2	1	1	2	2	1	1	2	1
3	2	2	1	1	2	2	1	1	2	2	1	1	2	3	1	1	2	3
4	2	1	1	1	2	1	1	1	2	1	1	1	2	2	1	1	2	2
5	1	1	1	1	1	1	1	1	1	1	1	1	2	2	1	1	2	2
6	2	2	2	2	2	2	2	2	2	2	2	2	2	2	2	2	2	3
7	2	2	2	2	2	2	1	1	2	1	2	1	2	3	1	1	2	3
8	1	1	2	2	1	1	2	2	1	1	2	2	2	2	1	1	2	2
9	2	3	2	1	2	3	2	1	2	3	2	1	2	2	1	2	2	2
10	3	2	1	1	2	3	1	1	2	3	1	1	2	3	1	1	2	3
11	2	1	2	2	2	2	2	2	2	1	2	2	2	2	2	2	2	2
12	2	2	1	1	2	2	1	1	2	2	1	1	2	3	1	1	2	2
13	2	3	1	1	2	3	1	1	2	3	1	1	2	2	2	2	2	2
14	1	1	2	2	1	1	2	2	1	1	2	2	1	1	2	2	1	1
15	1	1	2	2	1	1	2	2	1	1	2	2	1	1	2	2	1	1
HTU																		
16	2	1	1	1	2	2	1	1	2	2	1	1	2	2	1	1	2	1
17	2	1	1	1	2	2	1	1	2	2	1	1	2	2	1	1	2	1
18	2	2	1	1	2	2	1	1	2	2	1	1	2	2	1	1	2	2
19	2	1	2	2	2	2	2	2	2	2	2	2	2	2	2	2	2	1
20	2	2	1	1	2	3	1	1	2	3	1	1	2	2	1	1	2	2
21	2	2	2	2	2	2	2	2	2	2	2	2	2	2	2	2	2	2
22	2	2	1	1	2	2	1	1	2	2	1	1	2	2	1	1	2	2
23	2	1	1	1	2	1	1	1	2	1	1	1	2	2	1	1	2	2
24	2	2	1	1	2	3	1	1	2	3	1	1	2	3	1	1	2	3
25	1	1	1	1	1	1	1	1	1	1	1	1	2	2	1	1	2	2
26	2	2	2	2	2	2	2	2	2	2	2	2	2	2	2	2	2	2
27	2	2	1	1	2	2	1	1	2	2	1	1	2	2	1	1	2	2
28	1	1	2	2	1	1	2	2	1	1	2	2	1	1	2	2	1	1

Table 2. Continued.

	L 2		C 2				F 3				L 3				C 3			
	l1	l2	f1	f2	l1	l2	f1	f2	l1	l2	f1	f2	l1	l2	f1	f2	l1	l2
OTU																		
1	1	1	2	2	1	1	2	3	1	1	3	3	1	1	3	3	1	1
2	1	1	2	1	1	1	0	0	0	0	0	0	0	0	0	0	0	0
3	1	1	2	3	1	1	0	0	0	0	0	0	0	0	0	0	0	0
4	1	1	2	2	1	1	2	3	1	1	2	1	1	1	2	2	1	1
5	1	1	2	2	1	1	1	1	1	1	1	1	1	1	1	1	1	1
6	2	2	2	2	2	2	3	3	2	2	2	2	2	2	2	2	2	2
7	1	1	2	3	1	1	2	3	1	1	2	3	1	1	2	3	1	1
8	1	1	2	2	1	1	2	3	1	1	2	3	1	1	2	3	1	1
9	1	2	2	2	2	2	2	3	1	1	2	3	1	1	2	3	1	1
10	1	1	2	3	1	1	2	3	1	1	2	3	1	1	2	3	1	1
11	2	2	2	2	2	2	2	3	2	2	2	3	2	2	2	3	2	2
12	1	1	2	2	1	1	2	3	1	1	2	3	1	1	2	3	1	1
13	2	2	2	2	2	2	2	2	2	2	2	2	2	2	2	2	2	2
14	2	2	1	1	2	2	1	1	2	2	1	1	2	2	1	1	2	2
15	2	2	1	1	2	2	1	1	2	2	1	1	2	2	1	1	2	2
HTU																		
16	1	1	2	1	1	1	1	1	1	1	1	1	1	1	1	1	1	1
17	1	1	2	1	1	1	0	0	0	0	0	0	0	0	0	0	0	0
18	1	1	2	2	1	1	2	3	1	1	2	3	1	1	2	3	1	1
19	2	2	2	1	2	2	1	1	2	2	1	1	2	2	1	1	2	2
20	1	1	2	2	1	1	2	3	1	1	2	3	1	1	2	3	1	1
21	2	2	2	2	2	2	2	2	2	2	2	2	2	2	2	2	2	2
22	1	1	2	2	1	1	2	3	1	1	2	3	1	1	2	3	1	1
23	1	1	2	2	1	1	2	3	1	1	2	1	1	1	2	2	1	1
24	1	1	2	3	1	1	2	3	1	1	2	3	1	1	2	3	1	1
25	1	1	2	2	1	1	2	3	1	1	2	1	1	1	2	2	1	1
26	2	2	2	2	2	2	2	3	2	2	2	2	2	2	2	2	2	2
27	1	1	2	2	1	1	2	3	1	1	2	3	1	1	2	3	1	1
28	2	2	1	1	2	2	1	1	2	2	1	1	2	2	1	1	2	2

transforming one species into another and vice versa (Table 2, Fig. 4). In all species, the body surface is divided into a top (1), a middle (2), and a bottom (3) region. In each region, frontal (F), lateral (L), and caudal (C) sides are distinguished. The frontal side is the side exposed when the sac-like male is still in the pouch of the hermaphrodite (female). The lateral and caudal sides are in the corresponding positions.

There are good reasons to assume *Calantica* and *Smilium* as outgroup taxa: in both OTUs, the males display an ancestral condition regarding the organisation of their body into a capitulum and a peduncle, the number of plates on the capitulum, the six pairs of cirri, the extensible penis and in having a functional alimentary canal (Klepal 1987). As a consequence, the tree can be rooted and HTU 19 may be assumed to represent an ancestral form (Fig. 4). This 'species' (HTU 19) is likely to have long cuticular structures that were partly fused and partly free in body regions 1 and 2 and totally free in region 3. From this form, three ways of differentiation can be postulated. 1) All long structures become free on the outer surface of regions 1, 2

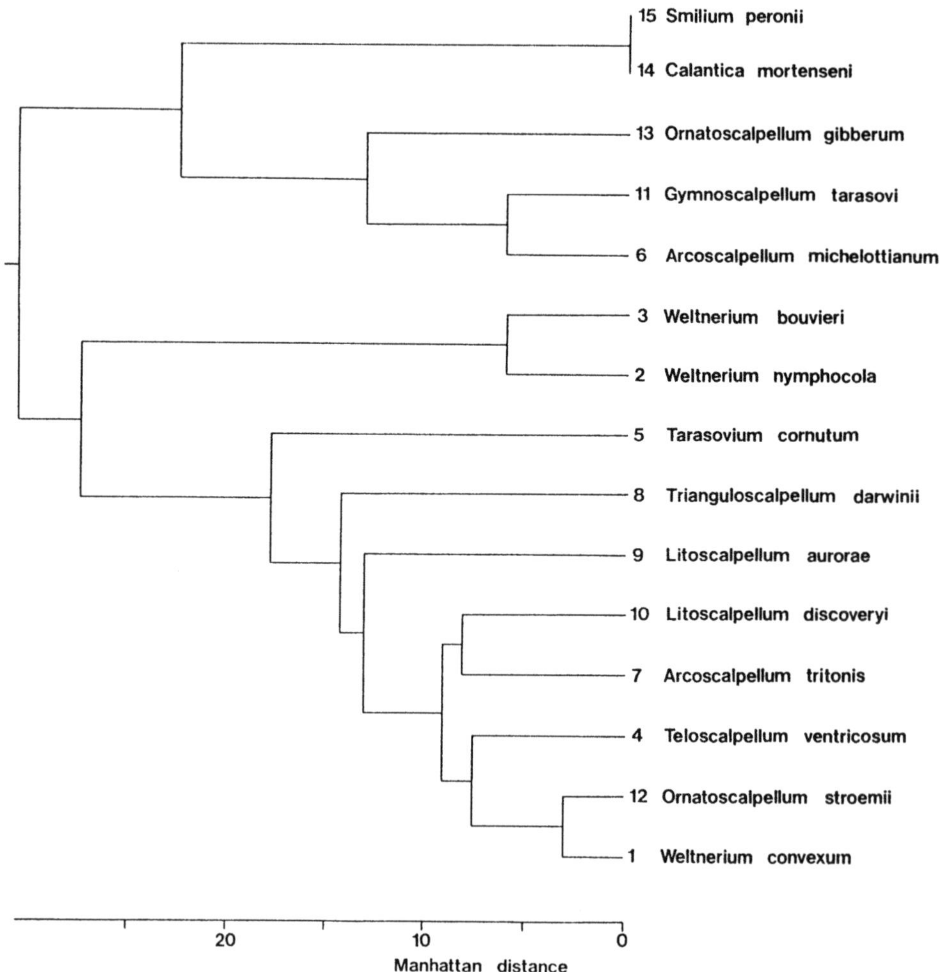

Figure 3. UPGMA-cluster analysis of a Manhattan distance matrix between the OTUs (numbers 1-15), and of all cuticular characters (Program routines written in AMIGA-Basic by HLN).

and 3 as in the calanticines *Calantica mortenseni* and *Smilium peronii*. 2) There was an increase in the degree of fusion in regions 1, 2 and 3. First, all structures fused basally (as in HTU 21). From there two ways could have been taken. In the first, some of the structures totally fused and all of them are short in region 1, as in the scalpellines *Ornatoscalpellum gibberum*. In the second, some of the still long structures were totally fused on the frontal side of region 3 (HTU 26). In the next step, either all of them were totally fused on the frontal side of region 3 and some of them were totally fused on the lateral side of region 2 (as in the arcoscalpelline *Arcoscalpellum michelottianum*), or some of the cuticular structures were in addition totally fused on the lateral and caudal sides of region 3 and, at the same time, some of the structures became free on the frontal and caudal sides of region 1, as in the meroscalpelline *Gymnoscalpellum tarasovi*. 3) There was a differentiation from HTU 19 that consisted in the shortening of the cuticular structures.

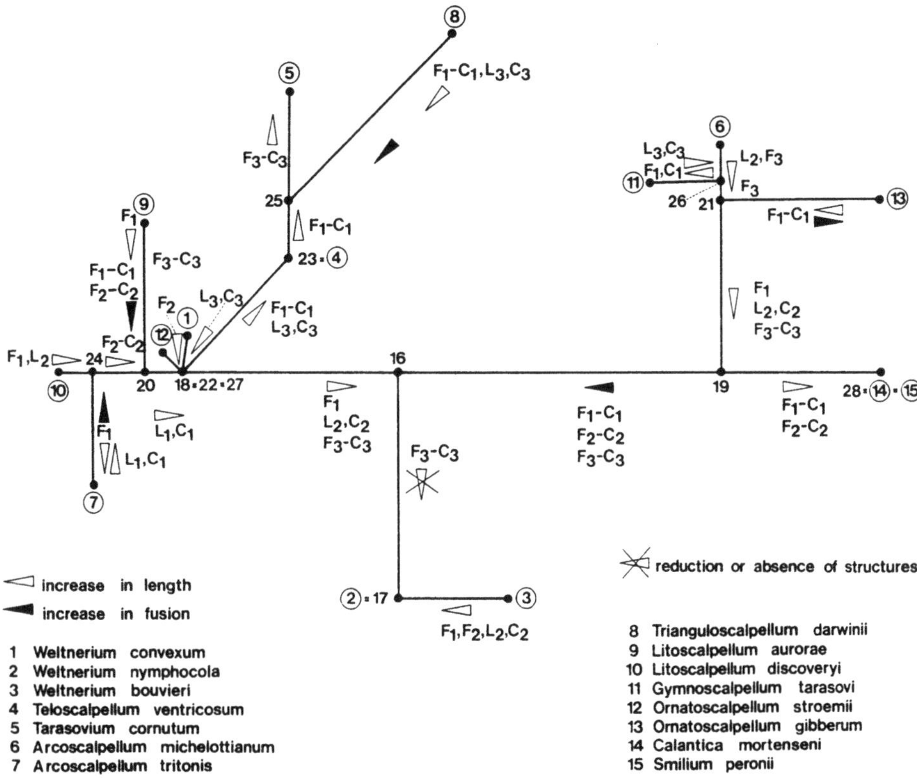

Figure 4. Wagner network of characters and scalpellid taxa. For character states see Table 2. The nodes on the tree represent recent species (encircled) and/or hypothetical ancestral species. The distances between immediately linked nodes are evolutionary distances which represent the numbers of changes in character states. (Program routines written in AMIGA-Basic by HLN).

Another hypothetical center of species differentiation was HTU 18, in which all the structures were short and basally fused in regions 1 and 2, some of them were basally, and some of them totally fused in region 3. From there, five ways of differentiation could have occurred. 1) Some of the structures totally fused on the frontal side of region 2 and on the frontal, lateral and caudal sides of region 3, as in the scalpelline *Ornatoscalpellum stroemii.* 2) All structures were totally fused on the lateral and caudal sides of region 3, as in the arcoscalpelline *Weltnerium convexum.* 3) Some of the structures became free on all sides of region 1, some of them became free on the lateral side, and some of them only basally fused on the caudal side of region 3, as in the arcoscalpelline *Teloscalpellum ventricosum.* As a next step of differentiation, the cuticular strucutures became free on all sides of region 1 (as in HTU 25). Then either all of them became free on all sides of region 3, as in the arcoscalpelline *Tarasovium cornutum,* or the structures became longer and basally fused on all sides of region 1 and some of them totally fused on the lateral and caudal sides of region 3, as in the arcoscalpelline *Arcoscalpellum tritonis.* 4) There was a differentiation from HTU 18 that consisted in some of the structures becoming free on the frontal side of region 1, the lateral and caudal sides of region 2 and on all sides of region 3, as in HTU 16. From there the differentiation proceeded towards a reduction of the structures on all sides of region 3, as in the arcoscalpelline *Weltnerium nymphocola*. In comparison with this species, all structures were basally fused on all sides of region 1 and some of them totally fused on all sides of region 2 in *Weltnerium bouvieri*. 5) There was a differentiation when some of the structures totally fused on the lateral and caudal sides of region 1, as in HTU 20. Beginning from that species, some of the structures became longer on all sides of region 1 and on the frontal and lateral sides of region 2, and all of them became longer on the caudal side of region 2. In addition, some of the structures totally fused on the frontal side of region 1. All theses changes were implemented in the meroscalpelline *Litoscalpellum aurorae*.

Alternately, the differentiation of HTU 20 possibly proceeded towards total fusion of the structures on all sides of region 2 as in HTU 24. From there, some of the structures totally fused on the frontal side of region 1 and on the lateral side of region 2, as in *Litoscalpellum discoveryi*. As an alternative, the structures became long on the frontal side of region 1, some of them basally fused on the lateral side, and some of them became free on the caudal side of region 1, as in the arcoscalpelline *Arcoscalpellum tritonis*.

The analysis of the Wagner network implies parallel trends in the character transformation of the length of the cuticular structures and in their degree of fusion. Long and free cuticular structures on all sides in all regions of the body are a primitive condition (see above arguments for the outgroup status of *Calantica* and *Smilium*; see also Klepal & Kastner 1980). This is in agreement with the taxonomy of the Scalpellidae. The Calanticinae are assumed to be more primitive in aspects of their systematic position, the kind of cuticular structures and the mode of their distribution on the body. Thus the Calanticinae, as more primitive subfamily of the Scalpellidae (Newman 1987; Zevina 1980), has the original cuticular structures in the original mode of distribution. No other subfamily has free structures distributed in the same way. The species investigated within this subfamily can be distinguished by the length and the density of these structures.

Within the Scalpellinae, two species were studied. In the Wagner network, they are far apart. In *Ornatoscalpellum gibberum*, the cuticular structures are longer in regions 2 and 3 than in *O.stroemii*. An increase in the degree of fusion of the structures affects only region 1 in *O.gibberum* and regions 2 and 3 in *O.stroemii* (see Table 2). From this it may be deduced that *O. gibberum* is the more original and *O.stroemii* the derived species within the genus (and possibly within the subfamily).

Within the Meroscalpellinae, the two species of *Litoscalpellum* are more similar to each other than to *Gymnoscalpellum tarasovi*. On the whole the cuticular structures are longer in *Gymnoscalpellum* than in *Litoscalpellum*, and the degree of fusion is less on the frontal and caudal sides of region 1 in *Gymnoscalpellum* than in *Litoscalpellum*. This lower degree of fusion and the relatively long cuticular structures all over the body make the genus *Gymnoscalpellum* more primitive than *Litoscalpellum* as regards cuticular ornament. The two species of *Litoscalpellum* may be distinguished by the length of the cuticular structures. *Litoscalpellum aurorae* has longer cuticular structures than *Litoscalpellum discoveryi*. This is presumably a secondary increase in length in body regions 1 and 2, and this makes *L.aurorae* a derived species in comparison with *L.discoveryi*.

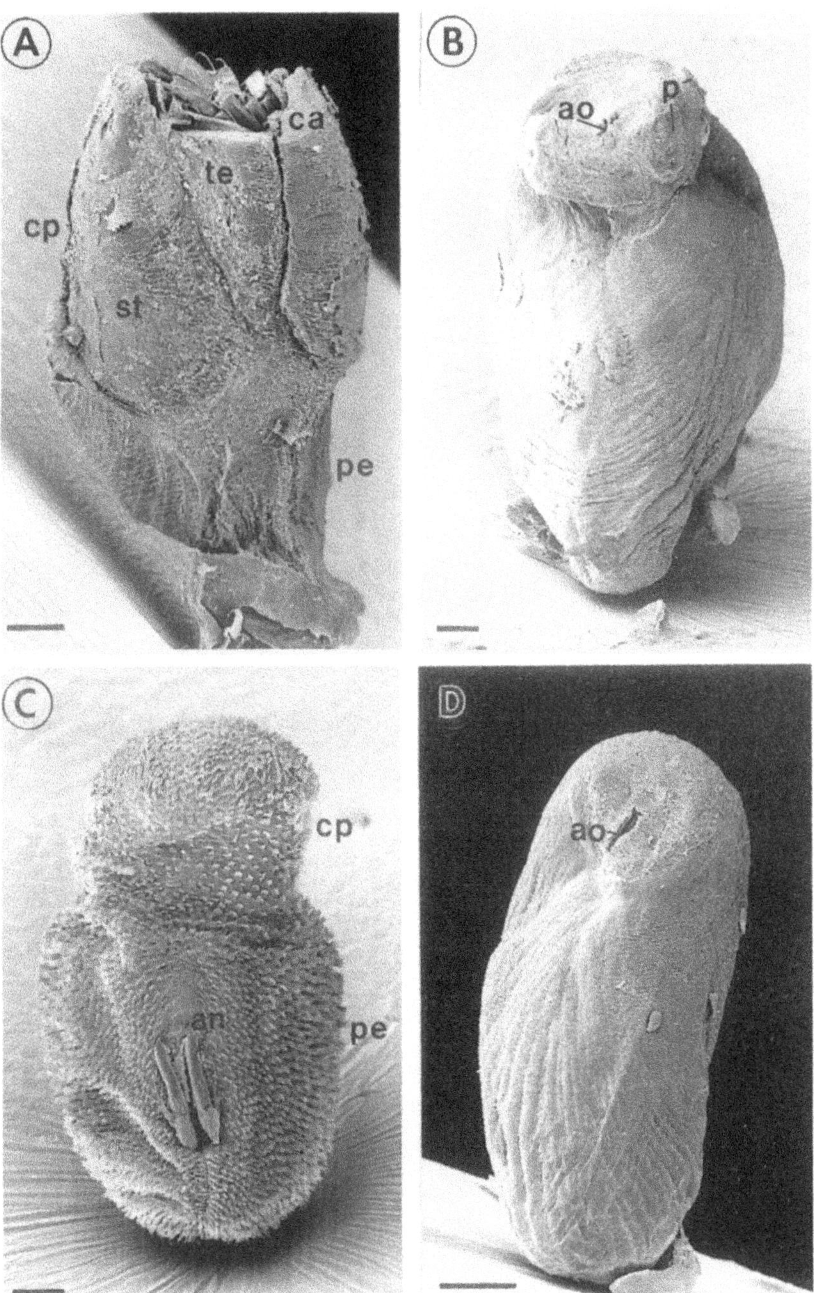

Plate 1. Examples of males of Scalpellidae. (A) *Calantica mortenseni* with distinct capitulum (cp) and peduncle (pe). Plates on the capitulum: scutum (st), tergum (te) and carina (ca) broken; for cuticular structures on outer surface of male see Plate 2A. (B) *Litoscalpellum discoveryi*, sac-like with four reduced plates (p) and an apertural opening (ao) on the capitulum; for cuticular structures see Plate 2D. (C) *Arcoscalpellum michelottianum*, sac-like without any plates, peduncle with antennulae (an); for cuticular structures see Plates 2C and 3B. (D) *Weltnerium convexum*, sac-like without any plates, apertural opening (ao) on capitular pole; for cuticular structures compare Plate 4C. Scale bars: 100 µm.

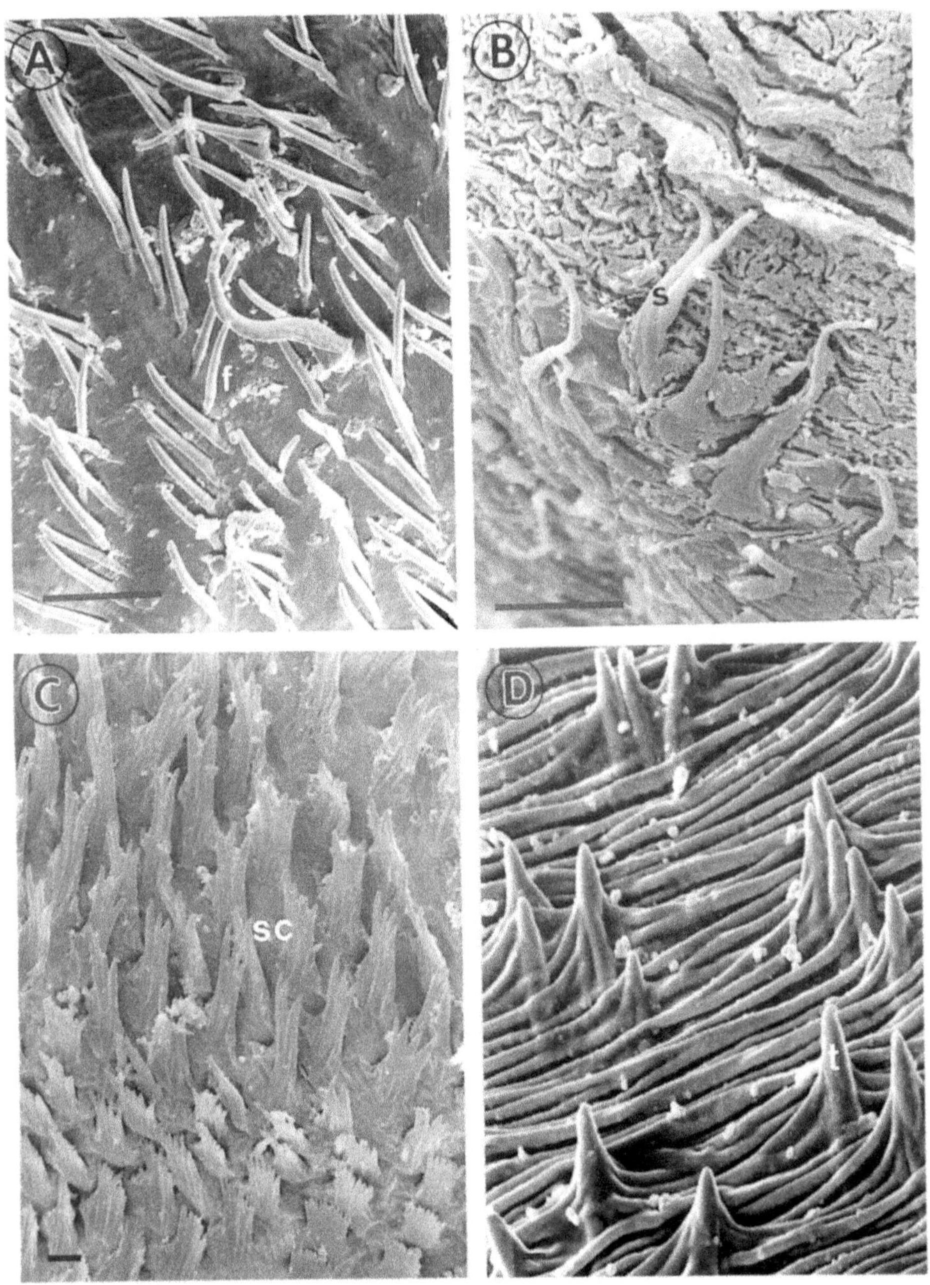

Plate 2. Cuticular structures on the outer surface of male scalpellids. (A) Fringes (f), (B) Setae (s), (C) Scales (sc), (D) Teeth (t). Scale bars: A, B, C 10 μm, D 1μm.

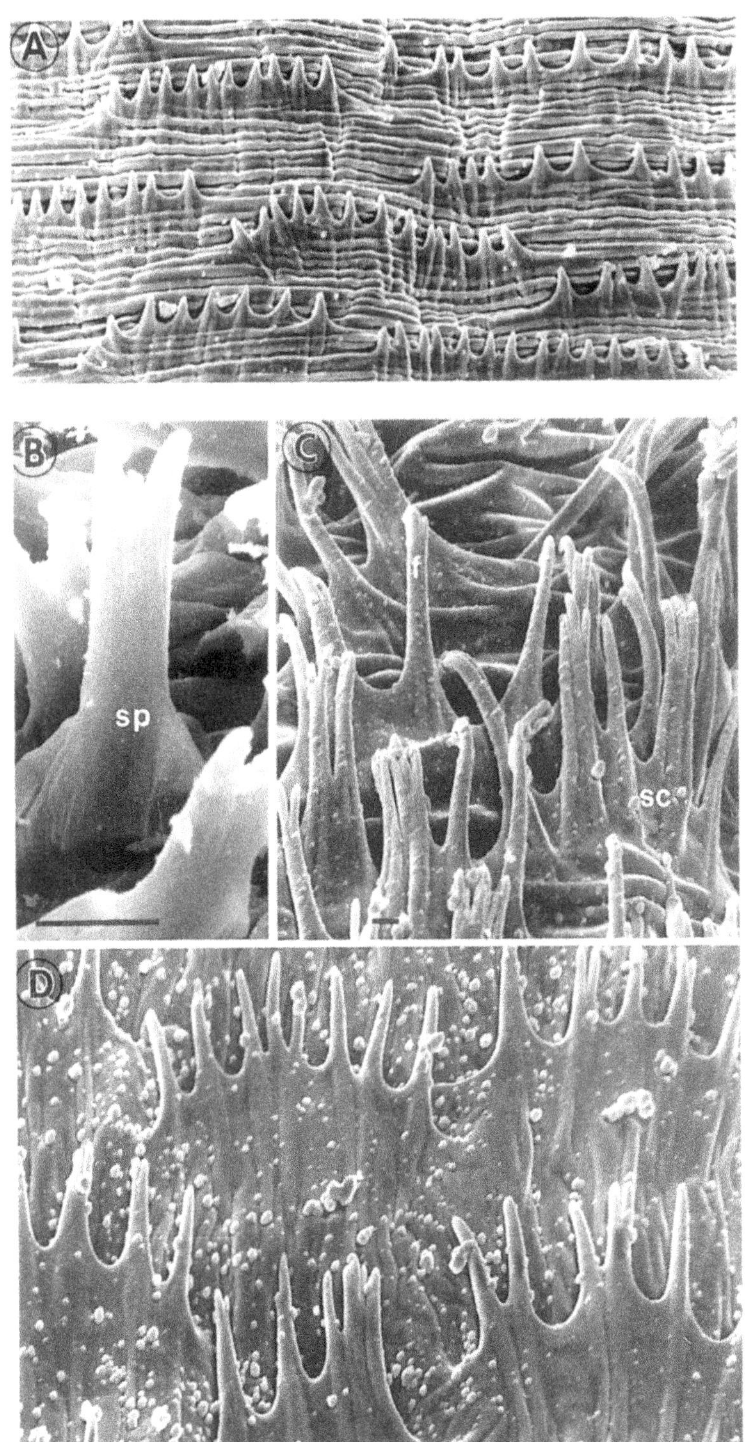

Plate 3. Cuticular structures on the outer surface of male scalpellids. (A) Combs; and their transitional stages, (B) Fringes-spine (sp), (C) Fringes (f)-scale (sc), (D) Scale-comb. Scale bars: A, C, D 1μm, B 10 μm.

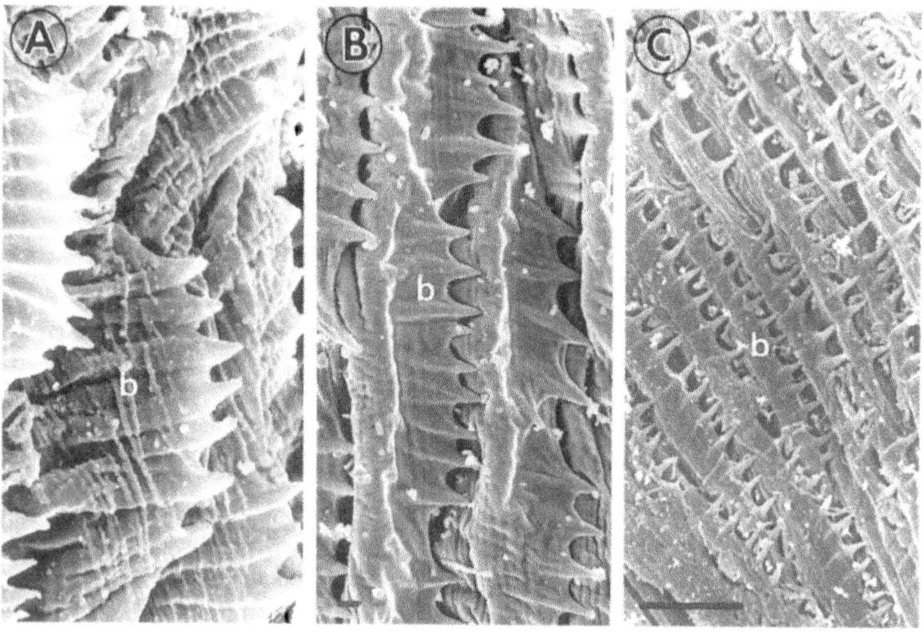

Plate 4. The comb basis (b) is widest in the Scalpellinae (A), narrower in the Meroscalpellinae (B) and narrowest in the Arcoscalpellinae (C). Scale bars: A, B 1 µm, C 10 µm.

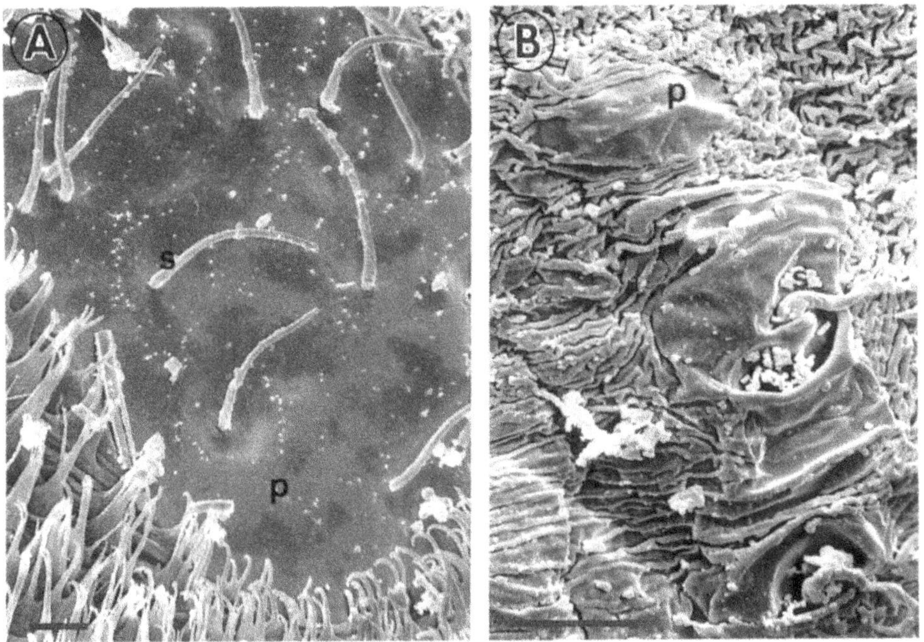

Plate 5. (A) Plate (p) fully developed in *Gymnoscalpellum tarasovi*. Several long setae (s). (B) Plate (p) reduced in *Weltnerium nymphocola* with single short seta (s). Scale bars: 10 μm.

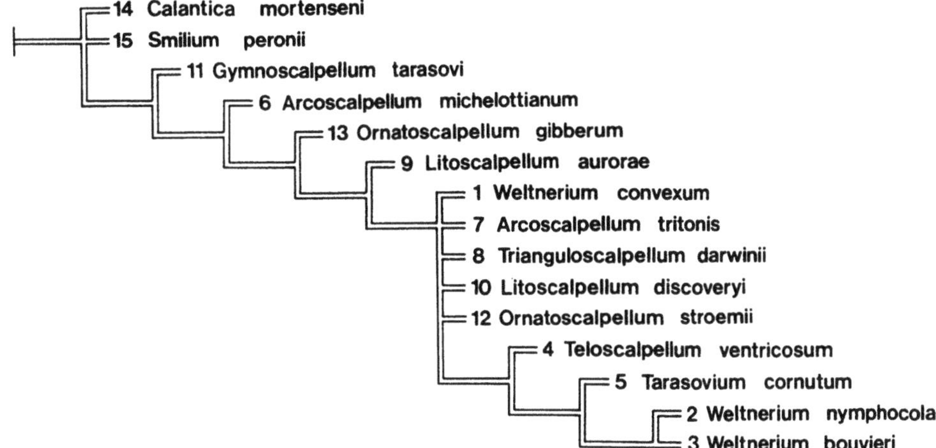

Figure 5. Nelson consensus tree of characters and scalpellid taxa (ie*, length = 110, ci = 56, ri = 73, trees = 10, outgroup = OTU 14, 15; HENNIG 86, vers. 1.5).

Various species of Arcoscalpellinae show parallel trends in the evolution of lengths and degrees of fusion of the cuticular structures. There are parallel increases in length in *Arcoscalpellum tritonis* and *Trianguloscalpellum darwinii.* With regard to fusion of the cuticular structures there are parallel increases in *Weltnerium convexum, Trianguloscalpellum darwinii* and partly in *Arcoscalpellum tritonis,* but parallel reductions in *Teloscalpellum ventricosum* and *Tarasovium cornutum.* These reductions are assumed to be derived conditions. These character transformations allow the distinction of the genera. Within the genus *Weltnerium, Weltnerium convexum* is distant from the other two species *W. nymphocola* and *W. bouvieri* in the Wagner network. *W. convexum* has cuticular structures on all 3 regions of the body and is thus more primitive than the other two species in this regard. In *W. bouvieri,* fusion of cuticular structures has progressed further in regions 1 and 2. Therefore *W. bouvieri* is presumably more derived than *W. nymphocola.* The two species of the genus *Arcoscalpellum* are far apart in the Wagner network because of differences in length and in the degree of fusion of cuticular structures. In *Arcoscalpellum tritonis,* some of the structures are shorter and more completely fused than in *A. michelottianum.* This makes *A.michelottianum* the more primitive species and *A. tritonis* the more derived one within the genus in this regard.

According to the principles of phylogenetic systematics, an outgroup was chosen and a cladistic analysis was made with HENNIG 86 (Fig. 5). Provided that the cuticular structures as a single character system are accepted as sufficient for a phylogenetic reconstruction, the analysis suggests a revision of the taxonomy. In the current state of analysis, however, the present paper exclusively emphasizes the discussion of possible taxonomic solutions but not final taxonomic statements. The following established genera (of the ones examined by us) are revealed as paraphyletic/polyphyletic units: *Weltnerium, Arcoscalpellum, Litoscalpellum* and *Ornatoscalpellum.* An inspection of the Wagner network as well as the phenetic analysis similarly recommend a splitting of those genera (Figs 3 and 4). The following (possible) proposals for a revision are made on the basis of the cladogram (Fig. 5):

The genus *Weltnerium* should include the species *W. nymphocola* and *W. bouvieri* as well as the former *Tarasovium cornutum*, or as an extension, additionally including the former *Teloscalpellum ventricosum*.

On the basis of the phenetic analysis (Fig. 3), the following assemblages suggest themselves as possibly new genera: 1) *Weltnerium nymphocola* and *W. bouvieri*, 2) *Arcoscalpellum michelottianum* and *Gymnocsalpellum tarasovi*, 3) *Arcoscalpellum tritonis*, *Litoscalpellum discoveryi*, *Weltnerium convexum*, *Ornatoscalpellum stroemii*, and *Teloscalpellum ventricosum*.

On the basis of the Wagner network (Fig. 4), the following units can be established: 1) *Weltnerium nymphocola* and *W. bouvieri*, 2) *Arcoscalpellum michelottianum* and *Gymnoscalpellum tarasovi*, 3) *Arcoscalpellum tritonis*, *Litoscalpellum discoveryi*, and *Ornatoscalpellum stroemii*. With the exception of the inclusion of *Teloscalpellum ventricosum*, the results on the basis of the Wagner network agree with the outcome of the phenetic analysis. Moreover, in the case of the connection between *Weltnerium nymphocola* and *W. bouvieri*, the two last analyses also agree with the phylogenetic systematic approach. These results coincide with a number of ecological factors. The cuticular structures are present both to maintain the distance between neighbouring cuticular surfaces (either between the mantle and the body, or between the males and their partners, or the males and the pocket in which they sit) and to avoid debris on the surface. In the first case, keeping the distance allows an exchange of substances between the body and the surrounding water via pores in the cuticle. In the second case, the cuticular structures prevent the settlement of fouling organisms on the surface as well as accumulation of detritus. The cuticular structures may then act as combs or fences and, by being moved passively in the hermaphrodites/females and the hardly reduced males, also as cleaning devices (Klepal & Kastner 1980; Lavalli & Factor 1992).

There are a number of possible ways to accomplish the required function (compare the combs in the Peracarida, Klepal & Kastner 1980). In the great majority of cases, species diversity is predominantly caused by the process of adaptive radiation. Mechanisms avoiding competition, such as character displacements, are an important driving force for adaptation and for the transformation of characters between species. Basically the male cuticular structures are a character like those of any arthropod cuticle, and they are as old as the cuticle and as the arthropods themselves (see also Müller & Walossek 1988). The existence of cuticular structures is thus plesiomorphic, but their structural details represent several (aut-) apomorphies, that can allow the distinction of lower taxonomic units.

4 CONCLUSION

It is obvious that the cuticular structures of the Scalpellidae resemble those found in other Crustacea (Klepal 1981; Klepal & Kastner 1980). Fringes are assumed to be the most primitive cuticular structures. First, they are the main structure in the Calanticinae, the most primitive of Scalpellidae. Second, they are the substructure of all other cuticular structures in the remaining subfamilies of Scalpellidae. Third, they occur on the shell of newly metamorphosed Balanomorpha (Glenner & Høeg 1993). Fourth, they are on the antennulae of all Cirripedia s.str. (e.g. Nott 1969; Nott &

Foster 1969). In the Calanticinae (Scalpellidae), fringes are plesiomorphic, both in their expression and distribution. The few serially arranged fringes on the wall plates of the Balanomorpha (Glenner & Høeg 1993) and the occasional ones on the males of the prothoracican *Ibla* (Klepal 1983) represent a derived condition. Cuticular structures are highly functional. Thus selection pressure acts on the flexible fringes to create by fusion more robust and stiff structures like the characteristic spines on the outside of the hermaphrodites (females) of *Ibla* and teeth, spines, scales and combs on the surface of the sac-like males in the Scalpellidae. Selection pressure does not affect the capitulum of the males of the Calanticinae presumably because this is protected by plates. Fringes disappear completely from the shell of the Balanomorpha once the animal is firmly attached to the substratum. Because of the high functionality and obvious cases of parallel evolution, the cuticular structures cannot be used for systematics at higher taxonomic levels. Their differing character states make them, however, valuable for systematics in the lower categories.

ACKNOWLEDGEMENTS

The help of the curators of the British Natural History Museum, the Museum of Stockholm and the National Museum of Natural History in Washington D.C. by lending the material investigated to one of us (WK) is gratefully acknowledged. It is a pleasure to thank D. Gruber who helped collecting the data and Dr. S. Neulinger who made the drawings.

REFERENCES

Crisp, D.J. 1983. *Chelonobia patula* (Ranzani), a pointer to the evolution of the complemental male. *Mar. Biol. Letters* 4: 281-294.

Darwin, C. 1851. *A Monograph on the sub-class Cirripedia, with figures of all the species.* The Lepadidae or Pedunculated Cirripedes. London: Ray Society.

Dayton, P.K., Newman W.A. & J. Oliver 1982. The vertical zonation of the deep-sea Antarctic acorn barnacle, *Bathylasma corolliforme* (Hoek). Experimental transplants from the shelf into shallow water. *J. Biogeogr.* 9: 95-109.

Farris, J.S. 1970. Methods for computing Wagner trees. *Syst. Zool.* 19:83-92.

Foster, B.A. 1983. Complemental males in the barnacle *Bathylasma alearum* (Cirripedia: Pachylasmidae). *Aust. Mus. Sydney. Mem.* 18: 133-139.

Glenner, H. & J.T. Høeg 1993. Scanning electron microscopy of metamorphosis in four species of barnacles (Cirripedia Thoracica Balanomorpha). *Mar. Biol.* 117: 431-439.

Henry, D.P. & P.A. McLaughlin 1965. Unique occurrence of complemental males in a sessile barnacle. *Nature* 207: 1107-1108.

Henry, D.P. & P.A. McLaughlin 1967. A revision of the subgenus *Solidobalanus* Hoek (Cirripedia Thoracica) including a description of a new species with complemental males. *Crustaceana* 12: 43-58.

Hui, E. & J. Moyse 1984: Complemental male in the primitive balanomorph barnacle, *Chionelasmus darwinii. J. Mar. Biol. Assoc. U.K.* 64: 91-97.

Jacques, F. 1989. The setal system of crustaceans: Types of setae, groupings, and functional morphology. *Crustacean Issues* 6: 1-13.

Klepal, W. 1981. Nicht-sensorische Cuticularstrukturen von Crustacea – eine rasterelektronenmikroskopische Studie. *Beitr. elektronenmikrosk. Direktabb. Oberfl.* 14: 611-618.

Klepal, W. 1983. Morphogenesis and variability of cuticular structures in the genus *Ibla* (Crustacea, Cirripedia). *Zool. Scr.* 12: 115-125.

Klepal, W. 1987. A review of the comparative anatomy of the males in cirripedes. *Oceanogr. Mar. Biol. Ann. Rev.* 25:285-351.

Klepal, W. & R.T. Kastner 1980. Morphology and differentiation of non-sensory cuticular structures in Mysidacea, Cumacea and Tanaidacea (Crustacea, Peracarida). *Zool. Scr.* 9:271-281.

Lavalli, K.L. & J.R. Factor 1992. Functional morphology of the mouthparts of juvenile lobsters, *Homarus americanus* (Decapoda: Nephropidae), and comparison with the larval stages. *J. Crust. Biol.* 12:476-510.

Müller, K.J. & D. Walossek 1988. External morphology and larval development of the upper Cambrian maxillopod *Bredocaris admirabilis. Fossils and Strata.* 23:1-70.

Newman, W.A. 1987. Evolution of cirripedes and their major groups. *Crustacean Issues* 5: 3-42.

Nilsson-Cantell, C. A. 1921. Cirripedien -Studien. *Zool. Bidr. Uppsala,* 7: 75-390.

Nilsson-Cantell, C. A. 1932. Cirripeds from the Indian Ocean and Malay Archipelago in the British Museum (Nat. Hist). *London. Ark. Zool.* 23A (18): 1-2.

Nott, J.A. 1969. Settlement of barnacle larvae: surface structure of the antennular attachment disk by scanning electron microscopy. *Mar. Biol.* 2: 248-251.

Nott, J.A. & B.A. Foster 1969. On the structure of the antennular attachment organ of the cypris larva of *Balanus balanoides* (L). *Phil. Trans. R. Soc. Lond.*(B) 256:115-134.

Pilsbry, H.A. 1908. On the classification of scalpelliform barnacles. *Proc. Acad. Nat. Sci. Philad.* 60: 104-111.

Sneath, H.A & R.R. Sokal 1973. *Numerical Taxonomy.* San Francisco: Freeman.

Zevina, G.B. 1980. A new classification of Lepadomorpha (Cirripedia). *Zool. Zh.* 59:689-698.

Sex and the single cirripede: A phylogenetic perspective

Jens T. Høeg
Department of Cell Biology and Anatomy, Institute of Zoology, University of Copenhagen, Copenhagen, Denmark

ABSTRACT

In Cirripedia, the traditional scenario considers hermaphroditism as part of the ground pattern, and various theories have tried to explain how and why a separate male sex evolved secondarily in numerous taxa. As an alternative I map the distribution of dioecy, androdioecy, and hermaphroditism onto a tentative phylogeny of the Thecostraca and the Cirripedia. This indicates that both the urthecostracan and the urcirripede had separate sexes and that hermaphroditism represents an apomorphy for the Thoracica, only, while the Ascothoracida, the Acrothoracica, and the Rhizocephala displays the plesiomorphic pattern with separate sexes. I emphasize that a separate male sex may well have evolved convergently several times in the Cirripedia, but we cannot forward such conclusions in a phylogenetic vacuum using adaptionist arguments only.

1 INTRODUCTION

The Cirripedia exemplify a most amazing variety of sexual systems, not only compared with other Crustacea but with the Metazoa at large. This, and the sessile mode of life seen in all cirripedes, makes them ideally suited for population level studies and explains why cirripede sexual biology has attracted the attention of so many biologists since the influential work of Darwin (1852, 1854).

The majority of cirripedes are hermaphrodites, a very uncommon strategy in Crustacea. Many species live in dense populations, which insures that each individual will have one or several partners for mating within reach of the long penis. However, many species live either strictly solitary or at much lower population densities, where a hermaphroditic strategy would seem very risky unless the species employ self-fertilization, which usually they do not (Ghiselin 1969). However, with the most astonishing insight, Darwin (1852) unveiled that many cirripede species have one or several minute, dwarf males associated with either larger hermaphrodites (androdioecy) or pure females (dioecy).

Among other crustacean taxa we find hermaphroditism as a ground pattern feature only in the Remipedia and the Cephalocarida. Competing hypotheses on phylogeny consider either of these groups as sister groups to all remaining Crustacea (Schram 1986; Hessler 1992). This could indicate that the stem species to all Crustacea had hermaphroditism and raises the question whether the hermaphroditism seen in the Cirripedia constitutes an advanced trait or a very ancestral trait taken over from the urcrustacean. Cirripedes themselves admittedly deviate much from the crustacean ground pattern, but the closely related Ascothoracida have numerous plesiomorphic features, including a flagellated, motile sperm that represents the most original type seen in all Crustacea (Grygier 1981, 1982; Healy & Anderson 1990). Moreover, the cirripede lineage may go back as far as the Cambrian (Collins & Rudkin 1981). It therefore seems worthwile to discuss what arguments exist concerning the sexual system in the urcirripede.

2 A PHYLOGENETIC PERSPECTIVE ON CIRRIPEDE SEX

Table 1 summarizes key aspects of cirripede biology, especially with regard to the sexual system. Recently we have seen some important accounts that attempt to raise the discussion of cirripede sex above the anecdotal level and construct testable sets of hypotheses. Crisp (1983) provided a thought-provoking account on how dwarf males could have evolved from hermaphrodites, while Charnov (1987) went further

Table 1. The sexual systems of the Cirripedia and closest relatives. (•) Denotes variations in a feature for a taxon. (1) Separate sexes, gonochorism. (2) Only a single family have hermaphroditism as an apomorphic feature. (3) Males and hermaphrodites. (4) Cryptogonochorism *sensu* Bresciani & Lützen (1972).

Taxon	Lifefor	Habitat	Sexual system	Male
TANTULOCARIDA	Parasitic	• Copepoda • Tanaidacea • Ostracoda	Dioecy and parthenogenesis	Almost size of sexual female
ASCOTHORACIDA	Parasitic	• Echinodermata • Anthozoa	• Dioecy (1) • Hermaphroditism (2)	• Almost size of female • Dwarf male
CIRRIPEDIA				
Acrothoracica	Setose feeders	Calcareous substrata (mostly shells of Mollusca)	Dioecy	Lecithotrophic dwarf male
Rhizocephala	Parasitic	• Decapoda • Peracarida • Stomatopoda • Cirripedia	Dioecy	Dwarf male, integrated in and nourished by female (4)
Thoracica	• Setose feeders • Parasitic (few)	Various substrata e.g.: • Hardbottom • Ships • Coelenterata • Turtles • Whales • Crustacea	• Hermaphroditism • Androdioecy (3) • Dioecy	• Feeding dwarf male • Lecitotrophic dwarf male

and constructed rigorous mathematical models for the various possible reproductive strategies and how they would evolve under natural selection.

Almost all accounts on cirripede sexual biology, including those of Ghiselin (1969), Crisp (1983), and Charnov (1987), seem more or less implicitly to assume that the cirripede stem species had hermaphroditism. This view will obviously bias ideas on how the sexual systems evolved, since cirripede males, where they exist, must accordingly have evolved secondarily from hermaphrodites. The view that the ancestral cirripede had pure hermaphroditism does not stem from a phylogenetic analysis but rather has a historical explanation. First, the majority of cirripede species have pure hermaphroditism and, although this confers no information by itself upon the plesiomorphic character state, it may well have biased the view of specialists not familiar with phylogenetic systematic arguments. Second, we have long considered the Thoracica as a 'core group' from which evolved both the boring Acrothoracica and the parasitic Rhizocephala (Newman et al. 1969). The urcirripede thereby became identical with the urthoracican, which was assumed to be a hermaphrodite. Finally, while we have long recognized the gonochoristic nature of the Acrothoracica, the realization that this also holds true for all Rhizocephala is fairly recent (Høeg 1991).

Klepal (1985, 1987) provided a very scholarly treatment of the various theories concerning the evolution of the cirripede sexual system. She reviewed the arguments for and against both a hermaphroditic and a gonochoristic ancestor for the Cirripedia and rightly emphasized the importance of a phylogenetic approach. However, lacking a well-established phylogeny she could not arrive at a hypothesis for the sexual status of the urcirriped.

Since Klepal's (1985, 1987) studies, evidence from sequencing of genes and from larval ultrastructure combined with the advent of phylogenetic systematics to cirripedology have for the first time provided us with a well argued gross phylogeny for the Cirripedia and their nearest relatives (Grygier 1987a; Spears et al. 1994; Høeg 1992; Jensen et al., 1994). In addition, we now have important new information concerning the sexual biology of the Tantulocarida and the Ascothoracida, both close relatives of the Cirripedia (Grygier 1987b,c, Huys et al. 1993).

We can employ characters in two ways in a phylogenetic discussion. First, we can use characters as primary input in a phylogenetic analysis. This is what we habitually do in a cladistic analysis using the principle of parsimony. Second, once we have confidence in a certain phylogeny, we can map characters, including ones not originally used in constructing that scheme, onto the branching pattern and analyse how the states changed throughout evolution. The latter would seem a sound principle for characters for which we have reason to suspect a high degree of homoplasy and which we therefore excluded from the initial analysis leading to the phylogeny at hand. However, we should eventually include as many characters as possible in the matrix since we cannot claim homoplasy a priori, but only *a posteriori* following a fully fledged phylogenetic analysis.

In the following, I will first discuss the various characters involved in the sexual systems seen in Cirripedia and then map some of these characters onto a tentative phylogeny for the taxon and its nearest relatives. With this approach and using the parsimony principle in Maddison et al. (1984), I will forward a hypothesis on what was the sexual status of the urcirripede.

3 CHARACTERS OF THE SEXUAL SYSTEM

The sexual system involves a multitude of characters, not all of which concerns us here. Although important functional links undoubtedly exist between the characters listed below, they do not evolve strictly in parallel and this allows us to consider each of them separately in a phylogenetic analysis. Such separation of complex traits into separate characters based on new biological insight provides a much more objective method of 'character weighting' than putting a subjective weight number onto a single trait in a matrix (Glenner et al. 1995).

3.1 *Females/hermaphrodites*

In most cirripedes the large, feeding organism is a hermaphrodite. However, some taxa instead have pure females. This raises the question whether such females represent a plesiomorphic condition or whether they evolved secondarily from hermaphrodites. For instance, in scalpellid thoracicans, hermaphrodites and females do not differ morphologically except that the latter lack both testes and a penis.

3.2 *Males present/absent*

This character denotes the presence or absence of a separate male sex irrespective of its morphology or whether it coexists with females (dioecy) or hermaphrodites (androdioecy). From the considerable difference in the morphology and mode of sex determination seen in males throughout the Cirripedia one might argue that they evolved convergently. However, statements on convergent evolution can only result a posteriori following a phylogenetic analysis. Unless opposed by incongruence in other characters, we must obviously consider the presence of a separate male sex as a valid and homologous character.

3.3 *Hermaphroditism/androdioecy/dioecy*

The combination of the states in characters (Sections 3.1 and 3.2) determines whether the taxon has pure hermaphroditism (no separate male sex), androdioecy (hermaphrodites and a separate male sex), or dioecy (separate male and female sexes). Dioecy may also be called gonochorism or separate sexes.

3.4 *Dwarf male*

Cirripede males, wherever present, occur as 'dwarf males', i.e. male individuals much smaller than and permanently associated with their female or hermaphrodite partner. Many of them lack calcareous plates and appendages, and usually also a functional gut (Klepal 1987). Rhizocephalan and acrothoracican males even lack any signs of body segmentation.

Foster (1983) opines that dwarf males evolved convergently numerous times in the Cirripedia. Admittedly, cirripede males often resemble each other in no other features than having a reduced size and morphology, but we must keep in mind that differences alone represent a poor argument against homology. Thus Høeg (1991) and Høeg & Lützen (1993) consider males of the Rhizocephala as homologous throughout the order despite considerable variation in structure and location, and we could forward a similar case for males of the Acrothoracica. In Thoracica, the Iblidae and the Scalpellidae both exemplify independent reduction series from more complete, multi-plated males to highly reduced, naked forms. We could take this as an indication males evolved along different evolutionary pathways in these two families. Nevertheless, such an argument, however reasonable, assumes that the Iblidae and the Scalpellidae are monophyletic, whence we already stand squarely in an *a priori* phylogenetic argument.

We usually characterize cirripede dwarf males by small size and simplified morphology, but Klepal (1985, 1987) have rightly emphasized that we cannot merely consider cirripede males as resulting from a simple process of reduction. In fact, most dwarf males in the Cirripedia exhibit several apomorphies not shared by their hermaphrodite or female partners. In both the Acrothoracica, the Rhizocephala and the Thoracica Scalpellidae *sensu stricto*, the ontogeny of males and females begins to deviate very soon after cyprid settlement and the adult male habitually represents a morphology never passed through in females (Turquier 1970, 1971; Høeg 1991; Glenner & Høeg 1994; Klepal 1987; Klepal & Høeg unpublished). Except in the Rhizocephala Akentrogonida (Høeg 1991) all cirripede males moult at least once following settlement (Klepal 1987), but they always cease to grow at a very small size and have no potential to grow into hermaphrodites or females. Thus we cannot explain their small size as due to lack of space or nutrients when settling on or inside a female or hermaphrodite.

Crisp (1983) outlined an elegant and simple model for the evolution of cirripede dwarf males in suggesting that they originated from hermaphrodites settling on conspecifics. In *Chelonibia* individuals settling within the aperture of larger conspecifics become functional males because they are arrested in growth and therefore never develop beyond the protandric stage of hermaphroditism. In agreement with Crisp's 'apertural male' model, the least specialized males of the Scalpellidae seem to differ little if at all from stages seen in female ontogeny, and some of them can apparently ingest food with the cirri. However, the observations outlined above shows that the evolution of most cirripede males must have involved more complex processes than Crisp's (1983) model would suggest.

3.5 *Dwarf or complemental male*

Darwin originally distinguished between dwarf males (associated with pure females) and complemental males (associated with hermaphrodites). In purely morphological terms we cannot uphold this distinction, since morphologically very similar males may be either 'dwarf' or 'complemental' (Klepal 1987). However, in terms of sexual strategy the distinction obviously retains importance. I find it best to reserve the term dwarf male as a morphological descriptor only, and employ the more informative terms dioecy and androdioecy (see Section 3.3) to designate the sexual state of the partner of the male (Charnov 1967).

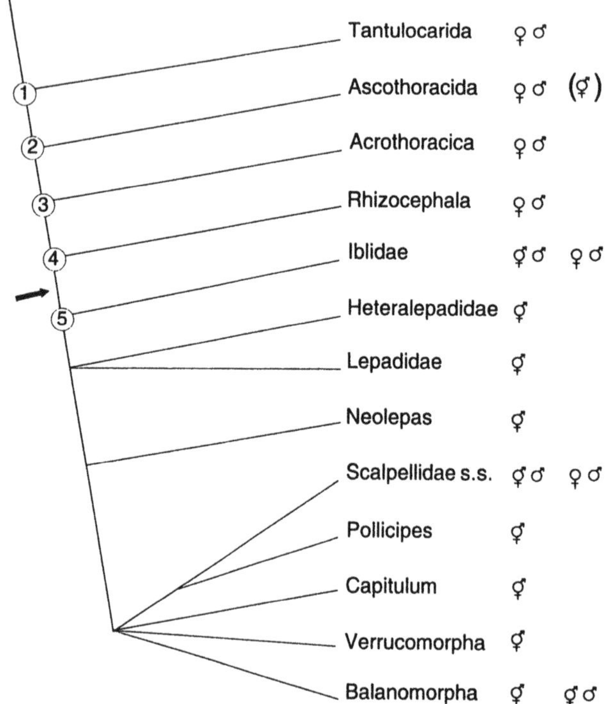

Figure 1. The distribution of sexual systems within the Cirripedia and closest relatives plotted onto a tentative phylogeny derived from Spears et al. (1994), Huys et al. (1993), and Glenner et al. (1995), but modified as explained in the text. Node (2): Thecostraca, node (3): Cirripedia; node (5): Thoracica. Below the Thoracica we find only separate sexes, indicating that this represent the ground pattern for both the Thecostraca and the Cirripedia. If so, hermaphroditism evolved between nodes (4) and (5) (*arrow*) and therefore represent an apomorphy for the Thoracica rather than for the Cirripedia as a whole. The Ascothoracida Petrarcidae do have hermaphroditism, but this represent an apomorphy not present in the ground plan (Grygier pers comm.). Males in the Thoracica, except possibly for the Iblidae, may have evolved secondarily from hermaphrodites, but the extent to which this has occurred must await future detailed analysis of male anatomy and ontogeny and the arrival of a more stable phylogeny for the Thoracica.

4 THE PHYLOGENY

The tentative phylogeny in Figure 1 and used below to map the distribution of males in the Cirripedia derives from several sources of information. Grygier (1987a) established the monophyly of the taxon Thecostraca, comprising the Ascothoracida, the Facetotecta, and the Cirripedia. Huys et al. (1993) provide data supporting a sister group relationship between the Tantulocarida and the Thecostraca. Within the Thecostraca, I have intentionally omitted the Facetotecta, since we know them from pelagic larvae only. The gross phylogeny of the remaining Thecostraca follows Høeg (1992), Jensen et al. (1994) and particularly Spears et al. (1994), who all agreed on a monophyletic Cirripedia in which the Acrothoracica split off first. Moreover, the gene sequencing data of Spears et al. (1994) support a true sister group relationship between the Rhizocephala and the Thoracica.

Within the Thoracica, Figure 1 illustrates the strict consensus tree in Glenner et al. (1995), except for the position of the Iblidae, which I favour as sister group to the remaining thoracicans (see discussion in Glenner et al. 1995). Moreover, due to their uncertain phylogenetic positions, the Poecilasmatidae and a few small families are omitted.

I have only mapped the presence of hermaphrodites/females (Section 3.1) and males (Section 3.2) onto the tree, but indirectly this naturally also covers the distribution of characters (Sections 3.3 and 3.5).

5 THE TAXA

5.1 *Tantulocarida*

The Tantulocarida, not recognized as a class until Boxshall & Lincoln (1983), parasitize other Crustacea. All species have separate sexes although the life cycle is more complex than first believed and also involves a parthenogenetic phase (Huys et al. 1983). Tantulocarids do not have dwarf males, and males and females apparently mate while swimming freely.

5.2 *Ascothoracida*

These thecostracans parasitize Anthozoa and Echinodermata. Some previous accounts (Newman 1974) suggested that ascothoracidans have protandric hermaphroditism as is also the case in most Thoracica. However, Melander (1950), unknown to Newman (1974), and Grygier (1987b,c) clarified their sexual biology. Most species have separate sexes and males and females differ morphologically already before settlement in the cyprid-like ascothoracid larvae. Only in the Petrarcidae do we find hermaphrodites, but Grygier (pers. com.) considers this as a secondary state derived from a ground pattern with separate sexes. We cannot classify the males as dwarf males, since in the Synagogidae they are but a little smaller than the females and organized the same way with regard to appendages and other morphological traits. In fact, synagogid males remain strong swimmers even following settlement, and we may speculate that they can seek out a new female partner if separated from the original one.

5.3 *Acrothoracica*

These boring and setose feeding cirripedes have separate sexes. Dwarf males with a much reduced and very specialized morphology attach, often in great numbers, to the attachment disc of the larger females. We principally owe our limited knowledge of their sexual biology to the accounts of Kühnert (1934) and Turquier (1972a,b).

5.4 *Rhizocephala*

All Rhizocephala have separate sexes. The females parasitize other Crustacea while the adult males lives as an integral part (pseudotestis) of the female. Bresciani &

Lützen (1972) created the term 'cryptogonochorism' for this fascinating phenomenon. Not surprisingly, the Rhizocephala were considered to be self-fertilizing hermaphrodites until the penetrating studies of Yanagimachi (1961). Bocquet-Vedrine (1961, 1972) claimed that hermaphroditism exists in the suborder Akentrogonida. However, subsequent studies have shown that also the Akentrogonida have separate sexes (Høeg 1991; Høeg & Lützen 1993), but their males have become reduced to mere spermatogonia, which cyprids inject directly into the connective tissue of the female parasite.

5.5 *Thoracica*

This suborder exhibits the most extraordinary variation in sexual systems varying from separate sexes, through androdioecy to pure hermaphroditism.

5.5.1 *Iblidae*

All species have dwarf males that associate with either hermaphrodites or females. All males of this small family have a highly advanced morphology, but it differs considerable between the species (Bathham 1945; Klepal 1985, 1987). Males of *Ibla cumingi* moult regularly and can ingest food with their cirri. Males of *Ibla idiotica* consist of little more than a bag of cuticle containing the testes and apparently float freely in the mantle cavity as do the male organs in some Rhizocephala Akentrogonida (Batham 1945; Høeg 1991). However, due to profound differences in ontogeny of the males the possibility for a homology seems very remote.

5.5.2 *Heteralepadidae and Lepadidae*

These two families consist of pure hermaphrodites. The individuals must accordingly occur gregariously and one may wonder what mechanisms insure that e.g. floating objects in the neuston almost immediately become populated by large numbers of settled lepadids.

5.5.3 *Anelasma and Rhizolepas*

These taxa constitute the only true parasites in the Thoracica. Their systematic position remains highly uncertain but they hardly have any close relationship to the Rhizocephala, since *Rhizolepas* parasitize polychaetes while *Anelasma* parasitize sharks. It is surprising that these parasitic genera have pure hermaphroditism without dwarf males since they occur under conditions comparable to the gonochoristic Rhizocephala. Again one may wonder what mechanisms insure that these parasites occur two or more together despite the fact that locating and successfully infesting a host animal must be a very unlikely event.

5.5.4 *Scalpelloidea*

This taxon is most likely para- or polyphyletic (Glenner et al. 1995), whence I have separated it into components, whose monophyly seems more assured. The Scalpellidae *sensu stricto* (i.e. excluding *Neolepas*, *Capitulum*, and *Pollicipes*) all have dwarf males, associated either with hermaphrodites or with pure females. However, *Neolepas*, occurring in hydrothermal vent habitats, *Lithotrya* (not entered in Fig. 1), *Pollicipes*, a candidate sister group to the Scalpellidae sensu stricto and *Capitulum*, a possible close relative of the balanomorphans, lack a separate male sex. The males of the Scalpellidae s.s. vary from feeding individuals that largely resemble early stages of females-hermaphrodite ontogeny to non-feeding males, which have lost all or almost all calcareous plates and consist of little but the testicular tissue.

5.5.5 *Verrucomorpha and Balanomorpha*

No males exist within the Verrucomorpha, but we find them in several families of the Balanomorpha, where they always associate with hermaphrodites (McLaughlin & Henry 1972; Gomez 1973, 1975; Dayton et al. 1982; Foster 1983; Hui & Moyse 1984). We have very little detailed information on the morphology of balanomorph males.

6 MALE OR HERMAPHRODITE ANCESTOR?

6.1 *Phylogeny and sex*

When we map the presence of hermaphrodites and males onto the tentative phylogeny of the Cirripedia in Figure 1 we see an interesting pattern. The two relevant outgroups both have separate sexes (dioecy), either exclusively (Tantulocarida) or predominantly and in the ground pattern (Ascothoracida). Moreover, within the Cirripedia itself, all species of the Acrothoracica and the Rhizocephala have separate sexes, and we have no indication that hermaphroditism ever existed in their lineages. It is only within the Thoracica that we meet hermaphroditism where the presence of males appear to exhibit a complicated pattern.

Using the principles for outgroup comparison developed by Maddison et al. (1984), we must therefore conclude that both the last common ancestor to the Thecostraca (Fig. 1: node 2) and to the Cirripedia (node 3) had separate sexes. Doing otherwise would amount to an unsupported *ad hoc* hypothesis. Separate sexes also emerge as our best guess for the last common ancestor to the Rhizocephala and the Thoracica (node 4). If, as depicted in Figure 1, the Iblidae represent the sister group to the remaining Thoracica we could also argue that the urthoracican in node (5) had a separate male sex.

Hermaphroditism therefore seems to have evolved somewhere along the lineage between nodes (4) and (5) (Fig. 1: arrow). If true, we arrive at the important hypothesis that hermaphroditism represent an apomorphy for the Thoracica, only, not for the Cirripedia as a whole.

Higher in Thoracica, the pattern becomes difficult to interpret, since the phylogenetic relations between scalpelloids, verrucomorphs, and balanomorphs remains very uncertain (Glenner et al. 1995). With regard to scalpellid males, both the phylogeny and the morphological reduction series would indicate that males evolved from larger females or hermaphrodites. For the Balanomorpha, I agree with Klepal (1987) that we must study many balanomorph males in fine detail before we can estimate their phylogenetic significance and whether they evolved convergently or from a dioecious ancestor. It remains thought-provoking, however, that *Chionelasmus* has dwarf males since in most phylogenetic schemes this species diverge very low on the balanomorph tree (Buckeridge & Newman 1992; Yamaguchi & Newman 1990; Glenner et al. 1995).

6.2 *Can we reconstruct the urcirripede?*

The omnipresence of a cypris larva shows that the last common ancestor to the Cirripedia had permanently sessile juvenile and adult stages. The fact that males and females associate more or less permanently in the Ascothoracida indicates that the cirripede stem species had a system where dwarf males associated with large. Alternatively, the sessile stem species had wholly separate and equal sized males and females, but this require either that the individuals settled gregariously or that the males shed sperm freely into the water column. The fact that both the Cirripedia and the Ascothoracida have a flagellated and motile sperm and that fertilization takes place in a sea-water filled brood chamber at least indicates that the thecostracan ancestors never passed through a stage with internal fertilization. We may therefore assume that internal fertilization does not form part of the crustacean ground pattern.

6.3 *Adaptation and phylogeny*

Future efforts may convince us that a large amount of convergent evolution has occurred in the evolution of cirripede sexual systems, but if so, this will only be because we have a robust phylogeny onto which we can map our characters. The mathematical models of Crisp (1983) and Charnov (1987) predict what frequency of males, females, and hermaphrodites will be optimal under given ecological conditions, such as frequency of settlement sites, reproductive output, and survival of larvae. However, phylogenetic constraints may more or less seriously modify the observed condition compared to that predicted by adaptationist theory. When we survey the sexual systems of the Cirripedia there exist numerous examples not readily explainable by adaptionist models. The Acrothoracica have separate sexes, which dovetails with their patchy distribution, in e.g. gastropod shells. On the other hand, acrothoracicans usually occur several together so we may also speculate why no hermaphrodites exist. In the Thoracica, the parasites *Rhizolepas* and *Anelasma* have pure hermaphroditism although most models would predict that they would benefit from having dwarf males as seen throughout the Rhizocephala. These examples do not invalidate models such as those in Crisp (1983) and Charnov (1987), but we must accept that their predictions represent evolutionary end points that may not yet have been or may never be realized.

7 CONCLUSION

Obviously, this paper can give no final answer to the sexual status of the urcirripede. Rather, I wanted to emphasize that approaches, such as the important papers of Crisp (1983) and Charnov (1987), cannot and should not take place in a phylogenetic vacuum. Only by integrating hypotheses on how and why characters change with the phylogenetic framework in which these events occur can we arrive at a truly integrated and testable set of hypotheses on such complex matters as the evolution of the cirripede sexual systems. We can never hope to understand adaptive radiation within a taxon before we have established the evolutionary starting point, viz., the ground pattern set of characters for the taxon under discussion. To obtain a clearer view of the evolution of the cirripede sexual system we must: 1) Study the morphology and ontogeny of males throughout many taxa and in much more detail so we can use characters from males as input in a phylogenetic analysis. As emphasized by Klepal (1985, 1987) the characters of the males must rank on par with those from females or hermaphrodites. Also, 2) We must obtain detailed knowledge concerning the mechanism employed in determination of sex throughout the taxa, and 3) Study the reproductive biology of a wide selection of species with different sexual systems and use this information to test the validity of the theoretical models.

ACKNOWLEDGEMENTS

The stimulus to write this paper stems largely from a discussion with Prof. R.R. Hessler concerning sex in the Cirripedia and in the Crustacea in general. I also owe much to inspiring discussions with Dr. H. Glenner, Dr. P.G. Jensen, Prof. W. Klepal, Prof. W.A. Newman, and Prof. F.R. Schram and I am especially indebted to Dr. M.J. Grygier for letting me in on his ideas concerning the ground pattern features of the Ascothoracida.

REFERENCES

Batham, E.J. 1945. Description of female, male and larval forms of a tiny stalked barnacle, *Ibla idiotica* n.sp. *Trans. R. Soc. N.Z.* 75: 347-356.

Bocquet-Védrine, J. 1961. Monographie de *Chthamalophilus delagei* J. Bocquet-Védrine, Rhizocéphale parasite de Chthamalus stellatus (Poli). *Cah. Biol. Mar.* 2: 455-593.

Bocquet-Védrine, J. 1972. Les Rhizocéphales. *Cah. Biol. Mar.* 13: 615-626.

Boxshall, G.A. & R. Lincoln 1983. Tantulocarida, a new class of Crustacea ectoparasitic on other crustaceans. *J. Crust. Biol.* 3: 1-16.

Bresciani, J. & J. Lützen 1972. The sexuality of *Aphanodomus* (Parasitic copepod) and the phenomenon of cryptogonochorism. Vidensk. *Meddr dansk naturh. Foren.* 135: 7-20.

Buckeridge, J.S. & W.A. Newman 1992. A reexamination of *Waikalasma* (Cirripedia: Thoracica) and its significance in balanomorph phylogeny. *J. Paleont.* 66: 341-345.

Charnov, E.L. 1987. Sexuality and hermaphroditism in barnacles: A natural selection approach. *Crustacean Issues* 5: 89-103.

Collins, D. & D.M. Rudkin 1981. *Priscansermarinus barnetti*, a probable lepadomorph barnacle from the Middle Cambrian Burgess Shale of British Columbia. *J. Paleont.* 55: 1006-1015.

Crisp, D.J. 1983. *Chelonobia patula* (Ranzani), a pointer to the evolution of the complemental male. *Mar. Biol. Let.* 4: 281-294.

Darwin, C. 1852. *A monograph on the sub-class Cirripedia, with figures of all the species. The Lepadidae; or, pedunculated cirripedes. Ray Society, London*: 1-400.

Darwin, C. 1854. *A monograph on the sub-class Cirripedia, with figures of all the species. The Balanidae, (or sessile cirripedes); the Verrucidae, etc., etc., etc.* Ray Society, London: 1-684.

Dayton, P.K., W.A. Newman & J. Oliver 1982. The vertical zonation of the deep-sea Antarctic acorn barnacle, *Bathylasma corolliforme* (Hoek): experimental transplants from the shelf into shallow water. *J. Biogeogr.* 9: 95-109.

Foster, B.A. 1983. Complemental males in the barnacle *Bathylasma alearum* (Cirripedia, Pachylasmidae). *Mem. Aust. Mus.* 18: 133-140.

Ghiselin, M.T. 1969. The evolution of hermaphroditism among animals. *Quart. rev. biol.* 44: 189-208.

Glenner, H., M.J. Grygier, J.T. Høeg, P.G. Jensen & F.R. Schram 1995. Cladistic analysis of the Cirripedia Thoracica (Crustacea: Thecostraca). *Zool. J. Linn. Soc.* 114: 365-404.

Glenner, H. & J.T. Høeg 1994. Metamorphosis in the Cirripedia Rhizocephala and the homology of the kentrogon and trichogon. *Zool. Scr.* 23: 161-173.

Gomez, E.D. 1973. The biology of the commensal barnacle *Balanus galeatus* (L.) with special reference to the complemental male-hermaphrodite relationship. Ph.D. dissertation. University of California San Diego:

Gomez, E.D. 1975. Sex determination in *Balanus (Conopea) galeatus* (L.) (Cirripedia Thoracica). *Crustaceana* 28: 105-107.

Grygier, M.J. 1981. Sperm of the ascothoracid parasite *Dendrogaster*, the most primitive found in crustacea. *Int. J. Inv. Reproduc. Devel.* 3: 65-73.

Grygier, M.J. 1982. Sperm morphology in Ascothoracida (Crustacea: Maxillipoda); confirmation of generalized nature and phylogenetic importance. *Int. J. Inv. Reproduc. Devel.* 4: 323-332.

Grygier, M.J. 1987. Classification of the Ascothoracida (Crustacea). *Proc. Biol. Soc. Wash.* 100: 452-458.

Grygier, M.J. 1987. Nauplii, antennular ontogeny, and the position of the Ascothoracida within the Maxillopoda. *J. Crust. Biol.* 7: 87-104.

Healy, J.M. & D.T. Anderson 1990. Sperm ultrastructure in the Cirripedia and its phylogenetic significance. *Rec. Aust. Mus.* 42: 1-26.

Hessler, R.R. 1992. Reflections on the phylogenetic position of the Cephalocarida. *Acta Zool.* 73: 315-316.

Hui, E. & J. Moyse 1984. Complemental male in the primitive balanomorph barnacle, *Chionelasmus darwini*. *J. Mar. Biol. Ass. U.K.* 64: 91-97.

Huys R., G.A. Boxshall & R.J. Lincoln 1993. The tantulocaridan life cycle: The circle closed. *J. Crust. Biol.* 13: 432-442.

Høeg, J.T. 1991. Functional and evolutionary aspects of the sexual system in the Rhizocephala (Thecostraca: Cirripedia). In R.T. Bauer & J.W. Martin (eds). *Crustacean sexual Biology*. pp. 208-227. New York: Columbia University Press.

Høeg, JT. 1992. The phylogenetic position of the Rhizocephala: Are they truly barnacles?. *Acta Zool.* 73: 323-326.

Høeg, J.T. & J. Lützen 1993. Comparative morphology and phylogeny of the family Thompsoniidae (Cirripedia, Rhizocephala, Akentrogonida), with description of three new genera and seven new species. *Zool. Scr.* 22: 363-386.

Jensen, P.G., J. Moyse, J.T. Høeg & H. Al-Yahya 1994. Comparative SEM studies of lattice organs: Putative sensory structures on the carapace of larvae from Ascothoracida and Cirripedia (Crustacea Maxillopoda Thecostraca). *Acta Zool.* 75: 125-142.

Klepal, W. 1985. *Ibla cumingi* (Crustacea, Cirripedia) - A gonochoristic species (Anatomy, dwarfing and systematic implications). *P.S.Z.N.I: Mar. Ecol.* 6: 47-119.

Klepal, W. 1987. A review of the comparative anatomy of the males in cirripedes. *Oceanogr. Mar. Biol. Annu. Rev.* 25: 285-351.

Kühnert, L. 1934. Beitrag zur Entwicklungsgeschichte von *Alcippe lampas* Hancock. *Z. Morph. Ökol. Tiere* 29:45-78.

Maddison, W.P., M.J. Donoghue & D.R. Maddison 1984. Outgroup analysis and parsimony. *Syst. Zool.* 33: 83-103.

McLaughlin, P.A. & D.P. Henry 1972. Comparative morphology of complemental males in four species of *Balanus* (Cirripedia Thoracica). *Crustaceana* 22: 13-30.

Melander, Y. 1950. Studies on the chromosomes of *Ulophysema oeresundense*. *Hereditas* 36: 233-255.

Newman, W.A. 1974. Two new deep-sea Cirripedia (Ascothoracica and Acrothoracica) from the Atlantic. *J. Mar. Biol. Ass. U.K.* 54: 437-456.

Newman, W.A., V.A. Zullo & T.H. Withers 1969. Cirripedia. In R.C. Moore RC (ed.) *Treatise on Invertebrate Paleontology, Part R, Arthropoda 4(1)*, pp. R206-R296. Lawrence: University of Kansas and Geological Society of America.

Schram, F.R. 1986. *Crustacea*. New York: Oxford University Press.

Spears, T., L.G. Abele & M.A. Applegate 1994. A phylogenetic study of cirripeds and their relatives (Crustacea Thecostraca). *J. Crust. Biol.* 14: 641-656.

Turquier, Y. 1970. Recherces sur la biologie des cirripèdes acrothoraciques. III. La métamorphose des cypris femelles de *Trypetesa lampas* (Hancock) et de *Trypetesa nassaroides* Turquier. *Archs Zool. Exp. Gén.* 111: 573-627.

Turquier, Y. 1971. Recherces sur la biologie des cirripèdes acrothoraciques. IV. La métamorphose des cypris mâles de *Trypetesa nassaroides* Turquier et de *Trypetesa lampas* (Hancock). *Archs Zool. Exp. Gén.* 112: 301-348.

Turquier, Y. 1972a. Recherces sur la biologie des cirripèdes acrothoraciques. VI. Contribution à l'étude de la reproduction et du déterminisme du sexe chez les Trypetesidae des côtes Françaises. *Archs Zool. Exp. Gén.* 113: 167-196.

Turquier, Y. 1972b. Contribution a la connaisance des cirripèdes acrothoraciques. *Archs Zool. Exp. Gén.* 113: 499-551.

Yamaguchi, T. & W.A. Newman 1990. A new and primitive barnacle (Cirripedia: Balanomorpha) from the North Fiji Basin abyssal hydrothermal field, and its evolutionary implications. *Pacific Science* 44: 135-155.

Yanagimachi, R. 1961. Studies on the sexual organization of the Rhizocephala. III. The mode of sex determination in *Peltogasterella. Biol. Bull. (Woods Hole)* 120: 272-283.

SEM-based morphology and new host and distribution records of *Waginella* (Ascothoracida)

Mark J. Grygier
Silver Spring, Maryland, USA

Tatsunori Itô†
Seto Marine Biological Laboratory, Kyoto University, Shirahama, Nishimuro, Wakayama, Japan

ABSTRACT

The first scanning electron microscopical (SEM) study of a morphologically generalized ascothoracidan crustacean is presented. The external morphology of a female *Waginella metacrinicola* (Okada), ectoparasitic on a pentacrinid stalked crinoid, *Metacrinus rotundus* Carpenter from Japan, is illustrated using SEM. Several kinds of gland openings on the flat, ventral side of the carapace are described. The inner wall of the large anteroventral pore on each carapace valve possesses lamellar ridges that bound a large number of small gland openings. Two anterior lattice organs (cardic organs) are found on each valve. The so-called second antenna or antenna-vestigial eyestalk complex does not arise from the cephalon proper, but from the mantle lateral to the antennule, and it most likely incorporates the external part of the organ of Bellonci complex. Records of *W. metacrinicola* and *W. axotremata* Grygier infesting metacrinine pentacrinids collected by recent French expeditions to the Philippines and New Caledonia are listed. The former species is reported from *Metacrinus musorstomae* Roux for the first time, and the latter from *M. levii* Cominardi, *M. serratus* Döderlein, and *Saracrinus nobilis* (Carpenter) for the first time. *Waginella axotremata* is also reported from northern Australia, infesting *S. nobilis*, and southeastern Australia, infesting *M. cyaneus* H.L. Clark. This species apparently uses its raspy, awl-like mandibles, drawings of which are presented herein, to drill holes in the cirri of its host; such drill-holes are proposed as potential trace fossils for studying the history of crinoid-ascothoracidan associations. The apparent absence of ascothoracidan parasites on other genera of Pentacrinidae suggests that the association may be no older than the Miocene. The possible synonymy of *W. metacrinicola* and *W. axotremata* is discussed on the basis of morphology, depth distribution, and biogeography, but is not resolved. *Crinoidoxenos* Blake, 1933 is revealed as a potential senior synonym of *Waginella*.

1 INTRODUCTION

Species of *Waginella* are among the ascothoracidan crustaceans that are least modified morphologically for a parasitic existence. Therefore, they are likely to be of value in phylogenetically directed comparisons of the Ascothoracida with their nearest relatives, the Cirripedia and Facetotecta. The latter group, comprising Hansen's y-larvae (nauplius y and cypris y), is still enigmatic due to ignorance about the adults. Several workers have suggested they are larvae of ascothoracidans or of other, unknown crustaceans closely related to the latter (Bresciani 1965; Schram 1970; Grygier 1984; Itô 1987, 1989, 1990). In this respect, there is no doubt that thorough comparative studies between facetotectans and ascothoracidans, such as that by Grygier (1987b), are of particular importance. Nevertheless, our current knowledge about the Ascothoracida is inadequate for this aim. For example, the identity and composition of even such prominent structures as the so-called second antennae or antenna-vestigial eystalk complexes (see Grygier 1984) is still debated (Grygier 1987b). The present paper takes some steps to relieve this paucity of information on the detailed morphology of ascothoracidans, based on observations of *Waginella metacrinicola* (Okada) using scanning electron microscopy (SEM). Until now, only one ascothoracidan, *Baccalaureus falsiramus* Itô & Grygier (1990), has been examined closely with SEM, but the females of that species are too highly modified to be of much use in the present context.

Species of *Waginella* for which the hosts are known are ectoparasitic on crinoids. Three species have been described. *Waginella metacrinicola* infests a stalked metacrinine pentacrinid, *Metacrinus rotundus* Carpenter in Japan (Okada 1926, 1938; Grygier 1983a, 1990a); we re-evaluate Grygier's (1990a) tentative Philippine record of *W. metacrinicola* herein (Section 5.1). *Waginella axotremata* Grygier has been found on several nominal species of *Metacrinus* and consubfamilial *Saracrinus* in the Philippines-Indonesia-Australia region (Grygier 1983a, 1990a). The present paper includes a listing of new records of both of these species of *Waginella* on additional metacrinine hosts in the New Caledonia area, southern Japan, the Philippines, and northern Australia. The third species, *W. sandersi* (Newman, 1974; q.v.), host unknown, comes from abyssal depths off Patagonia in the south Atlantic (see also Grygier 1983a, 1987a). An additional, as yet undescribed species of *Waginella* has been noted attached to the comatulid *Comactinia echinoptera* Pourtalès in the West Indies (Grygier 1988b).

Blake (1933: 97) very briefly announced his discovery of a primitive cirripede, *Crinoidoxenos halocypris*: 'This new genus (parasitic on a crinoid) is apparently the most primitive existing cirripede, since the adult is a true cypris 'larva.' The so-called antennal glands persist. The ventral sucker is unique.' It is possible that *C. halocypris* was a species of *Waginella*, a genus which has cypridiform adults, often parasitizes crinoids, and has a ventrally flattened carapace that might be considered sucker-like. Blake presented enough of a diagnosis to make *C. halocypris* an available name and *Crinoidoxenos* thus potentially a senior synonym of *Waginella*. He was affiliated with the Massachusetts Institute of Technology, but efforts to locate his material there, at the nearby Museum of Comparative Zoology, and at the Academy of Natural Sciences in Philadelphia have been unsuccessful (A.B. Johnston, personal communication). Therefore, we will continue to treat *Waginella* as a valid name.

2 MATERIALS AND METHODS

The SEM study reported here is mostly adapted from an unpublished manuscript written in 1988 by the late second author. Two adult females of *Waginella metacrinicola,* which were brooding eggs, were removed from an alcohol-preserved specimen of *Metacrinus rotundus* deposited in the Geological Institute, Faculty of Science, University of Tokyo. The crinoid had been collected by dredging at a depth of 100 m at a site 4 km west of Jogashima Island in Sagami Bay, Kanagawa Prefecture, Japan. The ascothoracidans, which were attached to the host's column opposite each other, were detached from the host and each was partially dissected to expose the main body inside the carapace. They were rinsed in distilled water, dehydrated through a graded ethanol series, transferred into isoamyl acetate, and desiccated in a critical point dryer using liquid CO_2. The dried specimens were mounted on brass stubs, sputter-coated with gold, and observed in a scanning electron microscope (JEOL T220) at an accelerating voltage of 10 kV. One of the specimens was fouled with a mucous or cement substance, so all the SEM micrographs in this paper are of the other specimen. In 1994, the first author borrowed the stubs from the Seto Marine Biological Laboratory and re-examined the carapace for lattice organs, not mentioned in Itô's manuscript, using a JEOL JSM-T20 scanning electron microscope.

The present faunistic records were compiled by the first author based on surveys of several museum collections. *Waginella metacrinicola* and *W. axotremata* were isolated from metacrinine crinoids in the Museum National d'histoire Naturelle, Paris (MNHNP) and the laboratory of Dr. Michel Roux, Université de Reims, France; all the parasites have been deposited in the MNHNP. These crinoids came from several recent French expeditions, mostly in the vicinity of New Caledonia and the Chesterfield Islands, but also in the Philippines (Forest 1989; Richer de Forges 1990). Some parasites were dissected and the parts mounted in glycerine jelly for light microscopical examination, and most were sexed by the morphological criteria developed for *W. sandersi* by Grygier (1987a). A specimen of *W. metacrinicola* was also found attached to a host *M. rotundus* in the Osaka Museum of Natural History (OMNH). Additional specimens of *W. axotremeta* were observed on *M. cyaneus* H.L. Clark in the Museum of Comparative Zoology (MCZ), and others parasitizing *Saracrinus nobilis* (Carpenter) were obtained on loan from the Northern Territory Museum of Arts and Sciences in Australia (NTMAS) and photographed.

3 *WAGINELLA METACRINICOLA* (OKADA)

3.1 *SEM observations*

3.1.1 *General observations*

The color in preserved condition was yellowish brown, the same as the color of the host, but the color in life is unknown. When the ascothoracidans were detached from the host, a white cement substance connected them with the host. This material extended between the valves and contacted the animal's cephalon. The cement was easily detached from the body and was accidentally lost during critical-point drying. [N.B., the cement pads for host attachment observed by Grygier (1983a, 1990a) conform to the flat, ventral sides of the valves and are colorless or amber, so the 'cement' observed by Itô may be a different, perhaps host-derived substance: M.J.G.]

The eggs were deposited around the dorsal part of the body inside the carapace valves and appeared to be loosely connected with each other and with the body. Most of the eggs became separated following dehydration; one is visible in Plate 1B. No hatched nauplii were found.

3.1.2 *Carapace*

The carapace is bivalved with a hinge and measures 3.5 mm long by 2.2 mm high. The ventral side of each carapace valve is almost flat and ornamented with a more or less regular array of hemispherical protuberances (Pl. 1A,B and 2B). These protuberances measure 9-10 μm in diameter and each is equipped with a circular collar around a collapsed tube (Pl. 2A). We refer herein to this kind of protuberance as a 'gland cone.' They are present only on the ventral surface of the carapace.

Other than the gland cones, numerous collapsed tubes, each of which arises within a circular pit, are distributed not only on the ventral surface of the carapace (Pl. 2A), but also on other external surfaces (Pl. 3A,C). We refer to them as 'pitted gland-tubes' herein. The pits measure 2-4 μm diameter. The pitted gland-tubes on the ventral face of the carapace are sunken within a mat of cuticular villi (Pl. 2A). Hair-like structures (possibly longer villi), which are at most 5 μm long, circumscribe the pitted gland-tubes at a distance. Prominent hair-like setae measuring about 50 μm long are scattered on the ventral surface of the carapace among the gland cones (Pl. 2B).

The ventral face of each carapace valve is armed along its medial margin with a dense pelage (Pl. 2B); the individual hair-like elements arise in tufts. Collapsed gland-tubes, each inserted into a prominent collar, occur within a narrow band of intermediate length pelage. We refer to them as 'collared gland-tubes' herein. They are arranged so that single ones and paired ones are positioned alternately (Pl. 2B). The collar of these gland-tubes appears to be a clear ridge measuring 4-6 μm in diameter.

The anteroventral carapace pore sensu Grygier (1983a) is located quite laterally, at the border between the ventral face and lower frontal face of each carapace valve (Pl. 1B and 3C). It is a round depression equipped with serrated, lamellar ridges. The ridges form a more or less square lattice all over the inner surface of the depression, and an oblong pore measuring about 3 μm along its major axis is located on the bottom of each square (Pl. 3D). Some tufts of fine setae arise from sockets along the rim of the depression, especially around the anterior half (Pl. 3C).

Part of the external surface of the carapace, but not the ventral face, is ornamented with low ridges which form a mesh-like structure that is most obvious ventrolaterally (Pl. 3C). The external surface also possesses a scattering of pitted gland-tubes (Pl. 3A,C), together with sparse hair-like setae of different sizes (Pl. 3C).

Each valve has two lattice organs (= cardic organs of Itô & Grygier 1990; see Jensen et al. 1994) close to the margin at the anterior end of the dorsal hinge; no posterior lattice organs were found. The lattice organs lie well away from the margin on both valves, among pitted gland-tubes (Pl. 3A). Each consists of a recumbent tube-like structure 12.5-13 μm long and 1.6-1.9 μm wide (the more anterior one being shorter) in a shallow trough 14.6-16.8 μm long and 3.3-3.8 μm wide, with distinct, radially grooved rims 2.2-2.7 μm wide (Pl. 3B). On the right valve of the cleaner specimen, the lattice organs are oriented at about a 50° angle to the valve margin, converging anteriorly toward the margin (Pl. 3A). On the left valve they are more nearly perpendicular to the margin (not illustrated). Due to adhering debris, it is difficult to make out the details of the tube in any of the lattice organs, but its surface may not be smooth and both tubes on the right valve seem to have a pore at the anteromedial end (Pl. 3B). On the right valve, a large pore is positioned between the lattice organs and the margin (Pl. 3A), while on the left valve, a crescent-shaped opening occupies the same position (not illustrated). Both openings might be artifacts.

The margins of each carapace valve, except for the anterodorsal portion that forms a hinge, are continuous with the mantle, the cuticular lining of the inner faces of the valves (Pl. 1A,B). The anteroventral portion of the mantle bears sparse, hair-like setae and ctenae composed of very short elements (not illustrated), whereas the posterodorsal portion is densely armed with long setae (70 µm or longer), together with a number of ctenae composed of separate, short elements (Pl. 2C). Simple pores of about 2 µm diameter are scattered sparsely on the mantle (Pl. 2C).

3.1.3 *Antennules*

Okada (1926, 1938) and Grygier (1983a) have already described or illustrated the antennules (first antennae) very well. The following observations can be added (Pl. 4A). Posterior to the claw guard, two setae insert on a disc-like plate, all comprising the proximal sensory process. The more medial seta is an aesthetasc and the more lateral one has a basal swelling from which a setula arises. The distal protrusion of the claw guard is delicate and hood-like. There are comb-like rows of short spinules along both margins of the sixth segment and the anterior margins of the preceding three segments (Pl. 6B). Similar comb-like rows are also present on the lateral face of the first segment.

3.1.4 *'Second antenna' or 'antenna-vestigial eyestalk complex'*

The so-called antenna or second antenna (Newman 1974; Grygier 1981, 1983a,b), or antenna-vestigial eyestalk complex (Grygier 1984), was found to occur on isolated carapace valves just below the large hole in the mantle where the adductor muscle was originally inserted before dissection (Pl. 1A). Hence, it is not located in the same plane of the body on which the first antenna arises. Instead, it is located in the deepest, dorsalmost part of the cavity that is formed between the anterior part of the body and the mantle of the carapace valve. This site is lateral to the base of the first antenna, rather than posterior to it.

In the female this structure is composed of a basal shaft and two attached branches (Pl. 5). The basal shaft is a papillary process with no proximal articulation; its distal half is the basal appendix sensu Grygier (1987b). It points to the rear, measures 150 µm long, and tapers apically. Two branches arise together from the midventral portion of the basal shaft and point respectively anteroventrally and posteroventrally. Neither of them has a clear articulation at the base. The anterior branch (proximal aesthetasc sensu Grygier (1987b)) is almost as long as, but markedly thinner than, the basal shaft. The posterior branch (ventral ramus sensu Grygier (1987b)) is about 250 µm long and is armed with a number of spirally arranged, ribbon-like filaments that are about 150 µm long and somewhat depressed toward their tips.

3.1.5 *Oral cone*

The apical part of the oral cone appears to be delimited from the rest by a circumscribing groove (Pl. 4B). The apical part is an almost conical tube, incised posteriorly, with a distal opening that is 40 μm in diameter. The sharp, undivided tips of the paired maxillae protrude from this opening (Pl. 4B).

3.1.6 *Female genital papilla*

A female genital papilla is located just posterior to each antennule, filling the gap between the antennule and leg 1 (Pl. 6B). It is a massive lobe directed posteriorly, and appears bipartite due to a deep, posterior incision. Some tufts of fine, setiform structures up to 100 μm long, and ctenae as well, arise from its ventral lobule, but the dorsal lobule is bare. [N.B., Itô identified this structure as a genital papilla, but another possibility, based on its position relative to the oral cone, is that it contains the end-sac of the maxillary gland; the gonopore may well be located farther dorsally: M.J.G.]

3.1.7 *General view of trunk*

The first thoracic somite is clearly articulated to the head along a deep anterior suture and is as long as the two succeeding somites combined (Pl. 6A). Comb-like rows of short spinules ornament its anterior half, and its posterodorsal margin is fringed with very long, thin setiform elements, mostly longer than 200 μm. The posterodorsal rims of the other thoracic somites are similarly fringed, and there are lateral tufts of similar hair-like elements on the second through fifth somites.

The lateral epaulets on the sixth thoracic somite have a proximal and an apical portion (Pl. 6A and 7A). The proximal portion is a cylindrical process with no basal articulation. Its surface is smooth, although a few small pores are present. The apical portion is a swelling delimited from the proximal portion by an anterior notch as well as by a clear difference in surface structure. Its surface appears to consist of thinner cuticle than that of the proximal portion, with a rough but soft texture and fine, mesh-like grooves that divide it into a number of hexagons. There is a trough-like depression on the posterolateral side.

The more posterior part of the body was heavily covered with cement or mucous, and could not be observed in detail. The seventh trunk somite (male genital somite) has a pair of lateral swellings or bosses, which face the epaulets on the preceding somite (Pl. 7A). The telson is armed ventrally with dense, comb-like rows of short hairs (Pl. 7B). Similar comb-like rows ornament the four preceding somites, but more sparsely, as well as the medial and lateral faces of the furcal rami (Pl. 7B). The fixed posteroventral spines on the telson have a basal articulation and are really a pair of large, spine-like setae.

3.1.8 *Thoracopods*

The thoracopods are markedly hairy and especially the margins of the protopods are densely fringed (Pl. 6B). The apical setae were fouled with mucous or cement, obscuring their detailed structure. The so-called filamentary appendage (Grygier 1984, 1990a) is a small lobe at the base of thoracopod I with a pointed, ventrolateral extension (Pl. 6B); its precise relation to the limb base could not be ascertained.

3.2 *List of other new records*

1. One female in OMNH on *Metacrinus rotundus* (File no. 77-60), collected by T. Imaoka, off W coast of Kagoshima Prefecture, Japan, 15 Nov 1977. The specimen is attached to the host's column facing downward. It is yellow and probably less than 3 mm long, and was not removed from the host for a more detailed examination; therefore, the specific identification based on the host is only tentative.

2. Three specimens in MNHNP (Ci-Asc 2351) on young *Metacrinus musorstomae* Roux, MUSORSTOM 3, CP 110, 2 Jun 1985, off Manila Bay, Philippines, 13°59.5'N, 120°18.2'E – 14°00.3'N, 120°17.2'E, 187-193 m. Two individuals were female, one brooding 111 late nauplii close to the molt into the ascothoracid-larva, the other not brooding; the third individual was a mature male. The females were attached to the host's column, one of them just below the 12th distinct whorl of cirri from the top, with its mouthparts over an inter-columnal joint; the other fell off before its position could be determined. The male was attached to a cirrus of the 14th whorl.

This find represents a new host for *W. metacrinicola* and the first unambiguous record outside Japan.

4 *WAGINELLA AXOTREMATA* GRYGIER

This chapter presents a clarification of some morphological details and a list of new distribution and host records, which are plotted together with older records in Figure 2. Much information concerning the position and orientation of the parasites on their hosts is compiled. The problem of discriminating species of western Pacific *Waginella* is outlined.

4.1 *Descriptive remarks*

4.1.1 *Mandible*

According to Grygier (1983a), the styliform mandible of *Waginella axotremata* is armed medially with retrorse spinules, i.e., ones that are directed back along the axis of the appendage. The material at hand allows a more detailed description. The distal mandibular armament is actually rasp-like, with arrays of basally directed spiniform elements (Pl. 8). These are rather small and numerous and arranged in a dense bed near the mandible's tip but become more robust and fewer in number proximally, where they occur in distinct clusters. Such a rasp allows the mandibles to act as awls for drilling the observed holes into cirral ossicles, presumably for feeding.

Among other ascothoracidans, only the members of the family Lauridae (mesoparasites of zoanthid cnidarians) have mandibles with medial 'retrorse spinules' and no other armament (Grygier 1985, 1990b, 1991; Itô & Grygier 1990). It would be useful to examine their mandibles and those of *Waginella* by SEM in order to assess their degree of similarity. Because of the sheath-like labrum, the oral cone of ascothoracidans must be dissected before critical point drying in order to expose the mouthparts to view. A planned SEM survey of additional ascothoracidan species by M.J. Grygier and W. Klepal will devote special attention to the mouthparts.

4.1.2 *Dorsal setae of seventh trunk somite*
Grygier (1983a) described a dorsal, transverse band of short setae on the seventh trunk somite (male genital somite) in both *Waginella axotremata* and *W. metacrinicola*, a characteristic also mentioned by Grygier (1990a). The present specimens of both species have tufts or rows of such setae dorsolaterally on this somite, but the mid-dorsal area is bare. Re-examination of the holotype of *W. axotremata* and the Japanese specimens of *W. metacrinicola* examined by Grygier (1990a) show that Grygier (1983a) was mistaken on this point, and all have just dorsolateral setae.

4.2 *New records*

1. Three specimens in MNHNP (Ci-Asc 2335) on *Metacrinus levii* Cominardi, CHALCAL 2, DW 82, 31 Oct 1986, SE of New Caledonia, 23°13.68'S, 168°04.27'E, 304 m. Each was found on a different specimen. One was a female brooding 22 nauplii that were near their molt to the ascothoracid-larva; it was attached to the host's column just above cirrus whorl 10 (counting from the top of the column). Another was a smaller female brooding nine unfertilized eggs, found on the side of a cirrus, on ossicles 2-4 from the end, with its mouthparts over a joint between two ossicles. The third specimen was a male found on the side of a cirrus, on ossicles 5-6 from the end, with no injury to the host discernible beneath the mouthparts. One of these last two specimens was attached to a cirrus of whorl 7, the other to a cirrus of whorl 13.
2. One specimen in MNHNP (Ci-Asc 2336) on *M. levii*, MUSORSTOM 5, DC 379, 20 Oct 1986, Chesterfield Islands, 19°53.20'S, 158°39.50'E, 370-400 m. The specimen was a female brooding eggs, six being found in the left valve for a total of about 12. It was attached to a cirrus, facing proximally on the oral side of ossicles 4-5 from the base.
3. One specimen in MNHNP (Ci-Asc 2337) on *M. levii*, MUSORSTOM 5, DW 338, 15 Oct 1986, Chesterfield Islands, 19°51.60'S, 158°40.40'E, 540-580 m. The specimen was a male attached to the side of cirral ossicles 2 and 3, facing proximally. No damage to the cirrus was observed.
4. Two specimens in MNHNP (Ci-Asc 2338) found loose in a jar with one *M. levii*, N.O. <<Vauban>> stn. 420, 24 Jan 1985, SE of New Caledonia, 22°44.2'S, 167°0.9'E, 345 m. One specimen was a brooding female which was not closely examined, and the other was a male. Bourseau et al. (1991) inadvertently omitted this record of *M. levii* from their report, noting only *Saracrinus nobilis* at this station (A. Crosnier, pers. comm.).
5. Two specimens in MNHNP (Ci-Asc 2339) on *M. levii*, MUSORSTOM 5, CP 352, 17 Oct 1986, Chesterfield Islands, 19°31.40'S, 158°37.70'E, 310-337 m. Both specimens were attached to cirri of the same whorl. One was a female brooding 21 eggs, attached laterally to cirral ossicles 8-9 and facing proximally. The other was a male, attached oral-laterally to ossicles 9-10 and facing proximally. In both cases there was a hole in the cirral ossicle beneath the mouthparts, that below the female being more obvious.
6. Six specimens in MNHNP (Ci-Asc 2340) on *M. levii*, MUSORSTOM 5, DW 339, 16 Oct 1986, Chesterfield Islands, 19°53.40'S, 158°37.9'E, 380-395 m. One specimen was attached to a cirrus of one host while the other specimens were on cirri

of whorls 7, 8, 9, 9, and 14 of another. There were two mature females, one of which was brooding, two immature females, and two males. Bourseau et al. (1991) recorded no crinoids from this station. This discrepancy may have been due to a misread label, but the problem remains unresolved (A. Crosnier, pers. comm.)

7. Two specimens in MNHNP (Ci-Asc 2341) on *M. levii*, BIOCAL, DW 38, 30 Aug 1985, SE of New Caledonia, 22°59.74-59.94'S, 167°12.31-12.42'E, 360 m. Both specimens were non-brooding females, attached to different cirri of whorl 6, one near the base and the other far distally.

8. Four specimens in MNHNP (Ci-Asc 2342) found loose in jar containing *M. levii* and *Saracrinus nobilis*; two were on cirrus fragments that matched the diameter of intact cirri of *M. levii*. BIOCAL, DW 37, 30 Aug 1985, SE of New Caledonia, 22°59.99'S, 167°16.65'E - 23°00.07'S, 167°16.34'E, 350 m. Three specimens were females, one non-brooding and two each with nine undeveloped eggs. The fourth specimen was a male.

9. Six specimens in MNHNP (Ci-Asc 2343) associated with three *S. nobilis*, two attached and four loose in the jar, MUSORSTOM 4, CP 192, 19 Sep 1985, NW of New Caledonia, 18°59.30'S, 163°25.00'E, 320 m. One attached specimen was near the tip of a cirrus of whorl 14, facing proximally; the other, a small specimen, was on the host's column below whorl 13. Two specimens were mature females brooding eggs, three were immature females, and one was a male.

Five specimens in MNHNP (Ci-Asc 2344) on one *S. nobilis*, MUSORSTOM 4, CP 192. Four specimens were found facing proximally on cirri of whorls 18, 19, 21, and 24 (three of them on the aboral side and at least two with a hole in the cirral ossicle below the mouthparts), the fifth on the column below whorl 23, facing upward. Three specimens were mature females, one brooding 30 eggs, one not brooding, and one not examined. Of the two smaller specimens, one was a male and the other was not examined.

Five specimens in MNHNP (Ci-Asc 2345), two found on and three loose in jar with *S. nobilis*, MUSORSTOM 4, CP 192. The loose specimens were a mature but non-brooding female, an immature female, and a male. One attached specimen was a female with eggs ready to hatch (not counted); it was on cirral ossicles 7-8, facing proximally. The final specimen was probably a male (not dissected), also on cirral ossicles 7-8, facing proximally. These last two specimens were on cirri of whorls 19 and 24, perhaps not respectively. In the case of the female, the host integument, but perhaps not the cirral ossicle itself, is damaged below the mouthparts; in the case of the supposed male, there may be a small hole in the cirral ossicle.

10. One specimen in MNHNP (Ci-Asc 2346) on *S. nobilis*, CHALCAL 2, CH 8, 31 Oct 1986, SE of New Caledonia, 23°13.36'S, 168°02.73'E, 300 m. One female was found on the side of the base of a cirrus of whorl 10, facing proximally with a well-developed glue pad. There was a hole in the most proximal cirral ossicle below the mouthparts. The specimen was brooding early nauplii, seven being counted on one side for a total of about 14.

11. One specimen in MNHNP (Ci-Asc 2347) found loose in jar with three fragments of *S. nobilis* (identified by M.J. Grygier based on fragments' 12 unsculptured internodal columnals), MUSORSTOM 4, DW 222, 30 Sep 1985, SE of New Caledonia, 22°57.6'S, 167°33.0'E, 410-440 m. This is a young specimen, the sex of which was not determined.

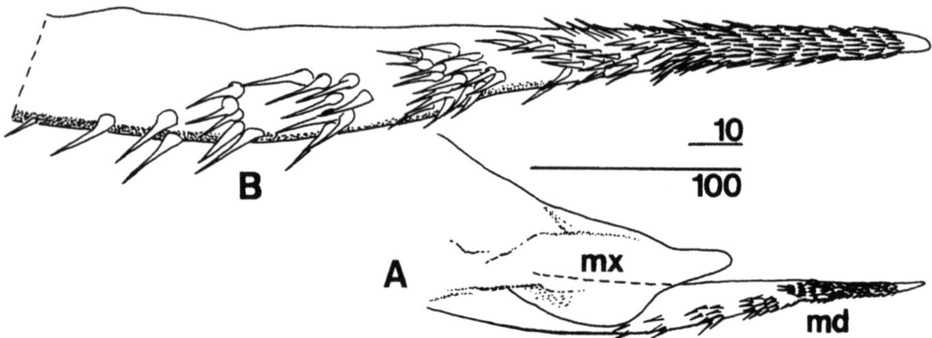

Figure 1. Photographs of two female *Waginella axotremata* (NTMAS Cr 009736) attached to cirri of *Saracrinus nobilis* from the Arafura Sea. (A) Lateral view; (B) Dorsal view; both with anterior end right, toward base of cirrus. Scales in mm.

Twelve specimens in MNHNP (Ci-Asc 2348), four attached to and eight loose in jar with *S. nobilis*, MUSORSTOM 4, DW 222. The attached specimens were found on three hosts, on a cirrus of whorl 11, cirri of whorls 9 and 15, and the column above whorl 16, respectively. Five were mature females, one with eggs, another with about 14 nauplii (seven counted in one valve), and a third with about 20 late nauplii (10 counted in one valve). One was an immature female, and the sex of the six smaller specimens was not determined.

12. Sixteen specimens in MNHNP (Ci-Asc 2349), 12 attached to and four loose in jar with *S. nobilis*, CHALCAL 2, DW 82, 31 Oct 1986, SE of New Caledonia, 23°16.68'S, 168°04.27'E, 304 m. Twelve specimens were adult females and six which were opened were all brooding. Two other specimens were a male and a very young female, and two remain undermined. Eleven specimens were found on the cirri of a single host, in whorls 8, 10, 12, 12, 13, 14, 14, 15, 15, 16, and 17, respectively.

13. Three specimens in NTMAS (Ref. no. Cr 009736) attached to two cirri of *S. nobilis*, collected by R. Williams, stn. RW 92-67, 20 Oct 1992, Arafura Sea between Arnhem Land and Aru Island, 9°00.08'S, 133°19.67'E, 193-195 m. All were females, and one was brooding 23 eggs. Two were attached to cirral ossicles 8 and 9, the other to ossicles 10 and 11. All were facing proximally, two sitting on the oral side of the cirrus (Fig. 1) and the other laterally next to one of them.

14. One specimen in Western Australian Museum (WAM 88-87) already reported by Grygier (1990a) as from an unidentified host, WNW of Lacepede Archipelago, Western Australia, 396-400 m. The host (WAM 30-87) has since been identified tentatively as *S. nobilis* (T. Oji & L. Marsh, pers. comm.).

15. Two specimens in MNHNP (Ci-Asc 2350) found loose in jar with four *Metacrinus serratus* Döderlein, MUSORSTOM 3, CP 105, 1 Jun 1985, off Manila Bay, Philippines, 13°52.6'N, 120°29.6'E-13°51.8'N, 120°30.2'E, 417-398 m. One specimen was a female brooding 13 late nauplii that were near the molt to the ascothoracid-larva. The other was a male.

16. Two specimens in MCZ attached to 'cotype' of *Metacrinus cyaneus* (No. 720), 'Endeavour' collection, locality not given but presumably SE of Cape Everard, Victoria, Australia (Clark 1916). The specimens were not examined, so the identifi-

cation is tentative. Both are of undetermined sex, perhaps a female/male pair, and are attached to the oral side of the basal parts of two cirri in adjacent whorls.

4.3 *Host relations, new hosts, and range extensions*

Section 4.2 presents a lot of data concerning the physical relationship between *Metacrinus levii* and *Saracrinus nobilis* and their parasitic ascothoracidans. Although this information is incomplete and somewhat anecdotal, some generalities can be mentioned. The depth range of infested *M. levii* is 304-580 m; that of infested *M. nobilis*, including Australian records, is 193-440 m. Up to 11 parasites were found on a single host. *Waginella axotremata* occurs mostly on the cirri; only four specimens were found on the column. The infested cirri were in whorls 6-14 from the top in *M.*

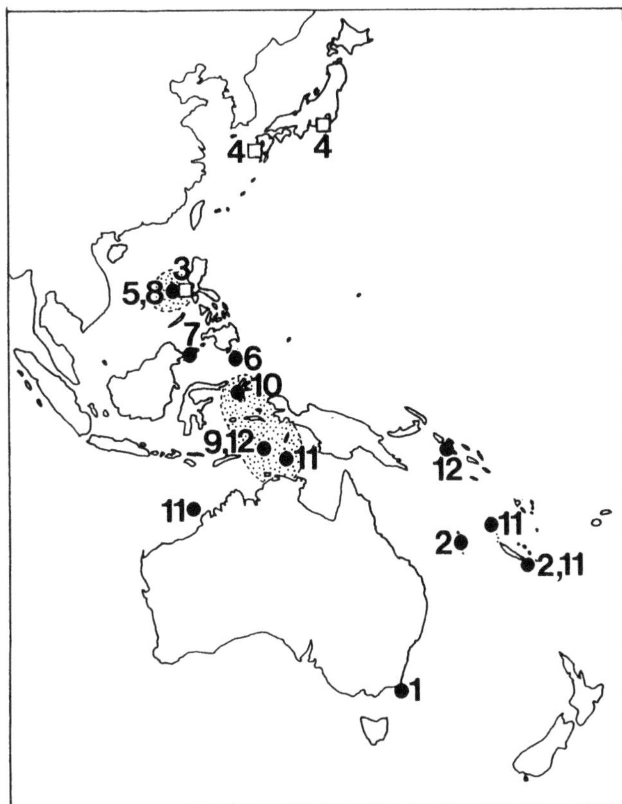

Figure 2. Distribution of *Waginella* spp. parasitizing Indo-West Pacific metacrinine pentacrinid crinoids: confirmed and supposed records of *W. metacrinicola* (.) and *W. axotremata* (•). Stippled areas include the localities where the latter species conforms most closely to the original description. Numbers at the various sites refer to the following nominal species of host crinoids: 1) *Metacrinus cyaneus*; 2) *M. levii*; 3) *M. musorstomae*; 4) *M. rotundus*; 5) *M. serratus*; 6) *Saracrinus varians*; 7) *M. zonatus*; 8) *S. acutus*; 9) *S. angulatus;* 10) *S. cingulatus*; 11) *S. nobilis*; 12) Unidentified. The following sets of synonymous host species have been proposed recently (see Bourseau & Roux 1989; Oji 1989): *M. cyaneus* (1) and *M. rotundus* (4); *S. varians* (7) and *S. nobilis* (11); *S. acutus* (8), *S. angulatus* (9), and *S. cingulatus* (10).

levii and whorls 8-24 in *S. nobilis*. While specimens could be found at any position along or around a cirrus, they always faced the column and were most often attached somewhere among the basal 10 cirral ossicles, occupying two adjacent ossicles. In seven cases, the cirral ossicle below the mouthparts, or its integument, was observed to be damaged, usually by a small hole leading into a larger cavity within the ossicle. A cirrus-by-cirrus search for such drill-holes in Recent crinoids in museum collections may provide a means for assessing the incidence of past parasitism on individual hosts. This type of drilling, causing a characteristic kind of skeletal damage to the host, is also potentially available as a trace fossil and could indicate ascothoracidan parasitism on fossil crinoids as well as the age of this association, a topic taken up in more detail below.

Metacrinus levii, M. serratus, and *Saracrinus nobilis* are recorded here for the first time as hosts of *Waginella*, and all are apparently hosts of *W. axotremata*. New Caledonia and the nearby Chesterfield Islands likewise represent a region from which *Waginella* has not previously been reported; they extend the range of this genus eastward from the nearest recorded localities in the Solomon Islands and southeastern Australia (Fig. 2; see Grygier 1990a). The parts of the range of *Metacrinus* and *Saracrinus* from which *Waginella* has not yet been reported include the Ogasawara Islands, East China Sea, western South China Sea, Palau, Indian Ocean off Sumatra, and New Zealand/Kermadec Islands region (from distribution map in Roux 1981). The present specimens from SE Australia on *M. cyaneus* represent the second record from that locality on that host (see Grygier 1990a). Grygier's (1990a) tentative identification of the host from the Solomon Islands as *M. cyaneus* is dubious, having been based solely on skeletal color. *Saracrinus nobilis* is apparently infested across the northern coast of Australia into the Indian Ocean, as well as around New Caledonia (Fig. 2). This extended distribution on recently collected *S. nobilis* is somewhat surprising, because the examination of many older lots of this crinoid in museums as part of the surveys reported by Grygier (1983a, 1990a) never turned up any infestations by *Waginella*. However, some hosts noted by Grygier (1990a) were unidentified, and Grygier (1983a) considered Carpenter's (1884) 'obscure larval Cirripede' found on a cirrus of *Saracrinus varians* (Carpenter) as possibly representing *W. axotremata; S. varians* is very likely synonymous with *S. nobilis* (see Bourseau & Roux 1989).

Metacrinus and *Saracrinus* comprise the Metacrininae, a monophyletic subfamily of probable Miocene age that is currently undergoing an adaptive radiation (Roux 1981). The family Pentacrinidae (= Isocrinidae) goes back to the Triassic and includes five additional extant genera (*Cenocrinus, Diplocrinus, Hypalocrinus, Teliocrinus, Neocrinus*) in three subfamilies of greatly varying antiquity. These genera are distributed variously in the western Pacific, Indian and Atlantic Oceans. Some of their species are well represented in museum collections but no ascothoracidan parasites are yet known from them. This suggests that the association of *Waginella* with Pentacrinidae may be of relatively recent origin (Miocene or later), a suspicion that is reinforced by the difficulty in judging whether more than one species of *Waginella* actually infest western Pacific metacrinines (see Section 5). The occurrence of an undescribed species of *Waginella* on a western Atlantic comatulid crinoid (see Section 1) is very interesting in this connection.

5 SPECIES DISCRIMINATION

5.1 *History of problem*

Even in the original description, Grygier (1983a) was at pains to diagnose *Waginella axotremata* as a distinct species from *W. metacrinicola*. The principal useful features, also discussed by Grygier (1990a), were the size and shape of the carapace, the proportions of the furcal rami, and a few quantitative differences in thoracopodal setation. Grygier (1990a) cast some doubt on the utility of some of these features when he described an immature specimen found on *Metacrinus zonatus* A.H. Clark from the Sulu Archipelago that was, in some respects, morphologically intermediate. In that paper, the immature specimen was assigned to *M. metacrinicola*, but, although the host is endemic to the Sulu Sea, its ascothoracidan parasite seems to be essentially identical to the present specimens from the New Caledonia region. The question is, should more than one species of *Waginella* continue to be recognized, and if so, do the New Caledonian specimens and the specimen from *M. zonatus* truly represent *W. axotremata*, or one or more new species? Some data pertaining to this question are given here.

5.2 *Color*

The normal color of *Waginella axotremata* is yellowish or yellow with black spots (Grygier 1983a, 1990a). However, many of the present specimens found attached to *Saracrinus nobilis* match the blue-green color of the host. This is most likely due to the uptake of pigment from the host tissue by the parasite while feeding, rather than a specific difference.

5.3 *Purportedly discriminative features*

A comparison of *Waginella* from the New Caledonian area with undoubted representatives of *W. metacrinicola* and *W. axotremata* from their type regions (Table 1) indicates that the former are intermediate between the latter two in several respects. Among these are the number of setae on the distal articles of both rami of thoracopod I, the shape of the furcal rami and the number of spines along their ventral margin. The New Caledonian specimens match *W. axotremata*, however, in the length and length/height ratio of the carapace, the number of medial setae on the basis in thoracopods II-V, and brood size. They match *W. metacrinicola* in the number of setae on the second exopodal article and third endopodal article of thoracopods II-V, but in both sites the lower end of the range overlaps that of *W. axotremata*. The habit of usually resting on the cirri is typical of *W. axotremata*, but neither species occurs exclusively on the column or the cirri.

5.4 *Depth distribution and biogeography*

Undoubted *Waginella metacrinicola* lives shallower than specimens assigned to *W. axotremata*. The depth range of the former species from Japan (Grygier 1990a) and the Philippines (herein) is 119-193 m. The depth range of all specimens referred

Table 1. Meristic and other features of *Waginella* spp. and comparison with New Caledonia populations. Most of the data are based on mature Japanese females of *W. metacrinicola* (Grygier 1983a, 1990a, herein), the original description of *W. axotremata* (see Grygier 1983a), and three females from New Caledonian *Saracrinus nobilis* and *Metacrinus levii*, but all available data were compiled concerning attachment site and brood size.

Character	Species/population *W. metacrinicola*	*W. axotremata*	New Caledonia
Carapace length (mm)	>3.25	<3.0	2.4-3.1
Carapace l/h	1.4-1.6	1.6-1.7	1.65-1.72
TI exopod 2 setae #	20	15-16	16-19
TI endopod 2 setae #	5-6	3	2?-5
TII-V basis setae #	7-9	4-6	4-6
TII-V exopod 2 setae #	10-12	8?-11	10-12
TII-V endopod 3 setae #	7-8	6-7	7-8
Furca l/h	2.8-3.4	2-2.5	2.3-2.7
Furca spines #	13-15	10-12	11-14
Attachment site	Column, seldom cirrus	Cirrus	Cirrus, sometimes column
Brood size	111	13-32	9-30

to the latter species (Grygier 1983a, 1990a, herein) is 193-580 m, except the unexamined specimen from *Saracrinus varians* (914 m).

The large number of geographical records now available for *W. metacrinicola* and *W. axotremata* provides an opportunity, which is very uncommon in the Ascothoracida, to discuss their biogeography with some degree of confidence. While the New Caledonian populations exhibit considerable morphological intermediacy between the two nominal species, biogeographical considerations make it impossible to recognize a cline in these features. *Waginella axotremata* per se occupies a region extending from northern Australia through eastern Indonesia into the Philippines (Fig. 2), and this area lies between Japan (the type locality of *W. metacrinicola*) and New Caledonia. The present find of *W. metacrinicola* off Manila Bay in the Philippines indicates that this species may be sympatric with, yet distinct from, *W. axotremata* there.

For now, the size and shape of the carapace of brooding females remain the most useful taxonomic characters to distinguish these two nominal species. This is an unsatisfactory state of affairs that does not help to distinguish the males and immature females. A final answer will require maintenance of specimens in the laboratory on a long-term basis in order to document allometry and increases in setal counts with growth. Such a rearing study may be possible in Japan, where *Metacrinus rotundus* is relatively accessible, harbors *W. metacrinicola*, and can be kept for long periods in aquaria (Amemiya & Oji 1982).

6 COMPARATIVE MORPHOLOGY

6.1 *Carapace pores*

The general structure of the anteroventral carapace pores was described by Grygier

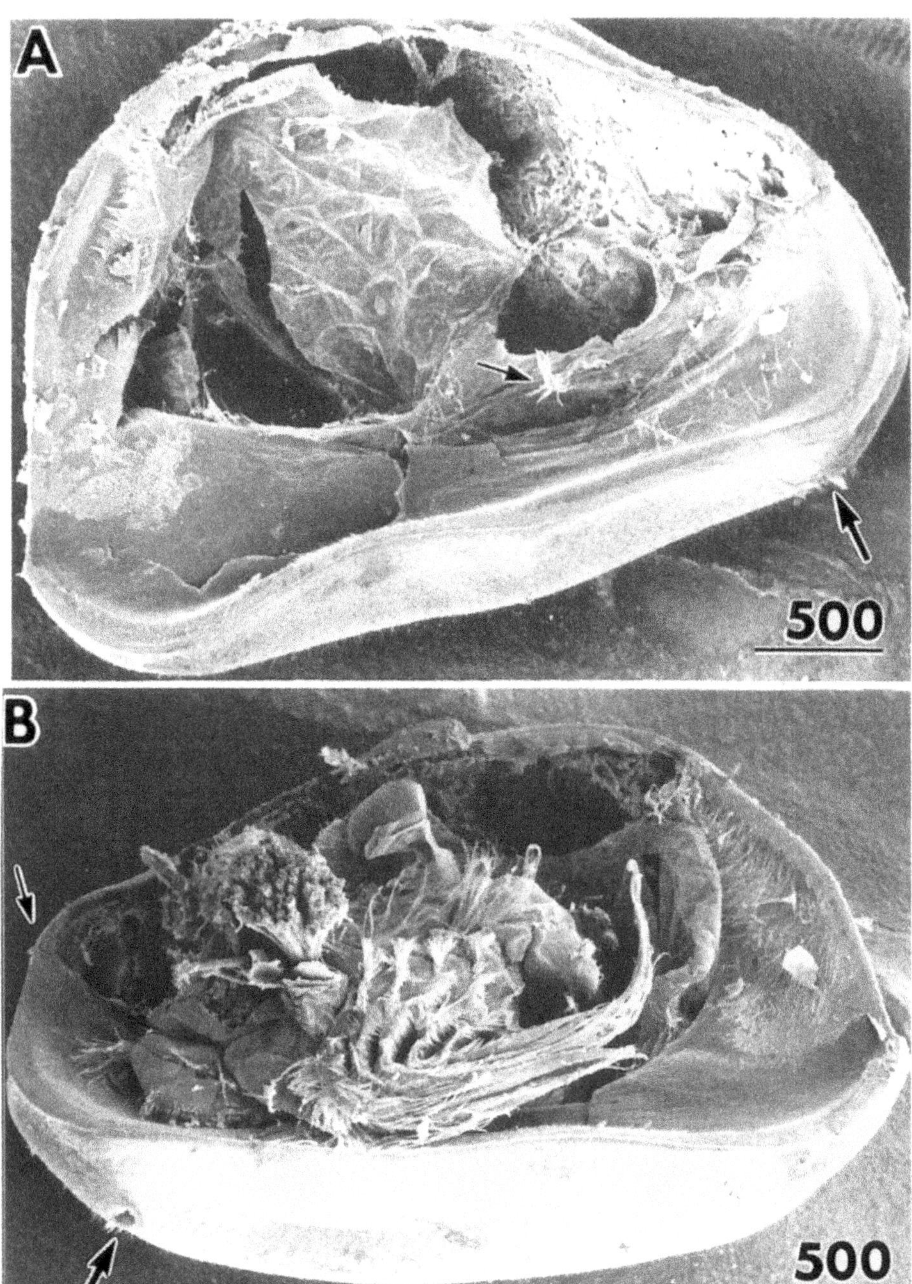

Plate 1. Scanning electron photomicrographs of *Waginella metacrinicola.* (A) Medial view of left carapace valve, anterior end right, showing so-called second antenna (small arrow; cf. Pl. 5) and site of anteroventral carapace pore (large arrow); (B) Ventrolateral view of main body and right carapace valve, anterior end left, with abdomen and left antennule removed, showing anteroventral carapace pore (large arrow; cf. Pl. 3C), approximate site of anterior lattice organs (small arrow), and one oval egg in dorsal brood chamber. Scales in μm.

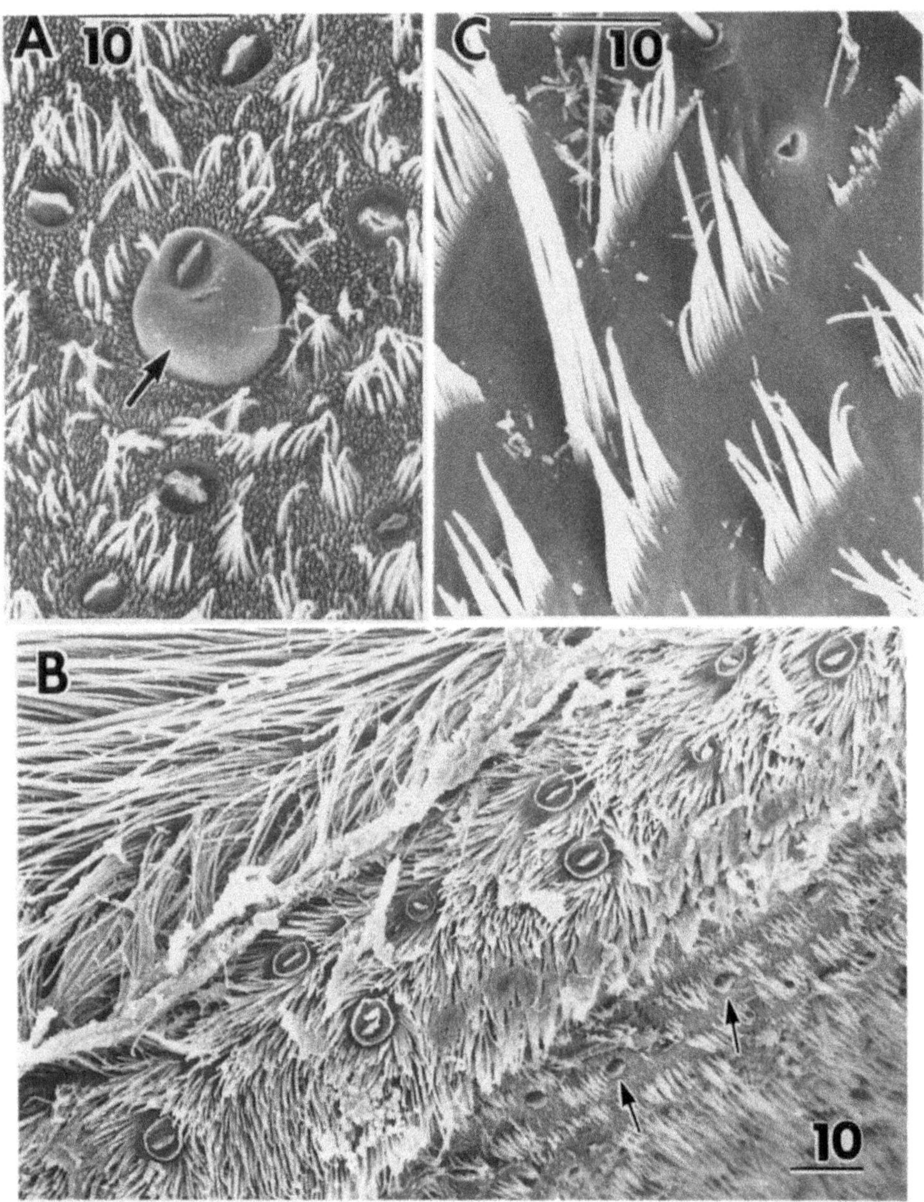

Plate 2. Scanning electron photomicrographs of *Waginella metacrinicola.* (A) Gland cone (arrow) and several pitted gland- tubes on ventral face of carapace valve among cuticular villi of two lengths; (B) Boundary of ventral face of carapace valve and mantle, showing ordered array of collared gland-tubes along medial margin proper (main diagonal), gland cones (arrows) and two broken setae on the ventral face (lower right), and dense pelage along the ventral margin of the mantle (upper left); (C) Setal base and simple integumentary pore among cuticular ctenae on posterior part of mantle. Scales in µm.

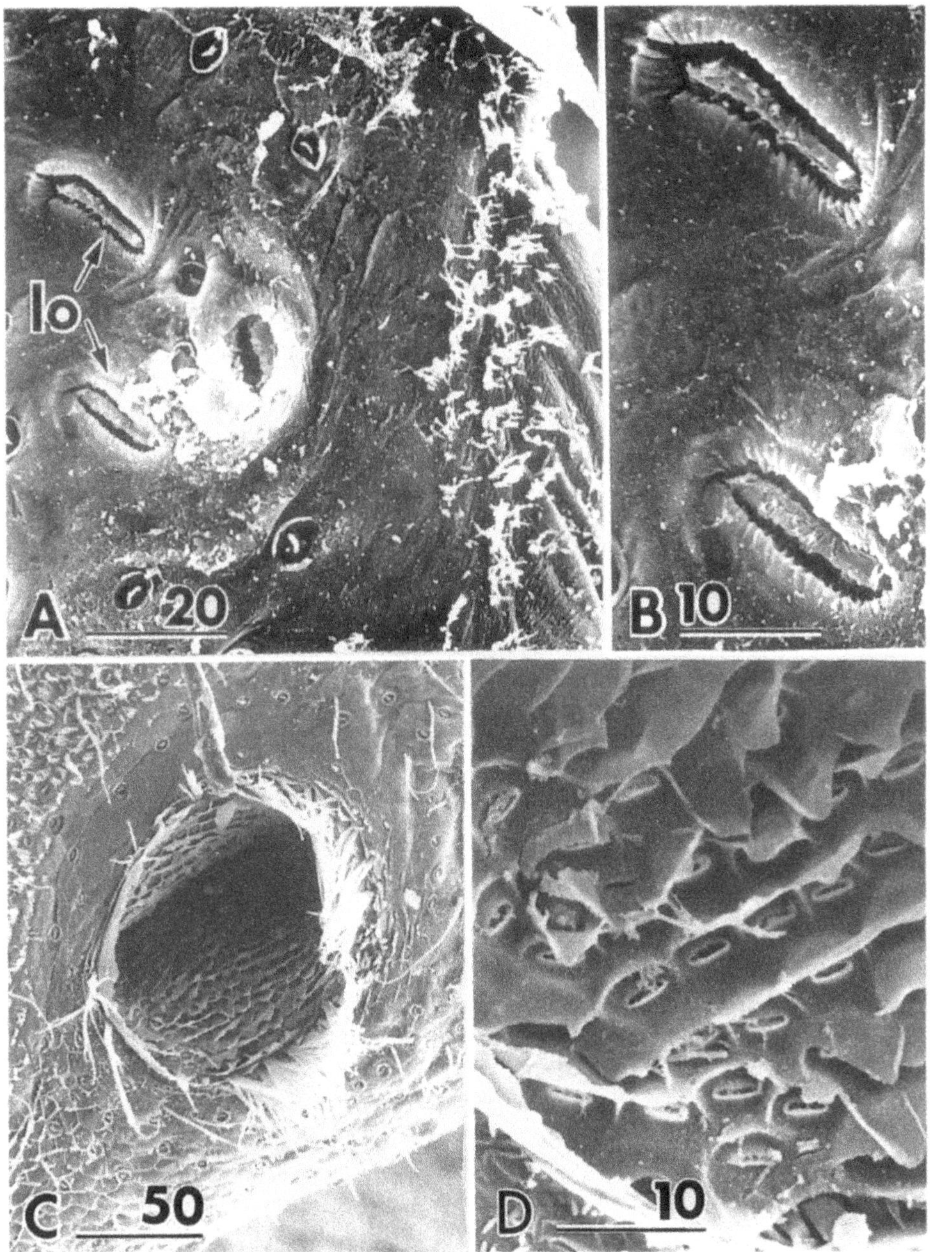

Plate 3. Scanning electron photomicrographs of *Waginella metacrinicola*. (A) Anterior lattice organs (lo), several pitted gland tubes and possible large pore (artifact?), and valve margin (right) with anteroventral direction towards bottom; (B) Detail of lattice organs; (C) Anteroventral carapace pore, anterior end right, lateral side below; (D) Lattice-like ridges and pores inside anteroventral carapace pore. Scales in μm.

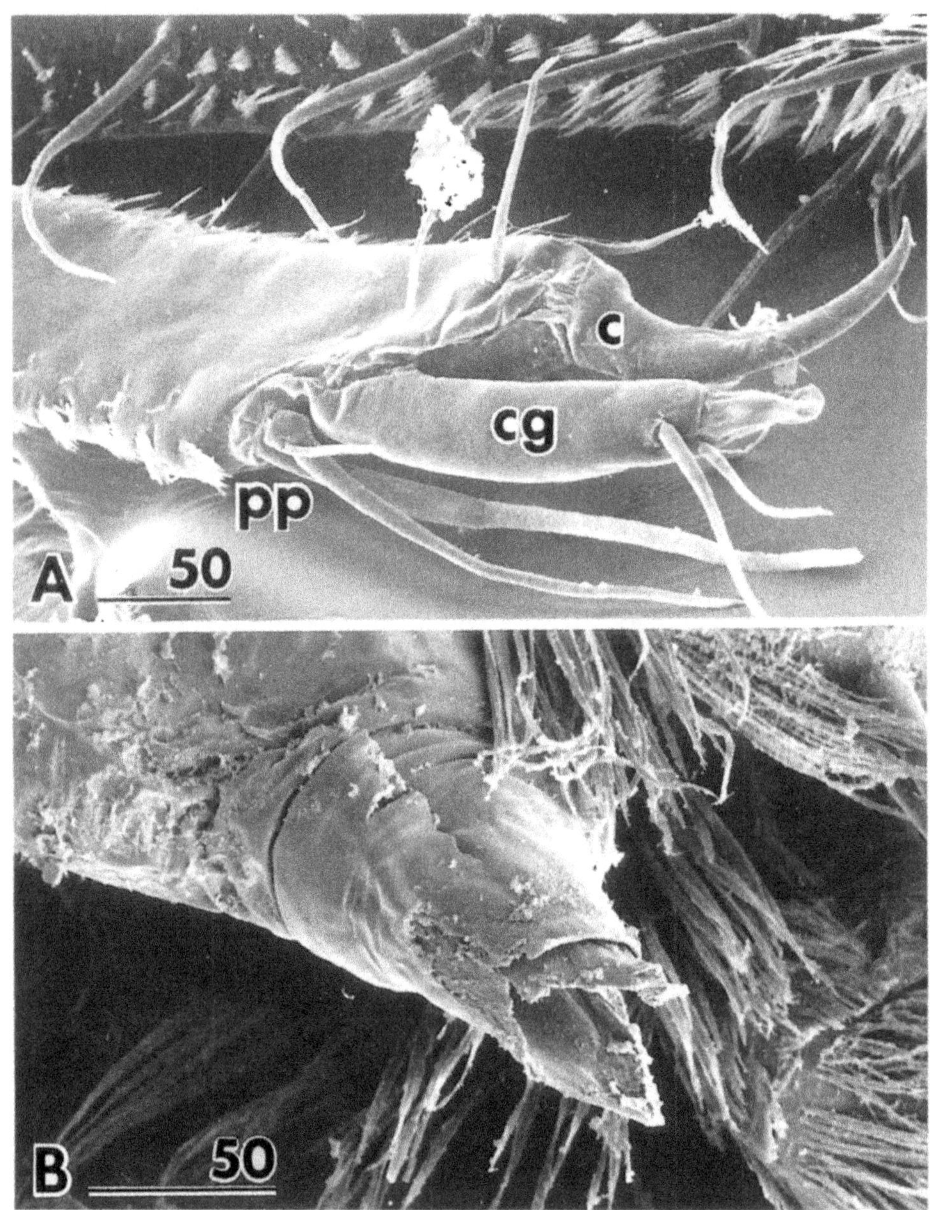

Plate 4. Scanning electron photomicrographs of *Waginella metacrinicola*. (A) Lateral view of distal half of sixth segment of antennule with claw (c), claw guard (cg), proximal sensory process (pp) and anterior margin of fifth segment (across top); (B) Apex of oral cone, anterior side left, with protruding tips of maxillae. Scales in μm.

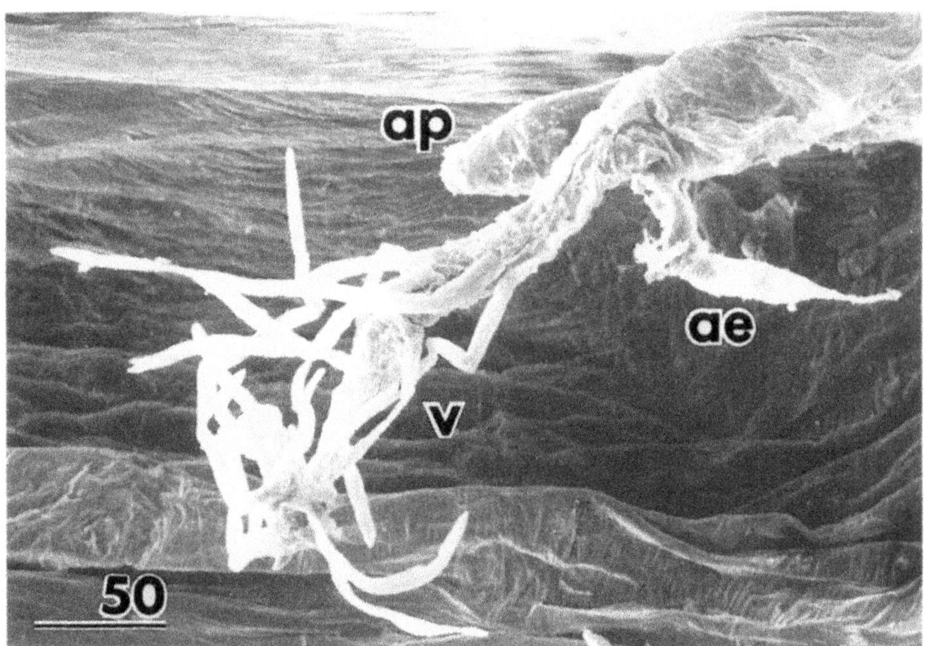

Plate 5. Scanning electron photomicrograph of so-called second antenna of *Waginella metacrinicola* dangling from mantle (cf. Pl. 1A), anterior end right. ae, proximal aesthetasc; ap, basal appendix; v, ventral ramus. Scale in μm.

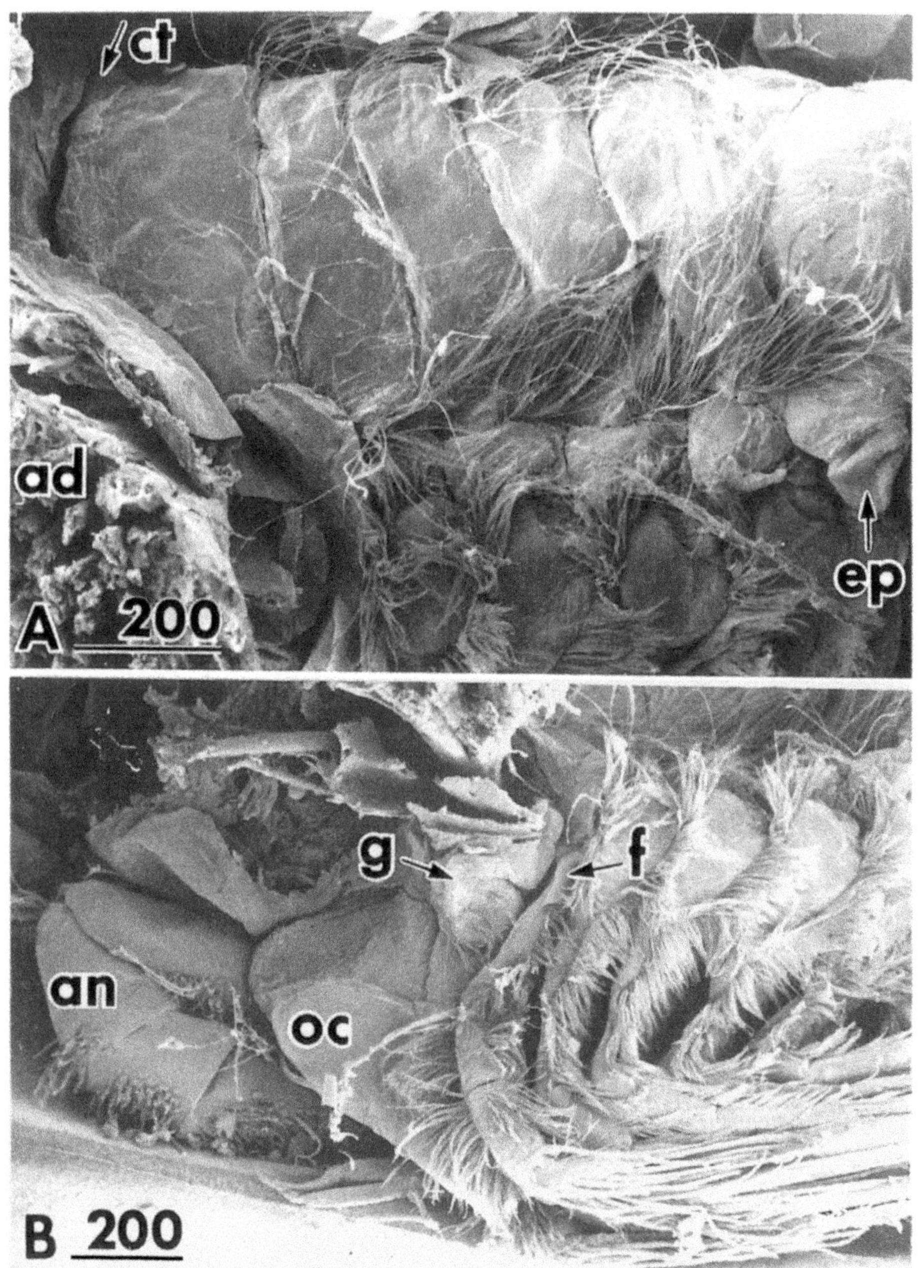

Plate 6. Scanning electron photomicrographs of *Waginella metacrinicola*. (A) Lateral view of thorax, anterior end left. (B) Ventrolateral view of cephalic and thoracic appendages, anterior end left. ad, adductor muscles; an, antennule; ct, suture between cephalon and first thoracomere; ep, epaulet (cf. Pl. 7A); f, filamentary appendage; g, genital papilla; oc, oral cone. Scales in μm.

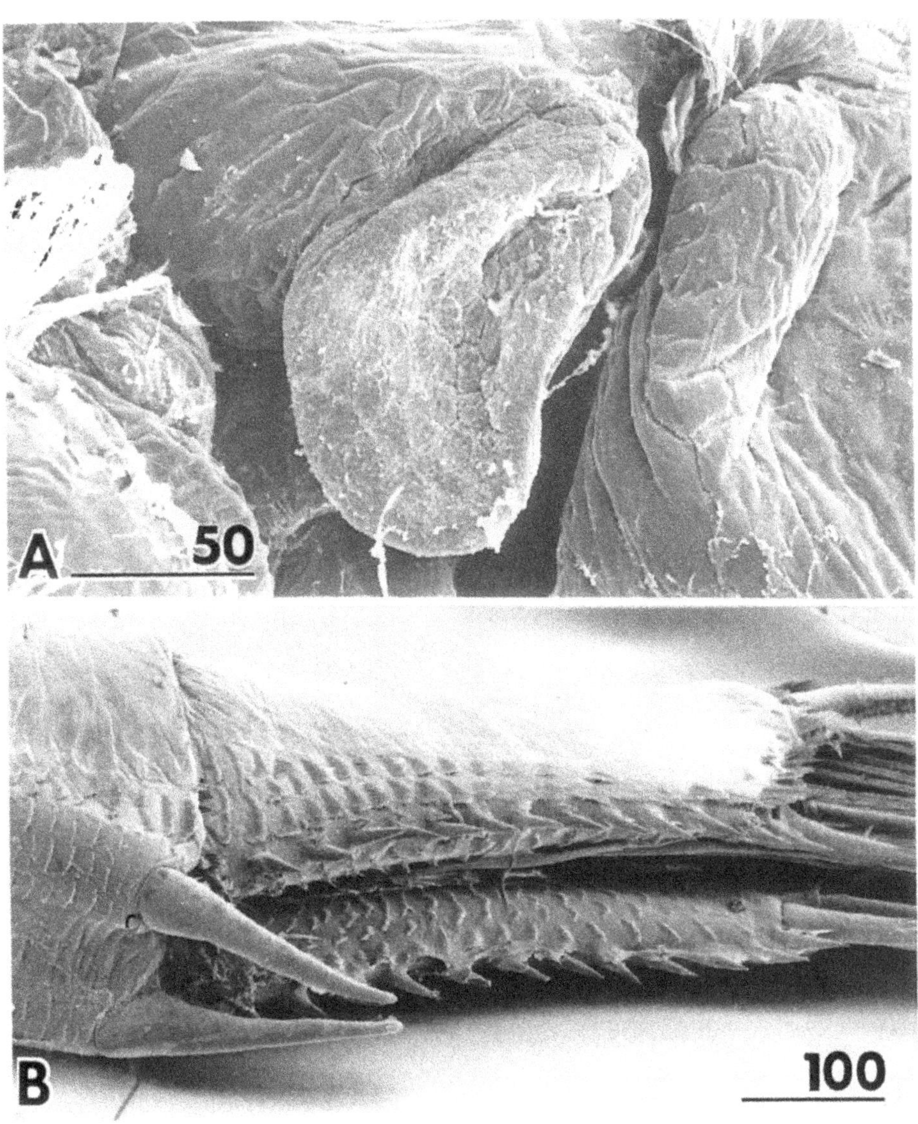

Plate 7. Scanning electron photomicrographs of *Waginella metacrinicola*. (A) Epaulet of sixth thoracomere (center) and lateral boss of seventh trunk somite (right); (B) Ventrolateral view of furcal rami and posterior part of telson with spine-like setae. Scales in μm.

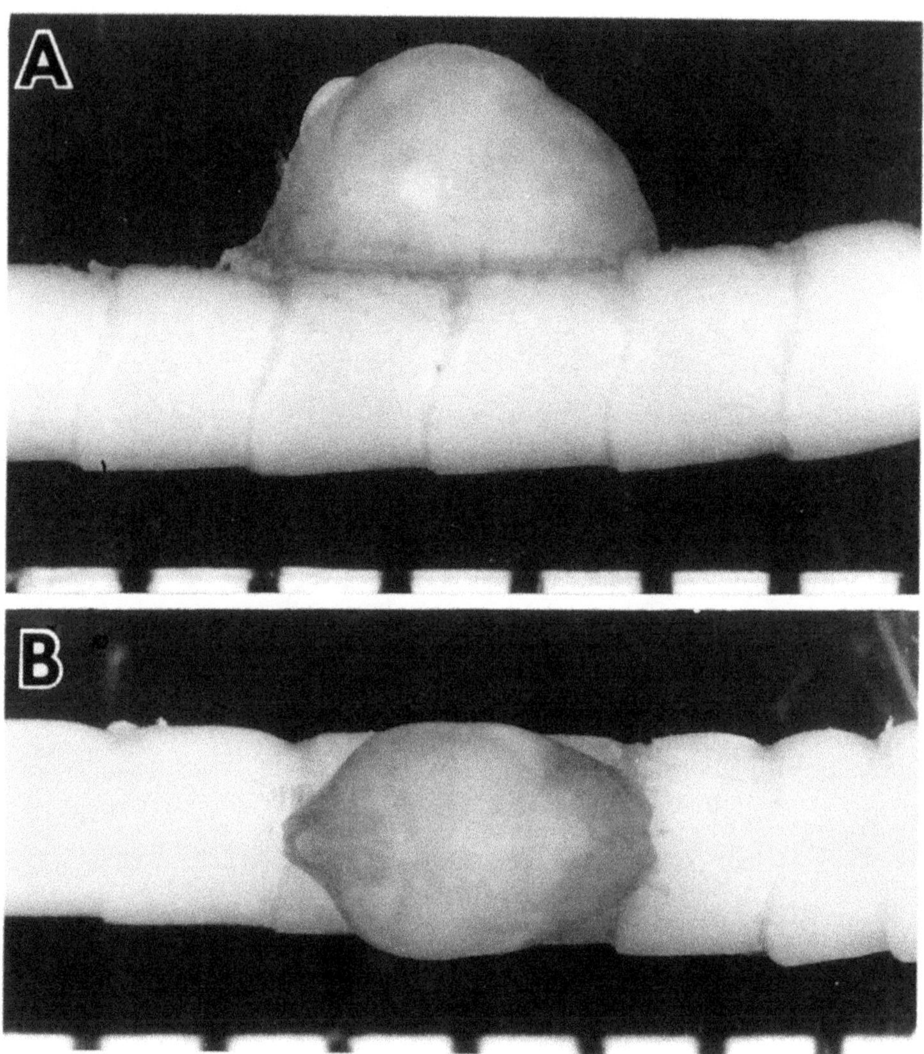

Plate 8. Mouthparts of *Waginella axotremata* from New Caledonia. (A) Mandible (md) and maxilla (mx); (B) Distal half of mandible. Scales in µm.

(1983a) based upon sections of *Waginella sandersi*, and the presence of small pores on the inner wall was clearly noted. Grygier did not mention, however, the lattice-like ridges or the distribution of pores among them, as demonstrated herein for *W. metacrinicola*.

Cirripede cyprid larvae have a prominent gland opening on each half of the carapace, i.e. the frontal horn pore, which is derived from the naupliar frontolateral horn (Walker & Lee 1976). Its situation on the carapace is similar to that of the anteroventral carapace pore in *Waginella*. However, this may be a coincidence, because the position of this structure is variable within the Ascothoracida; a similar and quite probably homologous pore is found more dorsally, opening on the inside of the carapace valve, in some species of *Synagoga*, another morphologically generalized genus of ascothoracidans (Grygier 1990b). Furthermore, the structural details are entirely different between *Waginella* and cirripede cyprids. The numerous small pores in *Waginella* presumably reflect a multiplicity of gland cells associated with the anteroventral carapace pores. The frontal horn pore of *Semibalanus* has neither a lattice-ridged wall nor small pores within, and it serves as the exit for just two large gland cells, just as in the nauplius (Walker 1973). The ontogenetic origin of the anteroventral carapace pore in *Waginella* is unknown, but the nauplii have no frontolateral horns (Grygier 1990a), so the process is probably different from that in the Cirripedia.

The anteroventral carapace pore is also somewhat similar to the so-called cone organ area described by Boxshall (1982) in misophrioid copepods such as *Benthomisophria palliata* Sars and *B. cornuta* Hulsemann & Grice. The cone organ area is located on both sides of the cephalon of these copepods and is equipped with a number of gland openings called cone organs, which are placed in a network of lamellar ridges.

Facetotectan cyprids have no prominent pair of depressions with aggregated pores on the anterior or anteroventral portions of their carapace, although typical integumentary pores are scattered in that region (Schram 1970; Itô & Ohtsuka 1984; Itô 1984, 1985, 1986, 1989).

Lattice organs of different morphologies are characteristic of all ascothoracid-larvae (in which they have been termed cardic organs) and cirripede cyprid larvae that have been examined for them to date (Itô & Grygier 1990; Grygier 1992; Jensen et al. 1994). However, in larvae studied by SEM, almost invariably five pairs have been found, two anterior and three posterior. Only the anterior two pairs are apparent in the present specimen of *Waginella metacrinicola*. Furthermore, Grygier (1983a) observed and illustrated only the anterior two pairs in *W. sandersi*, so perhaps the posterior three pairs are absent throughout the genus. In both species, the lattice organs are oriented at an unusually high angle to the valve margin. If their pores do truly open at the more anterior end in the present case, this matches the vast majority of cirripedes but not other ascothoracidans (Jensen et al. 1994).

The carapace of *W. metacrinicola* is equipped with gland cones, pitted gland-tubes, and collared gland-tubes, in addition to simple integumentary pores. These types of specialized gland openings have not so far been clearly characterized in ascothoracidans, although some pores are evident in SEM micrographs of the carapace of a larval form (Grygier 1988a) and on the body of adult female *Baccalaureus falsiramus* (see Itô & Grygier 1990). The few facetotectan and cirripede cyprids examined to date have no such specialized gland openings (e.g. Walker & Lee 1976; Elfimov 1986; Dineen 1987; Itô 1989; Jensen et al. 1994), only simple ones with a lip, but cyprid larvae are still very incompletely documented in this regard.

The gland cones are present only on the ventral faces of the carapace valves, which attach to the host crinoid. One might suppose that they are special organs for secreting the cement that keeps the carapace attached to the host. Structures that could be called 'compound gland cones' (many gland openings on a single protuberance) occur on the ventral side of the carapace valves in *Flatsia walcoochorum* Grygier (1991), a species of quite uncertain affinities within the Ascothoracida. Unfortunately, the single known specimen was found separated from its host and provides no information on the function of these structures. The collared gland-tubes that are especially prevalent on the medial margin of the ventral face of each carapace valve in *W. metacrinicola* also pose a question. Both the gland cones and the collared gland-tubes markedly differ from the simple apertures of typical integumentary pores in shape, so knowledge of their internal structures and their function would be of great interest.

6.2 *Supposed second antennae as organs of Bellonci*

The problematic processes located near the first antennae in *Waginella* spp. as well as in other ascothoracidans have been referred to by various names such as 'antenna' or 'second antenna' (Newman 1974; Grygier 1981, 1983a,b), 'antenna-frontal filament complex' or 'antenna-vestigial eyestalk complex' (Grygier 1984), or 'frontal filament complex' (Grygier 1990a). In *W. metacrinicola*, it was originally described as an 'appendice dont la fonction est incertaine' (Okada 1926). It is apparent from the many proposed names that subsequent authors have wished to emphasize a possible homology between the second antenna and this process, at least in part. Contrary to their views, Itô (1985) proposed a homology between the paraocular process of facetotectan cyprids and this ascothoracidan process, and supposed that these were all versions of the organ of Bellonci. Grygier (1987b) came back with the assertion that the posterior or dorsal ramus of the ascothoracidan organ is a separate process corresponding to the postocular filamentary tuft in facetotectan cyprids. Since that posterior ramus is present only in males of *Waginella* (see Grygier 1987a) and was not observed in the present female, we cannot address that point here.

By transmission electron microscopy we know that the paraocular process and postocular filamentary tuft of facetotectan cyprids are actually the external portion of the organ of Bellonci and the external extension of a gland, respectively (Itô & Takenaka 1988). Furthermore, the naupliar second antennae and mandibles remain as degenerate, papillary processes distinct from both the paraocular process and filamentary tuft. Therefore, the paraocular process is certainly not homologous with the second antenna.

The base of the paraocular process in facetotectan cyprids is situated lateral to the first antenna (Itô & Takenaka 1988). This matches the situation of the problematic process in the ascothoracidan here examined. Therefore, it is most reasonable to regard this ascothoracidan process as the homologue of the facetotectan paraocular process, namely, the external portion of the organ of Bellonci. What remains to be clarified is the identity of the basal appendix. It could be either a part of the organ of

Bellonci or a vestigial compound eye. Grygier (1987b) found that the paraocular process of his facetotectan cyprid had a similar basal appendix despite the occurrence of well-developed compound eyes, and was forced to reconsider his former view (Grygier 1984) that the basal appendix is a vestigial compound eye. The eye hypothesis recently gained new support from Grygier's (1992) discovery of a red pigment cup in the swollen base of the relatively simple frontal filament complex of a laurid ascothoracid-larva.

This issue could be resolved by a transmission electron microscopical study of an ascothoracidan like *Waginella*. Such a study could demonstrate eye-specific structures like rhabdomeres in the basal appendix and determine whether the vesicular part of the organ of Bellonci lies in the basal appendix or inside the cephalon nearby, the latter being true for a facetotectan cypris y (Itô & Takenaka 1988). It could also demonstrate the nervous connections of the basal appendix; a vestigial eye would be innervated from the optic center of the protocerebrum, but an organ of Bellonci from the lobus lateralis of the protocerebrum (= medulla terminalis of Malacostraca) (Walker 1974). As Itô & Takenaka (1988) have pointed out, such a comparative study will be incomplete without a detailed description of the ultrastructure of the frontal filament complex in barnacle cyprid larvae; only the naupliar frontal filament has been studied in the Cirripedia Thoracica until now (Walker 1974).

6.3 *Epaulets*

Epaulets are known to occur laterally on the sixth thoracic somite in various ascothoracidans, and their possible functions have been discussed (Grygier 1983a, 1984). The present study has revealed a previously unknown feature of this structure. The protrusive part of the epaulet appears to consist of soft, thin cuticle and, moreover, its rough surface is divided into numerous small polygons by a network of grooves. Such a surface structure occurs in the gills of some other crustaceans. The epipodites of the thoracic limbs in leptostracans are a case in point (Itô 1988). The surface of each epipodite is very rough due to delicate corrugations that increase the surface area available for gas exchange. The epaulets might thus also function as respiratory organs.

NOTE ADDED IN PROOF BY M.J. GRYGIER AND W. KLEPAL

The planned SEM study by Grygier and Klepal (Section 4.1.1) has been carried out and permits some additional description. Three females collected and indentified as *Waginella metacrinicola* by T. Itô early in 1990 were examined. The host crinoid from off Ushibuka, Amakusa Islands, western Japan, was not saved, but only *Metacrinus interruptus* Carpenter and *Saracrinus nobilis* are likely to occur there (T. Oji, pers. comm.); therefore, being rather small, the parasites might really be *W. axotremata*.

There are three posterior pairs of lattice organs as well as two anterior pairs (Section 3.1.2) on the carapace. The large pore near the anterior pairs is not an artifact (Section 3.1.2), but a pit with several pores inside; a similar pit is associated with the first two posterior lattice organs. On the rim of the anteroventral carapace pore, the supposed tufts of fine setae (Section 3.1.2) are actually tassel-like organs, possibly modified setae. There is a row of spiniform, bipectinate guard setae on the mantle, parallel to and well inside the edge of the posterior carapace aperture. Besides semi-articulated, restrorse spines along the inner side of the mandible (Section 4.1.1), there are widely-spaced, transverse rows of smaller, distally directed teeth on the outer side. The polygonal regions on the epaulets resemble bunches of 1.2-1.8 μm wide grapes at higher magnification. The dorsolateral tufts on trunk segment 7 (Section 4.1.2) and the lateral tufts on the thorax (Section 3.1.7) consist of non-articulated cuticular fringes, not setae. The outer magin of the protopod of thoracopod 2 is lined with a row of frayed, strap-like, compound fringes.

ACKNOWLEDGMENTS

This study was supported in part by Grant-in-Aid for Scientific Research No. 62540567 to T. Itô from the Japanese Ministry of Education, Science and Culture, and the following gentlemen at the University of Tokyo provided him with the ascothoracidan material for SEM: Drs. I. Hayami and K. Tanabe (Geological Institute), and S. Ohta and E. Tsuchida (Ocean Research Institute). In order to study the MUSORSTOM Ascothoracida, the first author was appointed temporarily to a position of Maître de Conférences in the Laboratoire de Biologie des Invertébrés Marins et Malacologie of the Muséum National d'Histoire Naturelle; thanks are extended to Drs. A. Crosnier, B. Métivier, and C. and F. Monniot for hospitality and assistance there. The first author also acknowledges a travel grant from ORSTOM to visit Dr. M. Roux (Université de Reims), and thanks Drs. R. Yamanishi (Osaka Museum of Natural History), A.J. Bruce (Northern Territory Museum of Arts and Sciences), and E. Harada (Seto Marine Biological Laboratory), and Mrs. A.B. Johnston (Museum of Comparative Zoology), for access to specimens, Dr. T. Oji (University of Tokyo Geological Institute) for identifying a crinoid, and Dr. S. Ohtsuka (Hiroshima University Fisheries Laboratory) for assistance with SEM and figure preparation during a visit to Japan that was supported by a fellowship from the Dan Charitable Trust Fund for Research in the Biological Sciences.

REFERENCES

Amemiya, S. & T. Oji 1992. Regeneration in sea lilies. *Nature* 357: 546-547.

Blake, C.H. 1933. A new parasitic cirripede (*Crinoidoxenos halocypris*). *Anat. Rec.* 57(4, supplement): 97.

Bourseau, J.-P., N. Ameziane-Cominardi & R. Avocat 1991. Echinodermata: Les Crinoïdes pédonculés de Nouvelle-Calédonie. *Mém. Mus. Natl. Hist. Nat. Sér. A Zool.* 151: 229-333.

Bourseau, J.-P. & M. Roux 1989. Echinodermes: Crinoïdes Pentacrinidae (MUSORSTOM 2 & CORINDON 2). *Mém. Mus. Natn. Hist. Nat. Sér. A Zool.* 143: 113-201.

Boxshall, G.A. 1982. On the anatomy of the misophrioid copepods, with special reference to *Benthomisophria palliata* Sars. *Phil. Trans. R. Soc. Lond. B* 297: 125-181.

Bresciani, J. 1965. Nauplius 'y' Hansen. Its distribution and relationship with a new cypris larva. *Vidensk. Medd. Dansk Naturh. For.* 128: 245-258, pl. XL.

Carpenter, P.H. 1884. Report upon the Crinoidea collected during the voyage of H.M.S. Challenger during the years 1873-76. Part 1. General morphology, with descriptions of the stalked crinoids. *Challenger Rep., Zool.* 11(32): 1-442.

Clark, H.L. 1916. Report on the sea-lilies, star-fishes, brittle-stars and sea-urchins obtained by the F.I.S. 'Endeavour' on the coasts of Queensland, New South Wales, Tasmania, Victoria, South Australia, and Western Australia. *Biol. Res. Fish. Exper. F.I.S. 'Endeavour', 1909-14* 4(1): 1-123, pls. I-XLIV.
Dineen, J. 1987. The larval stages of *Lithotrya dorsalis* (Ellis & Solander, 1786): a burrowing thoracican barnacle. *Biol. Bull.* 172: 284-298.
Elfimov, A.S. 1986. Morphology of the carapace in the cypris larva of *Heteralepas mystacophora* Newman (Cirripedia, Thoracica). *Biol. Morya* 1986(3): 30-34.
Forest, J. 1989. Compte rendu de la Campagne MUSORSTOM 3 aux Philippines (31 mai-7 juin 1983). *Mém. Mus. Natl. Hist. Nat. Sér. A Zool.* 143: 9-23.
Grygier, M.J. 1981. *Gorgonolaureus muzikae* sp. nov. (Crustacea: Ascothoracida) parasitic on a Hawaiian gorgonian, with special reference to its protandric hermaphroditism. *J. Nat. Hist.* 15: 1019-1045.
Grygier, M.J. 1983a. Revision of *Synagoga* (Crustacea: Maxillopoda: Ascothoracida). *J. Nat. Hist.* 17: 213-239.
Grygier, M.J. 1983b. Ascothoracida and the unity of Maxillopoda. *Crustacean Issues* 1: 73-104.
Grygier, M.J. 1984. Comparative morphology and ontogeny of the Ascothoracida, a step toward a phylogeny of the Maxillopoda. Ph. D. Thesis, University of California San Diego.
Grygier, M.J. 1985. Lauridae: taxonomy and morphology of ascothoracid crustacean parasites of zoanthids. *Bull. Mar. Sci.* 36: 278-303.
Grygier, M.J. 1987a. Reappraisal of sex determination in the Ascothoracida. *Crustaceana* 52: 149-162.
Grygier, M.J. 1987b. New records, external and internal anatomy, and systematic position of Hansen's y-larvae (Crustacea: Maxillopoda: Facetotecta). *Sarsia* 72: 261-278.
Grygier, M.J. 1988a. Larval and juvenile Ascothoracida from the plankton. *Publ. Seto Mar. Biol. Lab.* 33: 163-172.
Grygier, M.J. 1988b. Unusual and mostly cysticolous crustacean, molluscan, and myzostomidan associates of echinoderms. In R.D. Burke, P.V. Mladenov, P. Lambert & R.L. Parsley (eds.), *Echinoderm Biology*. Rotterdam: Balkema.
Grygier, M.J. 1990a. New records of *Waginella* (Crustacea: Ascothoracida) ectoparasitic on stalked crinoids from Japan and Australasia. *Galaxea* 8: 339-350.
Grygier, M.J. 1990b. Five new species of bathyal Atlantic Ascothoracida (Crustacea: Maxillopoda) from the equator to 50°N latitude. *Bull. Mar. Sci.* 46: 655-676.
Grygier, M.J. 1991. Additions to the ascothoracidan fauna of Australia and south-east Asia (Crustacea, Maxillopoda): Synagogidae (part), Lauridae and Petrarcidae. *Rec. Austral. Mus.* 43: 1-46.
Grygier, M.J. 1992. Laboratory rearing of ascothoracidan nauplii (Crustacea: Maxillopoda) from plankton at Okinawa, Japan. *Publ. Seto Mar. Biol. Lab.* 35: 235-251.
Itô, T. 1984. Another cypris y from the North Pacific, with reference to the bending behavior exhibited by a cypris y specimen of the formerly described type (Crustacea: Maxillopoda). *Publ. Seto Mar. Biol. Lab.* 29: 367-374.
Itô, T. 1985. Contributions to the knowledge of cypris y (Crustacea: Maxillopoda) with reference to a new genus and three new species from Japan. In, *Special Publication of Mukaishima Marine Biological Station*. Hiroshima: Faculty of Science, Hiroshima University.
Itô, T. 1986. A new species of 'cypris y' (Crustacea: Maxillopoda) from the North Pacific. *Publ. Seto Mar. Biol. Lab.* 31: 333-339.
Itô, T. 1987. Y-larvae – enigmatic crustacean larvae. *Ann. Rep. Seto Mar. Biol. Lab.* 1: 52-58 (in Japanese).
Itô, T. 1988. Location of the epipod on the thoracic limbs in *Nebalia bipes* (Crustacea: Leptostraca). *Ann. Rep. Seto Mar. Biol. Lab.* 2: 36-39 (in Japanese).
Itô, T. 1989. A new species of *Hansenocaris* (Crustacea: Facetotecta) from Tanabe Bay, Japan. *Publ. Seto Mar. Biol. Lab.* 34: 55-72.
Itô, T. 1990. [The true nature of y-larvae (Crustacea)]. Research grant 62540567 report, Faculty of Science, Kyoto University (in Japanese).

Itô, T. & M.J. Grygier 1990. Description and complete larval development of a new species of *Baccalaureus* (Crustacea: Ascothoracida) parasitic in a zoanthid from Tanabe Bay, Honshu, Japan. *Zool. Sci.* 7: 485-515

Itô, T. & S. Ohtsuka 1984. Cypris y from the North Pacific (Crustacea: Maxillopoda). *Publ. Seto Mar. Biol. Lab.* 29: 179- 186.

Itô, T. & M. Takenaka 1988. Identification of bifurcate paraocular process and postocular filamentary tuft of facetotectan cyprids (Crustacea: Maxillopoda). *Publ. Seto Mar. Biol. Lab.* 33: 19-38.

Jensen, P.G., J. Moyse, J. Høeg & H. Al-Yahya 1994. Comparative SEM studies of lattice organs: putative sensory structures on the carapace of larvae from Ascothoracida and Cirripedia (Crustacea Maxillopoda Thecostraca). *Acta Zool. (Stockh.)* 75: 125-142.

Newman, W.A. 1974. Two new deep-sea Cirripedia (Ascothoracica and Acrothoracica) from the Atlantic. *J. Mar. Biol. Ass. UK* 54: 437-456.

Oji, T. 1989. Distribution of the stalked crinoids from Japanese and nearby waters. *Univ. Mus. Univ. Tokyo, Nat. Cult.* 1: 27-43.

Okada, Y.K. 1926. Ascothoraciques. II. Note sur l'organisation de *Synagoga*. *Bull. Mus. Natl. Hist. Nat.* 50: 69-73.

Okada, Y.K. 1938. Les Cirripèdes Ascothoraciques. *Trav. Stat. Zool. Wimereux* 13: 489-514.

Richer de Forges, B. 1990. Les campagnes d'exploration de la faune bathyale dans la zone économique de la Nouvelle Calédonie. Explorations for bathyal fauna in the New Caledonian economic zone. *Mém. Mus. Natn. Hist. Nat. Sér. A Zool.* 145: 9-54.

Roux, M. 1981. Echinodermes : Crinoïdes Isocrinidae. *Coll. Mém. ORSTOM* 91: 477-543.

Schram, T.A. 1970. Marine biological investigations in the Bahamas. 14. Cypris y, a later developmental stage of nauplius y Hansen. *Sarsia* 44: 9-24.

Walker, G. 1973. Frontal horns and associated gland cells of the nauplii of the barnacles *Balanus hameri, Balanus balanoides* and *Elminius modestus* (Crustacea Cirripedia). *J. Mar. Biol. Ass. UK* 53: 455-463.

Walker, G. 1974. The fine structure of the frontal filament complex of barnacle larvae (Crustacea: Cirripedia). *Cell Tiss. Res.* 152: 449-465.

Walker, G. & V.E. Lee 1976. Surface structures and sense organs of the cypris larva of *Balanus balanoides* as seen by scanning and transmission electron microscopy. *J. Zool., Lond.* 178: 161-172.

New interpretations of South American patterns of barnacle distribution

Paulo S. Young
Museu Nacional/UFRJ, Depto de Invertebrados, Rio de Janeiro, Brazil

ABSTRACT

The shallow water (< 80 m) barnacle fauna occurring in the Southwestern Atlantic is divided in three groups: two comprising the tropical and subtropical species which occur along the stretch of the Brazilian coast under the influence of the Brazil Current and north branch of the Equatorial Current, and another by the temperate species which occur along the coast under the influence of the Falkland Current. The tropical species group related with the Brazil Current is subdivided in two subgroups: the first comprising the equatorial species which occur along the North coast of Brazil (0-4°), and the second comprising strict tropical species which occur from 4°S southwards and gradually disappears south of 18°S. The subtropical species are confined between 21-39°S. The temperate species group occurs from the southern tip of South America to 35-40°S along the Southwestern Atlantic. A Tropical-subtropical Transition Zone was detected between 21-24°S and another Subtropical-temperate Transition Zone between 35-42°S. The tropical and temperate groups have southern distribution limits related to a temperature gradient. A temperature gradient is also noted along the northward extension of the temperate species. The deep water species (> 80 m) can also be separated in three groups: tropical, subtropical and, temperate. The latitudinal patterns of species richness for the shallow water barnacles species shows a large number of species between 10-24°S, with a conspicuos decrease both to the equator and to higher latitudes. This pattern appears to be influenced by area and geologic history. The deep water species presented a greater richness at more southern latitudes. The depth associated patterns of barnacle diversity followed the general concept of a more balanomorph richness at shallow waters and a greater lepadomorph richness at greater depths. The tropical-subtropical barnacle fauna of the Southwestern Atlantic shows a great similarity with that of the Northwestern Atlantic, and no similarity with that of tropical West Africa. The temperate barnacle fauna, on the other hand, shows a similarity with the fauna of the southern oceans. Twelve generalized patterns of geographic distribution of South American barnacles were detected. Some of the uncommon patterns of distribution of the species may reflect changes of distribution due to antropic activities (introductions) or, alternatively, may reflect either cases of sibling species or misidentifications.

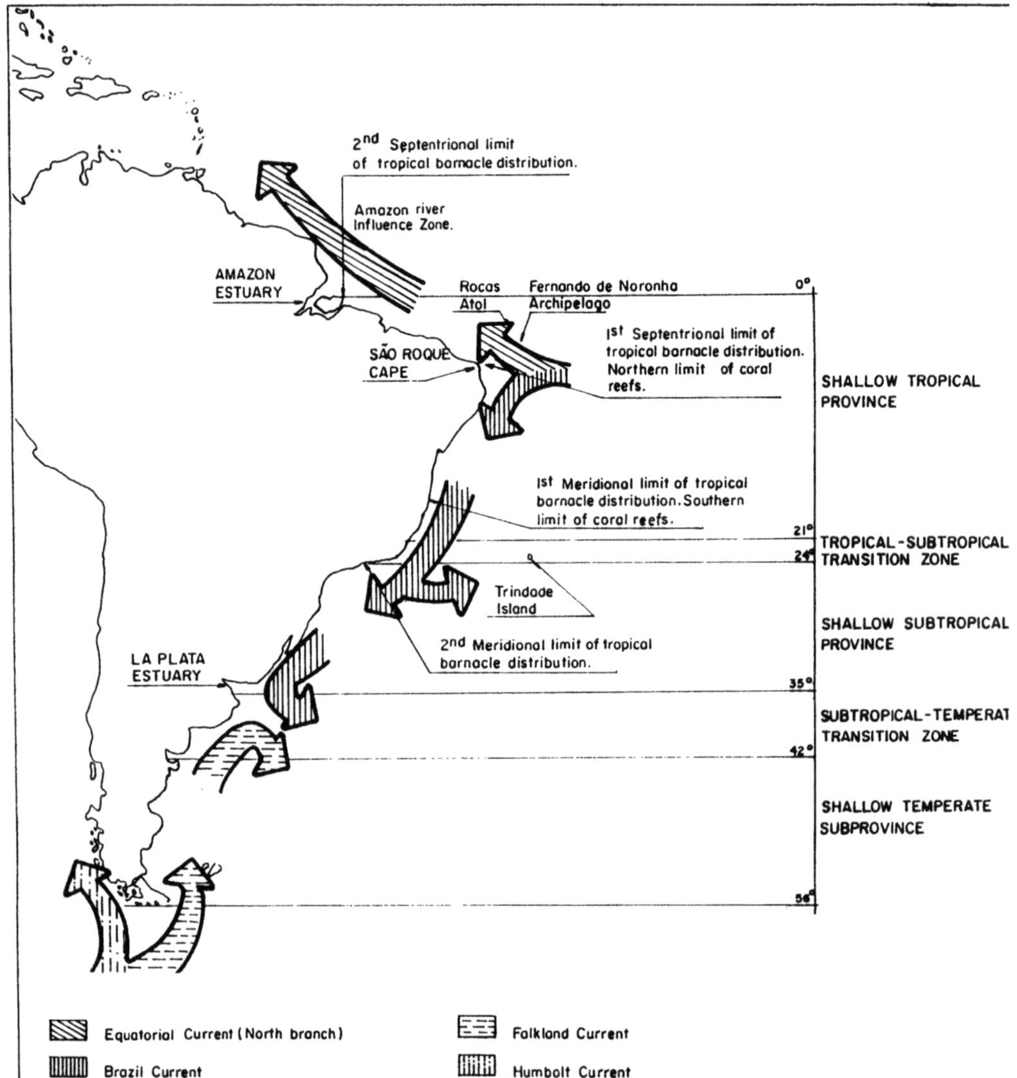

Figure 1. Depth range of the barnacle species from the Southwestern Atlantic. The strictly intertidal species were excluded. *W. aduncum* was not included due to the lack of data on its bathymetric distribution. 1) *C. amaryllis*; 2) *N. radiata*; 3) *B. reticulatus*; 4) *B. subalbidus*; 5) *B. amphitrite*; 6) *F. citerosum*; 7) *M. vesiculosus*; 8) *M. stultus*; 9) *A. psittacus*; 10) *M. tintinnabulum*; 11) *C. domingensis;* 12) *C. floridana*; 13) *C. paucicostata*; 14) *C. madreporarum*; 15) *A. cyathus*; 16) *M. coccopoma*; 17) *B. laevis*; 18) *B. improvisus*; 19) *C. patula*; 20) *C. galeata*; 21) *B. venustus*; 22) *B. trigonus*; 23) *O. hirtae*; 24) *O. lowei*; 25) *B. spongicola*; 26) *O. hoeki*; 27) *L. henriquecostai*; 28) *B. calidus*; 29) *P. kaempferi*; 30) *A. boubalocerus*; 31) *O. gibberum*; 32) *R. nexa*; 33) *H. rathbunae*; 34) *V. idioplax*; 35) *H. lankesteri*; 36) *P. martini*; 37) *V. flavidula*; 38) *H. cornuta*; 39) *A. portoricanum*; 40) *Balanus* sp.; 41) *W. scoresbyi*; 42) *A. gibbosa*; 43) *V. caribbea*; 44) *P. parallelogramma*; 44) *Altiverruca* sp; 46) *L. regina*; 47) *A. triangulare*; 48) *P. inaequilaterale*; 49) *B. bulata.*

1 INTRODUCTION

The distribution of barnacles in the Western Atlantic is well documented in the northern hemisphere. There is a complex pattern of distribution in Caribbean that appears to be determined by ocean circulation and temperature gradients (Southward & Newman 1977; Zullo 1979; Spivey 1981). The eastern coast of the USA and Canada presents a temperature gradient of fauna distribution (Briggs 1974).

Information on distribution in the Southwestern Atlantic, in contrast, is scarce due to the few collections available from this area. Nevertheless, the taxonomic studies on barnacles from the Brazilian and Argentinian coast (Aurivillius 1892; Hoek 1883; Weltner 1897; Lahille 1910; Pilsbry 1916; Calman 1918; Nilsson-Cantell 1927, 1930, Oliveira 1940, 1941; Weber 1960; Lacombe & Monteiro 1974; Lacombe & Rangel 1978; Spivak & L'Hoste, pers. comm.; Young 1988, 1989a, 1990, 1991, 1992, 1993, 1994; Young & Christoffersen 1984), complemented by new informations generated by recent deep dredging (personal observation) between 20-32°S, has suggested a general pattern of distribution for South American barnacles.

It should be noted that in the case of barnacles some species may have their geographical distribution altered by human activities (fouling and species associated with introduced fauna), thus breaking its natural barriers.

This study intends to discuss patterns of distribution of barnacles in the tropical, subtropical, and temperate Southwestern Atlantic. It also aims to compare the composition of the Southwestern Atlantic barnacle fauna with those of the Northwestern Atlantic and the Atlantic coast of Africa, as well as discuss the latitudinal and depth associated patterns related to species richness. Through the identification of global patterns of geographic distribution, it also intends to identify possible introduced and sibling species.

The taxonomic nomenclature used herein follows Newman & Ross (1976) for Balanomorpha, Zevina (1978a,b) for Lepadomorpha, and Zevina (1987a) for Verrucomorpha.

2 DISTRIBUTION PATTERNS OF THE BARNACLE FAUNA

The distribution of barnacles species along the Southwestern Atlantic, was displaceu along depth gradient (Fig. 1), to shows if there is a separation between shallow and deep water species. It was observed that 80 m was a limit between both species group. This depth is the zone where the smaller number of species occurred. These groups are discussed separetely after the characterization of the oceanographic circulation.

2.1 *The shallow water circulation*

South America presents along its coast the influence of two large currents in addition to the north branch of the Equatorial Current – the southward warm Brazil Current and the northward cold Falkland Current (Fig. 2).

The westward flowing Equatorial Current passes by Rocas Attol and Fernando de

Noronha Archipelago and reaches the coast of Brazil at São Roque Cape (5°S). At this site, the current splits into two branches, one of them running parallel to the northern coast of Brazil and forming the northern branch of the Equatorial current, which, off the Amazon river estuary, turns offshore and penetrates the north hemisphere. The second branch that flows southwards forms the southward Brazil Current.

Off the estuary of the La Plata river (38°S), the encounter of the southward flow of the warm Brazil Current with the northward flow of the cold Falkland Current occurs. At this latitude, these two surface currents deviate from the continental shelf and form a frontal confluence of subtropical and subantarctic waters (Legeckis & Gordon 1982, Gordon & Greengrove 1986) that will be part of the West Wind Drift.

The Brazil Current, a 100-150 m thick water layer, has a high salinity range (36,0 < S·% < 37,3) and warm temperatures (18,0 < T°C <27,0) (Pierre et al. 1991). This current flows southwards from the equator along the continental margin of South America. South of 18°S, it suffers the influence of lower temperatures during the cold seasons (cold fronts), and also of temporary upwellings along the coast generated by wind action. At about 22°S (Cape Frio), the Brazil current bifurcates, a branch of it flowing southward and the other moving eastwards, and forming a gyre (Tsuchiya 1985). As a consequence, a gradual decrease in temperature is noted southward to about the La Plata estuary. The Brazil Current turns offshore at a mean latitude of 35.8°S (s^2 = 1.1°, range 4.8°), more frequently at 36°S, with a secondary peak between 33-34°S (Olsson et al. 1988).

The Falklands Current is the northern branch of the Subantarctic Front. It is 500 m thick, with a salinity between 33,9-34,9·% during winter, but as low as 33·% during summer. Its temperature ranges between 4-10°C during winter and reaches 14°C in the summer (Picard 1979). It is generated at the Drake Passage, and flows northwards along the Argentinian coast to the La Plata estuary, where it turns offshore in the Brazil-Falkland Confluence. The Falkland Current flows seawards from the coast at a mean latitude of 38.8°S (s^2 = 0.9°, range = 4.4°), with higher frequency at its mean (Olson et al. 1988).

Both currents, when they flow offshore, present distinct latitude ranges during the

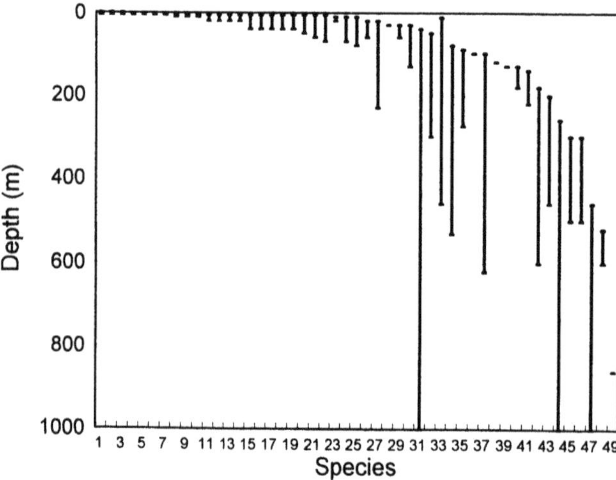

Figure 2. Shallow provinces of the Southwestern Atlantic, and Recent oceanographic process.

summer and winter seasons. During the winter, they flow offshore at smaller latitudes. During the summer such offshore movements are more southern. Furthermore, it was observed that the latitude of offshore movements of both currents are not coincident. Between the two currents there is a zone with waters of intermediate temperature with up to 300 km of extension (Legeckis and Gordon 1982; Olson et al. 1988; Pereira 1989).

Each of the two currents above hold a characteristic associated fauna, the southern related to cold waters and the northern with warm waters.

2.2 *Patterns of distributions of shallow water barnacles*

Based on the patterns of latidudinal distribution of the shallow water barnacle species, five faunal groups are observed. Of these, three show a pattern of short latitudinal range, and are geographically isolated (Fig. 3).

The first group indcludes species occurring between 0-24°S, all of which are restricted to the tropical zone (species 1-17 of Fig. 3). The second group is composed of species found along the subtropical zone, i.e., between 21-38°S (species 19-23). The third group is represented by the species from the temperate zone, occurring between 35°S and the southern tip of South America (56°S) (species 25-30). Besides the above groups, there is also a group of species occurring in both the tropical and subtropical zones (species 37-42), plus an additional one occurring in all the three zones (species 43).

The Tropical zone includes 17 species with distributions, in the Southwestern Atlantic, restricted to this zone. None of these species, however, is endemic to this zone. The distribution of these species in this area present some distinctions, probably due to the splitting of the Equatorial Current in its two branches and the influence of the discharges of rivers.

The Subtropical zone includes 5 species, all of which having their distributions at the Southwestern Atlantic restricted to it. Of the total, *Arcoscalpellum boubalocerus* and *Litoscalpellum henriquecostai* are endemic (Group II of Fig. 3). The composition of barnacle species from tropical continental shelf is very poorly known. Some species referred herein as subtropical may have a broader range to the north, specially *Octolasmis lowei*. This species occur in tropical zones of other provinces of the world. On the other hand, many disjunct patterns of distribution between Southwestern Atlantic and Northwestern Atlantic were detected for Decapoda Crustacea, probably determined by Pleistocene sea level changes (Melo 1985).

The Temperate zone presents 6 species, all of which with distributions at the Southwestern Atlantic restricted to this area. Five of these species also occur at the Southeastern Pacific in the area under influence of the Humbolt Current. *Hamatoscalpellum rathbunae* is endemic.

There are 12 species occurring also in the Tropical and Subtropical zones. Three of them, *Chthamalus bisinuatus, Fistulobalanus citerosum* and, *Megabalanus vesiculosus* are endemic from this area.

Only *Balanus improvisus* has a broad range, occurring throughout all three zones and in areas under the influence of both currents.

The patterns showed by the described groups suggest the occurrence of three biogeographic delimited areas in the Southwestern Atlantic. The arrangement, by lati-

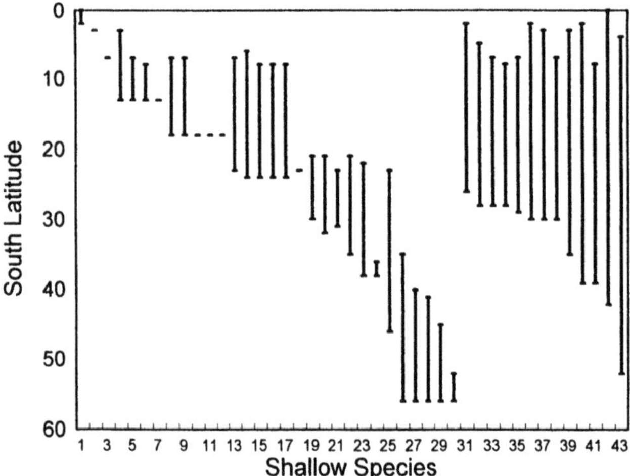

Figure 3. Latitudinal range of the barnacle species from the shallow (< 80 m) Southwestern Atlantic. Tropical province: 1) *B. calidus*; 2) *T. divisa*; 3) *B. subalbidus*; 4) *C. amaryllis*; 5) *L. dorsalis*; 6) *B. reticulatus*; 7) *M. declivis*; 8) *C. domingensis*; 9) *M. madreporarum*; 10) *O. hirtae*; 11) *A. cyathus*; 12) *C. paucicostata*; 13) *M. stultus*; 14) *C. floridana*; 15) *N. radiata*; 16) *B. eburneus*; 17) *O. hoeki*. Tropical-subtropical Transition Zone: 18) *T. floridana*. Subtropical province: 19) *O. lowei*; 20)*M. coccopoma*; 21) *L. henriquecostai*; 22) *B. spongicola*; 23) *A. boubalocerus*. Subtropical-temperate Transition Zone: 24) *B. glandula*. Temperate subprovince: 25) *H. rathbunae*; 26) *O. gibberum*; 27) *B. laevis*; 28) *A. psittacus*; 29) *E. kiingii*; 30)*C. scabrosus*. Tropical and Subtropical provinces: 31) *E. rhizophorae*; 32) *C. proteus*; 33) *C. galeata*; 34) *M. vesiculosus*; 35) *C. patula*; 36) *T. stalactifera*; 37) *M. tintinnabulum*; 38)*F. citerosum*; 39) *C. bisinuatus*; 40) *B. amphitrite*; 41) *B. venustus*; 42) *B. trigonus*. All three provinces: 43) *B. improvisus*.

tude of the barnacle species from each of the groups by latitudes (Fig. 4) illustrates the three faunal areas.

The tropical zone has a great species richness, but do not present any endemic species. All its components are also Northwestern Atlantic species. On the other hand, the tropical-subtropical group has three endemic species, which indicates some distinction of the Tropical South American fauna in relation to the Northwestern Atlantic. Due to this characteristic, I consider the Tropical zone as a faunal province for barnacles (Fig. 2).

The subtropical zone has fewer species than the above zone but has a rate of endemism of 40%, which also justifies the recognition of this area as a province (Fig. 2).

The temperate area of the Southwestern Atlantic has a barnacle fauna similar to that of the Southeastern Pacific (coasts of Chile and Peru), having a lower rate of endemism (17%). As a result, the Southwestern Atlantic temperate area is considered herein as a subprovince of the Magellean Province (Fig. 2).

The Northern limit of the distributions of species from the Tropical Province appears to be influenced by the distribution of the ecossystems. The Northern coast of Brazil (Fig. 2), which is under the influence of the northern branch of the Equatorial Current, is greatly influenced by the sediments discharged by the rivers. The dominant ecosystems in the province are mangroves and sand beaches, some of the latter

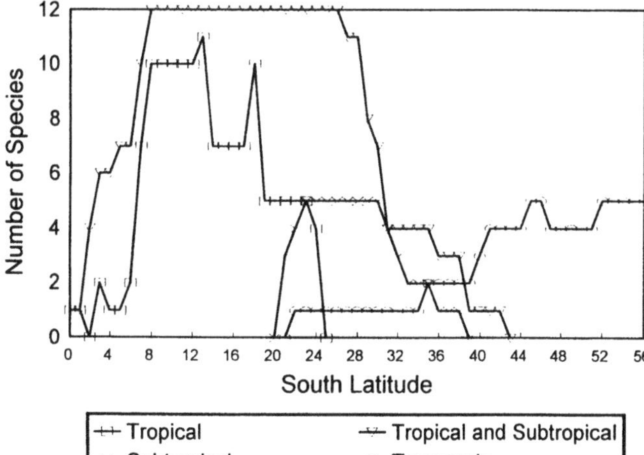

Figure 4. Number of barnacle species from shallow waters (< 80 m) of the Southwestern Atlantic by degree of latitude. The wide ranging *Balanus improvisus* is not included.

with arenitic deposits. There are few rocky coasts and no coral reefs. Few species of barnacles occur locally, and *Tetraclitella divisa* and *Balanus calidus* are characteristic of this area (Fig. 3). The first species has a circumtropical distribution, but is usually found at oceanic islands and the second is present untill North Carolina.

The tropical eastern coast of South America presents a greater diversity of environments such as mangroves, sand beaches, arenitic deposits, and coral reefs. In this area occur all the species related to reef corals, including cnidarian-associated barnacles (e.g. Pyrgomatidae, *Conopea*) (Fig. 3). *Chirona amaryllis* is the only species which also occurs at the northern and eastern tropical coast (Fig. 3). It is a recently introduced species (Young 1989b).

The northern limits of the distributions of tropical barnacle species appear to be divided in two areas. The first limit is between 5-7°S, where Recent coral reefs disappear together with the cnidarian-associated barnacles. The second limit appears to be between 2-3°S, where the influence of the discharge of the Amazon River is very strong, causing a reduction in salinity and an increase in sediment deposition (Fig. 2).

For the majority of the species, the southern distributional limit is at 18°S, although a southwards gradient from this latitude to 24°S is noted (Fig. 4). The latitude of 18°S is also the southern limit for coral reefs distribution, which in turn restrict a more southern distribution for some cnidarian-associated barnacles. Some coral species, however, have a more southern limit of distribution, thus allowing a southward range extension of their associated barnacles. All the cnidarian-associated barnacles, except *Ceratoconcha paucicostata*, have their southern limit of distribution superposed with their coral hosts.

The decrease of water temperature between 21-24°S, due to the presence of northward cold fronts and upwellings, characterizes this area as a transition zone between the tropical and the subtropical barnacle fauna (Fig. 2). In this area it is found the *Tetraclita floridana* a species at the present apparently restricted to this zone (Fig. 4).

The Fernando de Noronha Archipelago (4°S) and Rocas Atoll (4°S) have an impoverished tropical barnacle fauna. At the former occurs *Megabalanus tintinnabulum* and *Tetraclitella divisa*; at the latter, only the coral-barnacle *Ceratoconcha floridana* occurs. Trindade Island (20°S), situated off the coast of Espírito Santo State at the southern limit of the tropical fauna, has records of *Chthamalus bisinuatus* and *Lithotrya atlantica* Borradaile (= *L. dorsalis*).

The Northern limit of the distribution of subtropical barnacles is 21°S. To the north of such latitude, there is no upwelling and, therefore, water temperatures probably reach higher levels, a limiting factor for the development of subtropical species. The Subtropical province has strongly differentiated areas. In general terms, there is a gradient of lowering of water temperature of the Brazil Current as they flow southwards. In this way, a gradient in the distribution of the barnacle species from 28°S to the Subtropical Confluence, at 39°S, is shown.

The species occurring also in the Tropical and Subtropical provinces have their northern limits of distribution of coincident with the 1st and 2nd northern limits of the shallow tropical fauna. In the other hand, the southern limit of distribution is a little southern than the southern limit of the Shallow Subtropical fauna, reaching the 42°S (Figs 2 and 4).

The shallow tropical-subtropical and the subtropical faunas have their southern limits of distribution coinciding with the Subtropical Confluence. These species, therefore, are exclusively from waters of the Brazil Current in the Southwestern Atlantic (Fig. 2).

From 35-42°S, a transition zone between the Subtropical province and the Temperate subprovince is observed. This is due to the superposition of the subtropical, and the tropical-subtropical groups, with those species of the temperate group (Fig. 4). The introduced *Balanus glandula* (Spivak, pers. comm.) is found at this TZ, and has apparently not been able of expanding its range.

From 52-35°S, there is a northward gradient of the decrease in the number of species from Temperate subprovince, probably due the temperature increase of the Falkland Current as it flows northwards (Figs 2 and 4).

No species occurs in both the subtropical and temperate provinces.

The southern limits of distribution of barnacles of nearly all distributional groups are probably determined by temperature. Due to the cooling of the Brazil Current as it moves southwards, a southwards extinction gradient begining at 18°S and reaching the Subtropical Confluence (39°S) is noted. More stenothermic species have their southern limits of distribution at lower latitudes; those more resistent to cooler temperatures have a broader southwards range extension. Additionally, the northern limit of temperate and subtropical barnacle faunas may be determined by increasing water temperatures. Only the northern limit of distribution of tropical barnacle fauna may be due to other factors besides water temperature such as the ecosystems extinctions, high sedimentation and low salinities.

Both transition zones observed at the Southwestern Atlantic are very distinct from those observed by Newman (1979a) and Laguna (1994) in the Eastern Pacific. Newman (1979a) showed the presence of short-range endemics at the Californian Transition Zone, inferring that these endemic species were survives of a latitudinally compressed province or fugitive species which used the TZ as a refugium. Laguna (1994) reviewed the provincialism and transition zones in the Tropical Eastern Pacific and also observed several endemics, or restricted species, along the three transition zones he studied. When the distributional patterns of the barnacles from Southwestern Atlantic are compared to those from Eastern Pacific, two conspicuous distinctions can be noted: (1) There is a shorter latitudinal overlapping of the adjoining provinces, and (2) there is a small number of restricted species, none of them endemics.

Newman (1979a) based his hypothesis of a latitudinally compressed province on the high tectonic activity of the Californian coast. The South Atlantic continental margin, in contrast is a passive type and has a history of long geologic stability. Therefore, the Soutwestern Atlantic transition zones are probably not vestiges of extinct provinces. Furthermore, the absence of endemics shows that they were also inaccessible to serve as a refugium for fugitive species. The short latitudinal range of the transition zones in the Southwestern Atlantic may impede the occurrence of large number of restricted species. Besides, its long stable geologic history, and its consequent long time of physical stability, is probably the factor determining the small province species overlapping for those provinces.

Newman (1979a) used the percentual change and the Jaccard's coefficient of similarity to illustrate the change from the Oregonian to the Californian fauna, and

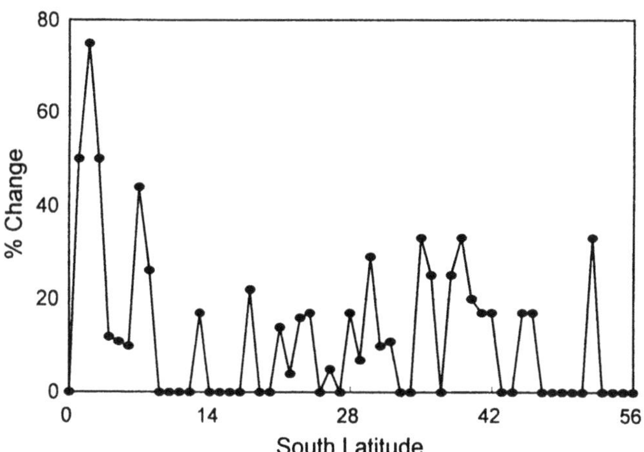

Figure 5. Percent of change of barnacle species from shallow waters (< 80 m) by degree of latitude.

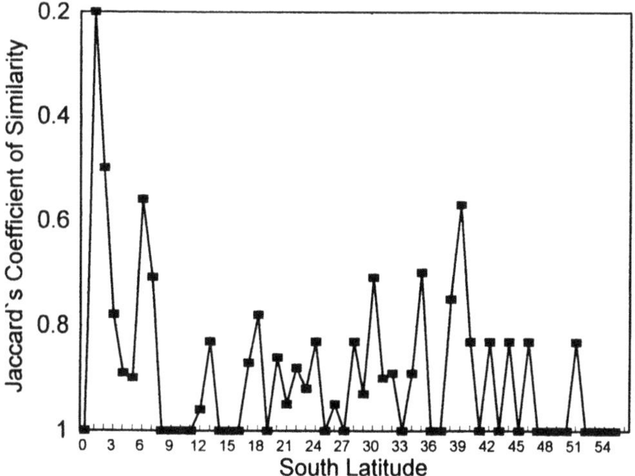

Figure 6. Jaccard's Coefficient of Similarity of barnacles from shallow water (< 80 m) by degree of latitude.

the TZ species pattern, along the latitudinal gradient. Both indices are presented in Figures 5 and 6 and are not much illustrative since both do not clearly shows the TZs. In relation to the accumulated number of species by latitude, the TZ are much better viewed in Figure 4. Only the Subtropical-Temperate Transition Zone along the 32-42°S may be evidenced by both indices. The gradient of extinction of the tropical and subtropical species is shown by the low and intermitent percent of change which is observed along the two provinces. The high index values for the low latitudes is due to the small number of species punctually distributed.

2.3 *The deep water circulation*

The behavior of the shallow water currents are relatively well elucidated (Lugeckis & Gordon 1982; Gordon & Greengrove 1986; Olson et al. 1988) but information on the deep-water circulation is scarce. Furthermore, the patterns of seasonal distribution of the currents exhibit great variability.

The encounter of the waters of the Brazil and Falkland currents near 38°S appears to have a distinct pattern at deeper waters. At such latitude, the Falkland Current penetrates beneath the waters of the Brazil Current and flows northwards reaching 29°S. This current provide waters for the subtropical temporary upwellings. North of 38°S, therefore, the cool deep waters occur at distinct depths determined largely by the influence of atmospheric circulation on the Brazil coast. The vertical and latitudinal distribution of deep water barnacle species of subtropical and temperate provinces exemplify the patterns of water submergence between 38-29°S. There is no information on deep water circulation north of 24°S.

2.4 *Distribution patterns of deep barnacle species*

Figure 7 shows the latitudinal distributions of the deep sea species. *H. rathbunae* and *O. gibberum*, both of which have been discussed before, are included because of their presence in deep waters as well. Lack of informations on the latitudinal distribution of its faunal components hinders the division of deep waters (80-1000 m) into provinces.

Despite the small amount of information, two interruptions on the faunal distribution may be observed between 23-24°S, and between 33-35°S. Such interruptions separate the tropical from the subtropical fauna, and the subtropical from the temperate one. The first transition zone (23-24°S) corresponds to the same latitude of the shallow Tropical-subtropical Transition Zone. Otherwise, the deep Subtropical-temperate Transition Zone (33-35°S) is only slightly displaced northwards when compared to the shallow one (35-42°S).

When the latitudinal and depth ranges of *H. rathbunae* are examined (Fig. 8) a peculiar pattern can be observed. It occurs at shallow water at the Temperate subprovince (44°S), also at 28-30°S, and again shallower at 24-26°S. This species appears to be a temperate species which occurrs along the area influenced by the Falklands current. It is found deeper at northern latitudes accompanying the northward submergence of this current beneath the warm Brazilian current. The shallower occurrences at 24-26°S is probably due to the emergence of the Falklands current as upwellings at these latitudes. Based on this hypothesis an overlap of subtropical and temperate fauna can be observed between 23-35°S.

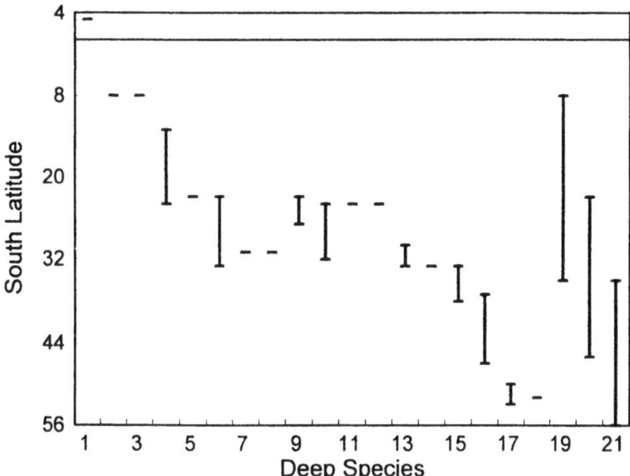

Figure 7. Latitudinal range of the barnacles species from deep (> 80 < 1000) Southwestern Atlantic. Depth distribution included after species. Tropical group: 1) *P. martini* (100 m); 2) *H. lankesteri* (80-275 m); 3) *V. flavidula* (120 m); 4) *W. aduncum* (without reference); 5) *P. kaempferi* (30-60 mm). Subtropical group: 6) *R. nexa* (50-300 m); 7) *A. portoricanum* (130 m); 8) *H. cornuta* (120 m). Subtropical or Temperate group!: 9) *V. idioplax* (80-540 m); 10) *P. inaequilaterale* (520-600 m); 11) *A. gibbosa* (180-600 m); 12)*Balanus* sp (130-180 m); 13) *V. caribbea* (200-460 m); 14)*Altiverruca* sp (300-500 m). Temperate group: 15) *A. triangulare* (460->1000); 16) *P. parallelogramma* (260->1000); 17) *W. scoresbyi* (140-200); 18) *B. bulata* (860); 19) *L. regina* (300-500); 20) *H. rathbunae* (10-460); 21) *O. gibberum* (40->1000).

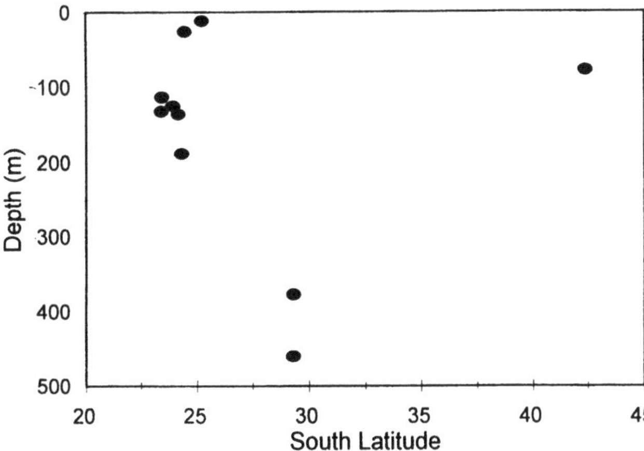

Figure 8. Distribution of *Hamatoscalpellum rathbunae* along latitude and depth gradients.

Some species with deep occurrences (species 9-14 of Fig. 7), and with distributions restricted to these latitudes cannot be placed without any assurance in either subtropical or temperate faunas.

At the present, it is premature to determine if these faunal groups reflects deep sea provinces because of both the small number of species and the small amount of information related to latitudinal distribution. All that should be emphasized is that two of the five species from the tropical group (species 1-5 of Fig. 7), none of the three species from the subtropical group (species 6-8), and three of the four from the temperate group (species 15-18) are endemics. Additionally, of the six species occurring exclusively between 23-35°S (species 9-14), only two are endemics. *L. regina* is the single species with a wide distribution, being a component of both the tropical and subtropical faunas (species 19).

Only from the southern tip of South America there are records of barnacles from more than 1000 m deep. Ten species were cited by Newman & Ross (1971) and Zevina (1990), all of them endemic to the area.

2.5 *Latitudinal patterns of species richness*

Abele (1982) reported a general inverse relationship between latitude and species richness for barnacles. This relationship appears not to occur along the Southwestern Atlantic. The data herein analysed were divided in two groups (one of shallow (< 80 m) and one of deep water species (80-1000 m)), to minimize the influence by the scarcity of deep samples at low latitudes. The latitudinal pattern of shallow water species shows a remarkable peak of species richness at 10-24°S (Fig. 9). From 10°S to the equator, species richness slowly decreases to about 8°S, from where species richness fastly decreases towards the equator. From 24°S to higher latitudes, there is the same decrease in number of species as that related by Abele (1982). Species richness decreases sharply from 20-34°S, from where, however, it decreases slower towards the southern tip of South America.

The greater species richness between 10-24°S appears to be a consequence of area and geologic history. At 18°S, the continental shelf shows a large extension, which supports a great habitat heterogeinity, including the more developed coral reefs of the South Atlantic. Many marine groups which occur in the Southwestern Atlantic have their greatest diversity at this area, e.g. corals and fishes (Laborel 1969; Nunan 1979, 1992). Besides the expansion of the continental shelf, this area is also characterized by having the continental break deeper than 130 m, a feature which made possible the maintenance of a relict population during Pleistocene sea level fluctuations (Laborel 1969; Melo 1985).

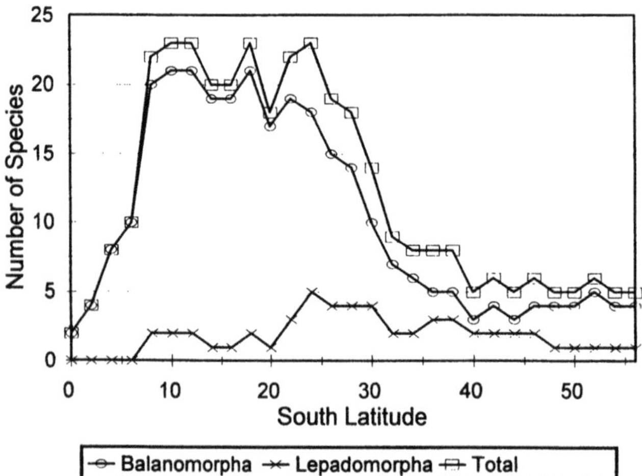

Figure 9. Patterns of latitudinal distribution in relation to species richness for shallow water species.

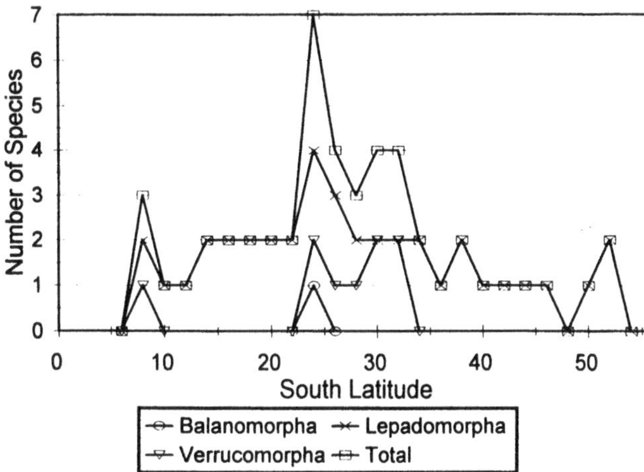

Figure 10. Patterns of latitudinal distribution in relation to species richness for deep water species.

Otherwise, the sharp decrease at low latitudes is probably also due to the influence of the Amazon river outflow, which causes a decrease in the available number of ecosystems. The inflexion of the graphic (Fig. 9) at 38°S represents the change of the tropical-subtropical fauna to the temperate fauna.

The deep water fauna of barnacles shows a greater species richness at 24°S (Fig. 10), the number of species presenting a similar decrease to higher latitudes as presented by the shallow water species. From 24°S to lower latitudes, the deep water curve is distinct from the shallow water group, with few species being found. The northern shelf of Brazil is poorly sampled, and very few records are known. On the other hand, the region between 20-30°S, with a large number of species, has the superposition of warm and cold water currents and presents a fauna related to both currents along a depth gradient, a characteristic which may account for the number of deep-sea species.

Newman & Ross (1971) and Spivey (1981) related a lepadomorph/balanomorph ratio decreasing markedly from high to low latitudes. Abele (1982) commented, however, that this was due to the increase in number of balanomorphh species instead of a change in number of lepadomorphs. The lepadomorph/balanomorph ratio showed a sharp decrease from 46°S to equator but also presented a decrease from 46 to 56°S (Fig. 11). In general, an expressive increase in number of species of balanomorph towards the equator was observed, except at 0°. More diversity is indeed expected (including the species richness component) in smaller than higher latitudes, a feature which is even more remarkable for balanomorph species. Attention has to be paid, however, to available habitat for this species. Balanomorphs are expressive in both the intertidal and shallow infralitoral zone, being present at all latitudes. At higher latitudes, however, they are influenced by the effects of low temperatures, and even freeze during the winter.

Otherwise, the lepadomorph species have an unimodal curve, with the greatest richness at 24°S. The majority of the lepadomorph species occurs along continental shelves and slopes showing small changes and having low climatic influence. In regard to the greater number of lepadomorphs at 25-35°S, it may be due to the superposing of the shallow warm and the deep cold currents, what has made possible the

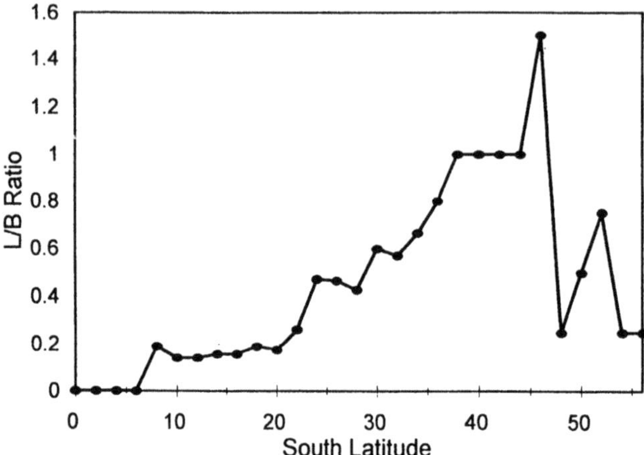

Figure 11. Lepadomorph/Balanomorph ratio along degree of latitude.

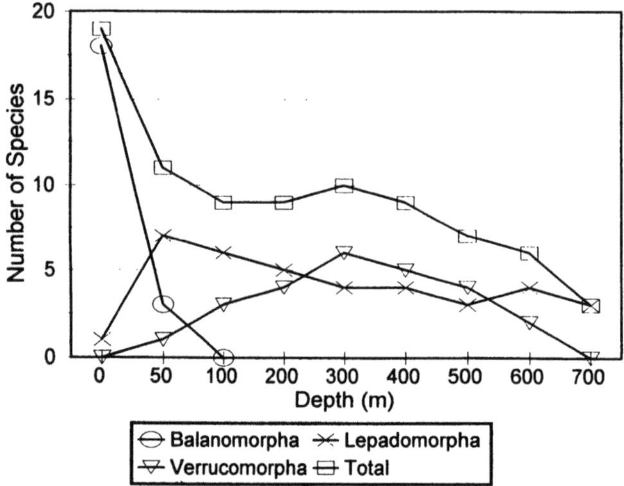

Figure 12. Patterns of depth distribution in relation to species richness for shallow water species.

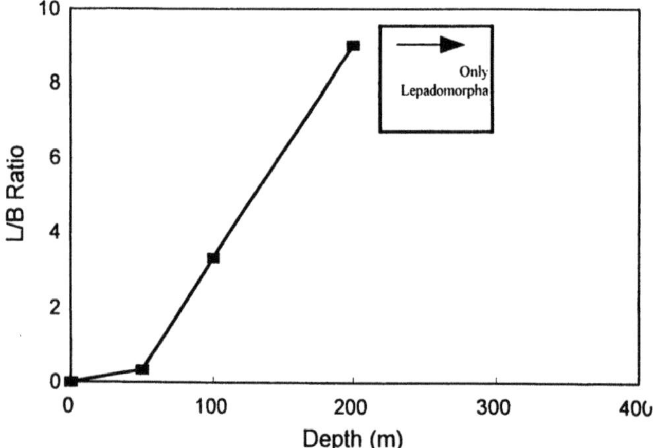

Figure 13. Lepadomorph/Balanomorph ratio along depth patterns.

presence of species associated with both currents. Otherwise, the variation in number of lepadomorph species, therefore, may be due exclusively to distinct sampling effort.

2.6 *Depth patterns of species richness*

There is an evident inverse relationship between depth and species richness, with the higher richness occurring at shallow litoral zone (0 m) (Fig. 12). This general pattern, however, is not observed for the three barnacle suborders, with only Balanomorpha following it. The Lepadomorpha and Verrucomorpha are poorly represented in shallow waters, their richness increasing, respectively, to 50 and 300 metres depth, from where they finally start to show a decrease characteristic of the general pattern. Due to these changes in composition along the depth gradient, it is remarkable the increase of the Lepadomorpha/Balanomorpha rate along depth gradient (Fig. 13).

3 BARNACLE FAUNA OF THE SOUTHWESTERN ATLANTIC

The information on the barnacle fauna of the Northwestern Atlantic and the Atlantic coast of Africa was compiled from the literature. For the Northwestern Atlantic, the data used were from Pilsbry (1953), Henry & McLaughlin (1975, 1986), Southward & Newman (1977), Zevina (1975, 1981, 1982, 1987b, 1990), Newman & Ross (1976), Weisbord (1977, 1979), Zullo (1979), Dando & Southward (1980), Spivey (1981), and Young (1989). For the Atlantic coast of Africa, the data was compiled from Barnard (1924), Stubbings (1967), Henry & McLaughlin (1975, 1986), Newman & Ross (1976), Dando & Southward (1979), and Zevina (1981, 1982, 1990).

In total, 163 species was compiled from the three areas (Table 1). The majority of them, 116 (71%), was recorded from only one of these areas. On the other hand, 15 species (9%) were recorded from all the three areas. Seven of the species are very common in fouling, and this recent pattern may be due to introductions made by man. Of the other eight species, half are shallow water forms, the other half being deep sea species wich are widely distributed throughout the Atlantic Ocean, some even in another oceans as well.

Table 1. Distribution patterns of barnacle species along Southwestern Atlantic, Northwestern Atlantic and Atlantic coast of Africa.

Species limited to the Southwestern Atlantic

Brochia bulata (Newman & Ross, 1971)
Compressoscalpellum smirnovi (Zevina, 1979)
Ornatoscalpellum gibberum (Aurivillius, 1892)
Litoscalpellum henriquecostai (Weber, 1960)
Hamatoscalpellum rathbunae (Pilsbry, 1907)
Meroscalpellum vinogradovae (Zevina, 1975)
Weltnerium aduncum (Aurivillius, 1894)
Weltnerium scoresbyi (Nilsson-Cantell, 1939)
Weltnerium speculum (Zevina, 1975)
Pilsbryiscalpellum parallelogramma (Hoek, 1883)
Amigdoscalpellum tortuosum (Zevina, 1975)
Arcoscalpellum acicularum (Newman & Ross, 1971)
Arcoscalpellum boubalocerus (Young, 1992)

Table 1. Continued.

Species limited to the Southwestern Atlantic

Arcoscalpellum imbricotectum (Newman & Ross, 1971)
Arcoscalpellum triangulare (Hoek, 1883)
Paralepas martini (Young, 1990)
Altiverruca quadrangularis (Hoek, 1883)
Altiverruca sp.
Chthamalus bisinuatus (Pilsbry, 1916)
Chthamalus scabrosus (Darwin, 1854)
Tetrachaelasma southwardi (Newman & Ross, 1971)
Chirona amaryllis (Darwin,1854)
Elminius kingii (Gray, 1831)
Balanus glandula (Darwin, 1854)
Balanus laevis (Bruguière, 1789)
Balanus sp.
Fistulobalanus citerosum (Henry, 1974)
Megabalanus coccopoma (Darwin, 1854)
Megabalanus vesiculosus (Darwin, 1854)
Austromegabalanus psittacus (Molina, 1782)

Species limited to the Northwestern Atlantic

Scalpellum gibbum (Pilsbry, 1907)
Litoscalpellum giganteum (Gruvel, 1902)
Verum hendersoni (Pilsbry, 1911)
Catherinum theorassi (Zevina, 1975)
Diceroscalpellum arietinum (Pilsbry, 1907)
D. diceratum (Pilsbry, 1907)
Anguloscalpellum gorgonophilum (Pilsbry, 1907)
Amigdoscalpellum mamillatum (Aurivillius, 1892)
A. svetlanae (Zevina, 1975)
Trianguloscalpellum pentacrinarum (Pilsbry, 1907)
T. pilosum (Zevina, 1975)
A. compositum (Zevina, 1975)
A. floccidum (Zevina, 1975)
Teloscalpellum antillarum (Pilsbry, 1907)
T. fedicovi (Zevina, 1975)
T. spicatum (Zevina, 1975)
Oxynaspis floridana (Pilsbry, 1953)
O. gracilis (Totton, 1940)
O. patens (Aurivillius, 1894)
Octolasmis americanum (Pilsbry, 1907)
O. antiguae (Stebbing, 1894)
O. aymonini (Lessona et Tapparone-Canefri, 1874)
O. dawsoni (Causey, 1960)
O. forresti (Stebbing, 1894)
O. geryonophila (Pilsbry, 1907)
O. muelleri (Coker, 1902)
Megalasma gracile (Hoek, 1883)
M. rectum (Pilsbry, 1907)
M. subcarinatum (Pilsbry, 1907)
Heteralepas belli (Gruvel, 1902)
H. luridas (Zevina, 1975)
Pagurolepas conchicola (Stubbings, 1940)
Verruca alba (Pilsbry, 1907)
V. calotheca (Pilsbry, 1907)
V. entobapta (Pilsbry, 1916)
V. floridana (Pilsbry, 1916)
V. xanthia (Pilsbry, 1916)
Cameraverruca euglypta (Pilsbry, 1907)
Metaverruca coraliophila (Pilsbry, 1916)
M. lepista (Zevina, 1985)
Altiverruca aves (Zevina, 1985)
A. bicornuta (Pilsbry, 1916)
A. darwini (Pilsbry, 1916)
A. gira (Zevina, 1985)
A. hoeki (Pilsbry, 1907)
A. longicarinata (Gruvel, 1902)
A. rathbuniana (Pilsbry, 1916)
Catophragmus imbricatus (Sowerby, 1827)
Chthamalus angustitergum (Pilsbry, 1916)
Aaptolasma americana (Pilsbry, 1916)
Armatobalanus circe (Kolosváry, 1947)
Chirona hameri (Ascanius, 1767)
Conopea merrili (Zullo, 1966)
Semibalanus balanoides (Linnaeus, 1767)
Balanus balanus (Linnaeus, 1758)
B. crenatus (Bruguiere, 1789)

Species limited to the Atlantic coast of Africa

Smilium acutum (Hoek, 1883)
Euscalpellum renei (Gruvel, 1902)
Pollicipes pollicipes (Gmelin, 1790)
Scalpellum scalpellum (Linnaeus, 1767)
Verum carinatum (Hoek, 1883)
Catherinum albatrossianum (Pilsbry, 1907)
C. trapezoideum (Hoek, 1907)
Arcoscalpellum kamenskae (Zevina, 1990)
A. michelottianum (Seguenza, 1876)
Oxynaspis celata (Darwin, 1851)
Octolasmis nierstraszi (Hoek, 1907)
O. tridens (Aurivillius, 1894)

Table 1. Continued.

Species limited to the Atlantic coast of Africa	
Poecilasma crassum (Gray, 1848)	*Tetraclita squamosa* (Bruguière, 1789)
Ibla atlantica (Stubbings, 1967)	*Solidobalanus fallax* (Broch, 1927)
Verruca striata (Gruvel, 1902)	*Acasta striata* (Gruvel, 1901)
A. galkini (Zevina, 1990)	*Conopea calceola* (Ellis, 1758)
Pachylasma giganteum (Philippi, 1836)	*Megatrema anglica* (Sowerby, 1823)
Euraphia aestuarii (Stubbings, 1963)	*B. perforatus* (Bruguière, 1789)
C. dentatus (Krauss, 1848)	*B. poecilotheca* (Krüger, 1911)
C. montagui (Southward, 1976)	*M. tulipiformis* (Ellis, 1758)
C. stellatus (Poli, 1791)	*M. zebra* (Darwin, 1854)
Species common to the Southwestern and Northwestern Atlantic	
Lithothrya dorsalis (Ellis & Solander, 1786)	*Newmanella radiata* (Bruguière, 1789)
Litoscalpellum regina (Pilsbry, 1907)	*Tetraclita floridana* (Pilsbry, 1916)
Verum idioplax (Pilsbry, 1907)	*Tetraclita stalactifera* (Lamarck, 1818)
Arcoscalpellum portoricanum (Pilsbry, 1907)	*Membranobalanus declivis* (Darwin, 1854)
Oxynaspis hirtae (Totton, 1940)	*Conopea galeata* (Linnaeus, 1771)
Poecilasma inaequilaterale (Pilsbry, 1907)	*Ceratoconcha domingensis* (Moullins, 1866)
Heteralepas lankesteri (Gruvel, 1900)	*Ceratoconcha floridana* (Pilsbry, 1931)
Verruca flavidula (Pilsbry, 1916)	*Ceratoconcha paucicostata* (Young, 1989)
Verruca caribbea (Pilsbry, 1916)	*Megatrema madreporarum* (Bosc, 1801)
Rostratoverruca nexa (Darwin, 1854)	*Balanus calidus* (Pilsbry, 1916)
Euraphia rhizophorae (Oliveira, 1940)	*Balanus subalbidus* (Henry, 1974)
Chthamalus proteus (Dando & Southward, 1980)	*Megabalanus stultus* (Darwin, 1854)
Species common to the Southwestern Atlantic and Atlantic coast of Africa	
Weltnerium campestrum (Zevina, 1975)	*Balanus spongicola* (Brown, 1844)
Amigdoscalpellum vitreum (Hoek, 1883)	
Species common to the Northwestern Atlantic and Atlantic coast of Africa	
Megalasma annandalei (Pilsbry, 1907)	*Chthamalus fragilis* (Darwin, 1854)
M. hamatum (Calman, 1919)	*Fistulobalanus pallidus* (Darwin, 1854)
Paralepas minuta (Philippi, 1836)	
Species common to the three studied areas	
Octolasmis hoeki (Stebbing, 1895)	*Balanus improvisus* (Darwin, 1854)
Octolasmis lowei (Darwin, 1851)	*Balanus trigonus* (Darwin, 1854)
Poecilasma kaempferi (Darwin, 1851)	*Balanus amphitrite* (Darwin, 1854)
Heteralepas cornuta (Darwin, 1851)	*Balanus eburneus* (Gould, 1841)
Altiverruca gibbosa (Hoek, 1883)	*Balanus reticulatus* (Utinomi, 1967)
Chelonibia patula (Ranzani, 1818)	*Balanus venustus* (Darwin, 1854)
Tetraclitella divisa (Nilsson-Cantell, 1921)	*Megabalanus tintinnabulum* (Linnaeus, 1758)
Acasta cyathus (Darwin, 1854)	

Based on the species present in two of the three areas studied, it was observed a greater similarity between the barnacle faunas of the Southwestern Atlantic and Northwestern Atlantic, with 24 (15%) species in common between the two areas. Six species are common from the tropical-subtropical intertidal zone and show a similar zonation at the South as much as the Northwestern Atlantic. There is only one species of *Chthamalus* endemic to the Southwestern Atlantic and an additional one to the Northwestern Atlantic (Dando & Southward 1980). Nine of the species are associated to reef organisms or restricted to coral reefs ecosystems, which are absent from the eastern Atlantic. The eastern Atlantic has an impoverished fauna of reef barnacles, with only *Megatrema anglicum* being recorded from ahermatypic corals in the area, despite a large number of fossil coral-barnacles (Newman & Ladd 1974, Newman & Ross 1970, Newman 1992). The remaining species are all deep water tropical species. A relationship, therefore, is evident between the barnacle faunas of the South and Northwestern Atlantic, and is valid for species from the intertidal to the deep sea zones.

The only 3 species (1,8%) shared between the Southwestern Atlantic and the western coast of Africa show no similarity between these two areas. There are no shared species from the intertidal zone, and two of the species are from the deep sea. Only one (*Balanus spongicola*) is a shelf species, but this is a taxon which has a characteristic pattern of distribution (see below).

The faunas of the Northwestern Atlantic and coast of West Africa also showed no similarity, since they only present five species (3,1%) in common. Of these, two are intertidal species which taxonomic status have been questioned: *Fistulobalanus pallidus*, previously recorded as an amphi-Atlantic, had its records from the Southwestern Atlantic proved by Henry (1974) to represent *F. citerosum*. *Chthamalus fragilis* has few records from West Africa (Broch 1927; Stubbings 1967) but is common in the Northwestern Atlantic (Dando & Southward 1980). Similarly, they may represent two closely similar but distinct species, which have been erroneously diagnosed, as is common with members of this genus (Dando & Southward 1979, 1980). The remaining three species are deep sea epizoants, of which few records are known.

As suggested above, there is a remarkable relationship between the tropical barnacle faunas of the Northwestern and Southwestern Atlantic but no relationship between either with that of the west coast of Africa. Furthermore, many of the Western Atlantic tropical species are closely related to species of the tropical East Pacific (Southward & Newman 1977), thus suggesting a closer relationship between the Western Atlantic and the eastern Pacific faunas. The closure of the Panamic isthmus was responsible for the differentiation of populations from both sides, and nowadays only *Conopea galeata* occurs both in the tropical Western Atlantic and eastern Pacific.

The majority of the Recent species appears to bear no relationship to the formation of the Atlantic ocean, even at the generic level. This observation, however, has to be regarded with caution since many genera, especially those of the Lepadomorpha, appears not to be monophyletic. Their diagnostic characters are very feeble, and do not represent synapomorphies. Many balanomorph genera, on the other hand, have a world wide distribution, with a few relict species showing a punctual distribution. Such feature makes it difficult to identify patterns of evolution.

The southern barnacle fauna of the Southwestern Atlantic is associated with the southern oceans, which in turn are under the influence of the West Wind Drift regime (Newman 1979b). All the nine genera limited to the Southwestern Atlantic occurs at the southern part of South America.

The North Atlantic has eleven genera which only occur there. One (*Semibalanus*) has a boreal distribution, but the other ten occur at the West Indies. One of these, *Cathophragmus*, is represented by a single species (*C. imbricatus*), which is considered as a relict species derived from an ancient and more generalized stock (Newman & Ross 1976). The large number of genera of the tropical Northwestern Atlantic that are absent from the Southwestern Atlantic is remarkable. Probably the greater number of islands, larger perimeter and more vicariant events may have lead to this richness and a greater possibility of survival of relict species.

Three of the six genera limited to the west coast of Africa (*Pollicipes, Ibla*, and *Pachylasma*) probably represent relict taxa which nowadays include species with remarkable allopatric distribution. The species of *Pachylasma* occur throughout the Indo-Pacific and elsewhere only at the southern coast of South Africa, a pattern which may represent a recent Atlantic invasion. The genus *Ibla* also has an Indo-Pacific distribution, being represented elsewhere only by *I. atlantica* in Sierra Leone (Stubbings, 1967). *Pollicipes* is a Tethyan relict, its species occurring at the eastern Pacific and eastern Atlantic (Newman & Killingley 1985; Newman 1991, 1992). The other genera only include deep sea species.

4 GENERAL PATTERN OF BARNACLES IN SOUTHWESTERN ATLANTIC

The barnacle fauna of the Southwestern Atlantic includes 62 species, plus ten more found only at depths greater than 1000 metres.

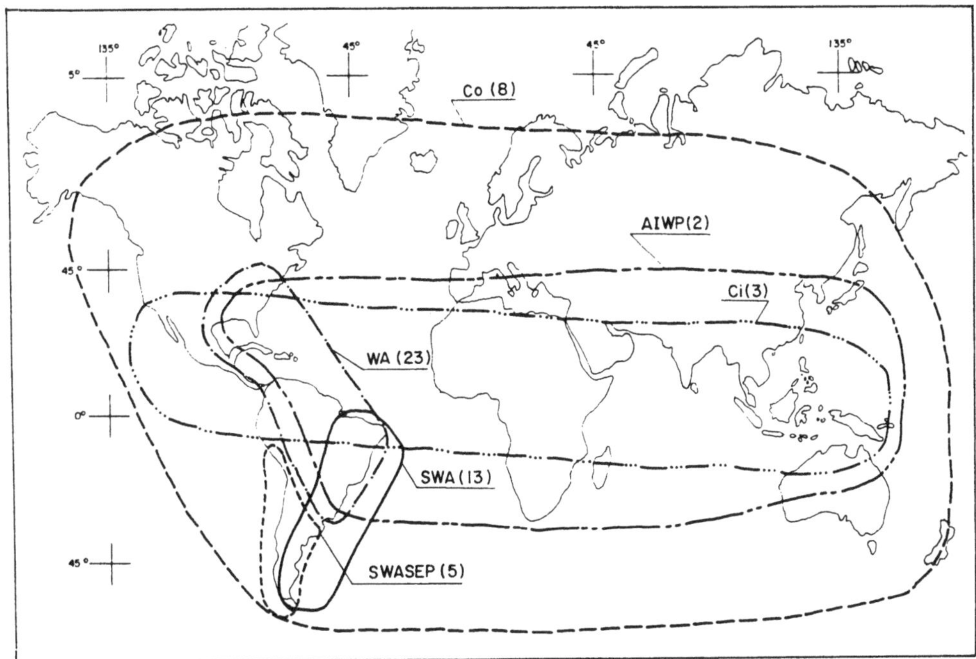

Figure 14. World distribution patterns of barnacles occurring at the Southwestern Atlantic. AIWP) Atlantic, Indo-West Pacific; Ci) Circumtropical; Co) Cosmopolite; SWA) Southwestern Atlantic; SWASEP) Southwestern Atlantic and Southeastern Pacific; WA) Western Atlantic.

4.1 *The general patterns*

Twelve generalized patterns were detected in relation to the South Atlantic barnacle fauna (Figs 14 and 15). The majority of the species had a distribution in five patterns: a) Western Atlantic, b) Southwestern Atlantic, c) Cosmopolitan, d) Southwestern Atlantic and Southeastern Pacific and, e) Circumtropical. Patterns a) and d) emphacize the relation of the tropical species of North and Southwestern Atlantic and the relation of temperate species with south oceans. The remaining patterns include less than two species each and may reflect introductions, cases of presence of sibling species, or misidentifications.

4.2 *Introduced species*

Many species which are found nowadays at the Southwestern Atlantic represent recent introductions. Others, however, may have been introduced in times prior to any knowledge of the local barnacle fauna. Such species, therefore, may present anomalous distribution patterns or, alternatively, may have broad ranges being cosmopolitan or even circumtropical species.

In relation to species with broad ranges (cosmopolitan or circumtropical), at the present it can be inferred that *Balanus trigonus, B. improvisus, B. amphitrite* and

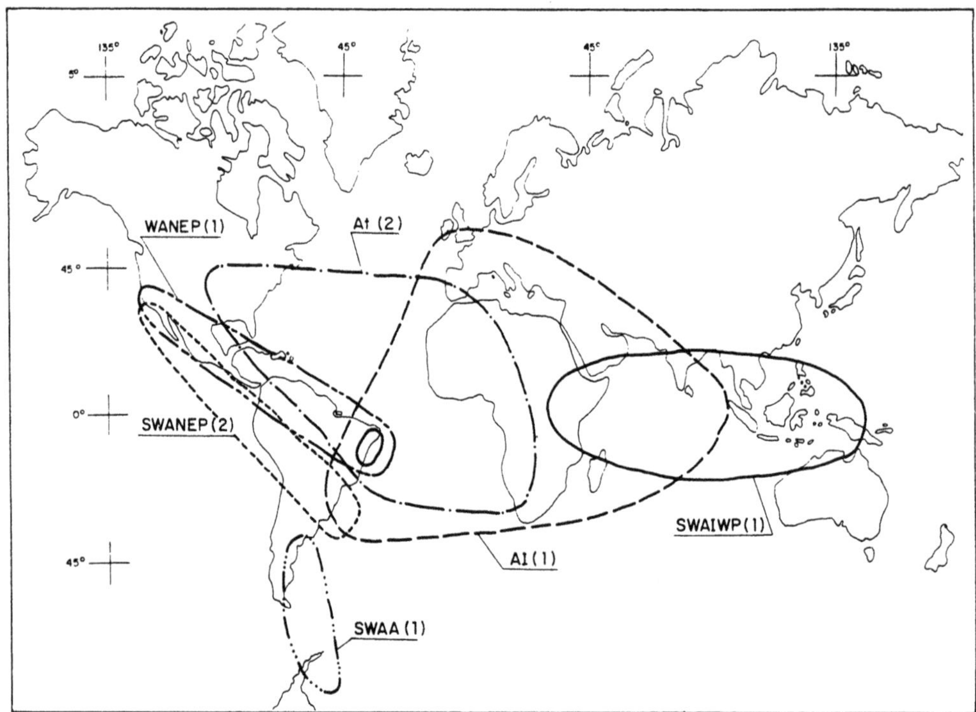

Figure 15. World distribution patterns of barnacles occurring at the Southwestern Atlantic (cont.). AI) Atlantic (except Northwestern Atlantic) and Indian; At) Atlantic; SWAA) Southwestern Atlantic and Antarctic; SWAIWP) Southwestern Atlantic and Indo-West Pacific; SWANEP) Southwestern Atlantic and Northeastern Pacific; WANEP) Western Atlantic and Northeastern Pacific.

Megabalanus tintinnabulum are cases of introduction in the Southwestern Atlantic. These species are commonly associated with fouling, and probably their ranges were expanded through human activities. Zullo (1992) commented on the probability of *B. trigonus* being introduced recently in the Northwestern Atlantic because of the lack of records of this species in the region. Similarly, *Balanus reticulatus* is a recent introduction to the Atlantic fauna that is, at the present, expanding its range in the Southwestern Atlantic. Hundreds of barnacle samples from the northeastern coast of Brazil collected between 1980-1984 have not recorded this species. Farrapeira-Assunção (1990) cited *B. reticulatus* from Recife (8°S) and by 1992 it was found at Todos os Santos Bay (13°S) at Bahia (pers. obs.). On the other hand, Newman & Ross (1976) cited *Balanus eburneus*, as originaly from Western Atlantic distribution and introduced at many new localities of the world. This species at Southwestern Atlantic have a very restricted distribution usually being found at polluted bays with harbors, which suggest to me to be an introduced species.

The Southwestern Atlantic/Northeastern Pacific pattern of barnacle distribution are represented by two undoubtly introduced species – *Balanus glandula* found around Rio de La Plata estuary (L'Hoste, pers. comm.), and *Megabalanus coccopoma*, nowadays occurring throughout the Subtropical province (Young 1994).

The species *Chirona amaryllis*, nowadays found along the tropical Brazilian coast and at the Indo-West Pacific, has been recently introduced to the former area. It was detected in 1982 at Piauí State (Young, 1989), in 1990 at Pernambuco State (Farrapeira-Assunção, 1990), and in 1992 at Bahia State (pers. obs.).

4.3 *Sibling species or misidentifications*

Some crustacean-associated barnacles have their specific status subject to debate, mainly due to descriptions usually based on variable external characters. Some of such species are found commonly at the Southwestern Atlantic and show uncommon patterns. Only *Octolasmis hoeki* and *Poecilasma kaempferi* show an Atlantic pattern of distribution. *Chelonibia patula* together with *B. venustus*, occurs throughout the Atlantic and the Indo-West Pacific.

The gorgonian-associated barnacle *Conopea galeata* is the only species which occurs throughout the Western Atlantic and Northeastern Pacific. Such pattern may reflect a vicariant event related to the closure of the Panama isthmus or, alternatively, may represent a group of sibling species. Recent studies of the *Euraphia rhizophorae* group, with identical distribution, proved to represent two species: one Atlantic – *E. rhizophorae* ss – and one Pacific – *E. eastropacencis* Laguna (Laguna, 1987). The detailed study of *C. galeata* specimens from Atlantic and Pacific may prove to be two sibling species.

The cosmopolitan species *Octolasmis lowei* and *Altiverruca gibbosa* have their specific status questionable, and may actually represent a group of sibling species, considering the variability of their diagnostic characters.

In another fashion, the barnacles associated with other animals may have their distribution patterns largely influenced by their hosts, a situation which may explain anomalous patterns of distribution.

Intriguing is the distribution of *Balanus spongicola*, found along the Eastern Atlantic and the subtropical part of the Southwestern Atlantic. The possiblity of introduction has to be discarded, since it only occurs at depths greater than 10 m. Furthermore, it is not found in fouling communities. This unexpected pattern is also observed in some Miocene taxa (e.g., corals of genus *Mussismilia*) which are extinct in Europe but are still represented at the Southwestern Atlantic (Laborel 1969). It may be a relict species which became extinct at Caribbean region. This hyothesis was discussed by Newman (1991) for same groups of Crustacea.

Finally, only a single species occurs at the Southwestern Atlantic and the Antarctic – *Arcoscalpellum triangulare* – which is a deep water species.

ACKNOWLEDGMENTS

I thank Ricardo Capitoli (Fundação Universidade do Rio Grande), Ana Maria Setúbal Vanin (Instituto Oceanográfico of Universidade de São Paulo), Yoko Wakabara (Instituto Oceanográfico of Universidade de São Paulo) and, Flavio Costa Fernandes (Instituto de Estudos Almirante Paulo Moreira), for the loan of some of the barnacle samples and, William Newman (Scripps Institution of Oceanography), Gustavo Nunan (Museu Nacional, Universidade Federal do Rio de Janeiro), Frederick Schram (Institute for Systematics and Population Biology, University of Amsterdam), Jens Hoeg (University of Copenhagen) and, an anonymous reviewer for bibliography and several useful comments on the manuscript. I also acknowledge the financial support from CNPQ.

REFERENCES

Abele, L.G. 1982. Biogeography: 241-304. In: L.G. Abele ed. *The Biology of Crustacea, 1, Systematics, the Fossil Record and Biogeography*. New York: Academic Press, New York.

Aurivillius, C.W.S. 1892. Neue Cirripedien aus dem Atlantischen, Indischen und Stillen Ocean. *K. Vet.-Akad. Forhandl. Stockholm* 3: 123-134

Barnard, K.H. 1924. Contributions to the crustacean fauna of South Africa. Cirripedia. *Ann. S. Afr. Mus.* 20: 1-103.

Briggs, J.C. 1974. *Marine Zoogeography*. New York: McGraw-Hill.

Broch,H. 1927. Studies on Moroccan Cirripeds (Atlantic coast). *Bull. Soc. Sci. nat. Maroc* 7: 11-38.

Calman, W.T. 1918. On barnacles of the genus *Scalpellum* from deep-sea telegraph-cables. *Ann. Mag. nat. Hist.* (9)1: 96-124.

Dando, P.R. & D.J. Crisp 1979. Enzyme variation in *Chthamalus stellatus* and *Chthamalus montagui* (Crustacea: Cirripedia): evidence for the presence of *C. montagui* in the Adriatic. *J. Mar. Biol. Ass. UK* 59: 307-320.

Dando, P.R. & A.J. Southward 1980. A new species of *Chthamalus* (Crustacea:Cirripedia) characterized by enzime electrophoresis and shell morphology: with a revision of others species of *Chthamalus* from the Western shores of the Atlantic Ocean. *J. Mar. Biol. Ass. UK* 60: 787-831.

Darwin, C. 1851. *A Monograph on the Subclass Cirripedia, with figures of all the species. The Lepadidae; or pedunculate cirripedes*. London: Ray Society.

Farrapeira-Assunção, C.M. 1990. Ocorrência de *Chirona (Striatobalanus) amaryllis* Darwin, 1854 e de *Balanus reticulatus* Utinomi, 1967 (Cirripedia, Balanomorpha) no Estado de Pernambuco. *Res. XVII Congr. bras. Zool.*: 7.

Gordon, A.L. & C.L. Greengrove 1986. Geostrophic circulation of the Brazil-Falkland confluence. *Deep-Sea Res.* 33: 573-585.

Henry, D.P. 1974 [1973]. Description of four new species of the *Balanus amphitrite* complex (Cirripedia, Thoracica). *Bull. Mar. Sci., Miami*, 23: 964-1001.

Henry, D.P. & P.A. McLaughlin 1975. The barnacles of the *Balanus amphitrite* complex (Cirripedia, Thoracica). *Zool. Verh., Leiden*, 141: 1-254, 22 pl.

Henry, D.P. & P.A. McLaughlin 1986. The Recent species of *Megabalanus* (Cirripedia: Balanomorpha) with special emphasis on *Balanus tintinnabulum* (Linnaeus) *sensu lato*. *Zool. Verh., Leiden*, 235: 1-69.

Hoek, P.P.C. 1883. Report on the Cirripedia colected by H.M.S. Challenger during the years 1873-76. *Rept. Sci. Res. Voyage H.M.S. Challenger, Zool.*, part 25, 8: 1-169, 13 pl.

Lacombe, D. & W. Monteiro 1974. Balanídeos como indicadores de poluição na Baía de Guanabara. *Revta bras. Biol.*, Rio de Janeiro, 34: 633-644.

Lacombe, D. & E.F. Rangel 1978. Cirripédios de Arraial do Cabo, Cabo Frio. *Publções Inst. pesq. mar.*, Rio de Janeiro, 129: 1-12.

Laborel, J. 1969. Les peuplements de madréporaires des côtes tropicales du Brésil. *Ann. Univ. Abidjan*, (E) 2: 1-260.

Laguna, J.E. 1988. *Euraphia eastropacensis* (Cirripedia, Chthamaloidea), a new species of barnacle from the Tropical Eastern Pacific: Morphological and electrophoretic comparisons with *Euraphia rhizophorae* (de Oliveira) from the Tropical Western Atlantic and molecular evolutionary implications. *Pac. Sci.* 41: 132-140.

Laguna, J.E. 1990. Shore barnacles (Cirripedia, Thoracica) and a revision of their provincialism and transition zones in the Tropical Eastern Pacific. *Bull. mar. Sci.* 46: 406-424.

Lahille, F. 1910. Los Cirripedios en la Argentina. *Revta Jard. zool. B. Aires*, (2)6: 69-89.

Legeckis, R. & A. Gordon 1982. Satellite observations of the Brazil and Falkland currents - 1975 to 1976 and 1978. *Deep-Sea Res.* 29: 375-401.

Melo, G.A.S. de. 1985. *Taxonomia e padrões distribucionais e ecológicos dos Brachyura (Crustacea: Decapoda) do litoral Sudeste do Brasil.* Dr. Sci. thesis, Universidade de São Paulo, 216 pp, 32 figs, 27 tabs.

Newman, W.A. 1979a. Californian Transition Zone: Significance of short-range endemics. Gray, J. & A.J. Boucot eds. *Historical biogeography, plate tectonics, and the changing environment*, pp. 399-416, Corvallis: OregonState Univ.

Newman, W.A. 1979b. On the biogeography of balanomorph barnacles of the Southern Ocean including new Balanid taxa; a subfamily, two genera and three species. *Proc. Int. Symp. Mar. Biogeogr. Evol. S. Hemi.* 1: 279-306, N.Zeal. DSIR Info. ser. 137.

Newman, W.A. 1991. Origins of southern hemisphere endemism, especially among marine Crustacea. *Mem. Queensl. Mus.* 31: 51-76.

Newman, W.A. 1992. Biotic cognates of Eastern Boundary conditions in the Pacific and Atlantic: Relicts of Tethys and climatic changes. *Proc. San Diego Soc. Nat. Hist.* 16: 1-7.

Newman, W.A. & J.S. Killingley 1985. The north-east Pacific intertidal barnacle *Pollicipes polymerus* in India? A biogeographical enigma elucidated by ^{18}O fractionation in barnacle calcite. *J. Nat. Hist.* 19: 1191-1196.

Newman, W.A. and H.S. Ladd 1974. Origin of coral-inhabiting balanids (Cirripedia, Thoracica). *Verhandl. Naturf. Ges. Basel* 84: 381-396.

Newman, W.A. & A. Ross 1971. Antarctic Cirripedia. *Antarct. Res. Ser.* 14: 1-257.

Newman, W.A. & A. Ross 1976. Revision of the balanomorph barnacles; including a catalog of species. *Mem. San Diego Soc. nat. Hist.* 9: 1-108.

Nilsson-Cantell, C.A. 1927. Some barnacles in the British Museum (Nat. Hist.). *Proc. Zool. Soc. Lond.* 1927: 743-790, 1 pl.

Nilsson-Cantell, C.A. 1930. Thoracic cirripeds collected in 1925-1927. *Discovery Rep.* 2: 223-260, 1 pl.

Nilsson-Cantell, C.A. 1938. Cirripedes from the Indian Ocean in the collection of the Indian Museum, Calcutta. *Mem. Ind. Mus.* 13: 1-81.

Nilsson-Cantell, C.A. 1939. Thoracic cirripedes collected in 1925-1936. *Discovery Rep.* 18: 223-238.

Nunan, G. 1979. *The zoogeographic significance of the Abrolhos area as evidenced by fishes.* M. Sc. thesis, University of Miami, viii + 146pp.

Nunan, G. 1992. *Composition, species distribution and zoogeographical affinities of the Brazilian reef-fishes fauna.* Ph.D. thesis, University of Newcastle upon Tyne, xx + 584pp.

Oliveira, L.P.H. de. 1940. Sobre uma nova espécie de crustáceo *Chthamalus rhizophorae* n. sp. *Mem. Inst. Oswaldo Cruz, Rio de Janeiro,* 35: 379-380, 1 pl.

Oliveira, L.P.H. de. 1941. Contribuição aos crustáceos do Rio de Janeiro: Sub-ordem 'Balanomorpha' (Cirripedia: Thoracica). *Mem. Inst. Oswaldo Cruz, Rio de Janeiro,* 36: 1-31, 11 pl.

Olsson, D.B., G.P. Podestá, R.H. Evans & O.T. Brown 1988. Temporal variations in the separation of Brazil and Malvinas currents. *Deep-Sea Res.* 35: 1971-1990.

Pereira, C.S. 1989. Seasonal variability in the coastal circulation on the Brazilian continental shelf (29°S-35°S). *Contin. Shelf Res.* 9: 285-299.

Picard, G.L. 1979. *Descriptive physical oceanography.* Oxford: Pergamon.

Pierre, C., C. Vergnaud-Grazzini & J.C. Faugeres 1991. Oxygen and carbon isotope tracers of the water masses in the Central Brazil Basin. *Deep-Sea Res.* 38: 597-606.

Pilsbry, H.A. 1907. The barnacles (Cirripedia) contained in the collections of the U.S. National Museum. *Bull U.S. Natn. Mus.* 60: 1-122, 11 pl.

Pilsbry, H.A. 1916. The sessil barnacles (Cirripedia) contained in the collections of the U.S. National Museum; including a Monograph of the American species. *Bull. U.S. Natn. Mus.* 93: 1-366.

Pilsbry, H.A. 1953. Notes on Floridan barnacles (Cirripedia). *Proc. Acad. Nat. Sci. Philad.* 105: 13-28, 2 pl.

Ross, A. 1975. *Heteralepas cornuta* (Darwin) in the Eastern Pacific abyssal fauna (Cirripedia, Thoracica). *Crustaceana* 28: 17-20.

Southward, A.J. & W.A. Newman 1977. Aspects of the ecology and biogeography of the intertidal and shallow-water balanomorph Cirripedia of the Caribbean and adjacent sea-areas. *Fish. Rep. F.A.O.,* Washington, 200: 407-425.

Spivey, H.R. 1981. Origins, distribution, and zoogeographic affinities of the Cirripedia (Crustacea) of the Gulf of Mexico. *J. Biogeogr.* 8: 153-176.

Stubbings, H.G. 1967. The cirriped fauna of tropical West Africa. ,*Bull. Br. Mus. Nat. Hist. Zool.* 15: 229-319, 1 pl.

Tsuchiya, M. 1985. Evidence of a double-cell subtropical gyre in the South Atlantic Ocean. *J. Mar. Res.* 43: 57-65.

Weber, L. 1960. Um novo Cirripédio Lepadomorfo da costa brasileira. *Scalpellum henriquecostai* n. sp. *Publçõe) Centro Est. zool.* 8: 1-7.

Weisbord, N.E. 1977. Scalpellid barnacles Cirripedia of Florida and of surrounding waters. *Bull. Amer. Paleon.* 72: 235-311.

Weisbord, N.E. 1979. Lepadomorph and Verrucomorph barnacles (Cirripedia) of Florida and adjacent waters, with an addendum on the Rhizocephala. *Bull. Amer. Paleont.* 76: 1-156.

Weltner, W. 1897. Verzeichnis der bisher beschriebenen recenten Cirripedienarten. Mit Angabe der im berliner Museum vorhandenen Species und ihrer Fundorte. *Arch. Naturgesch., Berlin* 1: 227-280.

Weltner, W. 1898. Cirripedien. *Hamburg. Magalhaens. Sammelreise* 1898: 1-15

Young, P.S. 1988. Recent cnidarian-associated barnacles (Cirripedia, Balanomorpha) from Brazilian coast. *Revta bras. Zool.* 5: 353-369.

Young, P.S. 1989a. *Ceratoconcha paucicostata,* a new species of coral-inhabiting barnacle (Cirripedia, Pyrgomatidae) from the Western Atlantic. *Crustaceana* 56: 193-199.

Young, P.S. 1989b. Establishment of an Indo-Pacific barnacle in Brazil. *Crustaceana* 56: 212-214.

Young, P.S. 1990. Lepadomorph Cirripeds from Brazilian coast. I-Families Lepadidae, Poecilasmatidae and Heteralepadidae. *Bull. Mar. Sci.* 47: 641-655.

Young, P.S. 1991. The Superfamily Coronuloidea (Cirripedia: Balanomorpha) from Brazilian coast, with redescription of *Stomatolepas* species. *Crustaceana* 61: 189-212.

Young, P.S. 1992. Lepadomorph Cirripeds from Brazilian coast. II-Family Scalpellidae. *Bull. Mar. Sci.* 50: 40-55.

Young, P.S. 1993. The Verrucomorpha and Chthamaloidea from the Brazilian coast (Crustacea: Cirripedia). *Revta bras. Biol.* 53: 255-267.

Young, P.S. 1994. The Balanoidea (Crustacea: Cirripedia) from the Brazilian coast. *Bolm. Mus. Nac., ser. Zool.* 356: 1-36.

Young, P.S. & M.L. Christoffersen 1984. Recent coral barnacles of the genus *Ceratoconcha* (Cirripedia, Pyrgomatidae) from Northeast Brazil (lat. 5°-18°S). *Bull. mar. Sci.* 35: 239-252.

Zevina, G.B. 1975. Cirriped Thoracica collected by r/v 'Academic Kurchatov' in the Atlantic sector of the Antarctic. *Trudy Inst. Okeanol.* 103: 183-193 [in Russian].

Zevina, G.B. 1978a. A new classification of the family Scalpellidae Pilsbry (Cirripedia, Thoracica). Part I. Subfamilies Lithotryinae, Calanticinae, Pollicipinae, Scalpellinae, Brochiinae and Scalpellopsinae. *Zool. Zh.* 57: 998-1006 [in Russian].

Zevina, G.B. 1978b. A new system of the family Scalpellidae Pilsbry (Cirripedia, Thoracica). 2. Subfamilies Arcoscalpellinae and Meroscalpellinae. *Zool. Zh. 57*: 1343-1352 [in Russian].

Zevina, G.B. 1981. Barnacles of the suborder Lepadomorpha of the world ocean. I. Family Scalpellidae. *Fauna U.S.S.R.* 127: 1-406. [in Russian].

Zevina, G.B. 1982. Barnacles of the suborder Lepadomorpha of the world ocean. II. *Fauna U.S.S.R.* 133: 1-222. [in Russian].

Zevina, G.B. 1987a. Deep-sea Verrucomorpha (Cirripedia, Thoracica) of the Pacific. 1. The North Pacific. *Zool. Zh.* 66: 1812-1821 [in Russian].

Zevina, G.B. 1987b. Abyssal Cirripedia Verrucomorpha (Thoracica) of the Atlantic and Indian Ocean. *Zool. Zh.* 66: 1304-1313.

Zevina, G.B. 1988. Deep-sea Verrucomorpha (Cirripedia, Thoracica) of the Pacific. 2. The South Pacific. *Zool. Zh.* 67: 31-40 [in Russian].

Zevina, G.B. 1990. Deep-sea Cirripedia Thoracica of the South Atlantic. *Trudy Inst. Okeanol.* 126: 80-89 [in Russian].

Zullo, V.A. 1979. Arthropoda: Cirripedia. Marine Flora and Fauna of the Northeastern United States. *N.O.A.A. Tech. Rep.* Circ. 425: 1-27.

Zullo, V.A. 1992. *Balanus trigonus* Darwin (Cirripedia, Balanidae) in the Atlantic basin: an introduced species? *Bull. Mar. Sci.* 50: 66-74.

Phylogeny and biogeography of the primitive Sessilia and a consideration of a Tethyan origin for the group

John S. Buckeridge
Department of Civil & Environmental Engineering, UNITEC Institute of Technology, Auckland, New Zealand

ABSTRACT

The Australian and New Zealand region currently provides the earliest stratigraphic records for the Pachylasmatidae, Tetraclitidae, Archaeobalanidae, and Verrucidae. These Sessilia are further characterised by both limited temporal distribution and high endemism. In this paper, the evolution and distribution of sessile cirripede species is analysed from a southern hemisphere perspective, with the proposal that although a significant number of taxa are certainly Tethyan relicts, the Australasian region was very likely also a centre for balanomorph evolution. An overview of cirripede biogeography is provided, and the phylogeny of the Sessilia is re-examined in light of recent observations. Consideration is given to the characteristics that have traditionally been used to evaluate balanomorph phylogeny, and it is concluded that many are in fact homoplasious, a result of convergence. A cladistic analysis of sessilian phylogeny is offered in response to these proposals. *Pachydiadema*, previously considered the earliest balanomorph is re-evaluated as a pedunculate.

1 INTRODUCTION

The study of biogeography, or the analysis of the distribution of organisms derived from a synthesis of their systematic disposition and ecology, is a comparatively recent phenomenon; the advancement and understanding of which relies heavily upon our appreciation of the earth's tectonic and surficial processes. The earliest biogeographic studies followed the pattern so evident in the advancement of all sciences: concerns were with *what* and *where*, (identification and distribution) rather than *how* and *why* (derivation of *explanations* for the patterns of both chronological and geographic distribution). They were also essentially terrestrial based, (e.g. Sclater 1858), with marine biogeography not being seriously considered until the mid 20th Century (Clements & Shelford 1939; Hedgpeth 1957; Thorson 1957; Knox 1963).

Although aspects of cirripede 'biogeography' were alluded to as early as the mid 1800s, in discussion on the distribution (Darwin 1854) and biostratigraphy of taxa (Darwin 1855), it was not until Newman (1976, 1979a) that aspects of cirripede phylogeny, stratigraphic distribution, and geographic distribution were integrated. Even in the monumental work by Newman et al. (1969), there is no attempt made to account for the geographic distribution of the group.

In a significant paper on short range endemics from the Californian Transition Zone, Newman (1976) used ecologic, biogeographic, stratigraphic, and climatologic factors to identify three cirripedes as Tethyan relicts. On the basis of current understanding of seafloor spreading, combined with knowledge of the average generic age of the three species concerned, Newman proposed that transition zone endemics exploited resources unavailable to species from faunal provinces to the north and south of the region. A recent paper by Newman (1992) reviews cognates amongst a range of invertebrates (including cirripedes) and concludes that the present amphitropical distribution of some taxa is best interpreted as relict faunæ, developed following the break-up of the Tethys, and accentuated by subsequent climatic changes.

In this paper, an approach similar to that employed by Newman (1976) is adopted for the Australasian region, and as well as the data sources employed by Newman, consideration will also be given to the phylogeny of taxa.

2 SESSILIA PHYLOGENY

Up until very recently, interpretation of the phylogeny of the Sessilia has been based upon what were perceived as 'primitive characteristics', which were retained in less derived genera as plesiomorphies. This reductionist approach, based on the approximation to a scalpellomorph ancestor for the Sessilia (i.e. with a heavily armoured

Table 1. Plesiomorphic and apomorphic characteristics of the Sessilia. Plesiomorphic characteristics are listed on the left, and are ranked in order of assumed phylogenetic age (most ancient at the top), modifications of each of these characters, recognised as derived features, are listed on the right. The ranking adopted here is used as the basis for developing the Taxon-Character matrix in Table 2.

Plesiomorphic features	Apomorphic features
Imbricating plates	
Parietes elevated above substrate by growth of imbricating plates	All wall plates in contact with substrate
Scutum and tergum with primordial valves	Scutum and tergum without primordial valves
Caudal appendages	No caudal appendages
8 Parietal plates	6, 4 or concrescent parietal plates
Rostrolatera not included in sheath	Rostrolatera included in sheath
Solid parietes	Parietes with interlaminate chitinous stringers
	Parietes with single row of tubes
	Parietes with multiple rows of tubes
Membranous basis	Calcareous basis
	Tubiferous basis
No radii	Radii
	Tubiferous radii
Tergum simple, triangular	Tergum with strong furrow
	Tergum with extended spur
	Tergum with beaked apex

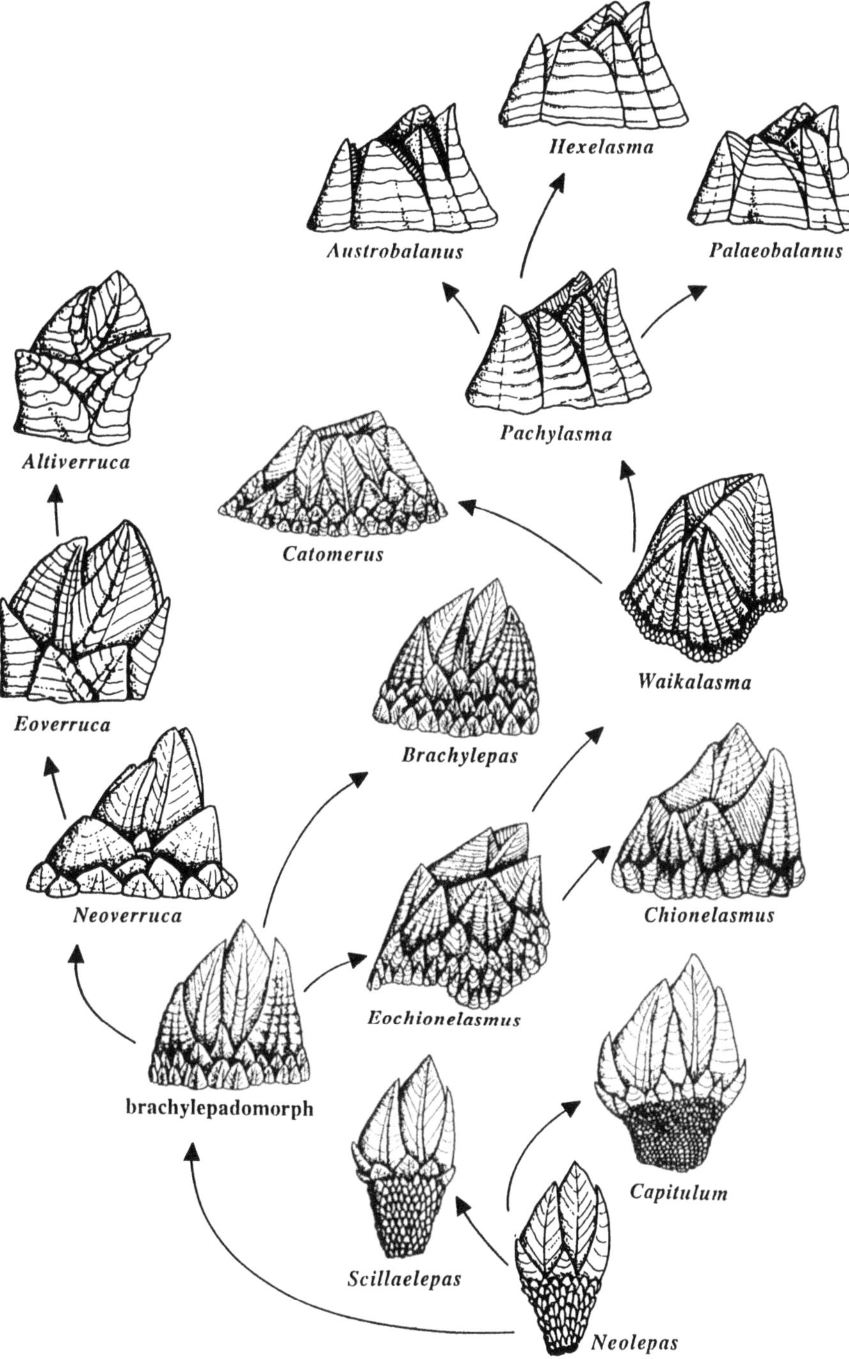

Figure 1. A traditional perspective of primitive Sessilia phylogeny. Lateral plates are absent in all Balanomorpha. *Waikalasma* and *Pachylasma* represent forms mid-stream in balanomorph evolution (i.e. they are cladistic outgroups). The Tetraclitidae, Archaeobalanidae (and Balanidae) have a *Pachylasma* rather than *Hexelasma* outgroup. The four plated condition of the post-*Austrobalanus* tetraclitids is seen as apomorphic rather than plesiomorphic. Modified from Yamaguchi & Newman (1990), Buckeridge & Newman (1992).

peduncle, a capitulum comprising paired terga, scuta, latera, carinolatera, rostrolatera, and with a rostrum and carina) has been the primary method for determining 'primitiveness,' and has resulted in the ranking of characteristics as listed in Table 1. The model only worked, however, if derived characteristics like radii, a tripartite rostrum and interlaminate chitin developed in the absence of convergent evolutionary mechanisms.

A number of problems exist when attempting to establish a phylogeny for the Sessilia, particularly if we accept the parsimonious view of a monophyletic origin for the suborder. A suitable outgroup candidate has not been agreed upon, primarily because of the difficulty in the capitular plate arrangement. Genera such as *Scillaelepas, Pollicipes* or *Capitulum* have long been considered as outgroups for the Sessilia (Darwin 1854). However, the capitulum of these pedunculates possess latera overlapped by carinolatera, the opposite of the condition observed in the Balanomorpha (Fig. 1).

Difficulties also arise in determining the intrafamilial relationships within the Balanomorpha. If *Pachylasma* is accepted as a valid outgroup in balanomorph phylogeny, and there are good reasons for this, (lack of specialisation, caudal appendages in some species, a number possess dwarf males, all have a tripartite rostrum, and all lack radii development), then *Pachylasma* has a strong claim as an outgroup for the Tetraclitidae, Coronulidae and Archaeobalanidae. However, a ranking of plesiomorphies (Table 1) does little to clarify the relationship between two very close families such as the Pachylasmatidae and the Tetraclitidae, i.e. if these families are closely related, and the Tetraclitidae are the more derived, then at what time did they develop from a pachylasmatid-like form?

The issue revolves around the point at which interlaminate chitin was incorporated into the parietes, a precursor for the development of interlaminate longitudinal tubes. The simplistic approach is that longitudinal tubes developed following the incorporation of interlaminate chitin in the parietes. This would infer a phylogeny through *Hexelasma* to the tetraclitids. However, *Austrobalanus*, the most primitive tetraclitid, possesses parietes without interlaminate chitin. As such it appears that the two families developed interlaminate chitin independently, but at approximately the same phyletic level. Two less likely alternatives are that the tetraclitids are diphyletic, with *Austrobalanus* in particular having developed from a *Bathylasma*-like form, and/or that *Austrobalanus* characteristics such as the six solid parietal plates, arborescent interlaminate figures, absence of chitin, and weakly developed radii are paedomorphic or convergent.

A similar problem exists when considering the relationship between the Pachylasmatidae and the Archaeobalanidae. The most primitive archaeobalanid, *Palaeobalanus,* is characterised by six parietal plates, (one of which is a compound rostrum), the development of radii and the inclusion of the rostrolatus into the sheath. On the basis of plate number and morphology, this infers a *Pachylasma* outgroup for *Palaeobalanus*, but as radii are (later) developed in the tetraclitids (again with a *Pachylasma* outgroup), the development of radii in balanomorph families must be homoplasious.

In Buckeridge & Newman (1992), a rather radical reinterpretation of capitular nomenclature was proposed. This was offered in response to new information concerning the morphology of *Waikalasma,* now considered one of the most primitive balanomorphs. In the new model, the latera (L) have been lost in the Balanomorpha, (the last sessile group to possess the latera being the Brachylepadomorpha), leaving a basic capitulum comprising rostrum, carina and paired carinolatera, rostrolatera, and 2nd carinolatera (RL-R-RL-CL_1-CL_2-C-CL_2-CL_1). The significance of this arrangement, if the primary wall of plates as R- CL_1-C-CL_1 is accepted, is the revisitation of the *Scillaelepas-Capitulum* model as a balanomorph outgroup (via a brachylepadomorph intermediate, and subsequent loss of the L). If the post larval *Chionelasmus* ontogeny, as figured by Newman (1987), is the model for all balanomorphs, ontogenetic development follows a primary R-CL_1-C-CL_1 arrangement, with the RL being subsequently added, overlapping both R and CL_1, thus achieving the standard balanomorph arrangement.

Based on an ontogenetic study of *Semibalanus*, Yamaguchi & Newman (1990) suggest that the addition of a further pair of lateral plates (CL_2) may be accomplished by replication.

There are clearly difficulties in determining a ranking of morphological characteristics when establishing a cirripede hierarchy, particularly when there is clear evidence for some of these characters being achieved through convergence, parallelism or reversal. Only certain characteristics can be safely utilized in evaluating phylogenetic position, those most valuable include:

- Presence of imbricating plates;
- Parietes elevated above substrate by growth of imbricating plates;
- Exclusion of the RL from the sheath;
- Primary wall in early ontogeny of R-CL_1-C-CL_1 (non-compound R);
- Scutum and tergum with primordial valves;

Table 2. Taxon-character matrix for selected genera of the Sessilia. A '1' indicates the presence of a character, '0' the absence of a character; '2' indicates that significant parietal chitin is found lining parietal pores; '3' indicates that although the RL-R-RL appear fused, fusion is incomplete, as a suture may be observed between the different plates. In some taxa, e.g. species of *Pachylasma*, the suture may only be observable in juveniles. Characters in bold '1' have a high frequency of convergence, those in italic '1' indicate that convergence has occurred. The characters are: 1 = tubiferous radii, 2 = loss of CL^2, 3 = parietal chitin, 4 = radii, 5 = tubiferous paries, 6 = RL fusion, 7 = RL sheath, 8 = imbricating scales, 9 = CL^1 replicated.

Taxa	1	2	3	4	5	6	7	8	9
Chionelasmus	0	0	0	0	0	0	0	1	0
Waiklalasma	0	0	0	0	0	0	0	1	1
Catomerus	0	0	0	0	0	0	1	1	1
Chelonibia	0	0	0	**1**	0	3	1	0	1
Pachylasma	0	0	0	0	0	3	1	0	1
Bathylasma	0	0	0	0	0	1	1	0	1
Austrobalanus	0	0	0	**1**	0	1	1	0	1
Epopella	0	1	**1**	**1**	0	1	1	0	1
Tetraclitella	*1*	1	2	**1**	**1**	1	1	0	1
Hexelasma	0	0	**1**	0	0	1	1	0	1
Palaeobalanus	0	0	0	**1**	0	3	1	0	1
Notobalanus	0	0	0	**1**	0	1	1	0	1
Balanus	0	0	**1**	**1**	**1**	1	1	0	1
Austromegabalanus	*1*	0	**1**	**1**	**1**	1	1	0	1

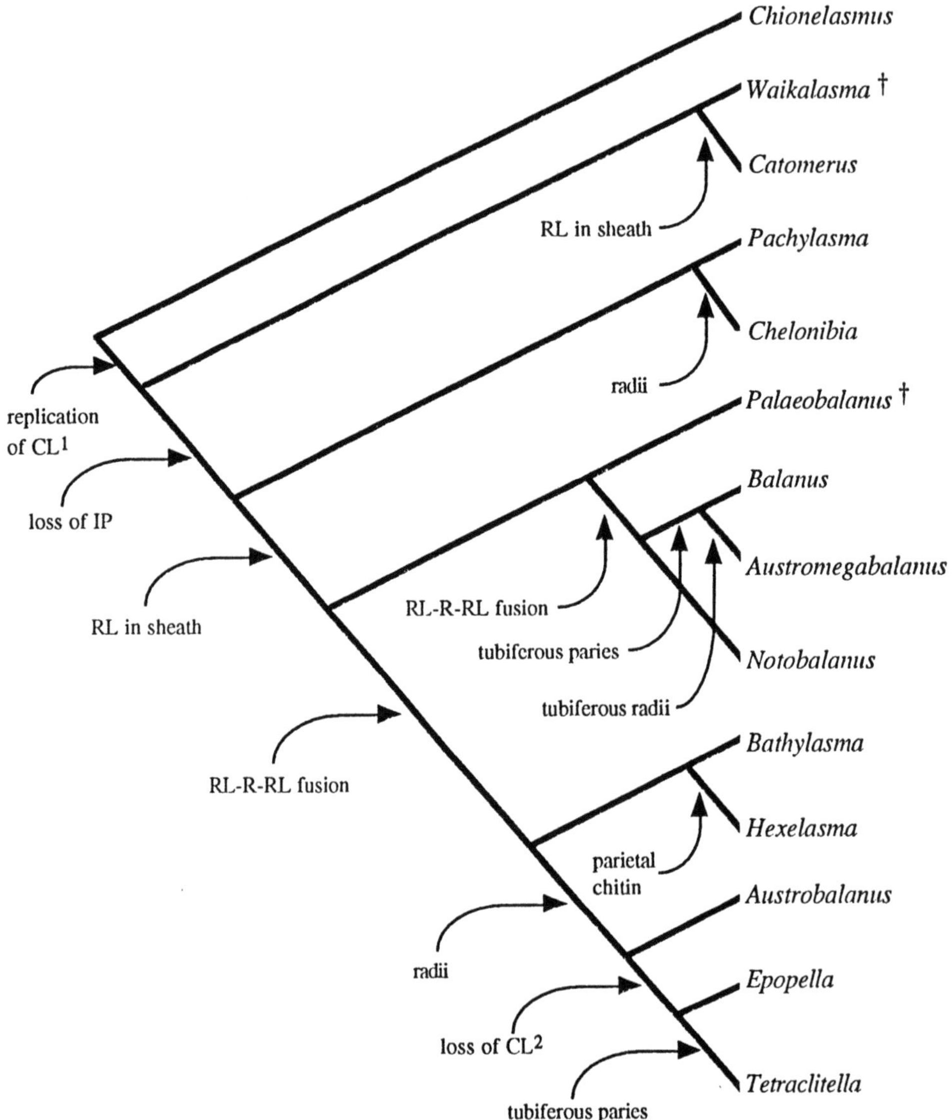

Figure 2. Cladogram of proposed Sessilian phylogeny. An analysis (manual) of sessilian phyletic relationships based upon a ranking of plesiomorphic characteristics as listed in the Taxon-Character matrix (Table 2). The presence of caudal appendages, an important indicator of antiquity, is not considered as *Waikalasma* and *Palaeobalanus* are known only as fossil (▪). Although features like radii and tubiferous parietes developed at differing times in different families, they may still be used in determining phyletic direction. Although not all balanomorph genera are recognised in the above figure, selected genera should be viewed as 'acting' for families, e.g. *Catomerus* for the Chthamalidae, *Palaeobalanus* for the Archaeobalanidae.

– Tubiferous radii (a useful character at familial level, although caution must be exercised in wider usage, as the condition is known to have developed independantly in *Tetraclitella* and the megabalanines).

Unfortunately key taxa in sessilian phylogeny are known only as fossil. In light of this, the presence (or absence) of caudal appendages is of limited use in any phylogenetic analysis, and has not been directly used in the cladistic model given in Figure 2. A number of the features listed in Table 1 as 'plesiomorphic' are not in themselves sufficient to bestow an early phylogenetic rank on a taxon. In particular, although the most primitive balanomorph species have solid parietes and a membranous basis, phylogenetically recent taxa, like *Elminius* also possess these. Further, the number of plates in the primary wall may be misleading, for apart from the Chionelasminae, the primitive balanomorph condition is for eight plates (*Waikalasma, Eolasma, Pachylasma*). Reversion to six, four and even concrescence is demonstrated in advanced species.

3 DISTRIBUTION OF SESSILE CIRRIPEDES

Recent studies on hydrothermal vent communities has resulted in the discovery of a remarkable number of new cirripedes: *Neolepas* (Newman 1979b; Jones 1993), *Neoverruca* (Newman & Hessler 1989), *Eochionelasmus* (Yamaguchi & Newman, 1990). What is particularly significant about these new species are the insights they have provided into cirripede phylogeny, each representing the most primitive living member of their respective suborders (Newman 1985, Jones 1993). Although the current distribution of the above taxa has a strong Pacific bias, this is almost certainly related to the availability of an appropriate niche: one that provides these less opportunistic taxa with protection from competition and predation. Unfortunately the fossil record of these genera is not great, with ?*Neolepas* known only from the Jurassic of New Caledonia (Buckeridge & Grant-Mackie 1985), plus *Eolasma* (Palaeocene to early Eocene) and *Chionelasmus*, a genus closely related to *Eochionelasmus*, only from Australia, New Zealand and the Chatham Islands (Buckeridge 1983). The impetus this may provide for proposing an austral origin for many cirripede families, is generally countered by a lack of suitably preserved facies in these other areas. It is fortunate however, that shallow, warm water biomicritic deposits of Palaeocene age are preserved in the Chatham Islands, and further that those depositional conditions favoured invertebrate life, including cirripedes (Buckeridge 1983, 1984b). More fortuitous still, is that this environment was penecontemporaneous with the explosive evolution of the Sessilia that marked the early Tertiary.

The Mesozoic *Pachydiadema*, which was interpreted by Withers (1935) as the earliest known balanomorph, is recorded only from Sweden. I have examined Withers' material, and concur that it probably lived in a high energy, shallow water environment like *Catomerus*. The imbricating plates of *Pachydiadema* are the most convincing balanomorph-like feature of the genus. The parietal plates however, differ from the chthamalids (and pachylasmatids), in possessing a very much simplified morphology. Surprisingly, no rostra are present in the collection, and amongst the more than 100 plates studied, no rostrolatera-like plates that could have entered the opercular sheath were identified. Further, I view the identification of the two carinal plates (Withers 1935: Pl. 50, Figs 4 and 5), as very subjective. The plates in question are very small, lack alae, and have a similar morphology to carinae of *Capitulum*. Withers places emphasis on the inner surface of these plates, which like *Capitulum*, are marked by transverse lines. Although he does not make the deduction that these lines form part of the sheath, he does conclude that:

'*C.* (*P.*) *cretaceum* approximates in the general characteristics of the shell with *Catophragmus*, especially in the form of the carinal compartment...'

The form of the carina provides the greatest obstacle to this genus as a candidate for the ancestral balanomorph. Primitive Sessilia like *Brachylepas, Waikalasma, Eolasma, Eochionelasmus* and *Chionelasmus* all possess large carinae (with alae), as an integral component for the sessile habit. Alae would certainly be one of the first pedunculate modifications *en route* to a sessile habit providing the necessary strengthening of the shell wall. Lateral extensions of the carina, or carinal parietes and intraparietes, which are not unlike alae, have already been achieved within a number of pedunculate genera like *Arcoscalpellum* and *Graviscalpellum.*

With the more complete appreciation of *Waikalasma* provided in Buckeridge and Newman (1992), a balanomorph phylogeny through brachylepadomorph, *Eochionelasmus* and *Waikalasma* stages is envisaged. *Pachydiadema* appears likely to be a late Mesozoic offshoot in cirripede phylogeny, with the form of the imbricating plates again demonstrating convergence. I am not at all confident that *Pachydiadema* is a balanomorph. If it is a sessile cirripede, it is most likely allied to the Brachylepadomorpha, but I believe it best represents a heavily armoured scalpellid.

3.1 *Verrucidae*

The Verrucidae, characterised by six calcareous plates, are the most derived family of the Verrucomorpha. Current knowledge suggests that they evolved from a multiplated brachylepadomorphan, via an intermediate form like *Neoverruca* (Newman & Hessler 1989). *Neoverruca*, extant on hydrothermal springs on the East Pacific Rise and the Mariana Trough (Newman 1989) is interpreted as a relic shallow water Mesozoic genus, escaping predation (and competition) pressures by adapting to the relatively inhospitable hydrothermal vent environment (Newman 1979b).

Schram & Newman (1980) have interpreted a possible puncture-like artifact on a stomatopod dorsum as *Verruca withersi*, and on the basis of this, extended the range of *Verruca* back to the middle Cretaceous. I am not convinced that this artifact was barnacle, let alone verrucid. As Schram and Newman state: 'The flattened and undecorated surfaces of the plates are unusual for a verrucomorph...'. Emphasis is placed on the '.. apparent asymmetry (of the orifice) that corresponds to an asymmetrical shell.' Unfortunately the orifice is incomplete, leaving questions about its asymmetry.

The earliest confirmed verrucid *Verruca tasmanica*, is found in the late Cretaceous of Western Australia. This species ranges through into the MacDonald Limestone (Oligocene) of New Zealand and the Batesford Limestone (Miocene) of Victoria (Buckeridge 1983, 1985), but interestingly is not known from the La Meseta Formation (Eocene) of Seymour Island, which possesses a similar associated fauna (including cirripedes) to that of the MacDonald Limestone (Zullo et al. 1988). In light of this it may be deduced that verrucid dispersal mechanisms are less effective than those of the balanomorphs. Certainly the high endemism characteristic of most modern verrucid species supports this interpretation. Living *Metaverruca* is represented by eight species, with 50% of these species being endemic to the West Pacific (Buckeridge 1994). The cosmopolitan *Metaverruca recta* (Aurivillius), also found in the West Pacific, is known from the early Miocene of New Zealand (Buckeridge 1983), suggesting that the southwest Pacific is either a site of relict faunae and/or a centre of verrucid speciation.

3.2 *Balanomorpha*

The earliest confirmed balanomorphans, *Eolasma maxwelli* and *Pachylasma veteranum*, are recorded along with *Verruca* and *Smilium* from several cirripede-rich horizons in the Chatham Islands Palaeocene. Not surprisingly, both the *Eolasma* and *Pachylasma* are relatively unmodified; *E. maxwelli* possesses opercular valves of very low relief, and has very weakly developed sheath ridges on all plates but the carina and rostra. *P. veteranum* retains compound rostral sutures, even in moderate sized specimens (unlike Holocene *Pachylasma* species, which tend to loose these sutures in early ontogeny). The only other known species of *Eolasma* occurs in the Miocene of Victoria (again associated with *P. veteranum*), but differing from *E. maxwelli* by a more clearly defined sheath and extended alae. During the Palaeocene, the Chatham Island region was clearly a site of explosive balanomorph evolution, as *Bathylasma* also occurs penecontemporaneously with *Eolasma* and *Pachylasma* (Buckeridge 1983, 1993). *Bathylasma rangitira* is unique amongst the bathylasmines in possessing very thick, heavy, buttressed parietal plates: adaptations for shallow water, high energy environments. Species of *Bathylasma* recorded from the later Cenozoic are characterised by thinner walls, and are associated with outer shelf and deep sea faunae.

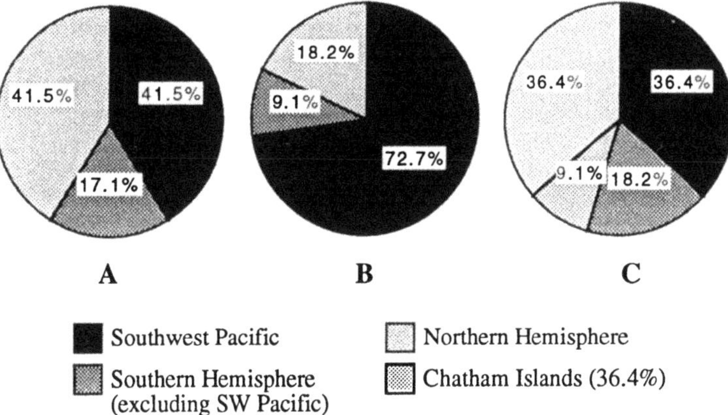

Figure 3. Antiquity index for primitive Balanomorpha and Verrucidae. Breakdown of the earliest geographic occurrence of genera as shown on Table 2. (A) Distribution of the Verrucidae, Pachylasmatidae, Tetraclitidae, Chthamalidae, Coronulidae and Archaeobalanidae (*Palaeobalanus* only). (B) Same families but only for genera with fossil records older than the Miocene. (C) Distribution of earliest records of the Pachylasmatidae only, with a further category showing the significance of the Chatham Islands. All values given as percentages of total genera considered.

Table 3. Antiquity of the Balanomorpha and Verrucidae. The high incidence of Austral endemism, coupled with the earliest known occurrence of primitive Balanomorpha and Verrucidae in the southwest Pacific infers an Austral origin to these sessile cirripedes. The Mesozoic *Pachydiadema*, interpreted variously as chthamalid (Withers 1935; Newman et al. 1969) and as a pachylasmatid (Foster & Buckeridge 1987; Buckeridge & Newman 1992) is here excluded from the Balanomorpha. Of the Archaeobalanidae, only the most primitive genus, the eight plated *Palaeobalanus* is considered in this chart. (L. Miocene = late Miocene, E. = early). A more comprehensive listing of the archaeobalanids and balanids is provided in Newman & Ross (1976) and Foster & Buckeridge (1987). [1]Only verrucids with a fossil record listed. [2]Only *Palaeobalanus*, is considered here. [3]Intimately linked with the development of cetaceans, the earliest of which are known from the Eocene.

Family	Genus	Earliest record	Age	Inferred antiquity
Verrucidae[1]	*Verruca*	Western Australia	L. Cretaceous	E. Cretaceous
	Metaverruca	New Zealand	E. Miocene	?L. Cretaceous
Pachylasmatidae	*Eochionelasmus*	North Fiji Basin	Recent	?E. Cretaceous
	Chionelasmus	Chatham Islands	Eocene	?M. Cretaceous
	Eolasma	Chatham Islands	Palaeocene	?M. Cretaceous
	Waikalasma	New Zealand	E. Miocene	?M. Cretaceous
	Pachylasma	Chatham Islands	Palaeocene	L. Cretaceous
	Bathylasma	Chatham Islands	Palaeocene	L. Cretaceous
	Mesolasma	Victoria	Oligocene	?Palaeocene
	Hexelasma	Victoria	E. Miocene	?Eocene
	Tesserolasma	Pakistan	Miocene	?Eocene
	Tetrachaelasma	South America	Recent	?Eocene
Tetraclitidae	*Austrobalanus*	Antarctica	L. Eocene	?Palaeocene
	Epopella	South Australia	L. Eocene	?Eocene
	Tesseropora	Italy	Oligocene	?Eocene
	Tetraclitella	New Zealand	E. Miocene	?Eocene
	Tesseroplax	?New Zealand	E. Miocene	?Eocene
	Tetraclita	France	Miocene	?Eocene
	Newmanella	Central America	Recent	?Oligocene
Chthamalidae	*Catomerus*	Eastern Australia	Recent	?Palaeocene
	Catophragmus	Central America	Recent	?Palaeocene
	Octomeris	Fiji & South Africa	Recent	?Palaeocene
	Euraphia	Warm Seas	Recent	?Palaeocene
	Chamaesipho	New Zealand	E. Miocene	?Palaeocene
	Chinochthamalus	Hong Kong	Recent	?Palaeocene
	Chthamalus	Chile	Miocene	?Palaeocene
	Jelius	Chile	Recent	?Palaeocene
	Tetrachthamalus	Indian Ocean	Recent	?Palaeocene
Coronulidae[3]	*Emersonius*	Florida	L. Eocene	?L. Cretaceous
	Chelonibia	France	Miocene	?Palaeocene
	Platylepas	Florida	Pleistocene	?Oligocene
	Coronula[3]	Italy	L. Miocene	?E. Miocene
	Cetolepas	California	Pliocene	?Miocene
	Stomatolepas	Cosmopolitan	Recent	?L. Miocene
	Cylindrolepas	West Indies	Recent	?L. Miocene
	Stephanolepas	Indo-Malaysia	Recent	?L. Miocene
	Cetopirus	Cool seas	Recent	?L. Miocene
	Cryptolepas[3]	California	Pleistocene	?L. Miocene
	Tubinicella[3]	Southern oceans	Recent	?L. Miocene
	Xenobalanus[3]	Cosmopolitan	Recent	?L. Miocene
Archaeobalanidae[2]	*Palaeobalanus*	New Zealand	M. Eocene	?Palaeocene

An analysis of the fossil record of the Verrucidae and primitive Balanomorpha (Table 3), results in a remarkably high incidence of earliest records from the Australasian-southwest Pacific region.

The significance of this southwest Pacific as a region for sessile cirripede expansion is confirmed using an Antiquity Index, which is determined as a percentage based on the earliest known occurrence of a genus (for earlier discussion on this see Buckeridge 1979, 1983; Newman 1979a). Figure 3 gives a representation of this antiquity index, where it is shown that of all the families considered, 41% of genera have their earliest record in the southwest Pacific. If however, only those with a known generic age of older than Miocene are considered, the southwest Pacific has 73% of the first occurrences. Due in part to the preservation of suitable facies, the Chatham Islands accounts for a remarkable 37% of this number.

4 INTERPRETATION OF PRIMITIVE BALANOMORPH DISTRIBUTION

Newman (1992) has interpreted a longitudinally disjunct distribution of several cirripede cognates as a consequence of reliction following the breakup of the Tethys. Two other alternatives were considered by Newman as explanations for this distribution, including chance dispersal and dispersal via higher latitudes, both being ruled out by lack of acceptable propagules. If the fossil record is considered, a disjunct distribution also exists for *Austrobalanus, Solidobalanus* and *Fosterella*. (Australasia and eastern Antarctic Peninsula), and *Notobalanus* and *Austromegabalanus* (Australasia and South America), although none of these occupy early phyletic positions in the Balanomorpha.

A significant number of plesiomorphic balanomorphs are however endemic to the Australasian region (*Palaeobalanus, Waikalasma, Eolasma*). In light of this, a model is proposed whereby during the late Cretaceous, the region was a centre for balanomorph evolution. The high regional endemism of early balanomorphs infers rapid phylogenetic change, with subsequently developing taxa, like *Austrobalanus, Solidobalanus, Fosterella*, *Notobalanus* and *Austromegabalanus* rapidly exploiting available niches. Following the breakup of Gondwana and the resultant closure of the Tethys, these more successful and derived taxa adopted their current disjunct distribution.

ACKNOWLEDGEMENTS

I wish to acknowledge the important mentoring rôle that Prof. W.A. Newman has provided throughout my cirripedological studies. The current understanding of the Cirripedia owes much to his foresight and enthusiasm, with his influence in cirripede biology and evolution being the most profound since the work of Charles Darwin. Thanks are also due to Prof. J.A. Grant-Mackie, Geology Department, University of Auckland, for helpful comments during the preparation of the manuscript and Ms J. Darrell, Palaeontology Department, British Museum of Natural History, London, who kindly and promptly provided specimens of *Pachydiadema* for examination.

REFERENCES

Buckeridge, J.S. 1979. Aspects of Australasian biogeography. Cirripedia: Thoracica. *Proceedings of the International Symposium on Marine Biogeography and Evolution in the Southern Hemisphere.* Auckland, New Zealand, July, 1978. *New Zealand Dept. Sci. Indust. Res. Inf. Ser.* 137(2): 485-490.

Buckeridge, J.S. 1982. The barnacle subfamily Elminiinae - Two new subgenera and a new Miocene species from Victoria. *J. Roy. Soc. New Zealand* 12: 353-357.

Buckeridge, J.S. 1983. The fossil barnacles (Cirripedia: Thoracica) of New Zealand and Australia. *New Zealand Geol. Surv. Pal. Bull.* 50: 1-151 + 14 pls.

Buckeridge, J.S. 1984a. A new species of *Elminius* from Pomahaka River, Southland, New Zealand. *New Zealand Jour. Geol. Geophys.* 27: 217-219

Buckeridge, J.S. 1984b. Two new Tertiary scalpellid barnacles (Cirripedia: Thoracica) from the Chatham Islands, New Zealand. *J. Roy. Soc. New Zealand* 14: 319-326.

Buckeridge, J.S. 1985. Fossil barnacles (Cirripedia: Thoracica) from the lower Miocene Limestone, Batesford, Victoria. *Proc. Roy. Soc. Victoria* 97: 139-150.

Buckeridge, J.S. 1991. *Pachyscalpellum cramptoni*: a new genus and species of lepadomorph cirripede from the Cretaceous of northern Hawke's Bay, New Zealand. *J. Roy. Soc. New Zealand* 21: 55-60.

Buckeridge, J.S. 1993. Cirripedia and Porifera. In Campbell H.J., et al. *Cretaceous-Cenozoic Geology and Biostratigraphy of the Chatham Islands, New Zealand. Institute of Geological* and *Nuclear Sciences Monograph.* 2: 1-240. (Released 1994).

Buckeridge, J. S. 1994. Cirripedia Thoracica: Verrucomorpha of New Caledonia, Indonesia, Wallis and Futuna Islands. MUSORSTOM 12. *Mém. Mus. natn. Hist. nat.* 1: 1-39.

Buckeridge, J.S. & J.A. Grant-Mackie 1985. A new scalpellid barnacle from the lower Jurassic of New Caledonia. *Geologie de la France.* 1: 77-80.

Buckeridge, J.S. & W.A. Newman 1992. A re-examination of *Waikalasma* (Cirripedia: Thoracica) and its significance in balanomorph phylogeny. *J. Paleo.* 66: 341-345.

Clements, F.L. & V.E. Shelford 1939. *Bio-ecology.* New York: John Wiley & Sons.

Darwin, C.H. 1854. *A monograph on the sub-class Cirripedia, with figures of all species.* The *Balanidae*, the *Verrucidae*. London: Ray Society.

Darwin, C.H. 1855. *A monograph on the fossil Balanidae and Verrucidae of Great Britain.* 44 pp. + pls 1-2. Palaeont. Soc. Mon., London (1854).

Foster, B.A. 1978. The marine fauna of New Zealand: Barnacles (Cirripedia Thoracica). *Mem. N.Z. Oceanog. Inst.* 69:1-160.

Foster, B.A. & J.S. Buckeridge 1987. Barnacle Palaeontology. *Crustacean Issues* 5: 43-62.

Hedgpeth, J.W. 1957. Marine biogeography, In Hedgepeth, J.W. (ed.) Treatise on Marine Ecology and Paleoecology: *Geol. Soc. Amer. Mem.* 67(1): 359-382.

Knox, G.A. 1963. The biogeography and intertidal ecology of the Australasian Coasts. Oceanogr. *Mar. Biol. Ann. Rev.* 1: 341-404.

Jones, D.S. 1993. A new *Neolepas* (Cirripedia : Thoracica : Scalpellidae) from an abyssal hydrothermal vent, southeast Pacific. *Bull. Mar. Sci.,* 52: 937-948.

Newman, W.A. 1976. Californian Transition Zone: Significance of Short Range Endemics. In Gray, J. & A.J. Boucot (eds). *Historical Biogeography, Plate Tectonics, & the Changing Environment.* Proc. Thirty-seventh Ann. Biol. Colloq. 32: 399-416. Corvallis: Oregon State Univ. Press.

Newman, W.A. 1979a. On the biogeography of balanomorph barnacles of the Southern Ocean including new balanid taxa: a subfamily, two genera and three species. Proc. Internl. Symp. Marine Biogeogr. & Evol. in the Southn. Hemisph. Auckland, New Zealand, July, 1978. *New Zealand Dept. Sci. Indust. Res. Inf. Ser.* 137(1): 279-306.

Newman, W.A. 1979b. A new scalpellid (Cirripedia); a Mesozoic relic living near an abyssal hydrothermal spring. *Trans. San Diego Soc. nat. Hist.,* 19: 153-167.

Newman, W.A. 1985. The abyssal hydrothermal vent invertebrate fauna: a glimpse of antiquity?. *Biol. Soc. Wash. Bull.,* 6 : 231-242.

Newman,W.A. 1987. Evolution of cirripedes and their major groups. *Crust. Issues* 5: 3-42.

Newman, W.A. 1989. Juvenile ontogeny and metamorphosis in the most primitive living sessile barnacle, *Neoverruca*, from an abyssal hydrothermal spring. *Bull. Mar. Sci.,* 45: 467-477.

Newman, W.A. 1991. Origins of Southern Hemisphere Endemism, especially among marine Crustacea. *Mem. Queensland Mus.*, 31: 51-76.

Newman, W.A., 1992. Biotic cognates of eastern boundary conditions in the Pacific and Atlantic: relicts of Tethys and climatic change. *Proc. San Diego Soc. Nat. Hist.*, 16: 1-7.

Newman, W.A. & B.A. Foster 1988. Southern Hemisphere endemism among the barnacles: explained in part by extinction of northern members of amphitropical taxa? *Bull. Mar. Sci.* 41: 361-377.

Newman, W.A. & R.R. Hessler 1989. A new abyssal hydrothermal verrucomorphan (Cirripedia: Thoracica): The most primitive living sessile barnacle. *Trans. San Diego Soc. Nat. Hist.*, 21: 259-273.

Newman, W.A. & A. Ross 1971. Antarctic Cirripedia. *Am. Geophys. Union Antarc. Res.* Ser. 14: 1-257.

Newman, W.A. & A. Ross 1976. Revision of the balanomorph barnacles; including a catalog of the species. *Mem. San Diego Soc. Nat. Hist.* 9: 1-108.

Newman, W.A., V.A. Zullo & T.H. Withers 1969. Cirripedia. In R.C. Moore (ed.), *Treatise on Invertebrate Paleontology*. Part R, Arthropoda 4(1): pp. R206-295. Lawrence: Geol. Soc. Am. & Univ. Kansas Press.

Schram, F.R. & W.A. Newman 1980. *Verruca withersi* (Crustacea : Cirripedia) from the middle Cretaceous of Columbia. *J. Paleo.* 54: 229-233.

Sclater, P.L. 1858. On the general geographical distribution of the members of the class Aves. *J. Linn. Soc., Zool.* 2: 130-145.

Thorson, G. 1957. Bottom communities (sublittoral or shallow shelf). In Hedgepeth, J.W. (ed.) *Treatise on marine ecology and paleoecology. Geol. Soc. Amer. Mem.* 67(1): 461-534.

Withers, T.H. 1935. *Catalogue of Fossil Cirripedia in the Department of Geology.* Vol. 2 Cretaceous. London: British Museum (Natural History).

Yamaguchi, T & W.A. Newman 1990. A New and Primitive Barnacle (Cirripedia : Balanomorpha) from the North Fiji Basin Abyssal Hydrothermal Field, and its Evolutionary Implications. *Pacific Sci.* 44: 135-155.

Zullo, V.A., R.M. Feldmann & L.A. Wiedman 1988. Balanomorph Cirripedia from the Eocene La Meseta Formation, Seymour Island, Antarctica. *Geol. Soc. Amer. Mem.* 169: 459-464.

Barnacle mitochondrial DNA: Determining genetic relationships among species of *Pollicipes*

Robert J. van Syoc
Department of Invertebrate Zoology and Geology, and Osher Laboratory for Molecular Systematics, California Academy of Sciences, San Francisco, USA

ABSTRACT

Polymerase chain reaction (PCR) and nucleotide sequencing techniques are developed for barnacle mitochondrial DNA (mtDNA) cytochrome c oxidase subunit 1 genes. Along with morphological evidence, nucleotide sequence data are analyzed to produce a phylogeny for the three species of edible goose barnacles in the genus *Pollicipes*. Genetic distances among these species are calculated to determine relative times of genetic divergence. The three extant species of *Pollicipes* have a Tethyan relict distribution. *Pollicipes* exhibits an eastern boundary distribution in the Pacific and Atlantic Oceans. Two species, *Pollicipes elegans* and *Pollicipes polymerus*, inhabit the eastern Pacific Ocean. The third, geographically isolated congener, *Pollicipes pollicipes,* lives in the eastern Atlantic Ocean. Molecular level data confirm the hypothesis, initially based on morphological similarities, that *Pollicipes pollicipes,* the eastern Atlantic species, and *Pollicipes elegans,* the amphitropical eastern Pacific species, are the most recently diverged of the three species in the genus. Application of average mtDNA sequence divergence rates for other marine poikilothermic taxa to the present data yields an estimated time of genetic divergence for these two species of about 36 million years before the present. This is near the time of the Eocene/Oligocene boundary when the Tethys Sea was uninterrupted and the Atlantic was significantly narrower that it is today.

1 INTRODUCTION

1.1 *Using mtDNA as a phylogenetic tool*

It is now generally recognized that the mitochondrial DNA molecule offers a relatively simple genetic marker for studies of evolution and biogeography (Wilson et al. 1985). The animal mitochondrial genome is small (often about 16,000 nucleotides), almost always circular and often can be quickly and easily extracted from a few micrograms of tissue. Although the arrangement of genes varies between major taxa, the molecule always contains genes for two rRNAs, 22 tRNAs, and thirteen proteins (Brown 1983; Wilson et al. 1985). Nuclear chromosomal DNA can also be extracted quite easily, but the size and number of the molecules involved may present researchers with a more complex task when attempting to isolate homologous sequences for comparison (Wilson et al. 1985).

Although PCR (see Section 2.3.2) allows homologous nuclear DNA sequences to be targeted with specific oligonucleotide primers and amplified to swamp out any other fragments, the mitochondrial genome copy number is much higher than most nuclear genes. This high copy number simplifies the process of successfully amplifying homologous fragments to generate nearly pure samples of template DNA for nucleotide sequencing or RFLPs. Furthermore, whereas nuclear DNA is represented in the diploid condition (usually) and is subjected to recombination during sexual reproduction, by crossing-over in meiosis, the haploid mtDNA molecule (an individual sequence of mtDNA is referred to as a haplotype rather than a genotype) is almost always acquired solely via egg cytoplasm (Birky et al. 1983; Wilson et al. 1985). In addition, there is evidence that mtDNA is more sensitive than nuclear DNA for tests of recent genetic divergence (Avise et al. 1979). The mitochondrial genes of most vertebrates evolve about 10 times faster than the nuclear genes (Brown et al. 1979; however, Martin et al. 1992 found that this rate was lower for sharks than for mammals). Despite the fact that the rate differential is only about two times faster in *Drosophila* (Powell et al. 1986; Satta et al. 1987), the generally higher rate of sequence divergence in mtDNA makes it especially appropriate for studies of population divergence or recent speciation events where relatively little time has elapsed to allow for nucleotide substitution events. For these reasons, mtDNA became popular for genetic studies particularly for those on endothermic vertebrates (e.g. Avise et al. 1983; Southern et al. 1988; Avise et al. 1989).

Much of what we know about mitochondrial DNA variation in natural populations comes from comparisons of restriction site maps or restriction fragment patterns on electrophoretic gels. This information allows for phylogenetic analysis and study of population dynamics (Brown & Simpson 1981; Ferris et al. 1981; Templeton 1983; Wilson et al. 1985; Wu & Li 1985; DeSalle et al. 1986; Hale & Singh 1987; Shields & Wilson 1987; Bowen et al. 1989). Although algorithms have been developed which allow analysis of RFLP patterns (Nei 1987), restriction length polymorphisms cannot demonstrate the details of nucleotide substitutions. These details are often of interest (for example, transitions vs. transversions and synonymous vs. nonsynonymous changes [explained more fully in Section 2.3.2 *Cytochrome c oxidase subunit 1 gene* and Section 2.3.6 *Analyses of DNA sequence data*]). In addition, because restriction endonucleases recognize only specific four or six base sequence sites, they only test for the presence or absence of those sites. Therefore, the actual number of nucleotides sampled may be small (four or six multiplied by the number of fragments per genome), and of ambiguous location.

As noted above, the sequencing of individual nucleotides from homologous DNA fragments offers larger data sets for analysis. It is also less susceptible than DNA hybridization or RFLPs to problems stemming from repetitive sequences (usually, but not always, in non-coding regions). However, repetitive sequences should be much less of a problem when using the mitochondrial genome rather than the nuclear genome.

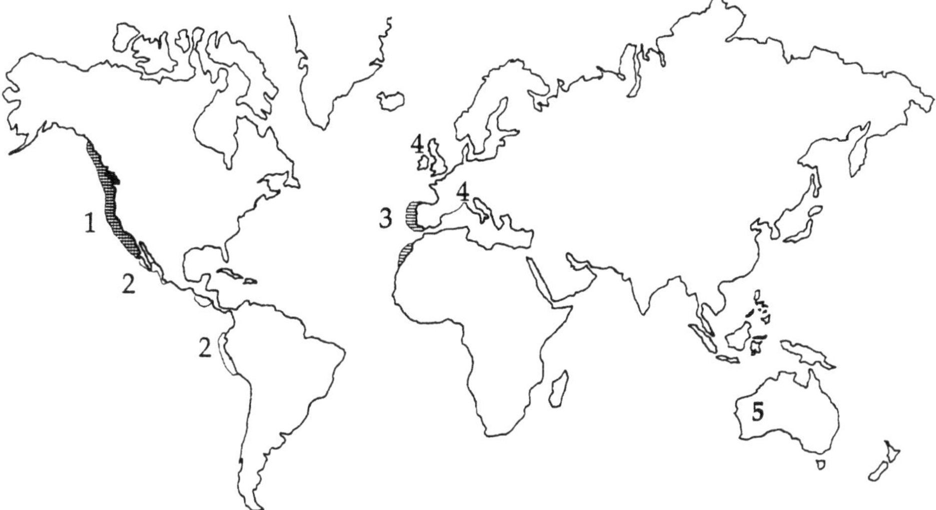

Figure 1. Distribution of living and fossil species in the barnacle genus *Pollicipes*. 1 = *P. polymerus*, extant, northeastern Pacific; 2 = *P. elegans*, extant, amphitropical eastern Pacific; 3 = *P. pollicipes*, extant, northeastern Atlantic; 4 = fossil species, England and Italy; 5 = fossil species, western Australia.

1.2 *Pollicipes biogeography: biological remnants of Tethys?*

Despite a rich fossil record in Europe (Darwin 1851; Withers 1953) and western Australia (Buckeridge 1983) dating from the Upper Cretaceous and perhaps as early as the Jurassic, only three species remain extant in the edible goose barnacle genus, *Pollicipes* (Foster 1978; Newman & Killingley 1985, Fig. 1 herein). Two of these species inhabit wave swept rocky intertidal zones along the open coast of the eastern Pacific margin. Specifically, *P. polymerus* ranges from Punta Abreojos, Baja California to British Columbia (Newman & Abbott 1980); *P. elegans,* the other eastern Pacific species, ranges from central Peru to Punta Abreojos, Baja California, with a distinct range gap along the Pacific coast of Central America, just north of the equator (Newman 1992; Kameya & Zeballos 1988); and the third species, *P. pollicipes,* is found in similar habitats in the eastern Atlantic from the Cape Verde Islands (northwestern coast of Africa) to the Iberian peninsula and, occasionally, southern England (Newman & Killingley 1985).

Darwin (1852) was the first to note that *P. elegans* of the eastern Pacific and *P. pollicipes* of the eastern Atlantic are morphologically very similar. It has since been hypothesized that *P. pollicipes* and *P. elegans* represent relict elements of the Tethys Sea fauna which have become restricted to the eastern boundary conditions of the Atlantic and Pacific Oceans (Newman & Killingley 1985; Newman & Foster 1987; Newman 1992). The present distributions of the three species therefore suggest that *P. polymerus* diverged from an ancestral pollicipedine prior to the divergence of *P. pollicipes* and *P. elegans*. To test this hypothesis, molecular phylogenies of the three species of *Pollicipes* and two other ecologically similar scalpellid barnacle outgroup species, *Capitulum mitella* and *Calantica villosa,* are presented. This molecular analysis is compared with a newly constructed phylogeny of the five scalpellid species based on a suite of morphological characters.

2 GENERAL METHODS

2.1 *Specimen acquisition and deposition*

P. polymerus from Vancouver Island, British Columbia were collected and sent to the author by Dr. William Austin. Prof. William A. Newman obtained specimens of *P. elegans* via Albertina Kameya Kameya, Instituto del Mar del Peru. Specimens of *P. pollicipes* were donated to the author by Teresa Cruz of the Universidade de Evora, and Professor Luiz Saldanha at the Universidade de Lisboa, Portugal. *Calantica villosa* specimens from New Zealand were collected and forwarded by Dr. Keith Probert. *Capitulum mitella* specimens were donated by Prof. Toshiyuki Yamaguchi of Chiba University, Japan. All specimens are now in the collections of the Invertebrate Zoology and Geology Department at the California Academy of Sciences (CAS). Each specimen used in the analyses is labeled with an 'RVS' number corresponding to the author's field notes. Most specimens were received preserved in ethyl or isopropyl alcohol, but some tissues were kept frozen prior to DNA extraction.

2.2 *Morphological characters and analysis*

Although many morphological characters have been previously used by others to describe or diagnose the species in the genus *Pollicipes*, the subfamily Pollicipedinae, and the other two scalpellids used in the present analysis (Darwin 1852; Foster 1978; Zevina 1981), many of these characters are either autapomorphies or otherwise phylogenetically uninformative. Therefore, five shell and appendage characters are described here for the first time and used in the cladistic analysis. These new binary characters, were examined in several specimens from each species, and are listed in the character matrix (Table 1) as numbers 4, 5 and 8-11. The remaining binary characters (numbers 1-3, 6 and 7) are the phylogenetically informative characters taken from previous analyses, as noted above. For each species a character is scored 1 if the state is as listed below, and 0 if not.

1. A single row/whorl of capitular plates below sublatera

Table 1. Morphological character data matrix for three species of *Pollicipes* and the two outgroup taxa, *Capitulum mitella* and *Calantica villosa*. See text for description of characters and character state definitions.

Species	Character number 11 12345678901
Capitulum mitella	00000001000
Calantica villosa	00101101000
Pollicipes pollicipes	11010010111
Pollicipes elegans	11010010111
Pollicipes polymerus	01111110100

2. Filamentary processes
3. Uni-articulate caudal appendages
4. Lower latera of different sizes
5. Peduncle scales point outward rather than up toward capitulum
6. Peduncle scales spine-like or spindle shaped
7. Tuft of spines at end of caudal appendages
8. Scutum triangular
9. Carina with sub-equal diamond shape
10. Peduncle scales flexed, sub-equal oval shape
11. Peduncle scales overlapping row above by about 1/2 scale length.

The most parsimonious tree was constructed from these characters using maximum parsimony analysis (Swofford 1993, PAUP 3.1.1). A branch-and-bound bootstrapping routine was repeated 1000 times to determine the stability of internal branch nodes.

2.3 *Molecular level methods*

2.3.1 *DNA extraction and PCR*

Total genomic DNA, both nuclear and mitochondrial, is easily extracted from whole organisms or small tissue samples. In the present study, whole specimens were either frozen or preserved in ethanol or isopropanol prior to subsequent DNA extraction. Tissues may be preserved by any of these methods or in saturated NaCl, 20% DMSO solution (see Rosel 1992) and yield DNA which is satisfactory for PCR. Small pieces (50-100 μg) of muscle tissue were dissected from specimen peduncles with a sterile razor blade or jeweler's forceps, and macerated briefly in a 1.5 ml Eppendorf tube with 400 μl homogenization buffer. Although sodium dodecyl sulfate (SDS) is customarily added (to a final concentration of 1%) to the homogenate, SDS is not necessary for effective digestion of such small amounts of tissue. Residual SDS in genomic DNA preps can interfere with PCR, so avoiding use of SDS in tissue digestions may be preferable. Finally, 4 μg of proteinase K and 4 μg RNase were added and the digestions were incubated at 55-65°C for two hours. Standard phenol/ chloroform extraction techniques were followed (Sambrook et al. 1989).

The quality of the template DNA is an important factor for successful PCR. Clean, intact, high molecular weight DNA template performs substantially better than 'dirty', fragmented, low molecular weight DNA. Also, purified clones of DNA fragments can yield PCR products from poorly matched primers, whereas genomic DNA preparations contaminated with proteins or polysaccharides may not give good products even from perfectly matched primer pairs.

The preparation of pure samples of specific DNA fragments has become commonplace in many laboratories. Clones can be made by ligating target DNA fragments into vector genomes which are then replicated by growing them out in bacterial colonies (see Sambrook et al. 1989). However, an increasingly popular technique is the more direct, labor- and time-saving method of the polymerase chain reaction (Mullis & Faloona 1987). PCR requires much less pure DNA than cloning. Moreover, dried, ethanol- or formalin-preserved material is acceptable (Pååbo 1985; Mullis & Faloona 1987). However, DNA from specimens preserved in formalin is nearly always broken into small pieces less than 100 base pairs in length. This restricts PCR to copies of short segments of DNA, thereby requiring a great deal more time and labor to accumulate sufficient sequence data. Therefore, if possible, use of frozen samples or initial preservation of tissue in alcohol, salt, or by drying is strongly recommended.

2.3.2 *Cytochrome c oxidase subunit 1 gene*

For practical reasons, the entire mitochondrial genome cannot be sequenced for each individual. Some portion of the molecule must be chosen. The gene for cytochrome c oxidase subunit 1 (CO1) has several properties which make it attractive for study. First, it has large regions of highly conserved amino acid sequence. These are necessary for PCR primer design and function. Second, there are also some regions of variable amino acid sequence, which may permit comparisons of non-synonymous base differences. This is most important for comparisons between highly divergent groups. Third, with several stretches of about 300 and 500 bases between highly conserved regions (used as primer sites), its size is appropriate for PCR, sequencing, and subsequent data analysis.

The CO1 gene has the additional advantage, shared by many protein coding genes with regions of conserved amino acid sequence, of relatively easy sequence alignment and analysis. Highly variable sections such as the control or origination region can be difficult to align and subsequent analysis may be compromised by misalignment or undiscovered gaps in sequence data. Although either mitochondrial or nuclear ribosomal genes (e.g., 18s rDNA) are more highly conserved, secondary structure differences may present a problem in aligning homologous sequences. The well known functional constraints of nucleotide placement within codon triplets in the protein coding regions result in conserved bases at the first and second positions that aid in aligning homologous sequences. Translating nucleotide sequence data to amino acid sequence can also aid in sequence alignment and reveal gaps and typographical or technical errors in the data set. The resulting amino acid sequences may also be useful in determining the phylogenetics of highly divergent taxa.

Another advantage of using protein coding regions rather than ribosomal genes is the presence of synonymous base substitutions (nucleotide changes which do not result in amino acid replacements) as well as non-synonymous base substitutions (which do result in amino acid changes). Synonymous changes occur at a more rapid rate than non-synonymous changes, presumably since they are not constrained by selection, and are therefore better markers for examining genetic relationships among closely related species or between populations within a species.

Two other mtDNA protein coding genes which are commonly used for phylogenetic studies are cytochrome b and NADH dehydrogenase (e.g., DeSalle et al. 1987; Meyer & Wilson 1990). In vertebrates, cytochrome b is slightly more variable than CO1 in amino acid sequence, while the NADH dehydrogenase subunits are extremely variable and thus lack many conserved regions of amino acid sequence for PCR primer sites. The CO1 gene was preferred for this study because it has sites (synonymous nucleotide positions) that evolve nearly as quickly as pseudogenes, which are the fastest changing sequences known. These positions can therefore be effective for determining population structure and building phylogenies for closely related species. The other nucleotide sites (non-synonymous nucleotide positions) that change very slowly can be used for PCR and DNA sequencing primer location, as noted above.

```
[----------K - N----------]   [---------M - A----------]
[----------K - J---------------]   [----M - Q-----]

K                  N J  M        Q         A
>                  < < >         <         <
```

SENSE STRAND PRIMERS (> DIRECTION)

		5' 3'
K	(1729)	GAGCTCCAGATATAGCATTCC
M	(2272)	GGAACATTAGGAATAATTTATGC

ANTI-SENSE STRAND PRIMERS (< DIRECTION)

		5′ 3′
N	(2234)	TGAGAAATTATTCCGAAGGCTGG
J	(2536)	CAATACCTGTGAGTCCTCCTA
Q	(2746)	GCTAATCCTAAGAAGTGTTGGGG
A	(2791)	AGTATAAGCGTCTGGGTAGTC

Figure 2. PCR and sequencing primers designed for crustacean cytochrome c oxidase subunit 1, mtDNA. Numbers for 5' end position are in reference to *Drosophila yakuba* sequence (Clary & Wolstenholme 1985). See text for details on relative effectiveness of the various primers. Primer A from *The Simple Fool's Guide to PCR* (Palumbi et al. 1991). All other primers designed by the author. PCR products for nucleotide sequencing reactions are shown above primer positions.

2.3.3 *PCR and sequencing primers for barnacle DNA*

'Universal' primers (e.g. Kocher et al. 1989; Palumbi et al. 1991) do not consistently result in PCR products from barnacle genomic DNA extractions. It therefore became necessary to design primers that are specific to barnacle DNA.

Much of the sequence data presented here were obtained using the two primer sets of K-N and M-A. Note that primer CO1 A was previously published (Palumbi et al. 1991) and was not designed by myself for barnacle PCR. A few species resisted PCR with either the K-N or M-A primer set. Therefore, additional primers CO1 O, CO1 P and CO1 Q were designed (Fig. 2). The O-P or O-N sets and M-Q set gave PCR products for the remaining species in the analysis. Although I have not tried these primers with non-barnacle crustaceans other than majid crabs (*Loxorhynchus*) and cloned mtDNA from brine shrimp (*Artemia*), the K-J set has worked with harpacticoid copepods (*Tigriopus californica*, Burton & Lee 1994).

Van Syoc (1994) details some specific problems with PCR primer design. In general, successful primer design involves the following steps:

1. Locate regions of conserved amino acid sequence. It is best to consider several points when selecting primer sites. The sites should be far enough apart to yield a sufficient amount of data (300-400+ bases), yet close enough together to allow for successful amplification from partially degraded DNA samples. That is, primers sited 400-500 bases apart may successfully amplify fragmented template DNA, when those 1000-1200 bases apart cannot. Similarly, badly degraded DNA may require positioning primers less than 100 bases from each other.

2. Examine the codon specificity of amino acids at the sites. Some amino acids offer very little flexibility in codon usage. This means that a more conserved nucleotide sequence will be required for the sequence of amino acids. Amino acids with high codon specificity are tryptophan (W), methionine (M), phenylalanine (F), asparagine (N), lysine (K), and cysteine (C). Those with very low codon specificity are leucine (L), arginine (R) and serine (S). Sequences with many high codon specific amino acids are more likely to possess more highly conserved nucleotide sequences and so make better sites for PCR primers.

A corollary of this point is that the codons of some amino acids have a higher guanosine or cytosine content than others. Guanosine and cytosine pair with three hydrogen bonds rather than the two hydrogen bonds which pair adenosine and thymine. Therefore, G-C pairs anneal at a higher temperature and melt at a higher temperature than A-T pairs. For this reason, it is advisable to look for sites with proline (P), glycine (G) or alanine (A) on the 3' end (see number 5 below).

3. Primer sequences of about equal A + T and G + C content are desirable.

4. Avoid sequences with many nucleotide repeats or palindromes.

5. It's best for the 3' end of the primer to have three or four guanosines or cytosines. This offers a better chance of tight annealing at the end of the primer where the DNA polymerase begins to add dNTPs to extend the complementary sequence of the template DNA (see Kwok et al. 1990).

6. Primers should be 20-23 bases in length. Shorter primers have a greater chance of mispriming and longer primers are generally not required and are more expensive to synthesize.

7. Primers sited on protein coding regions should end on the second base of a codon, as this is the most conserved position of any codon. If this is not possible, end the primer on the first base of the codon. Never end the primer sequence on the third base of the codon. This position is usually highly variable and will very likely result in a mismatch at the 3' end of the primer (see also number 5 above).

2.3.4 *PCR Buffers, cycle temperatures and other conditions*

It seems that there are nearly as many recipes for PCR buffers as there are labs using PCR and manufacturers of TAQ DNA polymerase. Some recommend adding non-ionic detergents (TritonX-100 or NP-40), gelatin and organic solvents (PEG or DMSO). Although, commercially available DNA polymerases may work better with buffers supplied by the manufacturer, a simple PCR buffer recommended by Kary Mullis (see Palumbi et al. 1991) works well with Perkin Elmer-Cetus TAQ DNA polymerase. Despite the variety of PCR buffers, all must contain $MgCl_2$. The recommended concentration of $MgCl_2$ in 10 buffer ranges from 1.5-10 mM. For nearly all of the PCR in the present study, 4 mM concentrations of $MgCl_2$ PCR buffer consistently produced good products.

PCR was run at the standard 94°C melting and 72°C extension temperatures on both the Perkin-Elmer Cetus and the Coy thermal cyclers. Annealing temperatures were varied from 42-50°C depending on condition of template and primer combination. If an accurate template to primer bond was assured, higher annealing temperatures were found to work better. If the template to primer match was uncertain or poor, lower annealing temperatures were most productive. Times at each temperature were 1.5 minutes at melting temperature, 2 minutes at annealing temperature and 2 minutes at extension temperature. Typically, PCRs were run for 35-40 cycles and ended with a 7 minute final extension step at 72°C before a 4°C hold.

2.3.5 *DNA sequencing*

Double stranded PCR products were made into single stranded DNA (ssDNA) in preparation for solid-phase sequencing reactions by using Biotin labeled primers in PCR (Hultman et al. 1989). Complementary strands were retained and prepared for sequencing reactions by washing with TE and double distilled water and filtering out primers and dNTPs with Centricon or Microcon tubes (available from Amicon Inc., see Lee & Vacquier 1993). Although Lee & Vacquier (1993) suggested Centricon tubes for this technique, Microcon tubes are preferable as they are cheaper and retain about 5 µl of retentate which is very close to the 6 µl required for sequencing reactions. This eliminates the need for the lyophilizing of 100 µl retentates from Centricon tubes down to 3-5 µl.

Dideoxy sequencing techniques (Sanger et al. 1977) with internal labeling were followed. US Biochemical's Sequenase version 2.0 enzyme kit and suggested protocols were employed with ^{35}S labeled dATP. Samples were heat denatured in heat blocks at 75-90°C and 2.5 µl loaded onto 8% polyacrylamide wedge sequencing gels, run at power settings from 60-80 watts for 1.5-4 hours depending on length and location of sequence desired. Gels were fixed in 10% methanol, 10% glacial acetic acid for 30-60 minutes and dried in a vacuum dryer for about 60-90 minutes, until dry. Dried gels were exposed on x-ray film for 12 hours.

2.3.6 *Analyses of DNA sequence data*

Sequences were initially aligned by eye. The sequence data were then entered into MEGA format and the program was used to translate them into amino acid sequence to check for errors in alignment, mistakes in the reading of autoradiographs, typographical errors when transcribing data from data sheets to word processing files used in sequence alignment, and gaps in the readable sequences. Amino acids sequences were not used to construct trees because only 3 of the 180 positions were phylogenetically informative. Therefore, only nucleotide sequence data were useful for tree building analyses.

DNA data analysis was performed using the computer programs contained in Molecular Evolutionary Genetics Analysis (MEGA ver. 1.0, Kumar et al. 1993). MEGA executes a neighbor-joining tree building algorithm using the individual genetic distance data. The Jukes-Cantor method was used to calculate the number of nucleotide substitutions per site between DNA haplotypes among the species. Confidence limits can be placed on genetic divergence by an analysis of co-variance (Nei 1987; Nei & Jin 1989). These variances are given as standard errors (SE) for all distance calculations.

Trees were also constructed by parsimony analysis (PAUP 3.1.1, Swofford 1993). Data for the construction of trees were derived from discrete character analysis with maximum parsimony methods (Swofford 1993, PAUP 3.1.1) as well as the genetic distance based neighbor-joining method which used distance data calculated from either the Juke-Cantor or the Tamura-Nei method (Kumar et al. 1993, MEGA 1.0). The Juke-Cantor method was chosen as the most generally appropriate method for calculating genetic distances for the present data. This decision was based upon recommendations made by Nei (1991) in his analysis of the accuracy of several distance estimating and tree-building methods under various circumstances. The Tamura-Nei method was used to select and analyze a reduced data set containing only transversional differences among the sequences compared.

Transversions were given double the weight of transitions in the parsimony (PAUP) analyses. It is now generally accepted that, when genetic isolation and divergence between species or populations first begins, the so-called transitional point mutations accumulate faster than transversional changes (DeSalle et al. 1987, mtDNA NADH gene nucleotide sequences in *Drosophila*; Irwin et al. 1991, mtDNA cytochrome b gene nucleotide sequences in mammals). Transitions, the changes from purine to purine (adenine and guanine) or pyrimidine to pyrimidine (thymine and cytosine) are favored over transversions (any other change) (Topal & Fresco 1976). This is because transitional changes allow for better base pairing with the bases in the complementary DNA strand. Although most base pairing follows the Watson and Crick adenine-thymine and guanine-cytosine pairing, 'mispairs' between a purine and a pyrimidine (A-C and G-T) are possible. If one of the paired bases is in a minor tautomeric form, the 'mispair' will conform precisely to the Watson and Crick model of DNA geometry. However, substitutions that form pyrimidine-pyrimidine pairs are not allowed (by model building) and purine-purine pairs require a 9° distortion of the glycosyl bond angle (Topal & Fresco 1976). Because early transitional changes tend to 'saturate' the positional differences, analysis of highly divergent sequence data is less biased if this difference in substitution rates of transitions and transversions is considered.

For each of the data sets analyzed with distance calculating methods, 500 bootstrapping replicates were performed for the calculation of each distance matrix and the construction of the accompanying tree. Gaps in sequences due to missing data were ignored in pair-wise comparisons.

3 RESULTS

3.1 *Phylogeny from morphological characters*

The morphological character data set gave trees that clearly link *P. elegans* and *P. pollicipes* to the exclusion of the other three species (Fig. 3). The most parsimonious

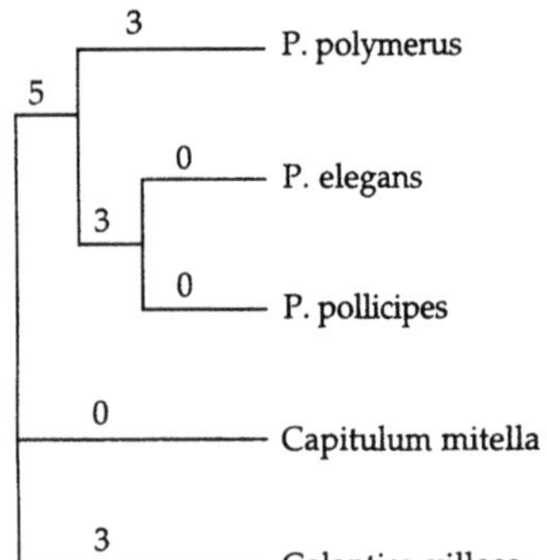

Figure 3. Most parsimonious tree (14 steps) for morphological character data set of the three species of *Pollicipes, Capitulum mitella* (outgroup 1) and *Calantica villosa* (outgroup 2). Numbers on individual branches indicate number of steps. See Table 1 for character data matrix.

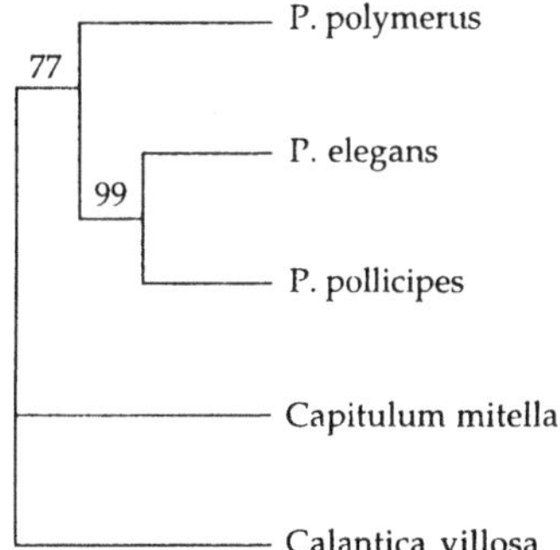

Figure 4. Branch-and-bound bootstrap tree of three species of *Pollicipes, Capitulum mitella* (outgroup 1) and *Calantica villosa* (outgroup 2), morphological data set (see text for character description, Table 1 for matrix and states).

tree places *P. polymerus* in the branch below the *P. elegans/P. pollicipes* node (Fig. 3). The bootstrap analysis of the morphological data gives very strong support to the node of *P. elegans* and *P. pollicipes* (Fig. 4, 99% bootstrap level, BL) but only moderate support for the node linking all three species of *Pollicipes* (Fig. 4, 77% BL).

3.2 *Molecular level comparisons*

3.2.1 *Sequence variation and statistics*

DNA sequence data from the CO1 gene were obtained from at least two individuals for each species. More than one individual was sequenced as a check against contamination of DNA samples. Intraspecific genetic variability was too low to alter tree structure, therefore, only one sequence for each species was used in the analyses. Homologous sequences of 540 nucleotides were obtained from every species (Appendix A). The five taxon scalpellid sequence comparison shows 198 variable positions (Appendix B), 74 of which are phylogenetically informative, 70 positions are two-fold degenerate positions (those positions for which either of two nucleotides will result in a codon for the same amino acid), and 80 are four-fold degenerate positions (those positions for which any of the four nucleotides will result in a codon for the same amino acid). Transition to transversion ratios are no more than 1.838 for each species pair (Table 2).

Sequences for each species have been submitted to GenBank. Accession numbers are: *Pollicipes polymerus* U12506, *P. pollicipes* U2824788, *P. elegans* U19273, *Capitulum mitella* U2824586, *Calantica villosa* U19483.

Table 2. Transition/transversion ratios for CO1 sequence data pair-wise comparisons among 5 scalpellid barnacle species. Ppy = *Pollicipes polymerus;* Pel = *Pollicipes elegans*; Ppl = *Pollicipes pollicipes*; Cmi = *Capitulum mitella* (outgroup 1); Cvi = *Calantica villosa* (outgroup 2).

OTU	Pel	Ppl	Cmi	Cvi
Ppy	1.390	1.838	1.585	0.885
Pel		1.545	1.349	1.016
Ppl			1.439	0.857
Cmi				1.053

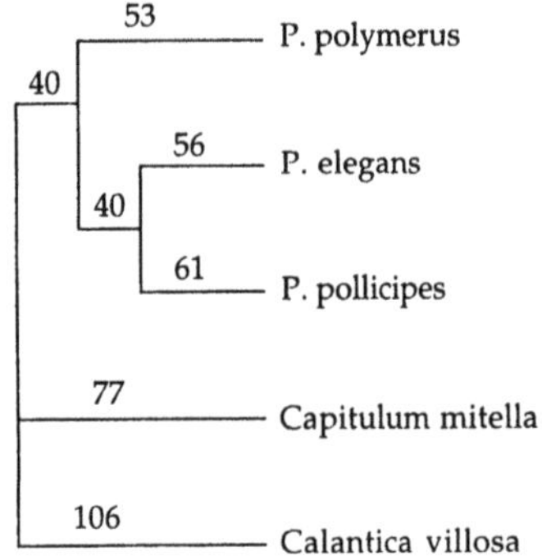

Figure 5. 433 step most parsimonious tree constructed from an exhaustive search of sequence data from CO1 gene of the three species of *Pollicipes*, *Capitulum mitella* (outgroup 1) and *Calantica villosa* (outgroup 2). Transversion weighted twice transitions, next shortest tree 435 steps. Numbers on individual branches indicate number of steps between nodes.

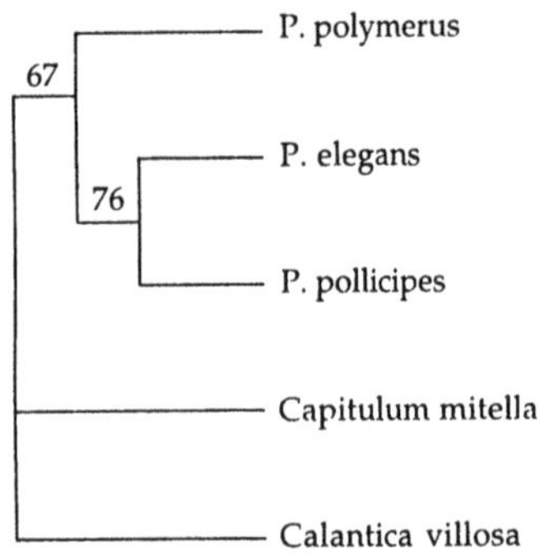

Figure 6. Branch-and-bound bootstrap parsimony tree constructed from sequence data from CO1 gene of the three species of *Pollicipes*, *Capitulum mitella* (outgroup 1) and *Calantica villosa* (outgroup 2). Transversion weighted twice transitions, 100 replicates.

3.2.2 *Parsimony trees*

The scalpellid barnacle haplotype phylogeny represented by a 433-step most parsimonious tree generated by an exhaustive search on PAUP (Swofford 1993) groups the three species of *Pollicipes*, with *P. polymerus* a sister taxon to the clade consisting of *P. elegans* and *P. pollicipes* (Fig. 5). The next shortest tree (435 steps) swaps the branches of *P. polymerus* and the first outgroup taxon, *Capitulum mitella. Calantica villosa,* the second outgroup, is an outlyer in both of these trees. A bootstrap analysis (Fig. 6) shows bootstrap level (BL) support for the *P. elegans/P. pollicipes* branch at 76% and all three species of *Pollicipes* at 67%. No other groups could be supported above the 50% BL by the data.

3.2.3 *Genetic distances and distance trees*

Distance values among the five scalpellid taxa are given in Table 3 (p-distance values) and Table 4 (Jukes-Cantor method). The Jukes-Cantor distance values between species pairs range from 0.1784 (*P. elegans* and *P. pollicipes*) to 0.2793 (*P. elegans* and *Calantica villosa*).

Pollicipes elegans and *P. pollicipes* are paired together in the tree generated from the genetic distance data (Fig. 7). The bootstrap of the distance data calculated by the Jukes-Cantor method supports the branch shared by these two species at the 81% BL (Fig. 7). The *C. mitella* (outgroup 1) linking node is supported at 54% BL (Fig. 7). The trees constructed by employing standard error calculations give higher levels of confidence, with 88% confidence probability (CP) for the *P. elegans* and *P. polli-*

Table 3. Genetic p-distance values for five scalpellid barnacle species analysis. Distances in the upper-right matrix, standard errors in lower-left matrix. Ppy = *Pollicipes polymerus*; Pel = *Pollicipes elegans*; Ppl = *Pollicipes pollicipes*; Cmi = *Capitulum mitella* (outgroup 1); Cvi = *Calantica villosa* (outgroup 2).

OTUs	Ppy	Pel	Ppy	Cmi	Cvi
Ppy		0.1835	0.1996	0.2103	0.2154
Pel	0.0168		0.1588	0.2000	0.2332
Ppl	0.0174	0.0159		0.2000	0.2216
Cmi	0.0182	0.0178	0.0179		0.2317
Cvi	0.0178	0.0183	0.0181	0.0188	

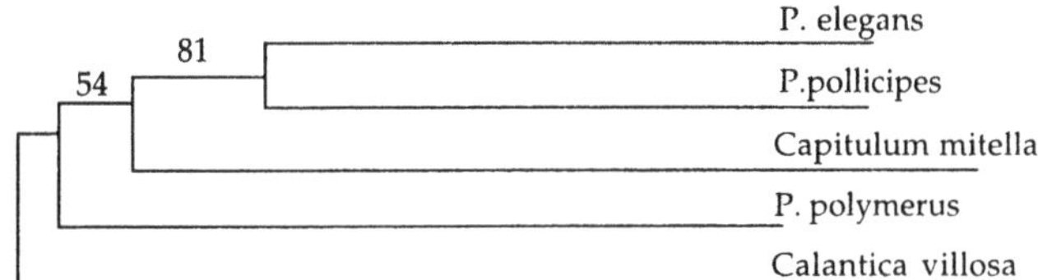

Figure 7. Branch-and-bound bootstrap tree for CO1 mtDNA sequence data from five scalpellid barnacle species, *Pollicipes elegans* (Pel), *P. pollicipes* (Ppl), *P. polymerus* (Ppy), *Capitulum mitella* (Cmi), and *Calantica villosa* (Cvi). Tree constructed from genetic distance data (Jukes-Cantor) using the neighbor-joining method. 500 bootstrap replications. 540 nucleotide positions compared. Bootstrap levels over 50% noted on branches. Branch-lengths reflect relative genetic distances. Scale: each cm approximately equal to genetic distance of 0.0136.

cipes node and 61% CP for the *C. mitella* node joining with the former two species.

The tree resulting from distance data comparisons groups *P. polymerus* and *Capitulum mitella* (outgroup 1) with *P. pollicipes* and *P. elegans* to the exclusion of *Calantica villosa* (outgroup 2) (Fig. 7). However, the tree offers a relatively weak link between *Capitulum* and the two most closely related species of *Pollicipes* with a very low BL value of 54% (Fig. 7). Therefore, the branching pattern between the *P. elegans*/*P. pollicipes* node, *P. polymerus* and *Capitulum* cannot be resolved with the present distance 'unfiltered' data set. For this reason, an attempt was made to filter some of the 'noise' from data set.

3.2.4 *Trees resulting from transversion data set analyses*

The transition/transversion ratios in the present data set (Table 2) are relatively low, none higher than about 1.8 with many near or below 1.0. This suggests that transitional changes can be considered at or near saturation levels, with transversional differences much more important for determination of branching order. Therefore, an analysis of genetic distance values was made using only transversional sequence data. This comparison used the Tamura-Nei method for calculating sequence divergence (Tamura & Nei 1993 as presented in Kumar et al. 1993, MEGA 1.0).

When only transversions are considered, *P. polymerus* groups with the other two species of *Pollicipes* (Fig. 8). *C. mitella* (outgroup 1) branches off below the *Pollicipes* clade. The BL values for *P. elegans* and *P. pollicipes* are still relatively high (Fig. 8, 85%). However, the *P. polymerus* connection is not convincing with BLs of

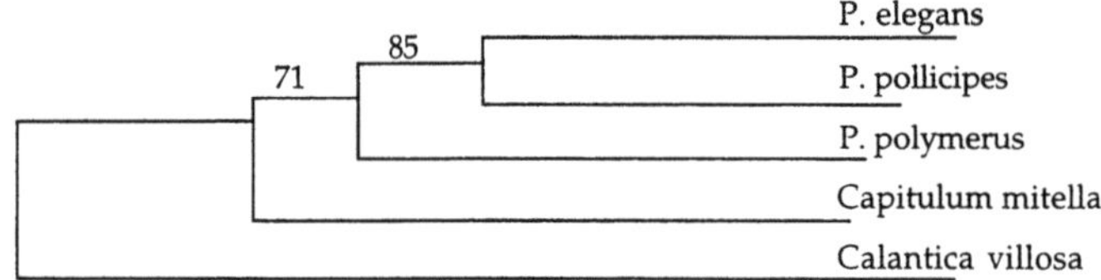

Figure 8. Transitional changes not included in analysis. Branch-and-bound bootstrap tree for CO1 mtDNA sequence data from five scalpellid barnacle species, the three ingroup species of *Pollicipes elegans* (Pel), *P. pollicipes* (Ppl), *P. polymerus* (Ppy) and two outgroup taxa, *Capitulum mitella* (Cmi, outgroup 1) and *Calantica villosa* (Cvi, outgroup 2). Tree constructed from genetic distance data (Jukes-Cantor) using the neighbor-joining method. 500 bootstrap replications. 540 nucleotide positions compared. Bootstrap levels over 50% noted on branches. Branch-lengths reflect relative genetic distances. Scale: each cm approximately equal to genetic distance of .00657.

Table 4. Genetic distance values for five scalpellid barnacle species analysis, Jukes-Cantor method. Distances in the upper-right matrix, standard errors in lower-left matrix. Ppy = *Pollicipes polymerus*; Pel = *Pollicipes elegans*; Ppl = *Pollicipes pollicipes*; Cmi = *Capitulum mitella* (outgroup 1); Cvi = *Calantica villosa* (outgroup 2).

OTUs	Ppy	Pel	Ppl	Cmi	Cvi
Ppy		0.2105	0.2321	0.2468	0.2539
Pel	0.0222		0.1784	0.2326	0.2793
Ppl	0.0238	0.0202		0.2326	0.2627
Cmi	0.0252	0.0243	0.0244		0.2771
Cvi	0.0250	0.0265	0.0257	0.0272	

only 71% (Fig. 8). Despite this low level of support, limiting the data analysis to transversional differences does produce a shift in the branching pattern between *P. polymerus*, *C. mitella* and the *P. elegans*/*P. pollicipes* node (Fig. 8).

3.2.5 *Genetic distances and sequence divergence*

The number of net nucleotide substitutions calculated with the Jukes-Cantor method between *P. elegans* and *P. pollicipes* is calculated as 17.84% with a standard error (SE) of 2.02% when the five scalpellid species are compared (Table 4). Levels of genetic distance between all other pairs of scalpellid species are all over 20% (Table 4). Examining the remaining distance values species by species, *P. polymerus* is closer to *P. elegans* than any species other than *P. pollicipes* (0.2105, Table 4). Distances between *P. polymerus* and the other species continue in ascending order from *P. pollicipes* (0.2321, Table 4), *C. mitella* (0.2468, Table 4), and *C. villosa* (0.2539, Table 4).

C. mitella (outgroup 1) is very slightly closer to *P. pollicipes* (0.2326, Table 4) and *P. elegans* (0.2326, Table 4) than *P. polymerus* (0.2468, Table 4). The distance values between *C. mitella* and the other two taxa are higher. *C. villosa* (outgroup 2) is clearly an outlier in the analysis.

4 DISCUSSION

4.1 *Phylogenetic relationships within the Pollicipedinae*

The present data strongly support the hypothesis that the amphitropical eastern sub-tropical Pacific species, *P. elegans*, and the eastern sub-tropical Atlantic species, *P. pollicipes*, are more closely related to each other than either is to their congener in the eastern warm-temperate Pacific, *P. polymerus*. All of the phylogenetic trees, whether based on molecular data or morphological data, as well as the calculated genetic distances, support this relationship among the three species. *Pollicipes elegans* and *P. pollicipes* sequence comparisons also show a higher ratio of transitions to transversions than those between either of these species and *P. polymerus*. This indicates a shorter time of genetic divergence for the *P. elegans* and *P. pollicipes* species pair as transitional differences are known to be present in higher proportions than transversions during the initial stages of genetic divergence (DeSalle et al. 1987).

Based upon the present molecular data sets and analyses, the closest living relative of the three species of the genus *Pollicipes*, the western Pacific goose barnacle *Capitulum mitella*, cannot be confidently removed from the branch containing the genus *Pollicipes*. It clusters next to or within the three species of *Pollicipes* in all trees. Despite this close relationship to *Pollicipes*, the shift in the branches of *P. polymerus* and *C. mitella* resulting from analysis of the 'filtered' data set (containing only transversions), does suggest that *Capitulum* is an outgroup to *Pollicipes*, as do the morphological data. The exact relationship of *C. mitella* to *Pollicipes* might be resolved by analyzing additional molecular data sets. This might be accomplished by either the addition of more sequence data from transversional changes within the CO1 mtDNA gene and other protein coding mtDNA genes, or by analysis of sequence data from a more slowly evolving gene, such as one of the ribosomal RNA genes. *Capitulum* is obviously very closely related to *Pollicipes*, both morphologically and genetically. In fact, it may be well to consider these two genera as phylogenetically indistinguishable taxa until we have additional evidence to allow for the clear separation of *Capitulum* from the *Pollicipes* clade.

4.2 *Levels of genetic divergence*

Despite the relative similarity of the DNA sequences of the two sub-tropical species of *Pollicipes*, they exhibit a very large percentage of divergent nucleotide sequence for congeners, about 18%. Although no mtDNA sequences have been previously published for barnacles, values for variation within genera are available for a variety of other taxa, including some crustaceans. Intra-generic differences in CO1 mtDNA gene sequences among some species within the snapping shrimp genus *Alpheus* from Panama are as high as those found among the barnacles in the present study (Knowlton et al. 1993). Trans-isthmian differences within and between species of these shrimp range from a low of 6.6% between *A. paracrinitus* in the Caribbean and *A. rostratus* in the eastern Pacific, to a high of 19.7% between Caribbean and Pacific populations of *A. cristulifrons*. The high degree of divergence found between populations of the same species on either side of the isthmus is quite remarkable and unique. It is quite likely that these populations represent morphologically conservative sibling species.

Levels of divergence above 10% are present within other crustacean genera. Palumbi & Benzie (1991) found that two congeneric species of penaeid shrimp (*Penaeus stylirostris* and *P. vannamei*) differed at 11% of the compared nucleotides in the CO1 mtDNA gene. The divergence between these species in 12s rDNA sequences was 9.6%. They noted that this level of divergence within a genus is very high and can be compared with levels of divergence among orders of mammals.

Indeed, most other available data comparisons come from within the Mammalia. The high level of divergence within crustacean genera in general and barnacle genera in particular may be due to greater time since evolution of 'genus level' morphological characters. For example, the barnacle genus *Pollicipes* is at least 55 million years old (Newman et al. 1969). The shrimp genus *Penaeus* dates to about 70 MYBP (Glaessner 1969). These dates coincide with the late Cretaceous/early Tertiary radiation of the placental mammalian orders (Carroll 1988).

More to the point, the current methods of ranking of taxa based on morphological characters can not be used as an absolute indicator of genetic diversity or age. This becomes obvious when we consider, for example, the disparity of comparing mtDNA genetic diversity within the barnacle genus *Pollicipes* (over 20%, present study) with that in the mammal genus of porpoises, *Stenella* (about 4%, Dizon et al. 1991). Clearly, the practice of using either species or genera as equivalents of a particular level of genetic diversity or divergence is troublesome and should be avoided. Rather, one should consider actual calculations of genetic diversity and divergence, preferably as determined from individual DNA sequence data.

4.3 *Estimating times of genetic divergence*

The calculated genetic distance (d_A) between *P. pollicipes* and *P. elegans* is about 18%. Converting this value into a date for genetic divergence requires an estimation of rate of nucleotide substitution over a relatively long period of time. Molecular clock variations have been well documented in several taxa; however, general smoothing of curves over long time scales may allow for an estimation of a general range for dates of genetic divergence. From the calculated d_A between the two species (17.84% ± 2%, Table 4) we can derive a time of effective genetic divergence using the formula presented by Nei (1987),

$$d_A = 2\lambda T$$

where T is time and l is the rate of nucleotide change in each sequence.

Nei (1987) lists the average rate of nucleotide change for synonymous positions in various protein coding nuclear DNA genes as $4.65 \cdot 10^{-9}$. A generally used average rate of net genetic divergence at all nucleotide positions in mtDNA is 2% per million years (Wilson et al. 1985; based on mammalian mtDNA restriction maps). Since the rate of net nucleotide divergence is a measure of total change in both diverging lineages, we must halve this value to determine the average within lineage rate of nucleotide change. This would equal a within lineage rate of nucleotide change (λ) of $10 \cdot 10^{-9}$. However, there is mounting evidence that in poikilotherm mtDNA, nucleotide sequences diverge somewhat more slowly (see Martin & Palumbi 1993 for review): rates of divergence are generally less than 1% per million years ($\lambda = 5 \cdot 10^{-9}$), many being less than 0.5% per million years ($\lambda = 2.5 \cdot 10^{-9}$). For these rates for nucleotide substitution, times of genetic divergence range from 7.92 MYBP (17.84% – 2 % divided by 2[10 · 10^{-9}]) to 35.68 MYBP (17.84% + 2 % divided by 2[2.5 · 10^{-9}]). Although this is a very broad range of time, it does allow us to search for a major vicariant event appropriate to the divergence of *Pollicipes* which might have occurred during that time frame.

4.4 *Potential causes of the genetic divergence of the Tethyan relict species of Pollicipes*

The geographically widespread and speciose fossil record of species reputed to be *Pollicipes* in Europe, and western Australia indicates a broad radiation during the Lower Eocene, about 55 MYBP. During this epoch, the warm waters of the Tethys Sea covered much of what is now southern Europe and Asia, stretching from a broad opening into the western margins of the tropical Pacific into central Europe and what is now the eastern Atlantic Ocean (Tozer 1989, Fig. 9 herein).

The opening of the southern Atlantic Ocean basin between South America and Africa began during the Cretaceous Period, about 125 million years ago (van Andel 1979). However, the northern margins of those continents remained relatively close to each other until the early Oligocene or late Eocene, about 30-40 MYBP. By that time the Atlantic may have become an effective barrier to gene flow via the plank-

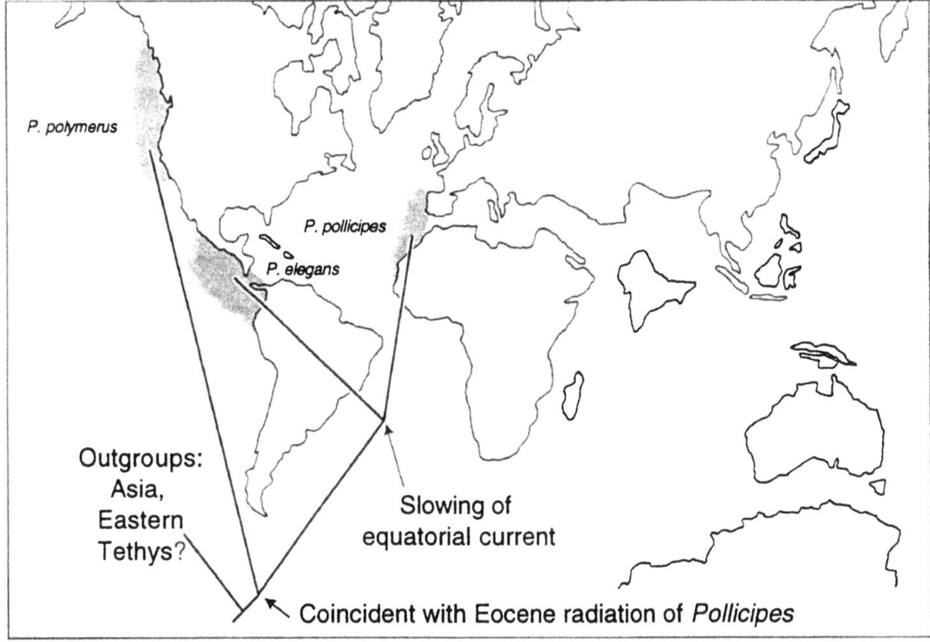

Figure 9. Summary of proposed biogeographic phylogeny of *Pollicipes*. The ancestors of the outgroups *Capitulum mitella* and *Calantica villosa* presumably originated in the eastern Tethys or Asian continental margins. Palaeogeographic reconstruction of the continents and oceans during early Tertiary time, about 50 MYBP (after Open University 1988). Note that continental margins are approximate present-day outlines, not meant to coincide precisely with ancient coastlines.

tonic larvae of shallow water coastal barnacle species. Van Andel (1979) suggests that a slowing of equatorial currents at the boundary of the Eocene/Oligocene epochs was the reason for a drop in productivity at that time along the equatorial tropical convergence zones. This slowing of the equatorial current across the widening Atlantic Ocean could correspond to a break in gene flow between populations of the ancestral stock of *P. pollicipes* and *P. elegans* (Fig. 9).

This stoppage of gene flow across the Atlantic correlated with the slowing of equatorial currents could be accompanied by extinction of *Pollicipes* populations in the western Atlantic or Caribbean coast of South America, from a drop in regional productivity at that time (Vermeij 1989). Unfortunately, there are no known fossils of *Pollicipes* from either the western Atlantic or the Caribbean. Therefore, there is no support from the fossil record for an Eocene/Oligocene extinction event in the western Atlantic/Caribbean.

Alternatively, we might consider a possible Tethyan origin of the genus *Pollicipes* and genetic divergence between *P. pollicipes* and *P. elegans* at, or near, the time of final closure of the Panamic seaway (mid to late Pliocene, about 3-3.5 MYBP, Saito 1976; Duque-Caro 1990). Although the final closure of the Isthmus of Panama is the most obvious physical barrier between the Pacific and Atlantic species of *Pollicipes,* the genetic divergence data presented here strongly support the suggestion that the physical separation of the eastern Pacific and eastern Atlantic species has existed for substantially more than 3 million years. In addition, this scenario implies a presence of *Pollicipes* in the western Atlantic just prior to the closure of the Panama seaway. The genus apparently has been excluded from the deep tropical areas surrounding Panama (Laguna 1985; Newman & Killingley 1985) and no fossil record of the genus in tropical America or the Caribbean exists (Donovan & Davis-Strickland 1993).

The foregoing analysis of divergence data between *P. elegans* and *P. pollicipes* suggests that *P. polymerus* diverged from their common ancestor at an earlier time, prior to the formation of the Atlantic and the complete destruction of the western Tethys. *P. polymerus* is presently restricted to waters with cooler average temperatures (8-20°C) than those inhabited by *P. elegans* and *P. pollicipes* (20-25°C). This is consistent with a scenario of *P. polymerus* emerging as a northeastern Pacific cold water tolerant form around the northwestern end of the Tethys, coincident with the great European radiation of *Pollicipes* in the Eocene (Fig. 9). In the absence of a good fossil record of *Pollicipes* in North or South America, it is not possible to independently assign a time to this hypothesized speciation event.

Including the fossil species currently known in a cladistic analysis with extant species may help to more fully develop a picture of the biogeography and evolution of the genus.

5 CONCLUSIONS

The eastern boundary distribution exhibited by the members of the genus *Pollicipes*, in both the Pacific and Atlantic Oceans, is relatively rare. This subtle pattern has been noted in but a few other organisms (Garth 1968; van den Hoek 1984; Vermeij 1986; crabs, red algae, and some mollusks, respectively). Newman (1992) has suggested that these are relicts of a once broader, Tethyan distribution. The present data tend to support this hypothesis. Specifically, the ranges of the extant species *Pollicipes* and their closest living relative, when combined with the fossil records for the genus, show a Tethyan distribution. *Pollicipes pollicipes*, the eastern Atlantic species, and *Pollicipes elegans*, the amphitropical eastern Pacific species, are the most recently diverged of the three species in the genus. From the current distribution of the two species and the amount of calculated genetic distance between them, it is estimated that they diverged from each other near the Eocene/Oligocene boundary. *P. polymerus* evolved at an earlier time on the northwestern margin of the Tethys, what is now the northeastern Pacific.

ACKNOWLEDGEMENTS

This project was funded in part by a grant from the National Sea Grant College Program, National Oceanic and Atmospheric Administration, US Department of Commerce, under grant number NA89AA-D-SG138, project number r/NP-1-18 M through the California Sea Grant College. The views expressed herein are those of the author and do not necessarily reflect the views of NOAA or any of its subagencies. The US Government is authorized to reproduce and distribute for governmental purposes.

Masatoshi Nei (Pennsylvania State University) provided a copy of the MEGA program. Bill Newman, Albertina Kamaya, Toshi Yamaguchi, Bill Austin, Teresa Cruz, Luiz Saldanha, and Keith Probert collected and/or sent specimens.

Jens Høeg, Frederick Schram, Rich Mooi and two anonymous reviewers offered thoughtful comments on earlier versions of this manuscript and I appreciate their help. Margo Haygood, Scripps Institution of Oceanography, supported my DNA lab work in the crucial early stages. Bill Newman provided the initial inspiration for this study and offered helpful advice and discriminating commentary throughout.

REFERENCES

Avise, J.C., C. Giblin-Davidson, J. Laerm, J.C. Patton & R. A. Lansman 1979. Mitochondrial DNA clones and matriarchal phylogeny within and among geographic populations of the pocket gopher, *Geomys pinetis*. *Proc. Natl. Acad. Sci. USA* 76:6694-6698.

Avise, J.C., J.R. Shapira, S.W. Daniel, C.F. Aquadro & R.A. Lansman 1983. Mitochondrial DNA differentiation during the speciation process in *Peromyscus*. *Mol. Biol. Evol.* 1:38-56.

Birky, C.W. Jr., T. Maruyama & P. Fuerst 1983. An approach to population and evolutionary genetic theory for genes in mitochondria and chloroplasts, and some results. *Genetics* 103:513-527.

Bowen, B.W., A. B. Meylan & J. C. Avise 1989. An odyssey of the green sea turtle: Ascension Island revisited. *Proc. Natl. Acad. Sci. USA* 86:573-576.

Brown, W.M. 1983. Evolution of animal mitochondrial DNA. In M. Nei & R.K. Koehn (eds.), *Evolution of genes and proteins*. Sunderland, Massachusetts: Sinauer Assoc. Inc.

Brown, G.G. & M.V. Simpson 1981. Intra- and interspecific variation of the mitochondrial genome in *Rattus norvegicus* and *Rattus rattus*: restriction enzyme analysis of variant mitochondrial DNA molecules and their evolutionary relationships. *Genetics* 97:125-143.

Brown, W.M., M. George & A.C. Wilson 1979. Rapid evolution of animal mitochondrial DNA. *Proc. Natl. Acad. Sci. USA* 76:1967-1971.

Buckeridge, J.S. 1983. Fossil barnacles (Cirripedia: Thoracica) of New Zealand and Australia. *NZ Geol. Survey. Paleontol. Bull.* 50:1-151.

Burton, R.S. & B.-G. Lee 1994. Nuclear and mitochondrial gene genealogies and allozyme polymorphisms across a major phylogeographic break in the copepod *Tigriopus californicus*. *Proc. Natl. Acad. Sci.* 91:5197-5201.

Carroll, R. L. 1988. *Vertebrate paleontology and evolution.* New York: W.H. Freeman & Co.

Clary, D.O. & D.R. Wolstenholme 1985. The mitochondrial DNA molecule of *Drosophila yakuba*: nucleotide sequence, gene organization, and genetic code. *J. Mol. Evol.* 22:252-271.

Darwin, C.R. 1851. *A monograph of the fossil Lepadidae, or, pedunculated cirripedes of Great Britain.* London: Palaeontographical Soc.

Darwin, C.R. 1852. *A monograph on the sub-class Cirripedia with figures of all the species. The Lepadidae; or, pedunculated cirripedes.* London: Ray Soc.

DeSalle, R., L. Val Giddings & K.Y. Kaneshiro 1986. Mitochondrial DNA variability in natural populations of Hawaiian *Drosophila*. II. Genetic and phylogenetic relationships of natural populations of *D. silvestris* and *D. heteroneura*. *Heredity* 56:87-96.

DeSalle, R., T. Freedman, E. M. Prager & A.C. Wilson 1987. Tempo and mode of sequence evolution in mitochondrial DNA of Hawaiian *Drosophila*. *J. Mol. Evol.* 26:157-164.

Dizon, A.E., S.O. Southern & W.F. Perrin 1991. Molecular analysis of mtDNA types in exploited populations of spinner dolphins (*Stenella longirostris*). In A.R. Hoelzel (ed.), *Genetic ecology of whales and dolphins.* Report of the International Whaling Commission. Special Issue 13. Cambridge: International Whaling Commission.

Donovan, S.K. & E.R. Davis-Strickland 1993. A possible lepadomorph barnacle from the Maastrichtian (Upper Cretaceous) of Jamaica, West Indies. *J. Paleont.* 67:158-159.

Duque-Caro, H. 1990. Neogene stratigraphy, paleoceanography and paleobiogeography in northwest South America and the evolution of the Panama Seaway. *Palaeog. Palaeocl. Palaeoecol.* 77:203-234.

Ferris, S.D., A.C. Wilson & W.M. Brown 1981. Evolutionary tree for apes and humans based on cleavage maps of mitochondrial DNA. *Proc. Natl. Acad. Sci. USA* 78:2432-2436.

Foster, B.A. 1978. The marine fauna of New Zealand: barnacles (Cirripedia: Thoracica). *Mem. NZ Oceanog. Inst.* 69:1-160.

Garth, J.S. 1968. *Globopilumnus xanthusii* (Stimpson), n. comb., a stridulating crab from the west coast of tropical America, with remarks on discontinuous genera of brachyrhynchous crabs. *Crustaceana* 15:312-318.

Glaessner, M.F. 1969. Decapoda. In R.C. Moore (ed.), *Treatise on invertebrate paleontology,* part R, Arthropoda 4:R399-R533. Lawrence: Geol. Soc. Am. & Univ. Kansas Press.

Hale, L.R. & R.S. Singh 1987. Mitochondrial DNA variation and genetic structure in populations of *Drosophila melanogaster*. *Mol. Biol. Evol.* 4:622-637.

Hultman, T., S. Stahl, E. Hornes & M. Uhlen 1989. Direct solid phase sequencing of genomic and plasmid DNA using magnetic beads as solid support. *NAR* 17:4937-4945.

Irwin, D.M., T.D. Kocher & A.C. Wilson 1991. Evolution of the cytochrome b gene of mammals. *J. Mol. Evol.* 32:128-144.

Kameya, A. & J. Zeballos 1988. Distribucion y densidad de percebes *Pollicipes elegans* (Crustacea: Cirripedia) en el mediolitoral Peruano (Yasila, Paita, Chilca, Lima). *Bol. Inst. Mar Peru-Callao* 12:6-22 (in Spanish).

Knowlton, N., L.A. Weigt, L.A. Solórzano, D.K. Mills & E. Bermingham 1993. Divergence in proteins, mitochondrial DNA, and reproductive compatibility across the Isthmus of Panama. *Science* 260:1629-1632.

Kocher, T.D., W.K. Thomas, A. Meyer, S.V. Edwards, S. Pååbo, F.X. Villablanca & A.C. Wilson 1989. Dynamics of mitochondrial DNA evolution in animals: Amplification and sequencing with conserved primers. *Proc. Natl. Acad. Sci.* 86:6196-6200.

Kumar, S., K. Tamura & M. Nei 1993. *MEGA: Molecular evolutionary genetics analysis, version 1.0.* University Park, PA: The Pennsylvania State University.

Kwok, S., D.E. Kellogg, N. McKinney, D. Spasic, L. Goda, C. Levenson & J.J. Sninsky 1990. Effects of primer-template mismatches on the polymerase chain reaction: human immunodeficiency virus type 1 model studies. *NAR* 18:999-1005.

Laguna,G.J. 1985. *Systematics, Ecology and Distribution of Barnacles (Cirripedia; Thoracica) of Panama.* Master's thesis, San Diego: University of California.

Lee, Y.H. & V.D. Vacquier 1993. A method for obtaining high-quality sequences from the non-biotinylated, free ssDNA remaining after solid-phase sequencing. *Biotechniques* 14:191-192.

Martin, A.P. & S.R. Palumbi 1993. Body size, metabolic rate, generation time, and the molecular clock. *Proc. Natl. Acad. Sci. USA* 90:4087-4091.

Martin, A.P., G.J.P. Naylor & S.R. Palumbi 1992. Rates of mitochondrial DNA evolution in sharks are slow compared with mammals. *Nature* 357:153-155.

Meyer, A. & A.C. Wilson 1990. Origin of tetrapods inferred from their mitochondrial DNA affiliation to lungfish. *J. Mol. Evol.* 31:359-364.

Mullis, K. & F.A. Faloona 1987. Specific synthesis of DNA in Vitro via a polymerase-catalyzed chain reaction. *Meth. Enzymol.* 155:335-350.

Nei, M. 1987. *Molecular evolutionary genetics.* New York: Columbia University Press.

Nei, M. 1991. Relative efficiencies of different tree-making methods for molecular data. In M.M. Miyamoto & J. Cracraft (eds.), *Phylogenetic analysis of DNA sequences:* 90-128. New York: Oxford University Press.

Nei, M. & L. Jin 1989. Variances of the average numbers of nucleotide substitutions within and between populations. *Mol. Biol. and Evol.* 6:290-300.

Newman, W.A. 1992. Biotic cognate of eastern boundary conditions in the Pacific and Atlantic: relicts of Tethys and climatic change. *Proc. San Diego Soc. Nat. Hist.* 16(1):1-7.

Newman, W.A. & D.P. Abbott 1980. Cirripedia: the barnacles. In R.H. Morris, D.P. Abbott & E.C. Haderlie (eds.), *Intertidal invertebrates of California:* 504-535. Palo Alto: Stanford University Press.

Newman, W.A. & B.A. Foster 1987. Southern hemisphere endemism among the barnacles: explained in part by extinction of northern members of amphitropical taxa? *Bull. Mar. Sci.* 41:361-377.

Newman, W.A. & J.S. Killingley 1985. The north-east Pacific intertidal barnacle *Pollicipes polymerus* in India? A biogeographical enigma elucidated by ^{18}O fractionation in barnacle calcite. *J. Nat. Hist.* 19:1191-1196.

Newman, W.A., V.A. Zullo & T.H. Withers 1969. Cirripedia. In R.C. Moore (ed.), *Treatise on invertebrate paleontology,* part R, Arthropoda 4 1:R206-295. Lawrence: Geol. Soc. Am. & Univ. Kansas Press.

Open University Course Team. 1988. *Ocean basins: Their structure and evolution.* UK: Open University, Milton Keynes. New York: Pergamon.

Pååbo, S. 1985. Molecular cloning of ancient Egyptian mummy DNA. *Nature* 314:644-645.

Palumbi, S.R. & J. Benzie 1991. Large mitochondrial DNA differences between morphologically similar Penaeid shrimp. *Mol. Mar. Biol. Biotech.* 1:27-34.

Palumbi, S.R., A. Martin, S. Romano, W.O. McMillan, L. Stice & G. Grabowski 1991. *The simple fool's guide to PCR.* Honolulu: University of Hawaii.

Powell, J.R., A. Caccione, G.D. Amato & C. Yoon 1986. Rates of nucleotide substitution in *Drosophila* mitochondrial DNA and nuclear DNA are similar. *Proc. Natl. Acad. Sci. USA* 83: 9090-9093.

Rosel, P.E. 1992. *Genetic population structure and systematic relationships of some small cetaceans inferred from mitochondrial DNA sequence variation.* Ph.D. dissertation, San Diego: University of California.

Saito, T. 1976. Geologic significance of coiling direction in the planktonic foraminifera *Pulleniatina. Geology* 4:305-309.

Sambrook, J., E.F. Fritsch & T. Maniatis 1989. *Molecular cloning. A laboratory manual. 2nd ed.* Cold Spring Harbor Laboratory Press, Cold Spring Harbor, NY.

Sanger, F., S. Nicklen & A.R. Coulson 1977. DNA sequencing with chain-terminating inhibitors. *Proc. Natl. Acad. Sci. USA* 74:5463-5467.

Satta, Y., H. Ishiwa & S.I. Chigusa 1987. Analysis of nucleotide substitutions of mitochondrial DNAs in *Drosophila melanogaster* and its sibling species. *Mol. Biol. Evol.* 4: 638-650.

Shields, G.F. & A.C. Wilson 1987. Calibration of mitochondrial DNA evolutions in geese. *J. Mol. Evol.* 24:212-217.

Southern, S.O., P.J. Southern & A.E. Dizon 1988. Molecular characterization of a cloned dolphin mitochondrial genome. *J. Mol. Evol.* 28:32-42.

Swofford, D.L. 1993. *PAUP: Phylogenetic Analysis Using Parsimony, Ver. 3.1.1.* Champaign, Illinois: Computer program distributed by the Illinois Natural History Survey.

Tamura, K. & M. Nei 1993. Estimation of the number of nucleotide substitutions in the control region of mitochondrial DNA in humans and chimpanzees. *Mol. Biol. Evol.* 10:512-526.

Templeton, A.R. 1983. Phylogenetic inference from restriction endonuclease cleavage site maps with particular reference to the evolution of humans and the apes. *Evolution* 37:221-244.

Topal, M.D. & J.R. Fresco 1976. Complementary base pairing and the origin of substitution mutations. *Nature* 263:283-289.

Tozer, E.T. 1989. Tethys, Thetis, Thethys, or Thetys - What, where, and when was it. *Geology* 17:882-884.

van Andel, T.H. 1979. An eclectic overview of plate tectonics, paleogeography, and paleoceanography. In J. Gray & A.J. Boucot (eds.), *Historical biogeography, plate tectonics, and the changing environment. Proc. of the Annual Biology Colloqium and selected papers:* 9-25. Corvallis: Oregon State University.

van den Hoek, C. 1984. World-wide latitudinal and longitudinal seaweed distribution patterns and their possible causes, as illustrated by the distribution of rhodophytan genera. *Helgolander Meeresuntersuchungen* 38:227-257.

van Syoc, R.J. 1994. *Molecular Phylogenetics and Population Structure Derived from Mitochondrial DNA Sequence variation in the Edible Goose Barnacle Genus Pollicipes (Cirripedia, Crustacea).* Ph.D. dissertation, San Diego: University of California.

Vermeij, G.J. 1986. Survival during biotic crises: The properties and evolutionary significance of refuges. In D.K. Elliot (ed.), *Dynamics of extinction*: 231-246. New York: John Wiley & Sons.

Vermeij, G.J. 1989. Interoceanic differences in adaptation: effects of history and productivity. *Mar. Ecol. Prog. Ser.* 57:293-305.

Wilson, A.C., R.L. Cann, S.M. Carr, M. George, U.B. Gyllenstein, K.M. Helmbychowski, R.G. Higuchi, S.R. Palumbi, E.M. Prager, R.D. Sage & M. Stoneking 1985. Mitochondrial DNA and two perspectives on evolutionary genetics. *Biol. J. Linn. Soc.* 26:375-400.

Withers, T.H. 1953. *Catalogue of Fossil Cirripedia in the Department of Geology. 3 (Tertiary)*:1-396 + pls. 1-64. London: Brit. Mus. (Nat. Hist.).

Wu, C.-I. & W.-H. Li 1985. Evidence for higher rates of nucleotide substitution in rodents than in man. *Proc. Natl. Acad. Sci. USA* 82:1741-1745.

Zevina, G.B. 1981. Barnacles of the suborder Lepadomorpha (Cirripedia, Thoracica) of the world ocean. Part 1. Family Scalpellidae. *Guides to the Fauna of the USSR* 127:1-406.

Appendix A: Nucleotide sequences for mtDNA CO1 gene. Nucleotide numbers 1-309 correspond to positions 1852-2160 in *D. yakuba*; numbers 310-540 correspond to positions 2491-2721 in *D. yakuba* (Clary & Wolstenholme 1985).

				111	111	111	122	222	222	223	333	333	333	444
	123	456	789	012	345	678	901	234	567	890	123	456	789	012
Pollicipes polymerus	TAC	CCA	CCT	CTA	GCC	AGC	AAT	ATC	GCA	CAC	TCA	GGA	GCC	TCT
Pollicipes elegans	..T	..G	..G	T..	...	..A	..C	..T	..C	..T	...	...	...	...
Pollicipes pollicipes	..T	...	..A	--.	...	..T	..C	..T	..T	...	..T	..G	..A	..C
Capitulum mitella	---	---	---	---	---	---	---	---	--T	...	.-G	...	.AT	.-A
Calantica villosa	..T	..T	...	..T	..A	..A	...	..T	..T	..T	..T	...	..T	...

	444	444	455	555	555	556	666	666	666	777	777	777	788	888
	345	678	901	234	567	890	123	456	789	012	345	678	901	234
Pollicipes polymerus	GTA	GAC	CTC	TCT	ATT	TTT	TCA	TTA	CAC	TTA	GCG	GGA	GCT	TCC
Pollicipes elegans	..T	..T	T.A	..A	...	...	..C	...	...	...	..C	...	..A	..T
Pollicipes pollicipes	...	..T	T.A	..A	..C	...	..C	C..	..T	...	..A	...	..A	..A
Capitulum mitella	...	..T	T.A	A.C	..C	..C	..C	C.T	...	..G	..T	..G	...	...
Calantica villosa	..T	..T	T.A	A..	...	...	..C	C.T	..T	...	..T	...	..G	...

						111	111	111	111	111	111	111	111	111
	888	889	999	999	999	000	000	000	011	111	111	112	222	222
	567	890	123	456	789	012	345	678	901	234	567	890	123	456
Pollicipes polymerus	TCT	ATT	TTA	GGA	GCT	ATC	AAC	TTC	ATA	TCC	ACA	GTA	ATT	AAT
Pollicipes elegans	...	...	C..	...	...	..T	...	...	...	..A	..G	..T	...	...
Pollicipes pollicipes	...	...	C..	..-	..-	..T	...	..T	...	...	...	.-.	...	...
Capitulum mitella	...	...	C.T	..G	...	..T	..T	..T	..G	...	...	...	...	..C
Calantica villosa	..A	...	...	...	...	..T	..T	...	...	..A	...	...	...	...

	111	111	111	111	111	111	111	111	111	111	111	111	111	111
	222	333	333	333	344	444	444	445	555	555	555	666	666	666
	789	012	345	678	901	234	567	890	123	456	789	012	345	678
Pollicipes polymerus	ATA	CGA	GCT	GAA	ACT	TTA	ACA	TTC	GAC	CGT	TTA	CCT	TTA	TTT
Pollicipes elegans	...	...	..C	...	...	C.G	...	..T	...	..A	C.C	...	C..	...
Pollicipes pollicipes	...	...	...	...	-..	C.T	...	..T	...	-.A	.A-	...	...	...
Capitulum mitella	...	...	..A	...	...	C..	...	..T	...	..A	.C-	...	.-.	...
Calantica villosa	...	..G	T.G	...	..A	...	...	..T	...	..A	A.T	..A	...	..C

	111	111	111	111	111	111	111	111	111	111	122	222	222	222
	677	777	777	778	888	888	888	999	999	999	900	000	000	001
	901	234	567	890	123	456	789	012	345	678	901	234	567	890
Pollicipes polymerus	GTA	TGA	AG-	GTA	TTT	GTG	ACG	GTG	ATT	CTT	CTA	TTA	TTA	TCT
Pollicipes elegans	. . .	. . .	. . A	. . T	. . .	. . C	. . C	. . A	. . .	T . A	T . .	C . G	. . G	. . .
Pollicipes pollicipes	. . T	. . G	. . A	. . C	. . C	. . .	. . C	. . A	. . .	T . A	T . .	C . T	. . .	. . C
Capitulum mitella	. . .	. . .	. . A	. . T	. . -	. . A	. . C	. . A	. . .	. . C	. . T	C . T	C . T	. . C
Calantica villosa	. A .	. . .	. . A	. . .	. . .	. . .	. . C	. . A	. . .	. . G	. . C	C . .	. . .	. . .

	222	222	222	222	222	222	222	222	222	222	222	222	222	222
	111	111	111	222	222	222	233	333	333	334	444	444	444	555
	123	456	789	012	345	678	901	234	567	890	123	456	789	012
Pollicipes polymerus	TTA	CCT	GTG	TTA	GCC	GGA	GCC	ATT	ACT	ATG	CTT	TTA	ACT	GAC
Pollicipes elegans	C . C	. . G	. . T	C . .	. . A	. . .	. . A	. . C	. . C	. . .	. . .	C . T	. . .	. . .
Pollicipes pollicipes	C . C	. . .	. . T	. . .	. . G	. . G	. . .	. . C	. . C	. . A	. . .	C . T	. . .	. . .
Capitulum mitella	C . T	. . .	. . T	C . C	. . A	. . .	. . .	. . .	. . A	. . A	. . .	C . T	. . C	. . T
Calantica villosa	C . .	. . A	. . T	. . .	. . A	. . T	. . T	. . .	. . .	. . A	T . A	C . .	. . A	. . .

	222	222	222	222	222	222	222	222	222	222	222	222	222	222
	555	555	566	666	666	667	777	777	777	888	888	888	899	999
	345	678	901	234	567	890	123	456	789	012	345	678	901	234
Pollicipes polymerus	CGG	AAT	CTT	AAT	ACA	TCA	TTT	TTC	GAC	CCT	ACA	GGG	GGA	GGA
Pollicipes elegans	. . A	. . C	. . .	. . .	. . .	. . T	. . C	. . T	. . T	. . A	. . C	. . A	. . .	. . G
Pollicipes pollicipes	. . T	. . .	. . A	. . .	. . .	. . C	. . C	. . .	. . T	. . .	. . T	T . .	. . .	. . G
Capitulum mitella	. . .	. . –	. . A	. . .	. . .	. . .	. . .	. . T	. . .	. . .	. . C	. . A	. . .	. . .
Calantica villosa	. . A	. . .	T . A	. . .	. . .	. . C	. . .	. . T	. . T	. . C	. . .	. . T	. . T	. . .

	222	223	333	333	333	333	333	333	333	333	333	333	333	333
	999	990	000	000	000	111	111	111	122	222	222	223	333	333
	567	890	123	456	789	012	345	678	901	234	567	890	123	456
Pollicipes polymerus	GAC	CCT	ATT	TTA	TAC	CTA	GGA	TTT	T - A	TTT	TTA	TTT	ACA	ATT
Pollicipes elegans	. . .	. . A	. . .	. . .	. . .	T . .	. . .	. . .	. T .	. . .	. . .	. . .	. . .	G . G
Pollicipes pollicipes	A . .	. . C	. . C	. . .	. . T	T . .	. . .	. . C	CT .	. . .	C . .	. . .	. . .	G . A
Capitulum mitella	. . .	. . .	. . .	C . T	. . T	T . .	. . T	. . .	. . .	. . .	C . T	. . .	. . G	G . G
Calantica villosa	A . .	. . C	. . C	. . G	. . .	T . .	. . .	. . .	CTT	. . .	C . T	. . –	. . T	G . A

	333	333	333	333	333	333	333	333	333	333	333	333	333	333
	333	444	444	444	455	555	555	556	666	666	666	777	777	777
	789	012	345	678	901	234	567	890	123	456	789	012	345	678
Pollicipes polymerus	GGA	GGA	TTA	-CA	G-A	GTA	GTC	TTA	GCC	AAT	TCG	TCC	CTA	GAT
Pollicipes elegans	...	...	G.G	A..	.G.	..T	..T	...	...	..C	..T	–.T	...	..C
Pollicipes pollicipes	..T	...	G.T	A..	.G.	..T	..T	...	...	..C	..A	..T	..T	..C
Capitulum mitella	..G	...	G..	A..	.G.	...	..T	C.C	...	..C	..T	..T	..C	..C
Calantica villosa	..T	..T	A.T	A..	–G.	...	..T	...	..T	...	..T	..T	T..	...

	333	333	333	333	333	333	333	444	444	444	444	444	444	444
	788	888	888	889	999	999	999	000	000	000	011	111	111	112
	901	234	567	890	123	456	789	012	345	678	901	234	567	890
Pollicipes polymerus	ATT	GTC	TTA	CAT	GAT	ACT	TAC	TAT	GTA	GTA	GCT	CAT	TTC	CAT
Pollicipes elegans	..C	CGT	...	...	..C	...	...	..C	...	...	..C	..C	...	...
Pollicipes pollicipes	..C	..T	.C.	...	..C	...	..T	...	...	..T	...	...	..T	...
Capitulum mitella	...	...	...	..C	..C	..C	...	..C	...	...	...	..C	...	..C
Calantica villosa	...	..T	...	..C	..C	...	..T	...	...	...	...	...	..T	...

	444	444	444	444	444	444	444	444	444	444	444	444	444	444
	222	222	222	333	333	333	344	444	444	445	555	555	555	666
	123	456	789	012	345	678	901	234	567	890	123	456	789	012
Pollicipes polymerus	TAC	GTT	TTA	TCA	ATA	GGA	GCA	GTA	TTT	GGA	ATT	ATA	GCT	GGA
Pollicipes elegans	..T	..G	...	..T	...	...	..G	...	...	...	..A	...	..A	...
Pollicipes pollicipes	..T	..C	...	..T	..G	..G	..G	..T	..C	...	...	...	..A	...
Capitulum mitella	...	..C	C.T	..T	...	..T	..C	..G	...	..T	...	...	..A	...
Calantica villosa	..T	..A	C..	...	...	...	...	..T	...	..T	...	...	...	..G

	444	444	444	444	444	444	444	444	444	444	444	444	455	555
	666	666	677	777	777	778	888	888	888	999	999	999	900	000
	345	678	901	234	567	890	123	456	789	012	345	678	901	234
Pollicipes polymerus	GCA	GTT	TAC	TGA	TTC	CCT	TTA	TTA	ACA	GGA	GTT	ACC	ATG	AAG
Pollicipes elegans	..T	...	...	...	..T	..A	C..	C.T	...	..–	...	...	..A	..A
Pollicipes pollicipes	..T	...	...	...	...	..C	...	..G	...	..–	–..	..T	..A	..A
Capitulum mitella	...	..C	...	...	..T	...	C..	...	..C	...	A..	..T	..A	..A
Calantica villosa	...	..A	..T	...	..T	...	...	A..	..T	...	A..	..A	T.T	..A

	555 000 567	555 001 890	555 111 123	555 111 456	555 111 789	555 222 012	555 222 345	555 222 678	555 233 901	555 333 234	555 333 567	555 334 890
Pollicipes polymerus	CCT	AAA	TGA	CTT	AAA	ATT	CAC	TTT	GGG	GCT	ATG	TTT
Pollicipes elegans	. . A	. . .	. . .	. . .	. . .	. . .	. . .	. . .	. . .	. . .	. . A	. . .
Pollicipes pollicipes	. . .	. . .	. . G	. . .	. . .	. . C	. . T	. . .	. . .	. . .	. . A	. . .
Capitulum mitella	. . A	. . .	. . G	. . A	. . .	. . C	. . .	. . .	. . .	. . .	. . .	. . C
Calantica villosa	. . .	. . .	. . .	T . A	. TT	. . C	. . .	. . C	ATT	T . A	. . A	. . .

Appendix B: Variable nucleotides in mtDNA CO1 gene.

	111	111	122	233	333	444	455	556	666	677	788	899	900	011
	369	025	814	703	689	258	912	470	346	925	814	713	625	814
Pollicipes polymerus	CAT	CAC	CTC	ACA	ACC	TAC	CCT	TTT	ATA	CAG	ATC	TTA	ACC	CAC
Pollicipes elegans	TGG	T..	ACT	CT.	...	.TT	TA.	A..	C..	..C	.AT	.C.	.T.	..A
Pollicipes pollicipes	T.A	-..	TCT	T.T	G.A	C.T	TA.	AC.	CC.	T.A	.AA	.C.	-T.	T..
Capitulum mitella	---	---	---	T.G	.AT	A.T	TAA	CCC	CCT	.GT	G..	.CT	GTT	TG.
Calantica villosa	TT.	.TA	A.T	TTT	..T	.TT	TAA	...	CCT	T.T	.G.	A..	.TT	..A

	111	111	111	111	111	111	111	111	111	122	222	222	222	222
	122	333	444	555	556	667	778	888	999	900	000	111	112	222
	706	235	124	067	892	380	140	369	268	912	457	013	690	258
Pollicipes polymerus	AAT	AGT	TTA	CTT	TAT	TTT	AAA	TGG	GCT	CAT	ATA	TTA	TGT	ACA
Pollicipes elegans	GT.	..C	.CG	TAC	.C.	C..	..T	.CC	ATA	T.C	G.G	.CC	GTC	.A.
Pollicipes pollicipes	...	...	.CT	TA.	A-.	...	TGC	C.C	ATA	T.C	T..	CCC	.T.	.GG
Capitulum mitella	..C	..A	.C.	TA.	C-.	...	..T	-AC	A.C	.TC	TCT	CCT	.TC	CA.
Calantica villosa	...	GTG	A..	TAA	.TA	.CA	...	..C	A.G	.CC	...	.C.	AT.	.AT

	222	222	222	222	222	222	222	222	233	333	333	333	333	333
	333	444	444	555	567	777	888	899	900	000	111	122	233	334
	147	013	469	258	910	369	256	814	503	469	058	915	734	692
Pollicipes polymerus	CTT	GCT	TAT	CGT	CTA	TCC	TAG	GAA	GTT	TAC	CAT	TAT	AAA	TAA
Pollicipes elegans	ACC	...	CT.	.AC	..T	CTT	AC.	A.G	.A.	...	T..	...	..G	G..
Pollicipes pollicipes	.CC	A..	CT.	.T.	.AC	C.T	.TT	..G	ACC	..T	T.C	C.C	..G	AT.
Capitulum mitella	..A	A..	CTC	T.-	.A.	.T.	.C.	A..	...	CTT	TT.	..C	TGG	GG.
Calantica villosa	T..	ATA	C.A	.A.	TAC	.TT	C..	TT.	ACC	.G.	T..	CTC	TTG	ATT

	333	333	333	333	333	333	333	344	444	444	444	444	444	444
	445	556	666	777	788	888	999	900	111	222	223	334	445	556
	354	780	369	235	812	346	036	928	147	036	792	581	470	392
Pollicipes polymerus	TAA	CTA	CTG	CCA	TTG	TCT	TTT	CTA	TTC	TCT	TAA	AAA	ATA	TTA
Pollicipes elegans	GGT	T..	.CT	T..	CCC	GT.	.C.	.C.	CC.	.TG	..T	..G	...	AA.
Pollicipes pollicipes	GTT	T..	.CA	T.T	CC.	.TC	.C.	T.T	..T	.TC	..T	GGG	TC.	.A.
Capitulum mitella	G..	TCC	.CT	T.C	C..	...	CCC	.C.	.C.	C.C	CTT	.TC	G.T	.A.
Calantica villosa	AT.	T..	T.T	TT.	...	.T.	CC.	T..	..T	.TA	C..	...	T.T	..G

	444	444	444	444	555	555	555	555	555	555
	667	788	888	999	000	111	112	222	333	334
	581	701	469	389	147	346	892	589	012	470
Pollicipes polymerus	ATC	CTT	TAA	GCA	GGT	ACT	AAT	CTG	GGG	TGT
Pollicipes elegans	T . .	TAC	CT .	. . .	AAA	. . .	. . .	. . .	. . .	. A .
Pollicipes pollicipes	T . .	. C .	. G .	– T .	AA .	G . .	. . C	T . .	. . .	. A .
Capitulum mitella	. C .	T . C	. . C	AT .	AAA	G . A	. . C	. . .	. . .	. . C
Calantica villosa	. AT	T . .	A . T	AAT	TA .	. TA	TTC	. CA	TTT	AA .

New frontiers in barnacle evolution

Frederick R. Schram
Institute for Systematics and Population Biology, University of Amsterdam, Amsterdam, Netherlands
Jens T. Høeg
Department of Cell Biology and Anatomy, Institute of Zoology, University of Copenhagen, Copenhagen, Denmark

ABSTRACT

We survey the state of knowledge in barnacle studies in several fields: use of modern technology in developing species descriptions, paucity of anatomical data, problems of over reliance on archetypal species, lack of a phylogenetic systematic framework for past work with lack of an adequate phylogeny. Several pathways for future research within these problem areas present themselves. We also identify 'hot spots' in cladistic studies of Maxillopoda that must be directly addressed if the new frontiers in barnacle evolution can move forward.

1 INTRODUCTION

How does one predict where a new frontier will lie? What determines a cutting edge of research in the future? How does one identify the lines that will emerge and define the future directions of study in a group? These questions often come up during the deliberations of granting agencies, but we rank and file scientists seldom have easy answers for them. Most research is Markovian in the sense that what might happen in the future grows directly from what happens right now, and what happens now grows directly out of what has just preceded it. In one sense, we can define 'innovation' as that which holds the greatest promise of fulfilling what is currently unfolding. Truly we should find it almost impossible to predict different lines of research.

This state of affairs certainly prevails for the field of cirripedology. In fact, the study of barnacles has directly grown from a rich heritage. As an excellent example of this, the CRUSTACEAN ISSUES volume 'Barnacle Biology' (Southward 1987), clearly tried to place current research in the context of the work that had gone on before. That book most particularly addressed how current work grew from the barnacle efforts of Charles Darwin, who serves as a sort of 'archetypal fountainhead' from which all barnacle studies have flowed.

We can see another example of this in the volume by Anderson (1994). Anderson produced an excellent summary of the current state of knowledge that exists in the realm of barnacle studies. However, we can discern from his review no clear direction as to how individual specialities within the field as a whole might further develop.

We want to examine in this paper just where the 'new frontiers' might lie and just which lines of research hold the greatest promise towards affording us some real quantum advances in our understanding barnacle evolution.

2 THE FRONTIERS WITHIN BARNACLE BIOLOGY

Several areas of research within the field of barnacle biology suggest themselves to us as ones prime for movement into new frontiers.

2.1 *The mechanics of species descriptions*

New techniques of handling large bodies of data can have immediate application in bringing about a revolution in how we describe species. Those of us who work on research problems involving the phylogeny of groups naturally must base our analyses on the descriptions of taxa that we find in the literature. Yet the one thing that immediately becomes obvious to a phylogenist when she or he enters the taxonomic literature is that taxonomists look at their materials differently. Taxonomists focus on characters that help identify species, while phylogenists want to deal with the nature of characters that might sort out the relationships among species. Certainly we found this to be true when we spearheaded a project that sought to address issues of cladistic relationships among some cirripede taxa (Glenner et al. 1995). We discovered that the standard literature diagnoses of species, genera, and families often did not contain equivalent information that would allow quick comparison among disparate taxa. We had either to mine the more detailed descriptions of specimens that occur in good taxonomic papers, or consult for ourselves actual specimens of species in question to uncover information that we could not find in the literature before we could build our character matrix.

Jones & Lander (1995) point this out quite effectively in their research concerning a revision of the calanticine scalpelomorphs. The selective use of a limited number of characters by previous workers left them unable to adequately evaluate all characters that had been employed by these people. Jones & Lander outline how a data processing system like DELTA can effectively act to overcome these short comings. This lack of across the board, in-depth examination of all characters in species descriptions in the past impinges on the ability of non-specialists to use the literature in identifying newly collected material. Sluys et al. (1995) summarize the LINNAEUS II package of programs that directly address user friendly interactions with taxonomic data bases.

Systems like DELTA and LINNAEUS II facilitate several things. 1) They allow non-experts to access taxonomic literature in ways that can serve purposes other than merely trying to describe species. 2) They actually can assist taxonomists in producing more effective and in-depth descriptions of new taxa by affording a technological framework to force looking at features other than those that provide an immediate delineation of one taxon from another. 3) They facilitate taking taxonomic information and interfacing it with other programs that could manipulate those data in other ways, e.g., programs like PAUP or Hennig86 that perform cladistic analyses.

The employment of such data handling techniques will actually take the rather staid science of taxonomy and thrust it into a frontier heretofore little explored, a frontier of direct practical application in other fields. These techniques will take taxonomy from being an 'art' performed by 'the authorities' to a real science with direct relevance and application to other sciences.

2.2 *Basic Data Paucity*

There often exists a distinct paucity of data or understanding for many basic aspects of adult or larval barnacle anatomy. One example of this comes from the field of barnacle reproduction. The penis figures as a crucial organ in the total array of reproductive structures available to cirripedes. In hermaphrodites, this organ plays a crucial role towards an individual literally 'groping' its way into the next generation's gene pool. Now, despite the fact that Darwin (1854) first outlined the general diversity evident in this structure, only since the work of Klepal et al. (1972), 118 years later, and Klepal (1990) have we had available to us any succinctly focused studies of the external and histological morphology of the penis.

Along these same lines, the recognition of the existence of dwarf males also stems from the work of Darwin, yet only now have Klepal & Nemeschkal (1995) begun to produce a detailed comparative morphologic analysis of these individuals. Even such mundane structures like carapaces still have secrets to yield as we can see with Elfimov's (1995) rather simple, straight forward comparison of cyprids; or the recent work of Jensen et al. (1994), where an entirely new organ has emerged from the SEM examinations of various maxillopodan carapaces; and this work may in turn prove quite helpful in discerning phylogenetic relationships within maxillopodans specifically and possibly even crustaceans as a whole.

So, in regard to addressing data paucity, the new frontier lies right in front of us whenever we seek to extend basic knowledge about animal anatomy (e.g., Moyse et al. 1995; Grygier & Itô 1995).

2.3 *The problem with archetypal species*

In trying to understand the biology of a group, one turns naturally to the most readily available species in order to begin ones research. This approach serves well in the initial phases of a field's development. However, we see an over reliance within the field of cirripedology on too few of these 'type' species to elucidate the forms and basic biology that supposedly apply to all cirripedes. As an example, many authors have expended a great deal of effort on trying to understand the phenomena of hatching, settling, and metamorphosis (e.g. Crisp et al. 1991; Yule & Walker 1987). Most of this work has focused on examination of the species such as *Balanus amphitrite, B. balanus,* and especially *Semibalanus balanoides.*

Yet one can see from studies such as that of Glenner & Høeg (1993) that some distinctly different patterns of settlement and metamorphosis occur in different taxa. Now, we do not want to imply that nothing of value can come from the use of typological species in the study of biology. However, we do believe that discoveries such as that of Glenner & Høeg should caution us not to focus exclusively on a few 'standard' types, but to spread our net farther when collecting information. Archetypic species can act as guide posts for where new information might be sought, but to apply any of these results in a broader evolutionary context must require a more broadly derived data base than that which we have used to date. Indeed, the admitted slow progress in fields of investigation such as discerning patterns in chemosensation, and trying to understand settlement and metamorphosis (e.g. see Clare 1995; Walker 1995) might be significantly speeded up if we started working with a wider array of taxonomic types.

In this regard, the new frontier lies in extending the range of organisms we work with and moving away from too much dependence on a few species. As with comparative anatomy, any direction we move in will extend the frontier.

2.4 *Lack of a adequate phylogenetic systematic context*

For a field that began with the research efforts of Charles Darwin, it may seem contradictory to maintain that we have a problem with the evolutionary foundations of cirripede studies. Nevertheless, there indeed exists a pervasive lack of phylogenetic context for what information we do have concerning barnacles.

For example, Anderson (1994) summarizes the great amount of work that he and others have performed to elucidate the patterns that they perceive in the function of cirri in feeding. He outlines not only strict anatomical facts in this regard, but also lays out the available knowledge concerning orientation of limbs and the active functioning of the components in the entire feeding apparatus. He has documented an interesting array of variations in form and function. However, he has framed almost none of this in any phylogenetic context. Despite a great deal of work, at this point we still do not know which patterns may be plesiomorphic or which apomorphic, where particular adaptive radiations among and between groups may lie, and what, if anything, any of this may tell us about the phylogenetic relationships among the cirripedes.

An analogous situation prevails in regard to reproductive biology. Klepal & Nemeschkal (1995) extend, as mentioned above, the frontier in regards to our anatomical knowledge of dwarf males. However, this information still has not reached the stage where they could do an effective cladistic analysis, although they do suggest that certain phenetic patterns exist that could guide continuing efforts to sample the diversity of form in yet-to-be sampled taxa.

Indeed, Anderson (1994) in a 357 page book on barnacles that in its subtitle purports to deal with 'structure, function, development, *and evolution*' (italics ours) devotes exactly 1½ pages to issues of phylogenetic relationships and that he summarizes in a single illustration (Fig. 1) presented in terms of evolutionary systematics. This approach is not unique among barnacle workers (e.g. see Newman 1987; Newman & Ross 1976).

Older cirripede workers commonly have not stated their conclusions as clearly outlined hypotheses about relationships, i.e., in a way that could allow the rest of us to easily test them. Nevertheless, a great deal of information exists in the literature such that anyone can gather the basic data that could frame more rigid hypotheses of phylogenetic relationships for barnacles. Only now have we taken the first tentative steps in this regard (Glenner et al. 1995; Buckeridge 1995), which means that only

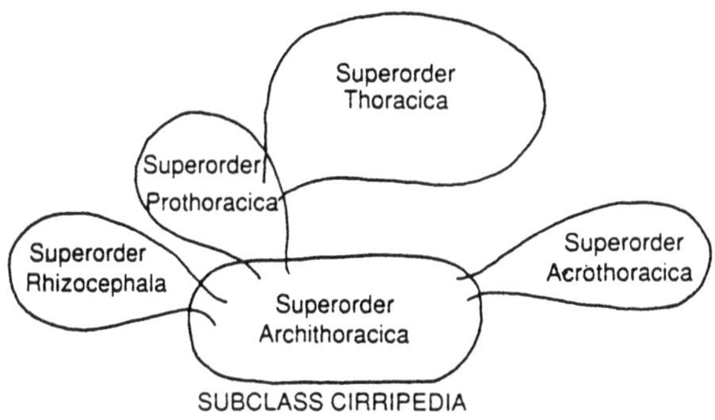

Figure 1. An Evolutionary Systematic presentation of relationships among major cirripede groups (taken from Anderson 1994).

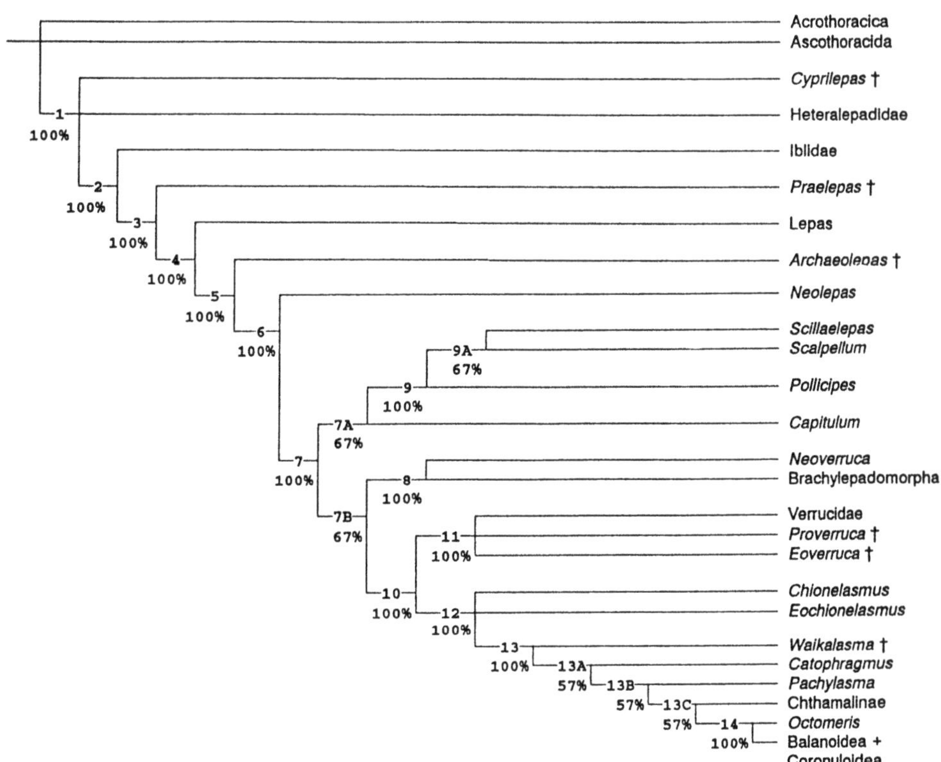

Figure 2. A phylogenetic systematic presentation of relationships among representative thoracican groups, using Acrothoracica and Ascothoracida as outgroups. This cladogram is a 50% Majority Rule consensus of 189 equally parsimonious trees (see Glenner et al. 1995 for details).

now has a whole new perspective opened up on the study of barnacle evolution.

Why is this important? The advantages of moving towards such a state should need no argument. As an example, Høeg (1995) takes up the available information concerning alternative sexual reproductive strategies in barnacles, re-organizes the facts at hand within the only cladistic framework (Fig. 2) available for barnacles (Glenner et al. 1995), and arrives at some interesting and remarkable conclusions.

Previous work in this field (viz., Crisp 1983; Charnov 1987) formulated mathematical models about reproductive strategies in barnacles completely divorced from any phylogenetic framework. These workers had assumed that the primitive condition in cirripedes had to be one of hermaphroditism simply because that is what past workers had thought, not because any phylogenetic analysis indeed had suggested that. Høeg's investigation, however, reveals that those earlier hypotheses may not stand up to what we can now discern about phylogenetic relationships among the Cirripedia as a whole. In Høeg's alternative scheme, hermaphroditism emerges as an apomorphic state for thoracican barnacles, and his conclusion clearly flows right out of the phylogeny. Even rigorous mathematical models emerge as seriously flawed when juxtaposed against a cladistic phylogeny. Previous workers couldn't consider any other alternatives because they had not posed their questions within the proper conceptual framework. Without a proper phylogenetic overview, conclusions concerning the evolution of any group of organisms, not just barnacles, become little more than just-so stories.

We can see another example of the efficacy of phylogenetics in the exchange that

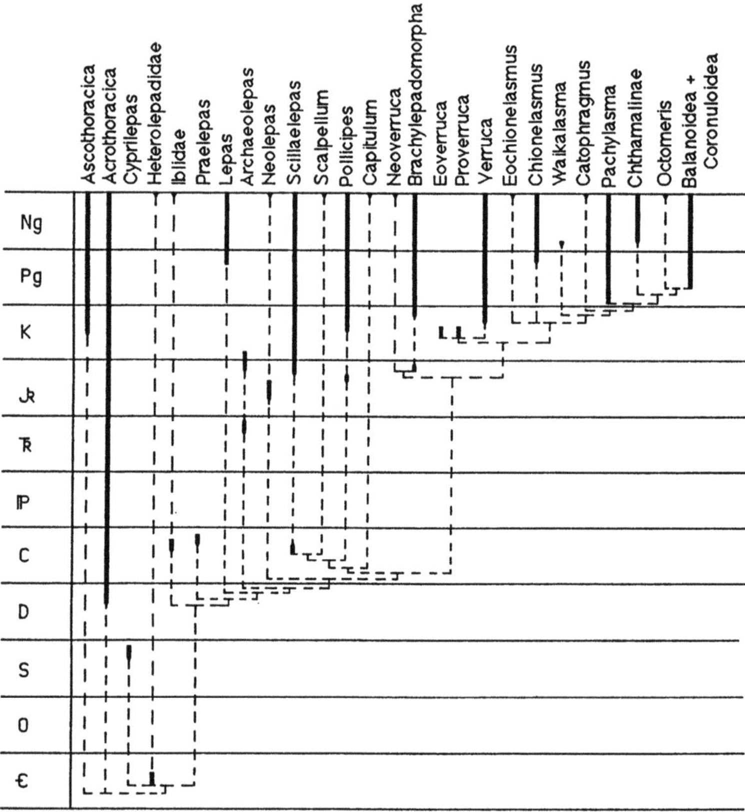

Figure 3. The cladogram of Figure 2 placed in a time-stratigraphic context, based on our current knowledge of the fossil record of the cirripede taxa (used in the analysis of Glenner et al. 1995) or their closest fossil relatives. The indicated times of the occurrence of particular branching points represent only approximations of the latest possible horizon for the event.

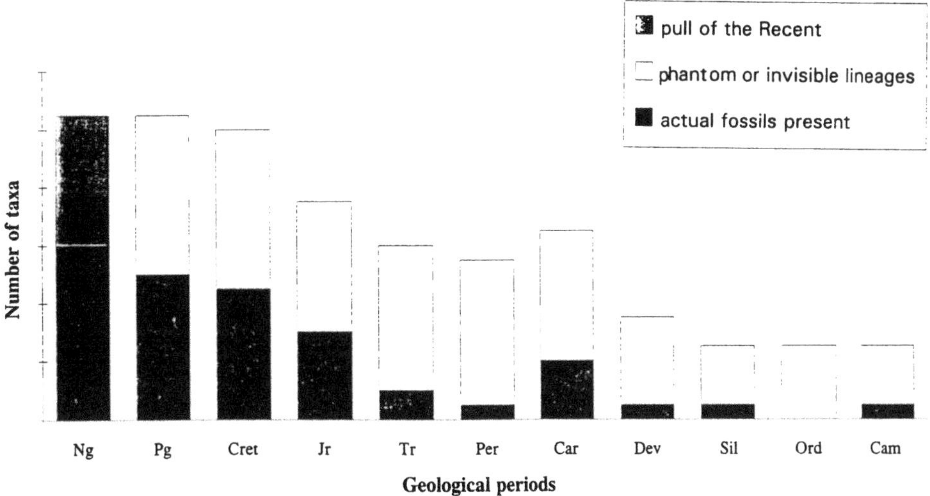

Figure 4. The diversity through time of the Cirripedia under different assumptions based on the taxa taken up by Glenner et al. (1995). The bottom, blackened portions indicate the diversity based on a purely taxic analysis of what specific taxa actually occur in each period. A possible scenario based on that data might indicate a minor early Paleozoic group, with a small radiation in the Carboniferous, followed by a steep decline due to the Permo-Triassic mass extinction event, and a subsequent, slow, steady recovery before an explosive Neogene radiation. The white boxes are those which include the ghost taxa and phantom lineages evident from the time-stratigraphic cladogram of Figure 3. A possible scenario based on that data would indicate a modest radiation in the early Paleozoic, with a striking Carboniferous radiation only little effected by the Permo-Triassic extinctions, a steady increase into the Cretaceous, with little change in diversity after that down to the present time. We can perceive that the so-called 'pull of the Recent' so common in taxic analyses is an artefact that results from ignoring the 'ghosts and phantoms' that the cladistic analysis tells us ought to be there.

occurred between Stanley & Newman (1980), Paine (1981), and Newman & Stanley (1981) concerning patterns of diversity in the past as indicators of the role of competition in the evolution of barnacles. Stanley and Newman believed that competition could explain the differences seen in the distributions in space and through time between chthamaloids and balanoids. Paine objected to this on several grounds, one of which was that he maintained there was no evidence for a chthamaloid decline through time. Newman and Stanley rebutted Paine's claim and produced some diversity charts that they believed buttressed their argument. However, their argument occurred within a purely taxic framework in which merely the presence or absence of groups was noted in the fossil record. Indeed, such an approach is often standard operation procedure in discussions of diversity seen in the fossil record (e.g. see Buckeridge 1983; Foster & Buckeridge 1987).

However, when you combine phylogenetics with time-stratigraphic distributions of taxa (Fig. 3) a substantially altered picture of the past history and timing of radiations within a group can emerge. One could count simply presence of taxa within a particular time period, in which case you would obtain one picture of the history of barnacle diversity through time (black shaded portions in Fig. 4). One, however, could also take into account the revelation of range extensions, the so called 'ghost taxa' and 'phantom lineages,' that the cladistic analysis allows us to recognize. In that case we could develop a completely different scenario about barnacle evolution. The debate of Paine, Stanley, and Newman almost completely looses its edge when one realizes that its conclusions could have been totally different if it had been adequately framed in a cladistic framework. One can clearly see also that the so-called 'pull-of-the-Recent', a problem inherent in all taxic analyses involving Recent groups, is an artifact that arises from ignoring the cladistic relationships. Incorporation of cladistic information into tallies of diversity through time thus provide a backup for Darwin's contention that the fossil record is inclomplete.

The cladogram of Glenner et al. (1995) is only a first step, but it clearly cannot adequately expand the cladistic frontier in a manner that will allow integration of physiologic, reproductive (Høeg 1995), larval (Korn 1995; Grygier 1995; Elfimov 1995), ultrastructural, biogeographic (Young 1995), and molecular information (van Syoc 1995). Barnacle workers will have to exert a concerted effort to do two things. First, we must expand the limited character data base of Glenner et al. (1995). In this regard, we do not mean only the use of morphological characters. Despite the difficulties already encountered (Spear et al. 1994) molecule sequencing must form an important component of this expansion. A total evidence approach, combining all lines of evidence into a single data base, should form the goal we strive toward when we do this. Second, we need to expand the range of taxa beyond those which Glenner et al. used. The only way we can address issues of establishing monophyletic clades within the barnacles or resolving issues of sister group status will be to expand the field of animals we are concerned with. When we do this, we should reconcile ourselves to the fact that in the short term we may induce a greater level of uncertainty into the analyses that we will get than we have had in the past. Nevertheless, it is the long term objective thast we must keep in mind, and the ability to use consensus and bootstrap analyses on the resultant data output affords us methods to handle that uncertainty.

3 LINGERING PROBLEMS IN MAXILLOPODAN CLADISTICS

In moving our horizons outward from barnacles per se, we also will have to tackle some vexing problems within the Maxillopoda as a whole. One could view these as separate issues, but we believe that they form a set of frontiers that we must scale before we can fully understand the barnacles.

3.1 *Cirripede monophyly and the concept of Thecostraca*

Before we can analyse the phylogeny of any group, we must first suggest a definition and come up with putative set of diagnostic apomorphies. Many modern accounts, including the cladistically oriented text book of Brusca & Brusca (1990) still treat the Ascothoracida as part of the Cirripedia proper. Other such texts, viz. that of Meglitsch & Schram (1991), view the relationships of the cirripedes and related groups within the context of a more inclusive taxon, the Thecostraca, and in that they followed in part Grygier (1987), who in turn emulated Wagin (1947), in considering as Cirripedia only the Rhizocephala, Acrothoracica and Thoracica. In a phylogenetic systematic framework, we would accordingly define the Cirripedia as comprising the stem species to these three taxa and all of its descendants. The very special type of sperm (Jamieson 1991) and numerous characters in their larvae support the monophyly of the Cirripedia so defined.

In fact, Grygier (1987) revived the taxon Thecostraca to comprise both cirripedes and ascothoracids, as well as the enigmatic y-larvae, which he isolated in the separate taxon Facetotecta. Several apomorphies support thecostracan monophyly, notably the unique position of the female gonopore on the first thoracomere and the possession of cyprids or cyprid-like larvae used in settlement.

The morphology of the cypris larva includes numerous apomorphies and therefore takes special importance in assuring cirripede monophyly (Høeg 1992; Jensen et al. 1994). Therefore, we most emphatically recommend that we reserve the terms cyprid or cypris larvae for the unique settling instar of the Cirripedia as defined above. This does not change the fact that we simultaneously accept a homology between the cirripede cyprid, the ascothoracid larva of the Ascothoracida and the 'y-cyprids' of the Facetotecta.

Grygier (1987) also left the basic phylogeny of the Thecostraca largely undecided. The attempts by Jensen et al. (1994) and Spears et al. (1994) broke new grounds in emphasizing the importance of cypris ultrastructure and gene sequence data, but since neither of these groups did not study facetotectans they also could not solve the dilemma left by Grygier's initial analysis. Adding to this confusion, Newman (1992) advocated that we include the Tantulocarida in the Thecostraca. We agree that the exciting new data brought forward by the studies of Boxshall & Lincoln (1987) and Huys et al. (1993) on these minute and intriguing parasites would seem to point to a close relationship with the Thecostraca. However, we also remain convinced that these two taxa can at best only be sister groups.

Concerning the Thecostraca, we would point to larval ultrastructure as a promising source of new data (Moyse et al. 1995). In addition, when sufficient amounts of facetotectan larvae become available for application of the PCR technique we should eventually employ data from the sequencing of genes to help elucidate relationships within the whole of the thecostracans. Until now, we have had a major problems in that studies, other than the pioneering efforts of Grygier (1987), Itô & Takenaka (1989), and Itô (1989), have almost unanimously ignored the existence of the Facetotecta in discussing the phylogeny of cirripedes and their closest relatives (but see Schram 1986). Indeed, this amounts to discussing mammalian phylogeny while ignoring marsupials.

3.2 *The quest for the 'Y' adult*

One reason why zoologists have largely neglected the facetotectans follows from our very limited knowledge of their biology. We have for a long time known of the existence of a special type of nauplii and cyprids, so called y-larvae (Schram 1986) and a close relation to the Cirripedia of these larvae was never in doubt. However, despite intensive research efforts in recent years by Grygier (1987), and especially the late Tatsunori Itô (e.g. 1986, 1987, 1989, 1990), the adults of the Facetotecta remain completely unknown.

Like the other Thecostraca, the adult facetotectans must undoubtedly have a sessile existence since their y-cyprids obviously have antennules highly adapted for attachment. Most specialists also anticipate that they are parasites, since free living adults would hardly have escaped detection. We remain wholly ignorant, however, concerning the taxonomic status of the putative host animals. Considering that thecostracans parasitize both Anthozoa and Echinodermata (Ascothoracida), Crustacea (Rhizocephala), and Vertebrata and Polychaeta (Thoracica), most metazoan phyla present possible candidates. In fact, some zoologist indeed may have found the 'adult y' already, but dismissed it as something like an 'uninteresting' parasitic copepod. Undoubtedly, therefore, we shall have to rely on sheer serendipity and await that unexpected moment when some especially observant biologists will recognize the characteristic 'y-nauplii' within a new parasite.

3.3 *Bredocaris, a Cambrian relative of the Thecostraca?*

Nothing has probably excited carcinologists in recent years as much as the discovery of minute, phosphatized animals from the Paleozoic, amenable for SEM analysis at a level of detail quite comparable to that common in extant species (e.g. see Walossek & Müller 1994). With admirable precision and perseverance, Müller and Walossek (1988) worked out the sequence of larval instars for *Bredocaris admirabilis* from the Swedish 'orsten'. They showed that it exhibits interesting similarities in ontogeny with the Thecostraca in developing directly from a nauplius instar with three appendages to one with a full fledged armament of thoracopods. We therefore are tempted to follow Walossek's (1993) suggestion that we consider *Bredocaris* as a very close relative of the Thecostraca. Nevertheless, investigations of crustacean phylogeny employing a master matrix for all Crustacea reveals some possibly alternative hypotheses (Schram & Hof, in press). For example, it is conceivable that all the orsten and some Burgess Shale arthropods actually fall within the Cambrian stem forms, either as a transition series leading to crown-group Crustacea, or as a distinct stem clade. In addition, the analysis of Schram and Hof, preliminary as it is, suggests that one could view Maxillopoda as a paraphyletic group and that in those circumstances forms like *Bredocaris Rehbachiella*, and *Skara* may not occur in the clade that contains thecostracans. Clearly the issue of the affinities of *Bredocaris* will need some careful attention.

The Tantulocarida have a very abbreviated ontogeny since they hatch directly at the settling stage, viz., the tantulus larva, which we and Huys et al. (1993) agree in considering as homologous to the thecostracan cyprid-like larvae. This impedes attempts to resolve the detailed relationships between *Bredocaris*, the Tantulocarida, and the Thecostraca. However, a Cambrian age for the origin of the Thecostraca (Glenner et al. 1995) can hardly surprise us since Collins & Rudkin (1981) described putative pedunculate cirripedes from that era. If true, this would indicate that tantulocarids, ascothoracidans, and facetotectans also existed as separate lineages at that time.

3.4 *The ancestral rhizocephalan*

Just as with the Facetotecta, no one has seriously questioned the cirripede nature of the Rhizocephala. However, this group has rarely entered into discussion on barnacle phylogeny despite the astonishing fact that this taxon presently comprises about one fifth to one fourth of all presently described species of Cirripedia. Schram (1986) and Schram & Hof (in press), in attempting an across the board analyses of all Crustacea with strict cladistic methods, had difficulty demonstrating that these obligate parasites belong to the Cirripedia based on the 'adult' characters immediately available for study. Only by forcing certain scorings of 'adult' morphology based on 'inferences' from the larval features could the rhizocephalans be brought into some proximity to the thecostracans. However, such results arise from fact that rhizocephalans have a highly reduced morphology and have therefore experienced much secondary loss of structures during evolution of their parasitism. Based on Høeg's detailed (1985) analysis, Schram (1986) and Newman (1987) advanced the unorthodox suggestion that the Rhizocephala diverged first from the cirripede lineage before the evolution of sessility. Both larval ultrastructure (Høeg 1992; Jensen et al. 1994) and rDNA sequence data (Spears et al. 1994) speak against Schram's and Newman's interpretations and argue instead for a sister group relationship between the Thoracica and the Rhizocephala. However, neither the studies on larval ultrastructure nor those on sequence data included the 'lepadomorph' families Iblidae nor Heterolepadidae, both of which compete for the status of sister group to all remaining Thoracica (see Glenner et al. 1995). This means that we cannot yet wholly dismiss the idea that the Rhizocephala diverged from 'within' the Thoracica. In fact, one might hope to find characters indicating such a position, because it would tell us that the Rhizocephala must have evolved from a stalked, setose feeding barnacle. In the more likely case that the Rhizocephala diverged outside the Thoracica, we shall perhaps never know the morphology of the ur-rhizocephalan. This exemplifies the often forgotten fact that the establishment of a convincing phylogeny does not necessarily also lead to a clear understanding of the morphology of the stem species or the evolutionary transformations in the ensuing pair of divergent lineages.

With respect to intrinsic rhizocephalan phylogeny, the cladistic analysis of Høeg & Lützen (1993) showed the way to go, but we nevertheless predict that a truly convincing phylogeny of these parasites must resort to data from gene sequencing.

3.5 *Tantulocarid-rhizocephalan relationships?*

Although never expressed in writing, many carcinologists have approached one of us (JTH) asking whether we could not contemplate a close relationship between the Rhizocephala and the Tantulocarida. The reasons for this seem straightforward. Both taxa parasitize Crustacea and have highly advanced life cycles including extreme processes of metamorphosis or possibly parthenogenetic reproduction (Høeg, in press). Moreover, set side by side it is easy to see some interesting similarities between the rhizocephalan kentrogon and the tantulus larva, notably the possession of an unpaired cephalic stylet used to penetrate into the hosts. However, these two larvae cannot be homologous for the simple reason that the settlement stage in rhizocephalans is a typical cirripede cypris, which following settlement metamorphoses into a kentrogon, while in tantulocarids the settling instar is the tantulus larvae itself. Nevertheless, the putative cladograms of Schram & Hof (in press) and Schram (1986) suggest the possibility of some relationship between these two groups.

If we could reject a tantulocarid-rhizocephalan relationship, that would leave us with another interesting scenario. Among the thecostracan taxa and their closest relatives, all but the setose feeding Acrothoracica and Thoracica have parasitic adults. Moreover, they apparently all evolved parasitism independently as evidenced by different host groups and non-homologous specializations in the invective larval stages. Taking this further, the ectoparasitic Branchiura also become candidates within the Maxillopoda for sitting in the 'thecostracan branch' (Grygier 1987; Walossek 1993). This leaves us with the impression, also noted by Newman (1987), that these groups, many of which combine parasitism with the possession of many plesiomorphic traits, survived only because they found a niche in the parasitic life style.

3.6 *The Acrothoracica – Conflicting evidence*

At the workshop at Kristineberg (Boxshall et al. 1992) one of the few major new lines of evidence, besides the fascinating 'orsten fauna', came from rDNA sequencing data. Among that were evidence from the group of Prof. Abele that put forth serious doubt concerning the cirripede affiliation of the Acrothoracica. The enigma remained, although somewhat tempered, in a subsequent paper (Spears et al. 1994). While sequencing data confirm a close relationship between the Thoracica and the Rhizocephala, the molecular data indicate that the Acrothoracica has closest cladistic affinity to the Ascothoracida. These authors acknowledge that morphological data almost unequivocally point to a monophyletic Cirripedia including the Acrothoracica. However, the study of cyprid sense organ by Jensen et al. (1994) revealed that the Acrothoracica had the same, putative, plesiomorphic character state as seen in the Ascothoracida, and most recently also in the Facetotecta (Jensen, pers. comm.). We must therefore seriously entertain the possibility of a phylogeny in which the Acrothoracica diverge first from the cirripede lineage and, like the issue of the heterolepadids and iblids mentioned above, this possibility has important impact on how to reconstruct the *Bauplan* of the urcirripede. For example, was such an ancestral form naked or shelled and did it have a stalk? Again, ideas about the *Bauplan* of a lineage must flow out from the phylogenetic analysis, and such analyses must always function as the home base from where we develop ideas concerning evolutionary events and their explanations.

4 VEXING FRONTIERS WITHIN THE PHYLOGENY OF BARNACLES

In addition to the problems outlined above that exist within the framework of maxillopodan phylogeny, we can also indentify two very specific probelm areas that will need attention within the thoracican cirripedes.

4.1 *The First Thoracican: adductor muscle vs. shell plates*

Based on the recent studies of Jensen et al. (1994), Spears et al. (1994), and Glenner et al. (1995), the monophyly of the Thoracica seems assured (with the above mentioned reservation concerning the Rhizocephala). However, we cannot hope to safely advance a branching sequence for the base of the Thoracica until we have data from larval ultrastructure, juvenile ontogeny, and gene sequences from both the Iblidae and the Heteralepadidae. In the first cladistic study of the Thoracica, Glenner et al. (1995) concluded that the heterolepadids diverged first from the thoracican lineage, a result which may conform with fossil data (Collins & Rudkin 1981). In such a scenario, the urthoracican completely lacked calcareous shell plates. However, cirripedologists have traditionally used the position of the carapace adductor muscle in the four plated Iblidae to argue that this family constitutes the sister group to all remaining Thoracica (Klepal 1985).

As often occurs in phylogeny, the importance of solving this issue goes well beyond the mere branching sequence and directly implicates the morphology of the thoracican, or even the cirripede, stem species. With heterolepadids diverging first we would conclude that both the urthoracican and most likely the urcirripede lacked shell plates altogether, thus their presence constitutes an intra-thoracican apomorphy evolved as late as the Late Palaeozoic (Glenner et al. 1995). However, accepting that the iblids could have diverged first entails a four plated urthoracican, and a reasonable, although not necessarily most parsimonious, scenario, that would postulate the plate absence seen in acrothoracicans and rhizocephalans as a secondary loss. It is conceivable that only gene sequence data will help us sort out the problem here or conceivably some new data from larval morphology.

4.2 *Neolepas, Neoverruca, and the elusive latera*

With the consideration of lateral plates we encounter a very complex frontier. Indeed this issue deals with the intricacies of scalpellid monophyly, the origin of sessile barnacles, and the astonishing phylogenetic impact of barnacles dredged from deep-sea vent habitats. Influenced by evolutionary systematic approaches, several specialists, even as late as the 1980s, argued that the 'balanomorphs' had evolved convergently from a 'pedunculate scalpelloid stock'. Presently, however, both Glenner at al. (1995), Yamaguchi & Newman (1990), Buckeridge & Newman (1992), and Buckeridge (1995) all agree on sessilian monophyly. Still, no phylogeny of the Sessilia can avoid some considerable amount of homoplasy, e.g., in the loss of imbricating whorls of shell plates and possibly also in the evolution of the various types of the so called sheaths (Anderson 1994). This, and the fact that the radiation of the advanced sessilians remains little less than a 'tangled web,' promises more than enough work for a generation of cirripedologists. Yet somehow this web must be unravelled.

Another compelling question concerns the phylogenetic position of various 'scalpelloids' and of the hydrothermal vent barnacles *Neolepas*, *Neoverruca*, and other such species presently under description by Newman (pers. comm.). An answer to these problems will also decide which taxon we should consider as the sister group to the Sessilia. Glenner et al. (1995) could confirm neither the monophyly of a Verrucomorpha that included *Neoverruca,* nor the monophyly of the Scalpelloidea in its widest sense.

The whole problem here arises from the fact that recent ontogenetic evidence from the vent barnacles has put into serious doubt our traditional concepts of homology among the so called lateral plates, viz., those encircling the capitulum below the wall of the five primary plates. For example, in the ontogeny of *Neoverruca*, and apparently also in *Neolepas*, the so called latus (L) develops heterochronously *after* other lateral plates, unlike in the 'traditional' scalpellids where the latus appear *before* the other latera right after the five primary plates develop. Obviously, we need detailed studies on early ontogeny along the lines set out by Glenner & Høeg (1994) to effectively deal with these issues. Until we have that kind of information, speculations about latera identities will remain just that, speculations.

5 CONCLUSIONS

We can see from the above, that there is no lack of new frontiers to attack in elucidating the patterns of barnacle evolution. In this effort, however, we should *absolutely not* become obsessed with obtaining a single TRUTH. The fact that there may exist multiple alternative hypotheses of relationships can only serve to strengthen the field as a whole. Heretofore, we cirripede workers have been too intimidated with the Darwinian orthodoxy.

Most of the progress in barnacle studies since the 1850s has been in describing new taxa. This is a task that is unending. Unless we move on to taking the data we now have and projecting broader scale generalizations from them, we will spend the next 150 years describing more taxa even while we continue to bewail the fact that we still do not adequately understand the mechanisms of hatching, settlement, metamorphosis, phylogenetic relationships, and other phenomena.

ACKNOWLEDGEMENTS

We owe an immense debt of gratitude to Prof. William A. Newman for his stimulating mentorship through the years. That we have been able to perceive any of the above has been due to his never flagging ability to synthesise new information and come up with new hypotheses. Hendrick Glenner helped review the draft manuscript and Giuseppe Fusco put together Figure 4.

REFERENCES

Anderson, D.T. 1994. *Barnacles: Structure, Function, Development, and Evolution*. London: Chapman & Hall.

Boxshall, G.A. & R. Lincoln 1987. The life cycle of the Tantulocarida (Crustacea). *Phil. Trans. Roy. Soc., Lond.* (B)315: 267-303.

Boxshall, G.A., J.-O. Strömberg & E. Dahl (eds.) 1992. The Crustacea: Origin and Evolution. *Acta Zool.* 73: 271-392.

Brusca, R. & G. Brusca 1990. *Invertebrates*. Sunderland: Sinauer Assc.

Buckeridge. J. 1983. Fossil barnacles of New Zealand and Australia. *N.Z. Geol. Surv. Paleo. Bull.* 50:1-151, 13 plates.

Buckeridge, J. 1995. Phylogeny and biogeography of the primitive Sessilia and a consideration of the Tethyan origin of the group. *Crust. Issues* 10: 255-267.

Buckeridge, J. & W.A. Newman 1992. A reexamination of *Waikalasma* and its significance in balanomorph phylogeny. *J. Paleo.* 66: 341-345.

Charnov, E.L. 1987. Sexuality and hermaphroditism in barnacles: a natural selection approach. *Crust. Issues* 5: 89-103.

Clare, A.S. 1995. Chemical signals in barnacles: old problems, new approaches. *Crust. Issues* 10: 49-67.

Collins, D. & D.M. Rudkin 1981. *Priscansermarinus barnetti*, a probably lepadomorph barnacles from the Middle Cambrian Burgess Shale of British Columbia. *J. Paleo.* 55: 1006-1015.

Crisp, D.J. 1983. *Chelonibia patula*, a pointer to the evolution of the complemental male. *Mar. Biol. Lett.* 4: 281-294.

Crisp, D.J., E.M. Hill & D.L. Holland 1991. A review of the hatching process in barnacles. *Crust. Issues* 7: 57-68.

Darwin, C.R. 1854. *A monograph on the subclass Cirripedia: The Balanidae; the Verrucidae, etc.* London: Ray Soc.

Elfimov, A. 1995. Comparative morphology of the thoracican cyprid larvae: studies of the carapace. *Crust. Issues* 10:137-152.

Foster, B.A. & J.S. Buckeridge 1987. Barnacle paleontology. *Crust. Issues* 5: 43-61.

Glenner, H. & J.T. Høeg 1993. Scanning electron microscopy of metamorphosis in four species of barnacles. *Mar. Biol.* 117: 431-439.

Glenner, H., M.J. Grygier, J.T. Høeg, P.G. Jensen & F.R. Schram 1995. Cladistic analysis of the Cirripedia Thoracica. *Zool. J. Linn. Soc.* 114: 365-404.

Grygier, M.J. 1987. New records, external and internal anatomy, and systematic position of Hansen's Y-larvae. *Sarsia* 72: 261-278.

Grygier, M.J. 1995. An unusual barnacle nauplius illustrating several hitherto unappreciated features useful in cirripede systematics. *Crust. Issues* 10: 123-136.

Grygier, M.J. & T. Itô 1995. SEM based morphology and new host and distribution records of *Waginella*, crinoid-associated ascothoracidan cristaceans. *Crust. Issues* 10: 209-228.

Høeg, J.T. 1985. Cypris settlement, kentrogon formation, and host invasion in the parasitic barnacle *Lernaeodiscus porcellanae*. Acta Zool. 66: 1-45.

Høeg, J.T. 1992. The phylogenetic position of the Rhizocephala: are they truly barnacles? *Acta Zool.* 73: 323-326.

Høeg, J.T. 1995. Sex and the single cirripede: a phylogenetic perspective. *Crust. Issues* 10: 195-206.

Høeg, J.T. in press. The biology and life cycle of the Cirripedia Rhizocephala. *J. Mar. Biol. Assc. UK.*

Høeg, J.T. & J. Lützen 1993. Comparative morphology and phylogeny of the family Thompsoniidae, with descriptions of three new genera and seven new species. *Zool. Scripta* 22: 363-386.

Huys, R., G.A. Boxshall & R.J. Lincoln 1993 The tantulocarid life cycle: the circle closed. *J. Crust. Biol.* 13: 432-442.

Itô, T. 1986. Three types of 'nauplius Y' from the North Pacific. *Publ. Seto Mar. Biol. Lab.* 31: 63-73.

Itô, T. 1987. Proposal of new terminology for the morphology of nauplius Y, with provisional designation of four naupliar types from Japan. *Zool. Sci.* 4: 913-918.

Itô, T. 1989. A new species of *Hansenocaris* (Crustacea:Facetotecta) from Tanabe Bay, Japan. *Publ. Seto Mar. Biol. Lab.* 34: 55-72.

Itô, T. 1990. Naupliar development of *Hansenocaris furcifera* Itô from Tanabe Bay, Japan. *Publ. Seto Mar. Lab.* 34:201-224.

Itô, T. & M. Takenaka 1988. Identification of bifurcate parocular process and postocular filamentary tuft of facetotectan cyprids. *Publ. Seto Mar. Biol. Lab.* 33: 19-38.

Jamieson, B.G.M. 1991. Ultrastructure and phylogeny of crustacean spermatozoa. *Mem. Qld. Mus.* 31: 109-142.

Jensen, J.G., J. Moyse, J. Høeg & H. Al-Yahya 1994. Comparative SEM studies of lattice organs: putative sensory structures on the carapace of larvae from Ascothoracida and Cirripedia. *Acta Zool.* 95: 125-142.

Klepal, W. 1985. *Ibla cummingi* – a gonochoristic species (anatomy, dwarfing, and systematic implications) *P.S.Z.N.I. Mar. Ecol.* 6: 47-119.

Klepal, W. 1990. The fundamentals of insemination in cirripedes. *Oceanogr. Mar. Biol. Ann. Rev.* 28: 353-379.

Klepal, W. & H.L. Nemaschkal 1995. Cuticular structures in the males of Scalpellidae: a character analysis. *Crust. Issues* 10: 179-194.

Klepal, W., H. Barnes & E.A. Munn 1972. The morphology and histology of the cirripede penis. *J. Exp. Mar. Biol. Ecol.* 10: 243-265.

Korn, O. 1995. Nauplius evidence for cirripede taxonomy and phylogeny. *Crust. Issues* 10: 87-121.

Meglitsch, P.A. & F.R. Schram 1991. *Invertebrate Zoology*, 3rd ed. New York: Oxford Univ. Press.

Moyse, J., J.T. Høeg, J.G. Jensen & H. Al-Yahya 1995. Attachment organs in cirripede cypris larvae: a comparative approach using mostly scanning electron microscopy. *Crust. Issues* 10: 153-178.

Müller, K. & D. Walossek 1988. External morphology and larval development of the Upper Cambrian maxillopod *Bredocaris admirabilis. Fossils & Strata* 23: 1-70, 16 plates.

Newman, W.A. 1987. Evolution of cirripedes and their major groups. *Crust. Issues* 5: 3-42.

Newman, W.A. 1992. Origin of Maxillopoda. *Acta Zool.* 73: 319-322.

Newman, W.A. & A. Ross 1976. Revision of the balanomorph barnacles: including a catalogue of the species. *Mem. San Diego Soc. Nat. Hist.* 9: 1-108.

Newman, W.A. & S.M. Stanley 1981. Competition wins out overall: reply to Paine. *Paleobiol.* 7: 561-569.

Paine, R.T. 1981. The forgotten roles of disturbance and predation. *Paleobiol.* 7: 553-560.

Schram, F.R. 1986. *Crustacea.* New York: Oxford Univ. Press.

Schram, F.R. & C.H.J. Hof in press. Fossils and the interrelationships of major crustacean groups. In: G. Edgecomb (ed), *Fossils & the Phylogeny of Arthropods*. New York: Columbia Univ. Press.

Southward, A. (ed.) 1987. Barnacle Biology. *Crust. Issues* 5: 1-443.

Spears, T., L.G. Abele & M.A. Applegate 1994. A phylogenetic study of cirripedes and their relatives. *J. Crust. Biol.* 14: 641-656.

Stanley, S.M. & W.A. Newman 1980. Competitive exclusion in evolutionary time: the case of the acorn barnacles. *Paleobiol.* 6: 173-183.

van Syoc, R.J. 1995. Barnacle mitochondrial DNA: determining genetic relationships among species of *Pollicipes*, a Tethyan relict. *Crust. Issues* 10: 269-296.

Wagin, V.L. 1947. *Ascothorax ophioctenis* and the position of the Ascothoracida in the system of the Entomostraca. *Acta Zool.* 27: 155-267.

Walker, G. 1995. Larval settlement: historical and future perspectives. *Crust. Issues* 10: 69-85.

Walossek, D. 1993. The Upper Cambrian *Rehbachiella kinnekullensis* and the phylogeny of the Branchiopoda and Crustacea. *Fossils & Strata* 32: 1-202.

Walossek, D. & K.J. Müller 1994. Pentastomid parasites from the Lower Paleozoic of Sweden. *Trans. Roy. Soc. Edinb.: Earth Sci.* 85: 1-37.

Yamaguchi, T. & W.A. Newman 1990 A new and primitive barnacle from the North Fiji Basin abyssal hydrothermal field, and its evolutionary implications. *Pac. Sci.* 44: 135-155.

Young, P.S. 1995. New interpretations of South American patterns of barnacle distribution. *Crust. Issues* 10: 229-253.

Yule, A.B. & G. Walker 1987. Adhesion in barnacles. *Crust. Issues* 5: 389-402.

Index

Milton Keynes UK
Ingram Content Group UK Ltd.
UKHW020828141024
449569UK00030B/1534